AF767792

Printed and Flexible Sensor Technology

Fabrication and applications

IOP Series in Sensors and Sensor Systems

The IOP Series in Sensors and Sensor Systems includes books on all aspects of the science and technology of sensors and sensor systems. Spanning fundamentals, fabrication, applications and processing, the series aims to provide a library for instrument and measurement scientists, engineers and technologists in universities and industry.

The series seeks (but is not restricted to) publications in the following topics:

- Advanced materials for sensing
- Biosensors
- Chemical sensors
- Industrial applications
- The Internet of Things
- Lab-on-a-chip
- Localization and object tracking
- Manufacturing and packaging
- Mechanisms, modelling and simulations
- MEMS/NEMS
- Micro- and nanosensors
- Non-destructive testing
- Optoelectronic and photonic sensors
- Optomechanical sensors
- Physical sensors
- Remote sensors
- Sensing for health, safety and security
- Sensing principles
- Sensing systems
- Sensor arrays
- Sensor devices
- Sensor networks
- Sensor technology and applications
- Signal processing and data analysis
- Smart sensors and monitoring
- Telemetry

Authors are encouraged to take advantage of electronic publication through the use of colour, animations, video and interactive elements to enhance the reader experience.

Printed and Flexible Sensor Technology

Fabrication and applications

Edited by
Subhas Chandra Mukhopadhyay
Macquarie University, Sydney, Australia

Anindya Nag
Faculty of Electrical and Computer Engineering, Technische Universität Dresden, Dresden 01062, Germany
and
Centre for Tactile Internet with Human-in-the-Loop (CeTI), Technische Universität Dresden, Dresden 01069, Germany

IOP Publishing, Bristol, UK

© IOP Publishing Ltd 2021

All rights reserved. No part of this publication may be reproduced, stored in a retrieval system or transmitted in any form or by any means, electronic, mechanical, photocopying, recording or otherwise, without the prior permission of the publisher, or as expressly permitted by law or under terms agreed with the appropriate rights organization. Multiple copying is permitted in accordance with the terms of licences issued by the Copyright Licensing Agency, the Copyright Clearance Centre and other reproduction rights organizations.

Permission to make use of IOP Publishing content other than as set out above may be sought at permissions@ioppublishing.org.

Subhas Chandra Mukhopadhyay and Anindya Nag have asserted their right to be identified as the editors of this work in accordance with sections 77 and 78 of the Copyright, Designs and Patents Act 1988.

ISBN 978-0-7503-3439-6 (ebook)
ISBN 978-0-7503-3437-2 (print)
ISBN 978-0-7503-3440-2 (myPrint)
ISBN 978-0-7503-3438-9 (mobi)

DOI 10.1088/978-0-7503-3439-6

Version: 20211201

IOP ebooks

British Library Cataloguing-in-Publication Data: A catalogue record for this book is available from the British Library.

Published by IOP Publishing, wholly owned by The Institute of Physics, London

IOP Publishing, Temple Circus, Temple Way, Bristol, BS1 6HG, UK

US Office: IOP Publishing, Inc., 190 North Independence Mall West, Suite 601, Philadelphia, PA 19106, USA

Contents

Preface

The intervention of sensing systems within the modern world has revolutionized the quality of human life. Almost every aspect of day-to-day activities in today's world involves the implementation of sensors in some way. Their presence has not only made our lives easier, but has also allowed humanity to progress on a technological level. The sensing systems as we use them today are being developed and characterized mostly in academic and industrial contexts. Following the sensorial pyramid, different kinds of prototypes have been fabricated, varying in terms of their processed materials, operating principles, and respective applications. Each of these prototypes is being specialized to have high efficiency in terms of sensitivity and robustness for the targeted application. The advancement in technology has allowed highly functionalized equipment to be developed that can be mass-produced for point-of-care sensing systems. Among the sensors that are currently available, the flexible sensors have been able to perform in a much more dynamic manner in comparison to the conventional MEMS-based silicon sensors. The additional wearable nature of flexible sensors has allowed them to be used for ubiquitous monitoring purposes. These flexible sensors have been fabricated using a range of printing techniques, each of which has the capability to form high-quality thin-film sensors. These prototypes have revolutionized the world of flexible sensors due to their simple fabrication techniques, quick roll-to-roll production, and simple operating mechanisms. The choice of printing technique primarily depends on the specifications of the sensors, which in turn depend on their applications.

In recent times, with the conjugation of nanotechnology and printing techniques, these flexible sensors have increased their dynamicity for real-time applications. The developed printed and flexible sensors have been exploited successfully in biomedical, industrial, and environmental applications. With the advent of nanotechnology these printed flexible sensors have attained further heights with respect to their performance. Nanomaterials have been included in various forms using the available printing mechanisms to enhance the resultant functionality. 1D and 2D conductive materials have been utilized thoroughly to enhance the sensitivity and selectivity of prototypes. Flexible printed sensors have allowed us to further formalize entire systems to determine the interfacing circuits for conditioning and transmitting the sensed data. Market surveys have also presented a great need for these sensors, not only to increase the multifunctional nature of the sensing systems but improve their impact on the chosen application. The exponential increase of these printed flexible sensors in the near future can be justified by the need in automated systems for detection in sectors of prioritized significance.

This book contains a structured overview of the fabrication techniques of the various printed flexible sensors that have been formulated and devised. It also contains some of the significant applications of these printed flexible sensors, conducted in controlled and real-time scenarios. This book contains chapters contributed by experts in their respective fields who have studied and operated printed flexible sensors via designing and employing them. It also presents the technological growth in the field of printed flexible sensors by presenting current trends. Each of the presented works will not only allow the reader to obtain an understanding of this field but will also assist the reader to improvise and innovate something of their own. This book is organized in the following manner.

Chapter 1 presents an overview of the sensors and fabrication techniques that are currently available. It also presents a brief overview of the sectors in which these prototypes have been deployed. This is followed by chapter 2, which exemplifies the utilization of printed flexible sensors for a range of research-based applications. It also categorizes the applications, providing the advantages and reliability related to each of them. Chapter 3 highlights the work done on the fabrication of printed flexible sensors in terms of their issues and resolution. It deals with the fabrication of printed flexible sensors via highlighting some of the challenges related to their fabrication techniques and possible corresponding solutions. Chapter 4 showcases the potential of 3D wax printing technology to create sophisticated biomedical devices. It also explains the application of printable devices for point-of-care testing and various biosensing purposes and details some recommendations and future directions for improving sensing accuracy and robustness. Chapter 5 summarizes some of the laser induced graphene preparation methods and their broad-spectrum applications, in particular in electrochemical and biosensing. It also focuses on the future outlook and research gaps in the reported literature. Chapter 6 provides an overview of the significant development in wearable microfluidic devices in terms of their fabrication and characterization. It also shows different categories for its use in physical property and body fluid sampling, manipulation, and detection. Chapter 7 reviews the recent developments in assembly techniques, including printing processes for forming films or networks of single-walled carbon nanotubes (SWNTs). The challenges related to the processing and performance of SWNTs, along with the potential research that can be performed with SWNTs, are also presented. Chapter 8 presents the fabrication and implementation of flexible strain sensors using graphene and its composites. It analyzes the capability of the remarkable material graphene in developing high-quality flexible strain sensors. Chapter 9 presents an overall review of screen-printed electrochemical and impedance biosensors. This is achieved by presenting different cases involving the estimation of environmental contaminants, food toxins, protein molecules, bacteria, and viruses. Chapter 10 deals with the properties of cellulose paper and its use for flexible electronic device applications. It explains the design and technology of flexible electronics produced using cellulose paper. Chapter 11 explains the synthesis and characterization methods for printing/fabricating graphene-based implantable electrodes for various neural recording and stimulation processes. This is followed by chapter 12 which shows the work done on screen-printed electrode-based sensors for the detection of biological and chemical species. It highlights the potential of electrochemical methods of screen-printed electrodes for accurate measurements of trace level biochemical species in a sample solution using a rapid and cheap analysis method. Chapter 13 provides a brief overview of the role of 3D printing in microfluidic enzymatic biofuel cells, along with related recent work in upcoming research areas. Chapter 14 presents the work done on the use of an array of interdigitated electrodes for developing novel incontinence sensing systems. These devices consist of simulation using ANSYS finite element analysis (FEA) and printing of the conductive polymer PEDOT:PSS on PET substrates. Chapter 15 presents an experimental study carried out to develop a transparent thin-film strain sensor (TFSS) using an unconventional substrate (PET with a pre-applied thin pressure-sensitive adhesive layer) and to investigate its response to static and dynamic strain environments. Insights into the fabrication of high figure of merit transparent thin films at room temperature and the methodology to develop a transparent TFSS, along with its performance evaluation, are described in this chapter. Chapter 16 reports the work done on high-performance radio frequency (RF) magnetron sputtered cerium oxide (CeO_2) thin-film-based oxygen sensors by optimizing the film thickness. Chapter 17 presents the fabrication of a metal oxide (MOX) trace moisture sensor on a flexible substrate. The sensors are parallel-plate capacitors with nanoporous alumina hydrophilic sensing films, deposited on polyimide using the solution method. Chapter 18 discusses the importance of droplet detection and the need for improvements in sensor geometry for precise and accurate detection of microdroplets. A novel type of printable and flexible sensor is introduced for droplet detection, which utilizes the capacitive transduction principle and the detection is contactless, thus increasing the lifetime of the detection system. Chapter 19 focuses on flexible and conformal antennas for wireless communication and sensing applications.

The chapters contributed to this book have been written with the utmost care and detail in order to present the work in a simple and efficient manner. The valuable research ideas imparted by the contributing authors have allowed us to present the high-quality research taking place in the microelectronics industries. Each of the presented research works will help the reader understand the functionality of printed flexible sensing systems at both the basic and advanced levels. The compilation of this interesting and useful book would not have been possible without the priceless contribution of the authors and their willingness to participate in this journey.

Subhas Chandra Mukhopadhyay
School of Engineering
Macquarie University
Sydney, NSW 2109, Australia
Subhas.Mukhopadhyay@mq.edu.au

Anindya Nag
Department of Electrical Engineering
CEMSE Division, King Abdullah University of Science of Technology
Thuwal 23955, Saudi Arabia
anindya1991@gmail.com

Editor biographies

Subhas Chandra Mukhopadhyay

Subhas Chandra Mukhopadhyay (Fellow, IEEE) holds a B.E.E. (gold medallist), M.E.E., Ph.D. (India) and Doctor of Engineering (Japan). He has over 31 years of teaching, industrial and research experience. Currently he is working as a Professor of Mechanical/ Electronics Engineering, Macquarie University, Australia and is the Discipline Leader of the Mechatronics Engineering Degree Programme. He is also the Director of International Engagement for the School of Engineering of Macquarie University. His fields of interest include Smart Sensors and sensing technology, instrumentation techniques, wireless sensors and network (WSN), Internet of Things (IoT), wearable sensors, medical devices, healthcare and environmental monitoring. He has supervised over 40 postgraduate students and over 100 Honours students. He has examined over 70 postgraduate theses.

He has published over 450 papers in different international journals and conference proceedings, written ten books and fifty two book chapters and edited eighteen conference proceedings. He has also edited thirty five books with Springer-Verlag and thirty two journal special issues. He has been cited so far **13933 times and has a h-index of 58**. He has received various awards, most notably: the Australian Research Field Leader in Engineering and Computer Science 2020; Distinguished Lecturer, IEEE Sensors Council 2020–2022; Outstanding Volunteer by IEEE R10, 2019; World Famous Professor by Government of Indonesia, 2018; Certificate of Distinction from IEEE Sensors Council, 2017; IETE R.S. Khandpur Award—India, 2016; Best Performing Topical Editor of *IEEE Sensors Journal* from 2013 to 2018, six years consecutively. He has organized over 20 international conferences as either General Chairs/co-chairs or Technical Programme Chair. He has delivered 389 presentations including keynote, invited, tutorial and special lectures.

He is a **Fellow of IEEE** (USA), a **Fellow of IET** (UK), a **Fellow of IETE** (India), a Topical Editor of IEEE Sensors journal, an associate editor of *IEEE Transactions on Instrumentation and Measurements*, and *IEEE Review of Biomedical Engineering*. He is the Editor-in-Chief of the *International Journal on Smart Sensing and Intelligent Systems* and Springer Nature on Computer Science. He is **a Distinguished Lecturer** of the IEEE Sensors Council from 2017 to 2022. He is the Founding Chair of the IEEE Sensors Council New South Wales Chapter.

More details are available at:
https://researchers.mq.edu.au/en/persons/subhas-mukhopadhyay
https://scholar.google.com.au/citations?hl=en&user=8p-BvWIAAAAJ
https://orcid.org/0000-0002-8600-5907

Anindya Nag

Anindya Nag (Member, IEEE) completed a B. Tech. degree from West Bengal University of Technology, India in 2013, M.S. degree at Massey University, New Zealand in 2015 and a Ph.D. degree from Macquarie University, Australia, in 2018. He worked as a lecturer in Dongguan University of Technology, China from February 2019 to August 2020. He has been postdoctoral fellow in King Abdullah University of Science and Technology (KAUST), Thuwal, Saudi Arabia and Shandong University, Jinan, China. He is currently working as a junior professor in Technische Universität Dresden, Dresden, Germany. His research interests are in the area of MEMS, flexible sensors, printing technology and nanotechnology-based smart sensors for health, environmental and industrial monitoring applications. His paper, 'Wearable flexible Sensors,' has been one of the top 25 downloaded papers in the *IEEE Sensor Journal* from June 2017–September 2018. Dr Nag has authored and co-authored over 75 research publications in the form of books, journal articles, international conference proceedings and book chapters.

List of contributors

Pablo Acedo
Universidad Carlos III de Madrid
Madrid
Spain

Jamil Akhtar
Department of ECE
Manipal University
Jaipur
India

Khairunnisa Amreen
Department of Electrical and Electronics Engineering
BITS Pilani
Hyderabad
India

Mohsen Asadnia
School of Engineering
Macquarie University
Sydney
Australia

Raghied Mohammed Helmy Atta
Taibah University
Al Madinah
Saudi Arabia

Mahroo Baharfar
Graduate School of Biomedical Engineering
University of New South Wales
Sydney
Australia

Manuel Bäuscher
Fraunhofer Institute for Reliability and Microintegration IZM
Berlin
Germany

Amin Beheshti
Computing Department
Macquarie University
Sydney
Australia

S Bhadra
Department of Electrical and Computer Engineering
McGill University
Montreal
Canada

Shreyas P Bhat
Centre for Nano Science and Engineering
Indian Institute of Science
Bengaluru
India

M S Bhat
Department of ECE
National Institute of Technology Karnataka
Surathkal
India

Gaozhe Cai
Key Laboratory of Agricultural Information Acquisition Technology
Ministry of Agriculture and Rural Affairs
China Agricultural University
Beijing
China

Shilun Feng
State Key Laboratory of Transducer Technology
Shanghai Institute of Microsystem and Information Technology
Shanghai
China
and
Nanyang Technological University
Singapore

Kamran Foroutani
Sahand University of Technology
Sahand
Iran

Kolla Lakshmi Ganapathi
Department of ECE
National Institute of Technology Karnataka
Surathkal
India

Sanket Goel
Department of Electrical and Electronics Engineering
BITS Pilani
Hyderabad
India

Chinthaka Gooneratne
Saudi Aramco
Darham
Saudi Arabia

Raheel M Hashmi
Faculty of Science and Engineering
Macquarie University
Sydney
Australia

Tarikul Islam
Department of Electrical Engineering
Jamia Millia Islamia
New Delhi
India

U S Jayapiriya
Department of Electrical and Electronics Engineering
Birla Institute of Technology and Science Pilani
Hyderabad
India

José Luis Jorcano
Universidad Carlos III de Madrid
Madrid
Spain

Vaishakh Kedambaimoole
Department of Instrumentation and Applied Physics
Indian Institute of Science
Bengaluru
India

Sara Khademi
Shenzhen Institutes of Advanced Technology
Chinese Academy of Sciences
Shenzhen
China
and
Sahand University of Technology
Sahand
Iran

Shakeb A Khan
Department of Electrical Engineering
Jamia Millia Islamia
New Delhi
India

Rajanna Konandur
Department of Instrumentation and Applied Physics
Indian Institute of Science
Bengaluru
India

Avinash Kothuru
Department of Electrical and Electronics Engineering
BITS Pilani
Hyderabad
India

Md Rahat Mahboob
Department of Electrical Engineering
Tezpur University
Assam
India

Naresh Mandal
Deptment of Electronics and Telecommunication Engineering
Indian Institute of Engineering Science and Technology
Shibpur
India

Dahiana Mojena Medinaa
Universidad Carlos III de Madrid
Madrid
Spain

Zizhen Ming
Department of Biomedical Sciences
Macquarie University
Sydney
Australia

Kalyan Kumar Mistry
CSIR—Central Mechanical Engineering Research Institute
Durgapur
India

Parisa Moazzam
School of Chemistry
University of New South Wales
Sydney
Australia

B Mohamadzade
Faculty of Science and Engineering
Macquarie University
Sydney
Australia

S Mohan
Centre for Nano Science and Engineering
Indian Institute of Science
Bangalore
India

Subhas Chandra Mukhopadhyay
School of Engineering
Macquarie University
Sydney
Australia

Anindya Nag
Faculty of Electrical and Computer Engineering
Technische Universität Dresden
Dresden 01062, Germany
and
Centre for Tactile Internet with Human-in-the-Loop (CeTI)
Technische Universität Dresden
Dresden 01069, Germany

M M Nayak
Centre for Nano Science and Engineering
Indian Institute of Science
Bengaluru
India

Ha-Duong Ngo
University of Applied Sciences
Berlin
Germany
and
Fraunhofer Institute for Reliability and Microintegration IZM
Berlin
Germany

Suresh Nuthalapati
Department of Instrumentation and Applied Physics
Indian Institute of Science
Bengaluru
India

B Pavithra
Centre for Nano Science and Engineering
Indian Institute of Science
Bengaluru
India

N Ramshanker
Centre for Nano Science and Engineering
Indian Institute of Science
Bangalore
India
and
Department of ECE
National Institute of Technology Karnataka
Surathkal
India

Amir Razmjou
School of Chemical Engineering
University of New South Wales
Sydney
Australia

Prakash Rewatkar
Department of Electrical and Electronics Engineering
Birla Institute of Technology and Science Pilani
Hyderabad
India

Chirasree RoyChaudhuri
Deptment of Electronics and Telecommunication Engineering
Indian Institute of Engineering Science and Technology
Shibpur
India

A S M Sayem
Faculty of Science and Engineering
Macquarie University
Sydney
Australia

Michael Schiffer
Fraunhofer Institute for Reliability and Microintegration IZM
Berlin
Germany

N N Sharma
Department of Mechanical Engineering
Manipal University
Jaipur
India

Vijay Shirhatti
Department of Instrumentation and Applied Physics
Indian Institute of Science
Bengaluru
India

Nireekshan Kumar Sodavaram
Faculty of Science and Engineering
Macquarie University
Sydney
Australia

Shaoxi Wang
School of Microelectronics
Northwestern Polytechnical University
Xi'an
China

Yupan Wu
School of Microelectronics
Northwestern Polytechnical University
Xi'an
China

Supriya Yadav
Department of Biosciences
Manipal University
Jaipur
India

Soma Zandi
School of Engineering
University of Kurdistan
Sanandaj
Iran

Zubair Hassan Zargar
Department of Electrical Engineering
Jamia Millia Islamia
New Delhi
India

Qianbin Zhao
School of Electrical and Electronic Engineering
Nanyang Technological University
Singapore

Chapter 1

Printed and flexible sensors: a review of products and techniques

Subhas Chandra Mukhopadhyay, Anindya Nag and Chinthaka Gooneratne

Sensors are playing an increasingly important role in our day-to-day lives and are helping us to lead safer and more secure lifestyles. Different types of sensors are used in different spheres in our lives, including healthcare, environmental monitoring, industrial applications, and many other fields. Printing is not a new technique in the sensor industry, in fact some types of sensors have always been printed. However, with the development of new and high quality materials and the advancement of technologies, a new generation of printed sensors is now emerging from various research and development activities with a wide range of applications. In recent times flexible sensors have emerged as another new paradigm of sensor research and application. This chapter will review the different available sensor products and will discuss the techniques used to fabricate such sensors.

1.1 Introduction

Humans are blessed with five excellent sensing organs—the eyes, ears, nose, tongue, and skin—along with the control centre, the brain, to organize and respond to all activities. To make our lives safer and more comfortable, the exploration, design, and development of different types of sensors are never-ending. Printing technology in the field of sensors began many decades ago and with time it has become smart, efficient, and productive. It allows many sensors to be printed within one fabrication step. Various types of sensors are manufactured partially by screen printing, in which the transducer is a printed layer of either a polymeric or ceramic material. This technology has been used in the sensor industry for many years. Progress in printed electronics now enables more sensors to be printed in their entirety. Since sensors have a much simpler structure than other electronic components such displays or logic circuits, the manufacturing learning curve is therefore less steep compared to many other printed electronics applications. In most cases, these new

doi:10.1088/978-0-7503-3439-6ch1 © IOP Publishing Ltd 2021

printed sensors can be fabricated on plastic substrates with the advantages of mechanical flexibility, thinness, and light weight [1–4]. Printed sensors are eco-friendly in nature and have the ability to fit onto various small, differently shaped electronic devices for different applications. Printed sensors eliminate the use of silicon and metal oxides, instead, with the help of innovative printing technologies, printed sensors can be manufactured on various flexible substrates such as paper, plastics, and foils [1–4].

Printed sensors and sensing devices find applications in various industries, including the automotive industry, environmental testing, consumer electronics, medical devices, smart packaging, building automation, and industrial equipment [1, 2]. The demand from some fields, in particular from smart packaging, is expected to grow at a high rate owing to the ability of sensors to enable manufacturers to keep track of inventory. The growing adoption of printed sensors in this industry to enhance the quality, visibility, hygiene, and safety of a product is anticipated to enhance the market growth of printed sensors. Printed sensors are being used increasingly in the food and beverage industries to monitor the temperature, gas, and humidity of sensitive products. With time many new exciting and challenging applications will find use for flexible and printed sensors.

Planar printed sensors can be of different shape, structure, and fabrication to cater for a certain application. In the early 2000s different types of sensors, mainly meander and mesh type configurations, were designed and developed [5–12]. The size and configuration were modelled and the experimental performance was evaluated. Planar printed sensors have been applied in many applications, for example, the determination of the quality of meat, in particular protein and fat content, is reported in [13], and consistent performance in the design of saxophone reeds is reported in [14]. The processing of leather, making use of planar electromagnetic sensors of interdigital type, is reported in [15] and the determination of the contamination of nitrate in water is described in [16]. The report in [17] describes the measurement and performance evaluation of novel planar interdigital sensors for different chemicals related to food poisoning. The contamination of phthalate leaching from plastic bottles is reported in [18]. The operating mechanism and preliminary results of using printed interdigital sensors to detect the leakage of liquefied petroleum gas are reported in [19] and the effects of particle size, composition, and coating layer thickness on its performance are reported in [20]. During the last few years, flexible sensors have been fabricated for different applications. Laser-ablated metallized PET films for tactile sensing are reported in [21], and a laser-induced graphene sensor has been designed and fabricated for salinity testing [22]. Printed sensors based on flexible materials as well as silicon based fabrication to detect water contamination and environmental monitoring are reported in [23] and [24], respectively.

A significant amount of research on printed and flexible materials is taking place in different parts of the world and many sensors are being developed. Although there remain many challenges, such as developing sensors with very small sized and adequate mechanical strength, selective detection, environmental friendliness, long lifetimes, and so on, the investigation of smart technology, new materials, and novel applications will continue in future.

1.2 Major manufacturers

Currently, many companies are involved in the design and development of printed and flexible sensors. Some of the major global players are included in the following with a short description, some elements of which have been quoted from their websites. There are also other manufacturers who are involved in the design, development, and production of printed and flexible sensors.

1.2.1 Interlink Electronics

Founded in the mid-1980s, 'Interlink Electronics has evolved into a leading provider of printed electronics, HMI devices, and sensor solutions' [25]. The company offers 'a full range of standard products as well as the ability to create custom solutions by leveraging the expertise in prototyping, materials science, firmware and software development, sensor fusion, system integration, and manufacturing' [25]. The company has developed sensor applications for numerous and diverse industries, such as the automotive, medical, industrial, and robotics fields. Interlink is also active within the burgeoning Internet of Things (IoT) sector, adopting the essential role of a full-service system integrator for IoT applications due to their membership in the LoRa Alliance ecosystem.

1.2.2 Tekscan

'Tekscan technology encompasses sensor technology, data acquisition electronics, and processing and analysis software. FlexiForce force sensors are ultra-thin and flexible printed circuits, which can be easily integrated into force measurement applications' [26].

1.2.3 PST Sensors

'Founded in 2010, PST Sensors is a Cape Town based technology company focussed on printed electronics and sensors. The letters PST stand for Printed Silicon Technology which forms the company's innovative base, and the word Sensors represents the team's passion for providing cutting edge temperature sensing solutions far beyond today's approaches' [27].

1.2.4 GSI Technologies

Established in the 1980s, the company has '40 years of experience with precision printing, use conductive inks to create durable, high-performance RFID antennae, electrodes, medical diagnostic sensors, drug delivery patches, organic photovoltaics, electro-chromatic and thermo-chromatic displays, smart cards, and many other applications across various industries. GSI is a leading provider of medical diagnostic electrodes, RFID antennas, printed heaters, printed sensors, EL lamps, and many other printed electronic applications' [28].

1.2.5 KWJ Engineering

KWJ Engineering 'builds products for detecting and monitoring a variety of gases including carbon monoxide, ozone, hydrogen sulfide, and other environmental pollutants. The developed instruments and modules are used in many industries including Smart Cities, Smart Homes, Personal Wearables, and IoT applications' [29].

1.2.6 Peratech Holdco

'Force-sensing HMI/MMI Solutions Company, founded in 1996 and inventor/developer of proprietary of Quantum Tunnelling Composites materials, providing next-generation Touch/Force-Sensing solutions. QTC® materials today are custom developed screen-printable inks, in both opaque and clear formulations, printing at only a few microns thick. These new incarnations of the material replace the legacy sheet materials, opening up many new opportunities for the technology as the key enabler. Peratech QTC® technology has been integrated in over one million devices, in areas such as smartphones, electronic whiteboards, cordless drills and even NASA robots' [30].

1.2.7 ISORG

'The company's core technology successfully integrates printed photodiodes on different substrates to enable large-area image sensors for the smartphone and security markets and extended applications in medical x-ray imaging, non-destructive testing and stock management. The company has strong knowledge in integration of printing-based photodiode on different substrate (glass, polyamide i.e. PI, ...) allows us to do image sensor of different size and shape dedicated to several applications' [31].

1.2.8 Fujifilm

'Starting in the 1980s, the use of digital technologies spread across a wide range of industries. From the beginning, Fujifilm was a digital pioneer, rapidly generating major digital advances in the fields of medicine, photography, printing, and more.

In the field of medicine, Fujifilm developed Fuji Computed Radiography (FCR), the world's first digital x-ray diagnostic system. In the field of photography, Fujifilm developed the DS-1P, the world's first digital camera, as well as the world's first digital minilab. In the field of printing, Fujifilm developed highly innovative computer-to-plate (CTP) systems.

To create all of these fundamental technologies, Fujifilm have developed strong fundamental technologies in electrical engineering, electronics, image analysis, imaging, and software. The core technologies that arose from these efforts remain one of Fujifilm's key strengths today' [32].

1.2.9 Canatu

'Canatu's solutions have brought the design freedom and user experience to the next level for 3D shaped touch devices. Canatu develops and manufactures innovative

3D formable and stretchable films and touch sensors, which are integrated into plastic, glass, textile or leather enabling 3D touch displays, smart switches and other intuitive user interfaces in automotive and consumer electronics' [33].

1.2.10 PolyIC

'PolyIC develops and markets products based on the platform technology printed electronics: In the course of this PolyIC focusses on individually manufactured transparent and flexible metal-mesh touch sensors. Touch sensors based on the PolyTC® technology offer transparent, conductive and flexible possibilities for touch screens and capacitive keys in any variants. The highlight is the possibility to combine decoration and function to achieve a maximum of design flexibility' [34].

1.2.11 MC10

'MC10's proprietary BioStamp® system wearable health tech creates comfortable, discreet sensors that can be applied anywhere on the body for targeted data collection. The wearable sensors are wireless and rechargeable via the included Link Hub. A dedicated mobile phone plus Link Application designed for remote data collection guides users through sensor application, prescribed activities, and eCOA' [35].

1.2.12 QUAD Industries

'Quad Industries uses its expertise in high-precision, fully automated screen printing-techniques to integrate functionality directly on various lightweight and flexible materials such as plastics, textiles, TPU and on paper. This allows the integration of a wide range of electronics—sensors, connectivity, heating—in any object, irrespective of its shape, size or material. Printed electronics not only enhance flexibility, they are also less harmful to the environment and more cost-effective to manufacture' [36].

1.2.13 Terabee

'Terabee develops and manufactures a wide range of sensor modules, including 2D infrared LED time-of-flight distance sensors, 3D time-of-flight depth cameras, and thermal cameras' [37]. The developed 'products are easy to use, compact, lightweight and offer great performance over a wide range' [37].

These major market players are involved in strategies, such as acquisitions, new product launches, and partnerships, to expand their reach in the market. For example, Thin Film Electronics ASA signed a distribution agreement with CymMetrik, China. This agreement focuses on sales expansion in India, China, and Taiwan. The printed sensors market ecosystem comprises raw material vendors, manufacturers, and end users. Some of the raw material vendors include T+ink, FlexEnable, Palo Alto Research Center, Brewer Science, and DuPont.

1.3 Materials for printed and flexible sensors

Manufacturing cost is an important consideration, but other than the cost of fabrication there are many other factors which need to be considered for the selection of materials for different parts of printed and flexible sensors. The size, mechanical strength, long-term performance, resistance to harsh environments, repeatability, reusability, and non-toxic and non-hazardous nature are the main requirements for the materials used in fabricating sensors.

Traditional printed sensors are based on silicon technology—an oxidized silicon wafer is spin coated with photoresist. A photoresist is a light-sensitive material used in several industrial processes, such as photolithography and photoengraving, to form a patterned coating on a surface. Different chemicals are used to give a material the desired permanent property variations. Materials such as poly(methyl methacrylate) (PMMA), poly(methyl glutarimide) (PMGI), phenol formaldehyde resin (DNQ/Novolac), and SU-8 are used. The materials are all applied as a liquid and are generally spin-coated to ensure uniformity of thickness.

Dry film stands alone amongst the other types in that the coating already exists as a uniform thickness, semi-solid film coated onto a polyester substrate and the user applies that substrate to the workpiece in question using lamination.

Gold sputtering is usually adopted on the side opposite to the sensing surface and it acts as a ground plane for single-sided measurement. The duration of sputtering depends on the density of the material sputtered and the thickness of the sputtered layer.

For flexible printed sensors the electrodes are usually printed on a substrate material. Different types of materials are currently available to be used as substrate materials. Although polydimethylsiloxane (PDMS) is extremely popular in research and development, other materials are also used. The comparative advantages and disadvantages of the different materials used as substrates are shown in table 1.1 [38–41].

Table 1.1. Comparative advantages and disadvantages of different substrate materials.

	Polydimethylsiloxane (PDMS)	Polyethylene terephthalate (PET)	Polyimide (PI)
Advantages	• Inert • Non-toxic • Non-flammable • Hydrophobic	• Inexpensive • Good chemical resistance • High resistance to temperature • High flexibility	• High flexibility • Good chemical and thermal resistance • High mechanical toughness
Disadvantages	• Difficult to integrate electrodes • Deposition needs to be carried out directly on its surface	• Very susceptible to heat degradation • Poor impact strength	• Expensive • Poor resistance to alkalis • Low impact strength

Table 1.2. Comparative advantages and disadvantages of different electrode materials.

	CNTs	Graphene	Aluminum
Advantages	• Better dispersion with a mixed polymer • Better compatibility • Higher flexibility	• High surface-to-volume ratio • Excellent electrical conductivity • High carrier mobility and density • High thermal conductivity	• Corrosion resistant • Strong in low temperatures
Disadvantages	• Low purity • Short lift time • Expensive growth process	• Does not have a band gap • High quality graphene is expensive and requires a complex process • Graphene exhibits some toxic qualities	• Growth of an oxide layer • More expensive than steel • Abrasive to tooling

The electrodes in flexible sensors are usually made of different materials, the most common among them are graphene, carbon nanotubes, aluminium, etc. The relative advantages and disadvantages of these materials are provided in table 1.2 [38–42].

The most common plastic substrates are polycarbonate (PC), polyethylene terephthalate (PET), polyethylene naphthalene (PEN), polyarylethersulfone (PES), polyamideimide (PAI), polyimide (PI), and polyethylene (PE).

1.4 Printing technologies

Many technologies are available and under development for printed and flexible sensors. In this section a few technologies will be described in brief. Some of the techniques will be used in combination to produce a printed sensor.

1.4.1 Thick-film technology

Thick-film technology is a common method used by many industries 'to produce electronic devices such as surface mount devices (SMD), hybrid integrated circuits, heating elements and sensors' [43, 44]. 'The complete method involves deposition of several successive layers of conductors, resistors and dielectric layers onto an electrically insulating substrate using a screen-printing process. The whole thick-film process usually consists of the following stages: lasering of substrates [45], preparation of ink [46], screen printing [47], drying/curing [48], firing [49], abrasive trimming [50] and laser trimming [51].

1.4.2 Thin film technology

Wikipedia defines a thin film as 'a layer of material ranging from fractions of a nanometer (monolayer) to several micrometers in thickness. The controlled synthesis of materials as thin films (a process referred to as deposition) is a fundamental step in many applications'. Thin film sensors 'are produced by directly depositing material onto fixtures in a vacuum deposition chamber by a process known as sputtering. A sputtering system allows for process control so that films may be produced with a high degree of repeatability' [52–54].

1.4.3 Inkjet printing

According to Wikipedia, the 'concept of inkjet printing originated in the 20th century, and the technology was first extensively developed in the early 1950s. The world first inkjet printer was invented by Ichiro Endo, who worked for Canon in Japan. In the late 1970s, inkjet printers that could reproduce digital images generated by computers were developed, mainly by Epson, Hewlett-Packard (HP) and Canon. In the worldwide consumer market, four manufacturers account for the majority of inkjet printer sales: Canon, HP, Epson and Brother' [55]. Inkjet printing is a type of digital or non-contact printing that recreates a digital image by propelling droplets of ink onto paper and plastic substrates [55].

1.4.4 Photolithography

Lithography is derived from the Greek words *litho* for stone and *graph* for drawing. Photolithography, also called optical lithography or ultra-violet (UV) lithography, is a process used in microfabrication to pattern parts on a thin film or the bulk of a substrate [56]. It uses light to transfer a geometric pattern from a photomask onto a photosensitive chemical photoresist on the substrate. The steps involved in the photolithographic process are wafer cleaning; barrier layer formation; photoresist application; soft baking; mask alignment; exposure and development; and hard-baking [57].

1.4.5 Masked photolithography

A photomask is made by exposing, or writing, the designer's pattern onto a resist coated chrome mask blank [58]. The latent image in the resist is then developed to form the required pattern. This resist image acts as a mask during the etching process. The pattern is transferred into the chrome film when the resist layer is removed. 'Photolithography uses three basic preparation steps to transfer a pattern from a mask to a wafer: coat, develop, expose. The pattern is transferred into the wafer's surface layer during a subsequent process. In some cases, the resist pattern can also be used to define the pattern for a deposited thin film. The wafer is exposed by UV (ultraviolet) from a light source traveling through the mask to the resist. A chemical reaction occurs between the resist and the light. Only those areas not protected by the mask undergo a chemical reaction' [58].

1.4.6 Maskless photolithography

The process of maskless photolithography, which does not use an intermediate static mask, utilizes methods that transfer the information directly onto the substrate. There are some advantages as well as disadvantages to this method compared to masked photolithography, which are detailed in [58, 59]. A key advantage of maskless lithography is the ability to change the lithography patterns from one run to the next, without incurring the cost of generating a new photomask [60].

1.4.7 Screen printing

The complete process of screen printing is explained in detail in [61]. Screen printing is a multidimensional process, and it becomes complex when micro- to nano-level electronics devices are to be printed precisely on a flexible or bendable substrates. The properties of substrates, such as surface roughness and chemical, mechanical, thermal, optical, thermo-mechanical, electrical, and magnetic properties, have to be studied in detail. The correct properties need to be chosen for optimum utilization. In the case of screen printed antennas, consistency in maintaining the required 'planarity' of the flexible or non-flexible underlying substrate will be of utmost importance.

1.4.8 Sputtering

According to Intlvac '[s]puttering can be described in a number of ways: cathodic sputtering, diode sputtering, RF or DC sputtering, ion-beam sputtering, reactive sputtering—but all of these are essentially describing the same physical process. Sputtering is a thin-film manufacturing process widely used across many industries including semiconductor processing, precision optics, and surface finishing. Sputtered thin films have excellent uniformity, density and adhesion making them ideal for multiple applications.

The target (source) material and substrate (destination) are placed into a vacuum chamber and a voltage is applied between them so that the target is the cathode and the substrate is attached to the anode. A plasma is created by ionizing a sputtering gas, usually an inert gas such as argon or xenon. Inert gases are typically employed as the sputtering gas because they tend not to react with the target material or combine with any process gases and because they produce higher sputtering and deposition rates due to their high molecular weight.

The sputtering process occurs when the target material is bombarded with the sputtering gas and the resulting energy transfer causes target particles to escape, travel and deposit on the substrate as a film. For the sputtering process to produce an effective coating, a number of criteria must be met. First, ions of sufficient energy must be created and directed towards the surface of the target to eject atoms from the material. The interaction of the ions and the target are determined by the velocity and energy of the ions. Since ions are charged particles, electric and magnetic fields can control these parameters. The process begins when a stray electron near the cathode is accelerated towards the anode and collides with a neutral gas atom converting it to a positively charged ion' [62, 63].

1.4.9 Direct laser writing

Prem Prabhakaran at l3dw.com defines direct laser writing (DLW) as '3D printing for the microscopic world. This techniques goes beyond the smallest shapes and sizes that can be accomplished by garden variety 3D printing. The highest resolution that can be achieved by DLW is typically [in the] few micron to sub-micron range. To put this into perspective, the average human hair has a cross section of 70–100 μm so we can imagine [...] the 3D structures [sitting] nicely [...] by the dozen across the cross-section of a hair. This technique allows 3D drawing of complex shapes with fine features (μ to few nm). The complexity of shapes that can be achieved by this technique are limited only by [the] photochemistry of [the] materials used for fabrication, and the limits of the optical setup' [64].

1.4.10 Direct dry printing of carbon nanotubes

As Binghao Liang *et al* explain 'in the direct dry printing of carbon nanotubes, a porous CNT block [is] used as both the seal and the ink; and Ecoflex film...serve[s] as an object substrate. Well-designed CNT patterns can be easily fabricated on the polymer substrate by engraving the target pattern on the CNT seal before the stamping process. Moreover, the CNT film can be directly used to fabricate [an] ultrathin (300 μm) strain sensor. This strain sensor possesses high sensitivity with a gauge factor (GF) up to 9960 at 85% strain, high stretchability (>200%) and repeatability (>5000 cycles)' [65].

1.4.11 Hybrid printed electronics

According to Holstcentre 'The hybrid printed electronics technology opens new possibilities for electronics applications. Combining printed circuits and devices with traditional electronic components like LEDs and chips, it enables large-area, flexible and freeform applications that can be manufactured in high volumes using roll-to-roll printing and assembly processes' [66, 67].

1.5 Conclusion

The chapter has described the progress of printed and flexible sensors over the last few decades. The demand for small sizes, smart devices, high performance, and critical applications is keeping the research and development of the field exciting. Many different sensors have been reported and a lot of work is in progress. With time new sensors will be available for use and will help us to live a better, healthier, and more comfortable life. New challenges, such as COVID-19, will be faced by humanity in the future and researchers will find ways to tackle these problems and find appropriate solutions.

Acknowledgement

Funded by the German Research Foundation (DFG, Deutsche Forschungsgemeinschaft) as part of Germany's Excellence Strategy—EXC 2050/1—Project ID 390696704—Cluster of Excellence 'Centre for Tactile Internet with Human-in-the-Loop' (CeTI) of Technische Universität Dresden.

References

[1] Global Market Insights Printed sensors market Size, COVID-19 impact analysis, regional outlook, application development potential, price trends, competitive market share & forecast, 2021–2027 *Report* GMI1621 https://gminsights.com/industry-analysis/printed-sensors-market

[2] https://marketwatch.com/press-release/global-printed-flexible-sensors-market-size-share-growth-analysis-forecast-to-2025-2019-11-26?mod=mw_quote_news

[3] Transparency Market Research 2021 Printed and flexible sensors market—global industry analysis, size, share, growth, trends and forecast 2019–2029 *Report* TMRGL270 https://transparencymarketresearch.com/printed-and-flexible-sensors-market.html

[4] https://idtechex.com/en/research-report/printed-and-flexible-sensors-2017-2027-technologies-players-forecasts/504

[5] Mukhopadhyay S C, Yamada S and Iwahara M 2002 Inspection of electroplated materials—performance comparison with planar meander and mesh type magnetic sensor *Int. J. Appl. Electromagnet. Mech.* **15** 323–29

[6] Mukhopadhyay S C, Yamada S and Iwahara M 2002 Evaluation of near-surface material properties using planar type mesh coils with post-processing from neural network model *Int. J. Appl. Electromagnet. Mech.* **23** 181–9

[7] Mukhopadhyay S C 2002 Quality inspection of electroplated materials using planar type micro-magnetic sensors with post processing from neural network model *IEE Proc.—Sci. Meas. Technol.* **149** 165–71

[8] Mukhopadhyay S C, Yamada S and Iwahara M 2002 Experimental determination of optimum coil pitch for a planar mesh type micro-magnetic sensor *IEEE Trans. Magn.* **38** 3380–2

[9] Mukhopadhyay S C, Gooneratne C P, Sen Gupta G and Yamada S 2005 Characterization and comparative evaluation of novel planar electromagnetic sensors *IEEE Trans. Magn.* **41** 3658–60

[10] Mukhopadhyay S C 2005 Novel planar electromagnetic sensors: modeling and performance evaluation *Sensors* **5** 546–79

[11] Mukhopadhyay S C 2004 A novel planar mesh type micro-electromagnetic sensor: part I—model formulation *IEEE Sensors J.* **4** 301–7

[12] Mukhopadhyay S C 2004 A novel planar mesh type micro-electromagnetic sensor: part II—estimation of system properties *IEEE Sensors J.* **4** 308–12

[13] Mukhopadhyay S C and Gooneratne C P 2007 A novel planar-type biosensor for non-invasive meat inspection *IEEE Sensors J.* **7** 1340–6

[14] Mukhopadhyay S C, Woolley J D M, Sen Gupta G and Demidenko S 2007 Saxophone reed inspection employing planar electromagnetic sensors *IEEE Trans. Instrum. Meas.* **56** 2492–503

[15] Mukhopadhyay S C, Deb Choudhury S, Allsop T, Kasturi V and Norris G E 2008 Assessment of pelt quality in leather making using a novel non-invasive sensing approach *J. Biochem. Biophys. Methods* **70** 809–15

[16] Yunus M A and Mukhopadhyay S C 2011 Novel planar electromagnetic sensors for detection of nitrates and contamination in natural water sources *IEEE Sensors J.* **11** 1440–7

[17] Rahman M S A, Mukhopadhyay S C, Yu P L, Chuang C H and Haji-Sheikh M 2011 Measurements and performance evaluation of novel interdigital sensors for different chemicals related to food poisoning *IEEE Sensors J.* **11** 2957–65

[18] Zia A I, Syaifudin Abdul Rahman M, Mukhopadhyay S C, Pak-Lam Y, Al-Bahadly I H, Gooneratne C P, Kosel J and Liao T-S 2013 Technique for rapid detection of phthalates in water and beverages *J. Food Eng.* **116** 515–23
[19] Nag A, Zia A I, Li X, Mukhopadhyay S C and Kosel J 2016 Novel sensing approach for LPG leakage detection: part I: operating mechanism and preliminary results *IEEE Sensors J.* **16** 996–1003
[20] Nag A, Zia A I, Xie L, Mukhopadhyay S C and Kosel J 2016 Novel sensing approach for LPG leakage detection: part II: effects of particle size, composition and coating layer thickness *IEEE Sensors J.* **16** 1088–94
[21] Nag A, Mukhopadhyay S C and Kosel J 2017 Tactile sensing from laser-ablated metallized PET films *IEEE Sensors J.* **17** 7–13
[22] Nag A, Mukhopadhyay S C and Kosel J 2017 Sensing system for salinity testing using laser-induced graphene sensor *Sens. Actuators* A **264** 107–16
[23] Alahi M E E, Xie L, Mukhopadhyay S and Burkitt L 2017 A temperature compensated smart nitrate-sensor for agricultural industry *IEEE Trans. Ind. Electron.* **64** 7333–41
[24] Azmi A, Azman A A, Kaman K K, Ibrahim S, Mukhopadhyay S C, Nawawi S W and Yunus M A M 2017 Electromagnetic sensing array to detect water contamination *IEEE Sensors J.* **17** 5244–51
[25] https://interlinkelectronics.com/
[26] https://tekscan.com/
[27] PST Sensors
[28] https://www.gsitech.com/
[29] KWJ Engineering
[30] Peratech
[31] ISORG
[32] Fujifilm
[33] Canatu
[34] Poly IC
[35] mc10
[36] QUAD Industries
[37] Terabee
[38] Yongzhao X, Xiduo H, Kundu S, Nag A, Afsarimanesh N, Sapra S, Mukhopadhyay S C and Han T 2019 Silicon-based sensors for biomedical applications: a review *Sensors* **19** 2908
[39] Han T, Nag A, Simorangkir R B V B, Afsarimanesh N, Liu H, Mukhopadhyay S C, Xu Y, Zhadobov M and Sauleau R 2019 Multifunctional flexible sensor based on laser-induced graphene *Sensors* **19** 3477
[40] Han T, Nag A, Afsarimanesh N, Mukhopadhyay S C, Kundu S and Yongzhao X 2019 Laser-assisted printed flexible sensors: a review *Sensors* **19** 1462
[41] Nag A, Alahi M E E, Feng S and Mukhopadhyay S C 2019 IoT-based sensing system for phosphate detection using graphite/PDMS sensors *Sens. Actuators* A **286** 43–50
[42] Nag A, Simorangkir B V B R, Valentin E, BJorninen T, Ukkonen L, Hashmi M R and Mukhopadhyay S C 2018 A transparent strain sensor based on PDMS-embedded conductive fabric for wearable sensing applications *IEEE Access* **6** 71020–27
[43] Wikipedia
[44] Thick Film Technologies

[45] Nag A, Mukhopadhyay S C and Kosel J 2016 Flexible carbon nanotube nanocomposite sensor for multiple physiological parameter monitoring *Sens. Actuators A* **251** 148–55
[46] Edberg J, Brooke R, Hosseinaei O, Fall A, Wijeratne K and Sandberg M 2020 Laser-induced graphitization of a forest-based ink for use in flexible and printed electronics *npj Flex. Electron.* **4** 1–10
[47] Maddipatla D, Narakathu B B and Atashbar M 2020 Recent progress in manufacturing techniques of printed and flexible sensors: a review *Biosensors* **10** 199
[48] Chlaihawi A A, Narakathu B B, Emamian S, Bazuin B J and Atashbar M Z 2018 Development of printed and flexible dry ECG electrodes *Sens. Bio-sens. Res.* **2** 9–15
[49] Kassem O., Saadaoui M., Rieu M. and Viricelle J.P. 2019 A novel approach to a fully inkjet printed SnO_2-based gas sensor on a flexible foil *J. Mater. Chem. C* **7** 12343–53
[50] Cardoso R M, Castro S V, Silva M N, Lima A P, Santana M H, Nossol E, Silva R A, Richter E M, Paixao T R and Munoz R A 2019 3D-printed flexible device combining sampling and detection of explosives *Sens. Actuators B* **292** 308–13
[51] Harper A F, Diemer P J and Jurchescu O D 2019 Contact patterning by laser printing for flexible electronics on paper *npj Flex. Electron.* **3** 1–6
[52] Wikipedia
[53] AZO Sensors
[54] https://grc.nasa.gov/WWW/sensors/PhySen/docs/TM-107418.pdf
[55] Wikipedia
[56] https://en.wikipedia.org/wiki/Photolithography
[57] https://lnf-wiki.eecs.umich.edu/wiki/Lithography_processing
[58] Nag A, Zia A I, Mukhopadhyay S C and Kosel J 2015 Performance enhancement of electronic sensor through mask-less lithography *Proc. of the Ninth Int. Conf. on Sensing Technology (Auckland, New Zealand)* (Piscataway, NJ: IEEE) pp 406–11
[59] Nag A, Zia A I, Babu A and Mukhopadhyay S C 2015 Printed electronics: present and future opportunities *Proc. of the Ninth Int. Conf. on Sensing Technology (Auckland, New Zealand)* (Piscataway, NJ: IEEE) pp 412–17
[60] https://en.wikipedia.org/wiki/Maskless_lithography
[61] Reid Print Technologies
[62] Intlvac Thin Film
[63] Nanografi
[64] Direct Laser Writing
[65] Liang B, Zang Z, Chen W, Lu D, Yang L, Yang R, Zhu H, Tang Z and Gui X 2019 Direct patterning of carbon nanotube via stamp contact printing process for stretchable and sensitive sensing devices *Nano-Micro Lett.* **11** 92
[66] Hybrid Printed Electronics
[67] Holst Centre

Chapter 2

Printed flexible sensors for academic research

Anindya Nag and Subhas Chandra Mukhopadhyay

The exploitation of printed flexible sensing prototypes in applications has been a great boon in the field of microelectronics. These sensors have been developed in research laboratories with enhanced attributes for their operation in specific sectors. Before the commercialization of these printed flexible sensors for real-time applications, these sensors are characterized and implemented in the research field to demonstrate their capabilities. This chapter exemplifies the utilization of printed flexible sensors for a range of research-based applications. These applications are categorized mainly under the biomedical, industrial, and environmental sectors. The spectrum of printed flexible sensors varies in certain factors such as fabrication techniques, processing materials, and working mechanisms. This chapter elucidates each of these categories, giving their advantages and viability. We conclude with some prospective ideas that can be executed in the future to improve the quality of these printed flexible sensors.

2.1 Introduction

The arrival of flexible sensors in sensing technology provided exponential improvement in terms of their fundamental attributes. When the use of smart sensors became popular among researchers in the late 1970s and early 1980s [1, 2] the types of sensors and their fabrication techniques were very different. At that time CMOS-based sensors were favored highly due to the available fabrication techniques. Most research was focused around the sensors available at that time. Then, with the growing interest in microelectromechanical systems (MEMS), silicon-based sensors were developed with higher quality. These sensors were applied mostly in industrial [3–5] and environmental [6–8] applications. Utilizing certain advantages of silicon sensors, low-cost sensing systems were developed with these prototypes. The great popularity of silicon sensors assisted in the commercialization of sensors in the real world. Researchers began developing sensors that can address applications in almost all sectors. With growing competition, work was done on optimizing the quality of

doi:10.1088/978-0-7503-3439-6ch2 © IOP Publishing Ltd 2021

the sensors employed for particular applications. However, due to certain disadvantages associated with silicon-based sensors, researchers had to seek alternatives that could serve their purposes.

Sensors with flexible mechanical characteristics began to be fabricated using novel processing materials. In addition to addressing the disadvantages faced by MEMS-based sensors, these flexible sensors have also been able to provide a wider spectrum of applications [9–11] in the real world. These sensors have been used in several sectors such as the military [12, 13], medicine [14, 15], sports [16, 17], etc. While scientists around the world synthesized the highly efficient materials carbon nanotubes (CNTs) [18, 19] and graphene [20, 21], others tried to use these materials to produce flexible sensors with high quality. Novel fabrication techniques were formulated that could process raw materials. Among the processed materials, different types of polymers were used to form the substrates of the prototypes. Some of them are organic, inorganic, and synthesized polymers. A few examples are polydimethylsiloxane (PDMS) [22–24], polyethylene terephthalate (PET) [25–27], polyimide (PI) [28–30], and poly(3,4-ethylenedioxythiophene) polystyrene sulfonate (PEDOT:PSS) [31–33]. Among the conductive materials, carbon-based allotropes and other metallic nanoparticles involving gold [34–36], aluminum [37–39], iron [40–42], and copper [8, 43, 44] have been employed to produce flexible sensing prototypes. In terms of fabrication techniques, certain printing techniques such as 3D printing [45–47], laser cutting [48–50], screen printing [51–53], inkjet printing [54–56], and gravure printing [57–59] have been used to develop sensors.

With the use of these materials and techniques, researchers have started developing a range of flexible sensors with enhanced electrical, mechanical, and thermal properties. The printed nature of flexible sensors provides additional advantages such as light weight, ease of production, high roll-to-roll (R-2-R) production, high sensitivity, and robustness. Furthermore, the inclusion of wearable properties in these prototypes has increased their applications greatly [60, 61]. These wearable sensors have been deployed for biomedical applications to a great extent [62, 63]. A market survey performed on the growth in the use of printed flexible sensors shows a steady increase over the next few years [64, 65]. These sensors come in the form of bio-, capacitive, piezoelectric, piezoresistive, temperature, humidity, optical, and gas sensors. The rise of printed sensors in the global market has been predicted to reach around USD 10.46 billion by 2022, with a compound annual growth rate (CAGR) of 7% [66]. These types of sensors will be used for both research and industrial purposes. The CAGR rate should further increase to 12.8% over the next few years, with the largest share accounted for by North America [67]. These facts indicate that the ground level work on these sensors will be performed at a large scale in future. This chapter elucidates the work related to the fabrication and implementation of printed flexible sensors for research purposes that has been done to date. It categorizes the use of sensors on the basis of three different applications, each of which is exemplified by a few significant works. These sensors have been chosen on the basis of their unique structure, working principles, and performance linked to the target application. These sensors have been developed by different research groups focusing on one or more of these categorized applications.

2.2 Printed flexible sensors

Printed flexible sensors have been popularized in the last decade [68, 69], particularly meeting the need for sensors with multifunctional capabilities. The prototypes developed with electrochemical properties have the additional feature of flexibility in these types of sensors. Using the available techniques, they can be fabricated on a large scale in a very short time. A combination of processing materials is used under controlled temperature and humidity conditions to form these sensors. Initially, the characterization and experiments were carried out on small samples, followed by testing with larger quantities. The chosen examples given below show the dynamic range of applications that have been addressed with these sensors. The operating mechanisms are electrochemical [70–72], strain [73–75], and electrical [76–78] sensing.

2.2.1 Biomedical applications of printed flexible sensors

The first category of applications is the most significant one for which flexible sensors have been introduced. Biomedical sensors have been used for measuring different kinds of samples with a wide range of concentrations, targeted for specific applications. Explanations regarding the operating principles of electrochemical and strain sensing are presented here. The quality of the sensors has been constantly optimized to improve the range of operation, the limit of detection (LOD), linear range, response time, and recovery time. One of the works [79] explains the development of flexible sensors using an inkjet printing technique for the electrochemical detection of lysozyme molecules. Selective operation was carried out using aptamer-based biosensors. The advantages of these sensors are their low-cost, high selectivity, and disposable nature. Cyclic voltammetry (CV) and electrochemical impedance spectroscopy (EIS) techniques were used to determine the responses of the sensors. Figure 2.1 [79] shows a schematic diagram of the fabrication and

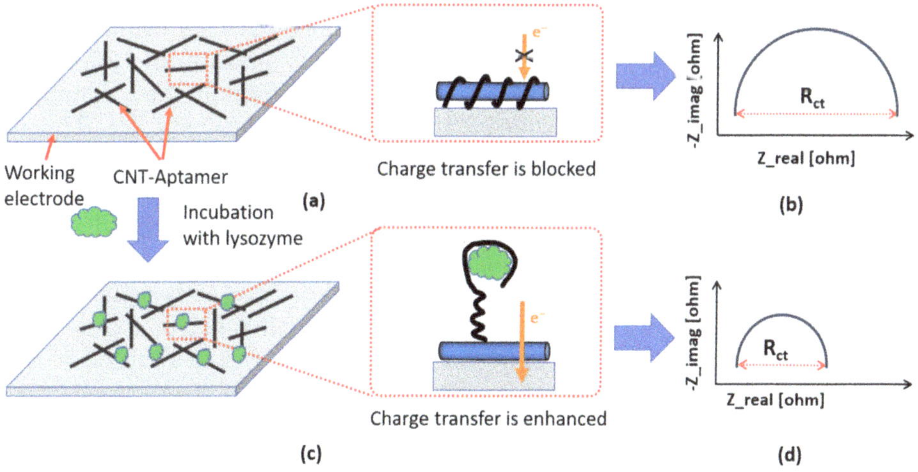

Figure 2.1. Schematic diagram of the (a), (c) operation and (b), (d) experimental results of the CNTs and aptamer-based complex for the detection of lysozyme molecule [79] CC BY 4.0.

working principle of the developed sensors. CNTs and aptamers-based complexes were formed and deposited on the working electrode. Printable inks of CNTs were deposited with inkjet printing on the electrodes. The inks were prepared by mixing 0.25 mg L^{-1} of multi-walled carbon nanotubes (MWCNTs) with 5 μM lysozyme in a 30% N-methyl pyrrolidone solution. The mixture was then sonicated, centrifuged, and filtered to form the ink cartridges for printing purposes. The inkjet printing was carried out with drop volumes of 10 pL, and an input voltage, temperature, and frequency of 40 V, 35 °C, and 5 kHz, respectively. Each prototype was printed with five layers of the CNT–aptamer ink. Strong π–π stacking interactions between the nucleotide bases of single-stranded DNA and the walls of the CNTs were exploited to form the complex. The formed sensors were used as point-of-care diagnostic tools and disposable testing kits. With the exposure of the target analyte, the bond between the CNTs and the aptamer was broken, while forming a bond between the aptamer and lysozyme. This led to the flow of electrons from the redox probes containing the aptamer–lysozyme complex to the working electrodes, which subsequently varied the charge-transfer resistance. The sensors were tested with different concentrations of 2.25, 0.5, 1, 2, 5, 10, and 20 μg mL^{-1} . The stability in the responses of the sensors was 35 days at room temperature, with a tolerance level of ±1.73% for the first 21 days. The LOD of the sensors for the tested analyte was 90 ng mL^{-1}.

The report in [70] explains the development of stretchable, screen-printed electrochemical sensors for the detection of glucose in human perspiration. The sensors were printed with platinum (Pt)-decorated graphite on a fabric glue polymer for the detection of physiological fluids. The measurement was performed using chronoamperometry via the immobilization of glucose oxidase on Pt-decorated graphite. Figure 2.2 [70] depicts the fabrication of Pt-decorated printed sensors. The quantification of glucose was achieved using the developed sensors to obtain high sensitivity. The reduction method was used to synthesize the graphite-supporting Pt nanoparticles. A value of 3 wt.% of Pt was optimized to produce the modified electrodes. Three different inks were used to develop the three working, counter, and pseudo-reference electrodes. The screen printing was performed with designs produced by AutoCAD software on flexible substrates. Initially, the printing and curing of Ecoflex were performed on temporary tattoo transfer paper. This was placed carefully on the PU sheets and cured to obtain proper connections between them. Then another layer of temporary transfer paper was used as a protective layer to cover the Ecoflex layer. Then two different types of devices were printed, with one consisting of working electrodes developed with Pt-graphite ink, and the other one consisting of Pt-decorated graphite and stretchable graphite as working and counter electrodes, respectively. Both types of sensors were then cured in a convection oven, followed by defining the electrode area and contact pads using an adhesive of silver epoxy ink. Finally, the samples were cured at 65 °C for 10 min. The immobilization of these sensors was performed using the GOx enzyme with the working electrode covered with a chitosan solution. The modified sensors were stored at 4 °C before their use. CV was used to analyze the output of the sensors. The sensitivity of the sensors was 105 ± 3 μA cm^{-2} mM^{-1}. The linear range was between 0 and 0.9 mM,

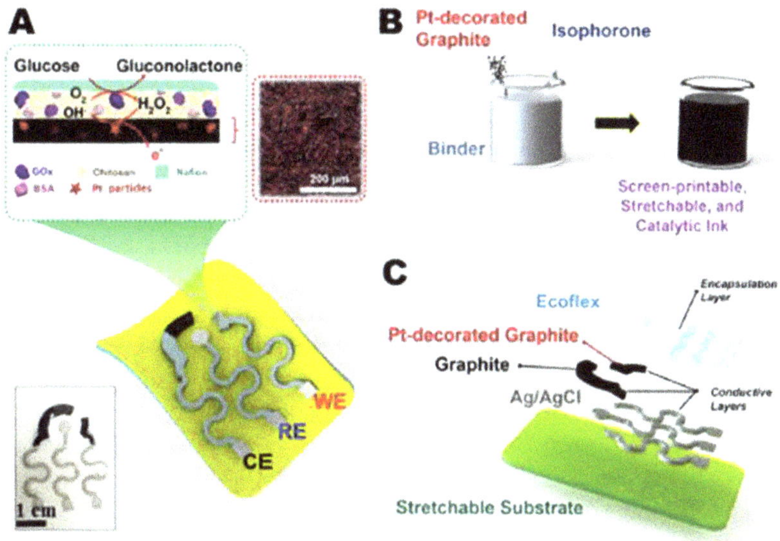

Figure 2.2. Schematic diagram of the development of Pt-immobilized graphite-based sensors. Reprinted from [70] Copyright (2016), with permission from Elsevier.

with an LOD of 6.6 mM. High repeatability of the responses was obtained, with recovery of the sensors between 71% and 84%.

Regarding the strain-sensing applications, an intriguing works is presented in [56], where inkjet printed pressure sensors were developed for monitoring of postural imbalance. The substrate of the sensors was formed using flexible printed circuit boards (FPCBs) using conventional technology. Two polymeric membranes made up of thermoplastic elastomers (TPU) were printed with PEDOT:PSS to form the electrodes of the prototypes. Four different flexible polymers, namely PDMS, PI, PET, and TPU, were used alongside two types of conductive inks to produce the sensors. The PEDOT:PSS conductive inks were printed on the TPU substrates to obtain proper quality adhesion. Drop-on-demand printing was performed to form the electrodes of the developed sensors. The printing was carried out on two separate flexible membranes to form the top and bottom electrodes of the sensors. The FPCBs were formed on PI substrates with copper connections with thicknesses of 125 and 35 μm, respectively. A multiplexing strategy was used to minimize the connection lines on the FPCBs. This was carried out by placing the connections for the sensing excitation and reading separately on the top and bottom of the electrode, respectively. Cross-coupling was avoided by using the same channel for excitation and reading for two sensors. The final step consisted of the bonding of the TPU membranes using a polyester-based, electrically conductive adhesive in certain zones of the substrates. This was done primarily to form the contacts of the capacitive sensors. The measurements were performed in terms of capacitance, followed by converting them into voltage using a graphical user interface developed in MATLAB. This helped in the real-time visualization of the collected data. The range of pressure varied from 100 to 180 kPa to obtain higher sensitivity and higher initial capacitance values. The sensitivity, noise, and stability of the developed capacitive sensors were 50 fF Pa^{-1}, 1.2 fF $\sqrt{Hz}^{-1}$.

Another significant work explaining the utilization of printed flexible sensors can be shown in the development of wearable, high-sensitivity graphene-based strain sensors [80]. The sensors operated in terms of piezoresistive sensing and were formed using graphene woven fabrics (GWFs) on polymers and medical tape composite films. The sensors had high potential for the fields of displays, robotics, fatigue detection, and body monitoring. Some of the advantages of these films are easy fabrication, ultralightess, superior physical robustness, high sensitivity, high reversibility, and retractability with respect to the human skin. The GWFs were fabricated using the chemical vapor deposition (CVD) technique via growing graphene on the surface of crisscross copper meshes. Followed by fabrication of GWFs with a sheet resistance of 400–500 Ω sq^{-1}, they were transferred to pre-treated films of PDMS and medical tapes. The PDMS substrates were cleaned, degassed, solidified, and peeled off from a Petri dish to form the substrates. Then, silver wires were connected with silver paste on both ends of the GWF–PDMS. Finally, another PDMS layer was used to cover the electrodes to protect them against exfoliation. This process was continued to develop multiple devices, where the devices were tailored into samples of appropriate sizes. The thickness of the substrates was around 0.5 mm. A strain of 0.1%–0.2% was applied to characterize the sensors to obtain a change in terms of relative resistance with respect to time. Some of the tested human activities include strain-induced motions such as hand clenching, phonation, expression changes, blinking, breathing, and pulse.

2.2.2 Industrial applications of printed flexible sensors

Even though the development of printed flexible sensors has primarily been focused on biomedical applications, the industrial uses of printed flexible sensors have also increased significantly. These sensors have been used mainly in robotics and related applications, where the dimensions and processed materials of these sensors have been varied to vary the sensitivity of the prototypes. One of the works explaining the use of printed flexible sensors for electrochemical sensing is the detection of hydrogen peroxide (H_2O_2) gas [81] using laser-induced sensors. Laser-inscribed graphene, decorated with silver nanoparticles, were used to form the electrodes of the sensors. Direct reduction of graphene oxide (GO) was performed to form electrochemical transducers for gas-sensing applications. The advantages of these sensors were light weight, small size, and high sensitivity. Figure 2.3 [81] shows the fabrication steps followed to develop GO, developed from Hummer's method, which was drop-cast on flexible PET sheets. These sheets were placed on CDs for the laser inscription process with a power of 5 mW. Single electrodes and triple electrodes were developed using laser inscription and transferred to the GO films. Then the PET sheets were removed from the disc. Finally, the electrodes were formed using Kapton tape and silver ink. The induced graphene was then decorated with silver nanoparticles with an electrodeposition technique using three-electrode systems. The sensors exhibited a high catalytic response towards the H_2O_2 gas,

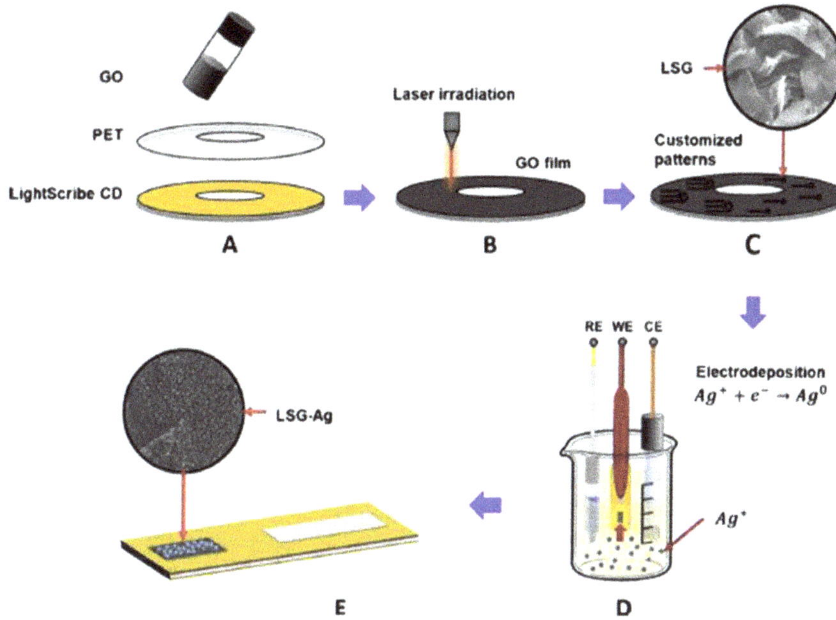

Figure 2.3. Schematic diagram of the fabrication of laser-inscribed graphene sensors decorated with silver nanoparticles. Reprinted from [81] Copyright (2019), with permission from Elsevier.

obtaining a fast response time of 3 s. Both the EIS and CV techniques were used to determine the responses of the sensors. The sensors operated within a linear range from 0.1 to 10 mM, having an LOD and R^2 value of 7.9 μM and 0.9982, respectively. Continuous bending of the sensors led to a loss of current of 13%. The repeatability and reproducibility of the sensors were 3% relative standard deviation (RSD) and 4.5% RSD, respectively.

Another work related to the use of printed flexible sensors for electrochemical sensing of industrial applications is [82]. This work was based on the use of solid polymer electrolyte (SPE), where a poly-vinylidene fluoride (PVDF) matrix was mixed in 1-methyl-2-pyrrolidone (NMP) for the immobilization of 1-ethyl-3-methylimidazolium tetrafluoroborate (EMIM-BF4) to develop electrochemical amperometric gas sensors. The SPE was coated on single electrodes of the sensors to produce a two-layered structured prototype. Interdigitated designs were formed on sensors using a screen printing technique. A carbon paste was used as the conductive ink to form the electrodes. Followed by pre-processing of the printing machine, the printing was carried out on flexible PET films. The SPE was developed on the PDVF matrix while optimizing the effects of temperature and blending ratio on the morphology and porosity of the SPE. The coating of the SPE layer was performed using an equal displacement method. The sensing area of the interdigitated electrodes was covered with uncooled SPE, followed by stabilization of the samples for 24 h before experimental use. The thickness of this SPE coating also varied the sensitivity towards the gas. Uniformity of the coating was achieved by magnetic stirring of the polymer with CJJ78–1. The sensors were tested in a controlled environment. The operating mechanism was based on the interconnection of the pore structures in the coated SPE. The sensors were tested for concentrations between 10 and 100 ppm of NO_2 gas. The best sensitivity was obtained when the sensors were treated at 80 °C for 2 min and the sensors were coated with SPE with a thickness of 1.25 mm and a weight ratio of 1:1:4.

The strain-sensing properties of printed flexible sensors were exploited to test them for soft-robotics applications [83]. The sensors were formed with elastic conductive thin films and carbon black composites. The advantages of these sensors were their high elasticity and robust nature. Conductive carbon black was mixed in an Ecoflex matrix with an optimized value of 8.33 wt.% to form the composites. The printing was performed using a customized printer where the thickness of the thin films was controlled by controlling the gap between the edge of the doctor blade and the substrate. The angle between the substrate of the sensors and the doctor blade was optimized to ensure uniform shearing of the composite material throughout the fabrication process. Then the samples were cured in an oven at 50 °C for 10 h, followed by creating the designated patterns of the electrodes. Additional dielectric support was given to the electrodes by attaching them on aluminum substrates, along with a calculated volume of Ecoflex cast over the composites. Two layers of composite-Ecoflex-based electrodes were bonded together and made normal to each other. The presence of Ecoflex as an adhesive to join the two layers of Ecoflex-coated composite-Ecoflex electrodes provided maximum shear and tensile strengths. An Ecoflex layer with a thickness of 0.05 μm joined the two active layers, followed by encapsulating them with another layer. The sensors were operated as tactile sensors, where the change in response with respect to pressure was detected in terms of capacitive values. The measurement was performed using an LCR meter to record the date against time. Load cells with a weight of 0.05 N were used to test the sensing arrays that consisted of 16 tactels. A sensitivity of 150 pF MPa^{-1} along with zero hystereses was obtained using these sensors.

Another work showcasing the use of printed flexible sensors for robotic applications was performed using 3D-printed sensors [84]. The sensors were developed using composites formed with conductive graphene pellets and flexible TPU pellets. 3D printing of the composite filaments was carried out and they were fused with each other to form an extensive 3D conductive network. The printing was performed on glass substrates with a nozzle diameter of 0.4 mm and a temperature of 70 °C. The sensors had a dimension of 2 × 1.5 cm, along with a layer depth of 200 μm. The sensors displayed good stretchability, high sensitivity, high durability, and reversibility of the results. These prototypes operated as resistive sensors, where the sensors were embedded inside a 3D-printed PLA-based finger hinge to detect bending feedback in certain applications such as robotics, actuators, and smart wearable devices. The initial testing was carried out manually by hanging a 250 gm load in the air with the help of paper clips. The initial resistance of the sensors was 100 kΩ, with a volume resistivity of 5 Ωm. The response time of the sensors was 200 ms, with high stability of the responses for over 6000 cycles. The gauge factor (GF) of the sensors was 11 and 80 at strains of 10% and 100%, respectively. The conductive network broke down when the strain was beyond 250%.

2.2.3 Environmental applications of printed flexible sensors

The third category in the use of these sensors is environmental applications, where prototypes have been employed to detect analyte present in the ambient at varied concentrations. These sensors have also been used to measure certain parameters such as temperature and humidity, whose optimization is critical for the fabrication, testing, and utilization of sensors. One interesting work is the development of laser-induced graphene sensors for the detection of saline concentrations in water bodies [85]. A single-step fabrication method was followed to form the conductive material, which was eventually transferred to tapes to form electrodes for the sensors. The advantages of these sensors are their low cost, quick fabrication, and high sensitivity towards the target material. Laser induction was performed on commercial PI films to photo-thermally convert the hybridization of carbon atoms from sp^3 in PI films to sp^2 in graphene. Figure 2.4 [85] shows a schematic diagram of the fabrication of laser-induced graphene sensors. After the PI films of 120 μm were attached to glass substrates for laser ablation, direct laser induction was performed to form the patterns, designed using CorelDraw software. Then the formed graphene was manually transferred to Kapton tapes by placing the tapes carefully over the formed graphene. Once the induced graphene was transferred, the sensors were employed as electro-chemical sensing prototypes for the testing of water samples from different water bodies. The devices operated as impedimetric sensors where the change in resistive and reactive values was studied with respect to the different concentrations of salinity of the samples. The sensors were connected carefully to the Kelvin probes of the LCR meter such that the bonding pads did not touch the water samples during experimentation. Concentrations ranging from 4 to 40 000 ppm were tested using the developed sensors. The resolution of the sensors was 1 ppm. Finally, the sensors were embedded in a microcontroller-based system to increase their potential for real-time applications.

Similar research related to the use of printed flexible sensors was shown in the detection of phosphate ions in water bodies [86]. 3D-printed mold-based sensors were developed using composites of graphite and PDMS to form the electrodes of the sensors. The advantages of these sensors lie in their low cost, simple operating principles, and ability to detect ions at low concentrations. 3D-printed molds were formed at controlled temperatures and humidity tolerances. Figure 2.5 [86] shows a schematic diagram of the fabrication steps for the 3D-printed mold-based sensors.

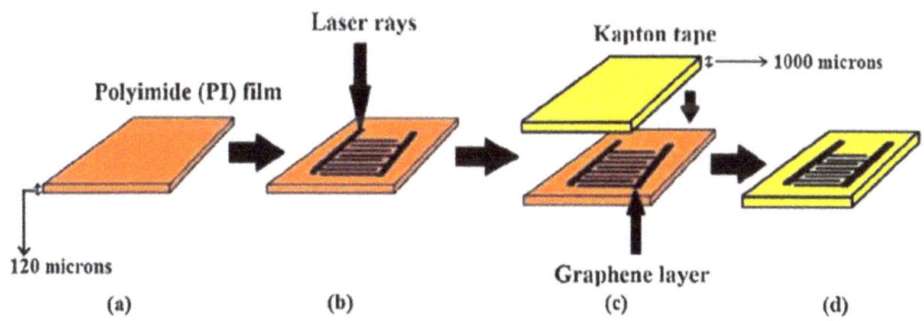

Figure 2.4. Schematic diagram of the fabrication steps for laser-induced graphene sensors. Reprinted from [85] Copyright (2017), with permission from Elsevier.

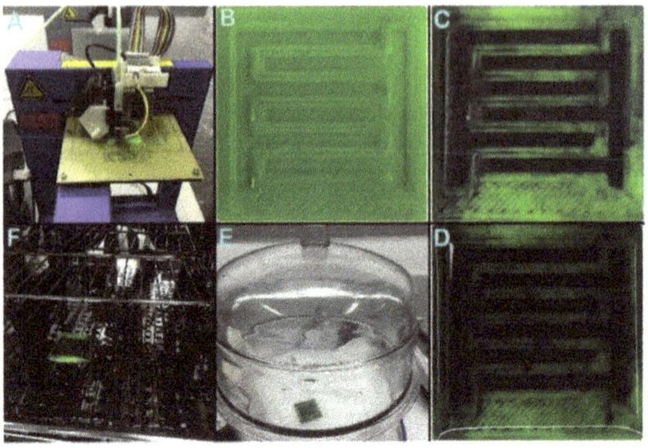

Figure 2.5. The fabrication steps that were followed to develop 3D-printed mold-based sensors are shown serially (A)–(F). Reprinted from [86] Copyright (2019), with permission from Elsevier.

Acrylonitrile butadiene styrene (ABS), with a diameter of 1.75 mm, was used as the printing filament to form the molds due to its high resilience to changes in temperature. The 3D-printed molds were formed with trenches having a depth of 500 μm. Then graphite was cast in the molds to fill up the trenches while removing the redundant graphite powder off the molds. Then PDMS was cast on the developed molds. The height of the cast PDMS was adjusted to 1000 μm using a casting knife. The samples were then desiccated to remove the trapped air bubbles, followed by oven curing at 70 °C for 2 h to solidify the substrates of the sensors. Each of the prototypes had a length and width of 18 and 2 mm, respectively. The overall sensing area of the patches was 900 mm^2. The sensors were used for testing water samples to determine the presence of phosphate ions at varying concentrations. The testing was performed for concentrations ranging from 30 to 0.03 ppm in order to analyze the changes in the reactance and impedance values of the sensors. A frequency sweep was performed between 1 Hz and 50 kHz, where an optimal response was obtained between 30 and 50 kHz. A particular frequency of 47 kHz was chosen to obtain a sensitivity of 0.0498 Ω ppm^{-1} . The response time and LOD of the sensors were 1 s and 0.03 ppm, respectively. Finally, the sensors were embedded with microcontroller-based systems for using these systems in real-time applications.

The use of printed flexible sensors for other environmental applications can be seen in their employment for temperature, humidity, and pH sensing. The report in [87] showed the fabrication of sensors on flexible substrates for use in temperature sensing. The sensors were fabricated using the inkjet printing process, particularly opting for a drop-on-demand technique. Silver ink and PI were used to form the electrodes and substrates, respectively. The Kapton substrates were initially cleaned to remove residual grease and organic contamination. The viscosity of the silver was 9–15 cps at 25 °C. These inks were sonicated to minimize nozzle clogging, remove large particles, and ensure stability of the suspension. The dispensing device used during the printing process had a capillary glass tube and an aperture coupled with a piezoelectric element. The printing process was carried out at a frequency of 400 Hz and a droplet velocity of 4.5 m s^{-1} . The silver lines formed were cured thermally at 130 °C and 150 °C for 10 and 30 min, respectively. After the process was completed, the resistivity and resistance of the two sheets were 5.9 $\mu\Omega$cm and 0.163 Ω sq^{-1}, respectively. The sensitive area, silver line width, and gap of the sensors were 6.2 cm^2, 300 μm, and 60 μm, respectively. The prototypes functioned as thermistors, with an input voltage ranging between 0 and 1 V. The sensors were tested for determining temperature variations ranging from 20 °C to 60 °C . The mean sensitivity of the sensors was 2.23×10^{-3} $°C^{-1}$. The responses of the sensors also displayed good linearity and hysteresis less than 5%.

Another work related to the use of printed flexible sensors for environmental applications is [88]. The sensors were formed using copper oxide (CuO) nanorods (NR) and nanoflowers (NF) on flexible PET substrates. The low-temperature hydrothermal synthesis technique of screen printing technology was employed to form the nanostructures for the electrodes. The size and morphology of the nanostructures were optimized by varying the solvent, reaction temperature, and duration of synthesis. Interdigitated electrodes were formed with an electrode finger length, width, and spacing of 10, 0.5, and 0.5 mm, respectively. The overall printed area was 18×13 mm^2. Then the developed interdigitated layers were dried at 120 °C for 20 min. This was followed by forming pastes by mixing a 40 wt.% of CuO NF and NR with co-polymers formed with poly(methyl methacrylate)—poly(butyl methacrylate) as a binder and butyl carbitol acetate as a solvent. The sensors were operated as electrochemical sensing prototypes that had great potential in certain areas suh as food, medicine, and agriculture. The CuO nanostructures formed on flexible substrates had high crystallinity and low surface roughness. The CuO-based sensors formed were employed for pH sensing, which was monitored using the EIS technique. The testing was performed within a frequency range between 20 Hz and 10 MHz. For a particular pH value, the capacitances decreased with an increase in frequency. The sensitivity of the sensors was 0.64 μF pH^{-1} within a pH range of 5–8.5. The sensors also showed negligible response towards other ions such as sodium, potassium and other ions.

2.3 Conclusion and future work

The chapter explains the work done on the development of printed flexible sensors and their utilization for different applications in academic research. Various kinds of printed sensors have been designed and fabricated, followed by characterizing and deployment in a range of applications. Some examples related to biomedical, industrial, and environmental uses have been shown above, in order to highlight the significance of these sensors. The advantages of these sensors include their high electrical conductivity, light weight, high flexibility, robustness, quick R-2-R production, multifunctional operation, and high selectivity and sensitivity toward the tested particles. Although the range of applications of these sensors is very large, there are certain sectors for which these sensors show great potential. The formation of composites, which can combine the individual properties of different materials, will not only improve the quality of the sensors, but will also increase their functionality. This increase in the functional nature of the sensors can reduce the number of sensors required for operation, thus reducing the total power required to run the sensors. The techniques used to develop the sensors can also be combined to generate sensors with optimized characteristics. For example, customization can be performed using 3D printing techniques to print thin-film sensors with high electrical conductivity and mechanical flexibility directly. The research into these sensors can also combine two superior elements for improved performance. For example, graphene and CNTs can be fused together to form sensors for electrochemical [89, 90] and strain-sensing [91, 92] applications. Signal-conditioning circuits can also be scaled up to where an array of printed sensors can operate together in an efficient manner. This would help to generate systems that can work under any environmental conditions. With the improvement in the quality of these sensors through the above-mentioned techniques, their prospects in academic research will increase to a great extent.

Acknowledgement

Funded by the German Research Foundation (DFG, Deutsche Forschungsgemeinschaft) as part of Germany's Excellence Strategy—EXC 2050/1—Project ID 390696704—Cluster of Excellence 'Centre for Tactile Internet with Human-in-the-Loop' (CeTI) of Technische Universität Dresden.

References

[1] Hampel D 1980 Application of very large scale integrated (VLSI) circuits to smart sensors *Proc. SPIE* **252** 2–9

[2] Breckenridge R and Katzberg S 1980 Smart sensors for the 80s—the status of smart sensors *Sensor Systems for the 80s Conf.* 1917

[3] Nag A, Zia A I, Li X, Mukhopadhyay S C and Kosel J 2015 Novel sensing approach for LPG leakage detection: part I—operating mechanism and preliminary results *IEEE Sensors J.* **16** 996–1003

[4] Beebe D J, Hsieh A S, Denton D D and Radwin R G 1995 A silicon force sensor for robotics and medicine *Sens. Actuators* A **50** 55–65

[5] Pérez R, Chaillet N, Domanski K, Janus P and Grabiec P 2006 Fabrication, modeling and integration of a silicon technology force sensor in a piezoelectric micro-manipulator *Sens. Actuators* A **128** 367–75

[6] Alahi M E E, Nag A, Mukhopadhyay S C and Burkitt L 2018 A temperature-compensated graphene sensor for nitrate monitoring in real-time application *Sens. Actuators* A **269** 79–90

[7] Mourzina Y G, Ermolenko Y E, Yoshinobu T, Vlasov Y, Iwasaki H and Schöning M J 2003 Anion-selective light-addressable potentiometric sensors (LAPS) for the determination of nitrate and sulphate ions *Sensors Actuators* B **91** 32–8

[8] Barhoumi L *et al* 2017 Silicon nitride capacitive chemical sensor for phosphate ion detection based on copper phthalocyanine–acrylate-polymer *Electroanalysis* **29** 1586–95

[9] Xu Y *et al* 2019 Silicon-based sensors for biomedical applications: a review *Sensors* **19** 2908

[10] Harraz F A 2014 Porous silicon chemical sensors and biosensors: a review *Sens. Actuators* B **202** 897–912

[11] Kumar S S and Pant B 2014 Design principles and considerations for the 'ideal'silicon piezoresistive pressure sensor: a focused review *Microsyst. Technol.* **20** 1213–47
[12] Becker L 2006 Influence of IR sensor technology on the military and civil defense *Proc. SPIE* **6127** 61270s
[13] Skorobogatov S and Woods C 2012 Breakthrough silicon scanning discovers backdoor in military chip *International Workshop on Cryptographic Hardware and Embedded Systems* (Berlin: Springer) pp 23–40
[14] Bisogni M G, Guerra A D and Belcari N 2019 Medical applications of silicon photomultipliers *Nucl. Instrum. Methods* A **926** 118–28
[15] Åstrand A P 2012 A flexible resonance sensor system for detection of cancer tissue: evaluation on silicon *Thesis* Umeå University
[16] Ishido H, Takahashi H, Nakai A, Takahata T, Matsumoto K and Shimoyama I 2015 6-axis force/torque sensor for spike pins of sports shoes 2015 28th IEEE Int. Conf. on Micro Electro Mechanical Systems (MEMS) (Piscataway, NJ: IEEE) pp 257–60
[17] Lightman K 2016 Silicon gets sporty *IEEE Spectr.* **53** 48–53
[18] Nag A, Mukhopadhyay S C and Kosel J 2016 Flexible carbon nanotube nanocomposite sensor for multiple physiological parameter monitoring *Sens. Actuators* A **251** 148–55
[19] Han T, Nag A, Mukhopadhyay S C and Xu Y 2019 Carbon nanotubes and its gas-sensing applications: a review *Sens. Actuators* A **291** 107–43
[20] Nag A, Mitra A and Mukhopadhyay S C 2018 Graphene and its sensor-based applications: a review *Sens. Actuators* A **270** 177–94
[21] Mehmood A *et al* 2020 Graphene based nanomaterials for strain sensor application—a review *J. Environ. Chem. Eng.* **8** 103743
[22] Nag A, Afasrimanesh N, Feng S and Mukhopadhyay S C 2018 Strain induced graphite/PDMS sensors for biomedical applications *Sens. Actuators* A **271** 257–69
[23] Lei K F, Lee K-F and Lee M-Y 2012 Development of a flexible PDMS capacitive pressure sensor for plantar pressure measurement *Microelectron. Eng.* **99** 1–5
[24] Nag A, Feng S, Mukhopadhyay S, Kosel J and Inglis D 2018 3D printed mould-based graphite/PDMS sensor for low-force applications *Sens. Actuators* A **280** 525–34
[25] Nag A, Mukhopadhyay S C and Kosel J 2016 Tactile sensing from laser-ablated metallized PET films *IEEE Sensors J.* **17** 7–13
[26] Li S *et al* 2018 The room temperature gas sensor based on Polyaniline@ flower-like WO_3 nanocomposites and flexible PET substrate for NH_3 detection *Sens. Actuators* B **259** 505–13
[27] Yaqoob U, Phan D-T, Uddin A I and Chung G-S 2015 Highly flexible room temperature NO_2 sensor based on MWCNTs-WO_3 nanoparticles hybrid on a PET substrate *Sens. Actuators* B **221** 760–68
[28] Han T *et al* 2019 Gold/polyimide-based resistive strain sensors *Electronics* **8** 565
[29] Yi Y, Ali S and Wang B 2019 An inkjet-printed strain sensor with a carbon-silverpolyimide topology 2019 IEEE Int. Conf. on Flexible and Printable Sensors and Systems (FLEPS) (Piscataway, NJ: IEEE) pp 1–3
[30] Qin Y *et al* 2015 Lightweight, superelastic, and mechanically flexible graphene/polyimide nanocomposite foam for strain sensor application *ACS Nano* **9** 8933–41
[31] Ramírez J, Rodriquez D, Urbina A D, Cardenas A M and Lipomi D J 2019 Combining high sensitivity and dynamic range: wearable thin-film composite strain sensors of graphene, ultrathin palladium, and PEDOT:PSS *ACS Appl. Nano Mater.* **2** 2222–29

[32] Bali C, Brandlmaier A, Ganster A, Raab O, Zapf J and Hübler A 2016 Fully inkjet-printed flexible temperature sensors based on carbon and PEDOT:PSS *Mater. Today Proc.* **3** 739–45

[33] Seekaew Y, Lokavee S, Phokharatkul D, Wisitsoraat A, Kerdcharoen T and Wongchoosuk C 2014 Low-cost and flexible printed graphene–PEDOT:PSS gas sensor for ammonia detection *Org. Electron.* **15** 2971–81

[34] Gong S *et al* 2014 A wearable and highly sensitive pressure sensor with ultrathin gold nanowires *Nat. Commun.* **5** 1–8

[35] Hu C, Bai X, Wang Y, Jin W, Zhang X and Hu S 2012 Inkjet printing of nanoporous gold electrode arrays on cellulose membranes for high-sensitive paper-like electrochemical oxygen sensors using ionic liquid electrolytes *Anal. Chem.* **84** 3745–50

[36] Xu M *et al* 2017 A novel flexible electrochemical glucose sensor based on gold nanoparticles/polyaniline arrays/carbon cloth electrode *Sens. Actuators* B **252** 1187–93

[37] Lamanna L *et al* 2019 Flexible and transparent aluminum-nitride-based surface-acoustic-wave device on polymeric polyethylene naphthalate *Adv. Electron. Mater.* **5** 1900095

[38] Meng X, Yang J, Liu Z, Lu W, Sun Y and Dai Y 2020 Non-contact, fibrous cellulose acetate/aluminum flexible electronic-sensor for humidity detecting *Compos. Commun.* **20** 100347

[39] Ueno N, Akiyama M, Ikeda K and Tateyama H 2002 A foil type flexible pressure sensor using nitelide aluminum thin film *Trans. Soc. Instrum. Control Eng.* **38** 427–32

[40] Bandgar D, Navale S, Naushad M, Mane R, Stadler F and Patil V 2015 Ultra-sensitive polyaniline–iron oxide nanocomposite room temperature flexible ammonia sensor *RSC Adv.* **5** 68964–71

[41] Wang Z, Yang S, Miao S, Shi Q, He T and Lee C 2019 A motion-balanced sensor based on the triboelectricity of nano-iron suspension and flexible polymer *Nanomaterials* **9** 690

[42] Villagrossi E, Cenati C, Pedrocchi N, Beschi M and Tosatti L M 2017 Flexible robot-based cast iron deburring cell for small batch production using single-point laser sensor *Int. J. Adv. Manuf. Technol.* **92** 1425–38

[43] Zhang Y *et al* 2020 A flexible non-enzymatic glucose sensor based on copper nanoparticles anchored on laser-induced graphene *Carbon* **156** 506–13

[44] Yang Y-J *et al* 2008 An integrated flexible temperature and tactile sensing array using PI-copper films *Sens. Actuators* A **143** 143–53

[45] Liu C *et al* 2018 3D printing technologies for flexible tactile sensors toward wearable electronics and electronic skin *Polymers* **10** 629

[46] Sajid M, Gul J Z, Kim S W, Kim H B, Na K H and Choi K H 2018 Development of 3D-printed embedded temperature sensor for both terrestrial and aquatic environmental monitoring robots *3D Print. Addit. Manuf.* **5** 160–69

[47] Muth J T *et al* 2014 Embedded 3D printing of strain sensors within highly stretchable elastomers *Adv. Mater.* **26** 6307–12

[48] Nag A, Menzies B and Mukhopadhyay S C 2018 Performance analysis of flexible printed sensors for robotic arm applications *Sens. Actuators* A **276** 226–36

[49] Han T *et al* 2019 Multifunctional flexible sensor based on laser-induced graphene *Sensors* **19** 3477

[50] Nag A and Mukhopadhyay S C 2018 Fabrication and implementation of printed sensors for taste sensing applications *Sens. Actuators* A **269** 53–61

[51] Lee C G *et al* 2018 Stretchable strain sensors fabricated by screen printing of silver paste on the surface modified transparent elastomeric polyurethane films *Mater. Sci. Appl.* **9** 1008

[52] Moorthi A, Narakathu B, Reddy A, Eshkeiti A, Bohra H and Atashbar M 2012 A novel flexible strain gauge sensor fabricated using screen printing *2012 Sixth Int. Conf. on Sensing Technology (ICST)* (Piscataway, NJ: IEEE) pp 765–8

[53] Chang W-Y, Fang T-H, Lin H-J, Shen Y-T and Lin Y-C 2009 A large area flexible array sensors using screen printing technology *J. Disp. Technol.* **5** 178–83

[54] Le D D, Nguyen T N N, Doan D C T, Dang T M D and Dang M C 2016 Fabrication of interdigitated electrodes by inkjet printing technology for apllication in ammonia sensing *Adv. Nat. Sci.: Nanosci. Nanotechnol.* **7** 025002

[55] Kisic M, Blaz N, Zlebic C, Zivanov L and Damnjanovic M 2015 Flexible inkjet printed sensor for liquid level monitoring *2015 38th International Spring Seminar on Electronics Technology (ISSE)* (Piscataway, NJ: IEEE) pp 472–76

[56] Cruz S, Dias D, Viana J C and Rocha L A 2015 Inkjet printed pressure sensing platform for postural imbalance monitoring *IEEE Trans. Instrum. Meas.* **64** 2813–20

[57] Bariya M *et al* 2018 Roll-to-roll gravure printed electrochemical sensors for wearable and medical devices *ACS Nano* **12** 6978–87

[58] Park J, Nam D, Park S and Lee D 2018 Fabrication of flexible strain sensors via roll-to-roll gravure printing of silver ink *Smart Mater. Struct.* **27** 085014

[59] Maddipatla D, Narakathu B, Turkani V, Bazuin B and Atashbar M 2018 A gravure printed flexible electrochemical sensor for the detection of heavy metal compoundsProceedings **2** 950

[60] Nag A *et al* 2018 A transparent strain sensor based on PDMS-embedded conductive fabric for wearable sensing applications *IEEE Access* **6** 71020–27

[61] Nag A, Mukhopadhyay S C and Kosel J 2017 Wearable flexible sensors: a review *IEEE Sensors J.* **17** 3949–60

[62] Hassannejad H, Matrella G, Ciampolini P, De Munari I, Mordonini M and Cagnoni S 2017 Automatic diet monitoring: a review of computer vision and wearable sensor-based methods *Int. J. Food Sci. Nutr.* **68** 656–70

[63] Chen H, Xue M, Mei Z, Bambang Oetomo S and Chen W 2016 A review of wearable sensor systems for monitoring body movements of neonates *Sensors* **16** 2134

[64] Printed and flexible sensors 2017–27: technologies, players, forecasts https://www.idtechex.com/en/research-report/printed-and-flexible-sensors-2017-2027-technologies-players-forecasts/504 (Accessed: 28 Sept)

[65] Printed and flexible sensors market worldwide https://www.databridgemarketresearch.com/reports/global-flexible-sensors-market (Accessed: 28 Sept 2020)

[66] Markets and Markets 2016 Printed sensors market worth *Report* SE 4468

[67] Esticast Printed and flexible sensor market revenue *Report* Esticast

[68] Khan S, Lorenzelli L and Dahiya R S 2014 Technologies for printing sensors and electronics over large flexible substrates: a review *IEEE Sensors J.* **15** 3164–85

[69] Han T, Nag A, Afsarimanesh N, Mukhopadhyay S C, Kundu S and Xu Y 2019 Laser-assisted printed flexible sensors: a review *Sensors* **19** 1462

[70] Abellán-Llobregat A *et al* 2017 A stretchable and screen-printed electrochemical sensor for glucose determination in human perspiration *Biosens. Bioelectron.* **91** 885–91

[71] Moya A, Gabriel G, Villa R and del Campo F J 2017 Inkjet-printed electrochemical sensors *Curr. Opin. Electrochem.* **3** 29–39

[72] Afsarimanesh N, Mukhopadhyay S C and Kruger M 2017 Molecularly imprinted polymer-based electrochemical biosensor for bone loss detection *IEEE Trans. Biomed. Eng.* **65** 1264–71

[73] Maddipatla D 2016 Development of fully printed and flexible strain, pressure and electrochemical sensors *Thesis* Western Michigan University, Kalamazoo, MI
[74] Lee H, Seong B, Moon H and Byun D 2015 Directly printed stretchable strain sensor based on ring and diamond shaped silver nanowire electrodes *RSC Adv.* **5** 28379–84
[75] Eshkeiti A *et al* 2014 A novel self-supported printed flexible strain sensor for monitoring body movement and temperature *SENSORS, 2014* (Piscataway, NJ: IEEE) pp 1615–18
[76] Nag A, Mukhopadhyay S and Kosel J 2017 Influence of temperature and humidity on carbon based printed flexible sensors 2017 Eleventh Int. Conf. on Sensing Technology (ICST) (Piscataway, NJ: IEEE) pp 1–6
[77] Aliane A *et al* 2014 Enhanced printed temperature sensors on flexible substrate *Microelectron. J.* **45** 1621–26
[78] Harada S, Kanao K, Yamamoto Y, Arie T, Akita S and Takei K 2014 Fully printed flexible fingerprint-like three-axis tactile and slip force and temperature sensors for artificial skin *ACS Nano* **8** 12851–57
[79] Khan N I, Maddaus A G and Song E 2018 A low-cost inkjet-printed aptamer-based electrochemical biosensor for the selective detection of lysozyme *Biosensors* **8** 7
[80] Wang Y *et al* 2014 Wearable and highly sensitive graphene strain sensors for human motion monitoring *Adv. Funct. Mater.* **24** 4666–70
[81] Aparicio-Martínez E, Ibarra A, Estrada-Moreno I A, Osuna V and Dominguez R B 2019 Flexible electrochemical sensor based on laser scribed graphene/Ag nanoparticles for non-enzymatic hydrogen peroxide detection *Sens. Actuators* B **301** 127101
[82] Luo R-B, Li H-B, Du B, Zhou S-S and Chen Y-H 2019 A printed and flexible NO_2 sensor based on a solid polymer electrolyte *Front. Chem.* **7** 286
[83] Devaraj H *et al* 2020 Highly elastic and flexible multi-layered carbon black/elastomer composite based capacitive sensor arrays for soft robotics *Meas.: Sens.* **2–4** 100004
[84] Gul J Z, Sajid M and Choi K H 2020 Retraction: 3D printed highly flexible strain sensor based on TPU–graphene composite for feedback from high speed robotic applications *J. Mater. Chem.* C **8** 2597–7
[85] Nag A, Mukhopadhyay S C and Kosel J 2017 Sensing system for salinity testing using laser-induced graphene sensors *Sens. Actuators* A **264** 107–16
[86] Nag A, Alahi M E E, Feng S and Mukhopadhyay S C 2019 IoT-based sensing system for phosphate detection using graphite/PDMS sensors *Sens. Actuators* A **286** 43–50
[87] Dankoco M, Tesfay G, Bènevent E and Bendahan M 2016 Temperature sensor realized by inkjet printing process on flexible substrate *Mater. Sci. Eng.* B **205** 1–5
[88] Manjakkal L, Sakthivel B, Gopalakrishnan N and Dahiya R 2018 Printed flexible electrochemical pH sensors based on CuO nanorods *Sens. Actuators* B **263** 50–8
[89] Radhakrishnan S and Mathiyarasu J 2019 Graphene–carbon nanotubes modified electrochemical sensors *Graphene-based Electrochemical Sensors for Biomolecules* (Amsterdam: Elsevier) pp 187–205
[90] Huang Q, Lin X, Tong L and Tong Q-X 2020 Graphene quantum dots/multiwalled carbon nanotubes composite-based electrochemical sensor for detecting dopamine release from living cells *ACS Sustain. Chem. Eng.* **8** 1644–50
[91] Shi J *et al* 2017 Graphene welded carbon nanotube crossbars for biaxial strain sensors *Carbon* **123** 786–93
[92] Pan S *et al* 2020 A highly stretchable strain sensor based on CNT/graphene/fullerene-SEBS *RSC Adv.* **10** 11225–32

Chapter 3

The fabrication of printed flexible sensors: challenges and possible outcomes

Anindya Nag and Subhas Chandra Mukhopadhyay

In the sensing world flexible sensors have proven to be a much better option in comparison to their counterparts. These flexible sensors have a dynamic range in terms of their electrical, mechanical, and thermal characteristics, which have been exploited in various forms for different kinds of applications. The utilization of printing technology to enhance the quality of flexible sensors has been revolutionary in the field of microelectronics. The attributes of printed flexible sensing prototypes have certain advantages that allow their deployment for multifunctional applications. This chapter deals with the fabrication of printed flexible sensors via highlighting some of the challenges of the fabrication techniques and possible corresponding solutions. The fabrication techniques include some of the major methodologies such as screen printing, inkjet printing, gravure printing, and flexographic printing, which have been used over the years to produce optimized printed flexible sensors. The conclusions are presented in the final section of the chapter.

3.1 Introduction

In the last few years there has been exponential growth in the use of sensors in the fields of science and technology. Sensors have been used in various sizes, shapes, and forms to meet the requirements of different applications. Earlier in the 1970s, when very large integrated circuits were popularized in the electronics industry [1, 2], these devices served very well in operating the early McIntosh and other computing systems. Although these systems formed the basis of modern micro-sized devices, certain factors had to be prioritized in order to improve their efficiency in terms of robustness, sensitivity, and longevity. The primary aims were the miniaturization of their size and optimizing their electrical and thermal characteristics. These changes not only assisted in integrating complex circuitries in a more compact manner, but also increased their commercial uses. These requirements were met by the

doi:10.1088/978-0-7503-3439-6ch3

© IOP Publishing Ltd 2021

introduction and popularization of semiconducting sensors in the late 1980s [3, 4], which were predominantly formed on silicon substrates. Some of the major advantages of these semiconducting sensors are their small size, excellent signal-to-noise ratio, low hysteresis, easy integration with conditioning circuits, high stability of responses, and capability to work under various environmental conditions [5, 6]. Thus, within a short span of time, these semiconducting sensors became widely used for a wide range of biomedical [7–9], industrial [10–12], and environmental [13–15] applications. These silicon-based sensors were fabricated using the microelectromechanical (MEMS) [16–18] technique. Some of the techniques involved in this process are photolithography [19, 20], spin-coating [21, 22], etching [23, 24], and sputtering [25, 26]. Each of these processes assists in achieving precise features for the formed prototypes. Although the large-scale implementation of these sensors was achieved for both commercial and research purposes, there were certain disadvantages that needed to be addressed. Some of these are a high fabrication costs, degradation and saturation in responses over a period of time, limitations in their use as wearable sensors due to their highly brittle nature, and the strong influence of the ambient temperature and humidity on the response of the sensors.

The limitations of silicon-based sensors led researchers to seek an alternative for ubiquitous sensing purposes. Sensors were then developed with flexible materials, where the degree of bending of the prototypes depended on the individual and combined mechanical attributes of the processed materials. The chosen materials were processed to form the electrodes and substrates of the flexible sensors, mainly based on the application of the sensors. The quality and performance of these flexible sensors are much better in comparison to the MEMS-based sensors in terms of sensitivity and efficiency. The substrates of these sensors are normally developed using different kinds organic, inorganic, and synthetic polymers. The conductive materials that have been considered to produce the electrodes include a range of nanomaterials and nanoparticles. Some of the common polymers that have been considered to form the substrates are polydimethylsiloxane (PDMS) [27–29], polyethylene terephthalate (PET) [30–32], polyimide (PI) [33–35], polyethylene naphthalate (PEN) [36–38], and poly(3,4-ethylenedioxythiophene) polystyrene sulfonate (PEDOT:PSS) [39–41]. Similarly, some of the nanomaterials that have been used to develop flexible sensors are carbon nanotubes (CNTs) [42–44], graphene [45–47], graphite [48–50], and metallic nanowires [51–53]. These materials have been processed to form sensors operating through different mechanisms to obtain the highest productivity. The working principles chosen to operate the sensors have been, either individually or in combination, resistive, capacitive, and inductive techniques. The designs of the electrodes have been based on interdigital [54–56], serpentine [57–59], and fractal [60–62] shapes, depending on the size, dimension, and processed materials.

The most frequent deployment of flexible sensors has been as wearable prototypes [63, 64]. These sensors have been embedded with different kinds of conditioning circuits to monitor patients with acute and chronic conditions. Wearable sensors have been used for monitoring patients with a range of conditions, where the durations of response of the sensors have been varied. The nature of flexible sensors varies depending on the technique used to fabricate them. One popular fabrication technique has been the printing process, which allows quick roll-to-roll production of high quality prototypes. Some of the common printing techniques are screen printing [65–67], inkjet printing [68–70], 3D printing [71–73], laser ablation [74–76], and gravure printing [77–79]. The attributes of the flexible sensors vary depending on the printing process used. Each of these techniques varies certain properties such as elasticity, aspect ratio, tensile strength, and thermal coefficients. Some of the specific advantages of these printed flexible sensors are their low cost, high flexibility, large sensing area, high degree of comfort, light weight, and biocompatibility [80, 81]. Due to these advantages there has been constant growth in the use of these printed flexible sensors in different sectors. These sensors have been used in different forms, including piezoelectric sensors, piezoresistive sensors, photodetectors, temperature sensors, gas sensors, and CMOS sensors [82]. Some of the major areas where these sensors are used are medical [83, 84], industrial [85, 86], and other automotive [87, 88] applications. For example, recently some of the major key players, such as Canatu, Fujifilm, ISORG, Interlink Electronics, etc, had a meeting related to the growth of the market in printed flexible sensors in the global scenario [89].

For the printed flexible sensors that have been developed using a range of fabrication techniques, a thorough review highlighting their advantages and disadvantages has not been carried out. A study of the challenges that are currently faced by the different fabrication techniques for developing printed flexible sensors is needed for a couple of reasons. First, for each of the above-mentioned fabrication techniques, the developed sensors have certain limitations that deter their multifunctional real-time applications. Second, sometimes the quality of the printed sensors degrades due to the shadowing effect. This generates an error in the output of the sensors. Thus it is important to deal with some of the problems related to the fabrication of flexible sensors using printing techniques. This chapter deals with some of the significant works linked to various printing techniques. It also describes the challenges related to current sensors, along with the possible solutions.

3.2 The fabrication of printed flexible sensors

The design and development of printed flexible sensors have been done in several steps, depending on the type of printing technique. Each of the sensors has been used for different types of applications, including electrochemical, strain and electrical sensing. Here are some of the examples of significant works for the three types of sensing, which have been fabricated using the five printing techniques.

3.2.1 Screen printing

This is the most popular type of printing, which was first popularized during the early 1900s [90, 91]. It has been considered for developing sensors due to its high quality of printing, high durability, and the ability to print on a range of substrates, including textiles, glass, and polymers. The sensors developed using this process have high sensitivity, high selectivity, and high stability of responses [92]. One interesting work is presented in [93], where stretchable electrochemical sensors were developed using a screen printing technique. These sensors were used for determining the concentration of glucose in human perspiration. The advantages of these sensors were their low-cost, all-printable nature, and high sensitivity toward the tested molecule. The sensors were immobilized with platinum (Pt) decorated graphite to increase the specificity of the target molecule. A wt.% of 4% of Pt was mixed with graphite using a reduction process at room temperature and atmospheric pressure. Three different types of inks were prepared for screen printing purposes on polyurethane (PU) sheets. The first type of ink was formed using 75% Pt–graphite and 25% ink. This composition was used for both the counter and working electrodes. The reference electrode was formed using a commercial silver/silver chloride ink, mixed with 10% of Ecoflex® to obtain extra stretchability. The designs of the sensors were created using AutoCAD software. The Ecoflex® was initially screen-printed on a temporary transfer tattoo paper and cured on the PU sheets for proper adhesion. Figure 3.1 [93] shows a schematic diagram of the generation of the inks, sensors, and their subsequent use for glucose detection. Cyclic voltammetry (CV) was used to test the sensors within a range between 33 μM and 0.9 mM. The performance of the sensors was much better after the immobilization of Pt-decorated graphite nanoparticles. The sensitivities obtained before and after the immobilization of graphite were $0.69 \pm 0.06\ \mu A\ cm^{-2}\ mM^{-1}$ and $105 \pm 3\ \mu A\ cm^{-2}\ mM^{-1}$. Similarly, the limit of detection (LOD) of the sensors after immobilization decreased from 6.6 mM to 10 μM. The linear range of the sensors was between 0 and 0.9 mM.

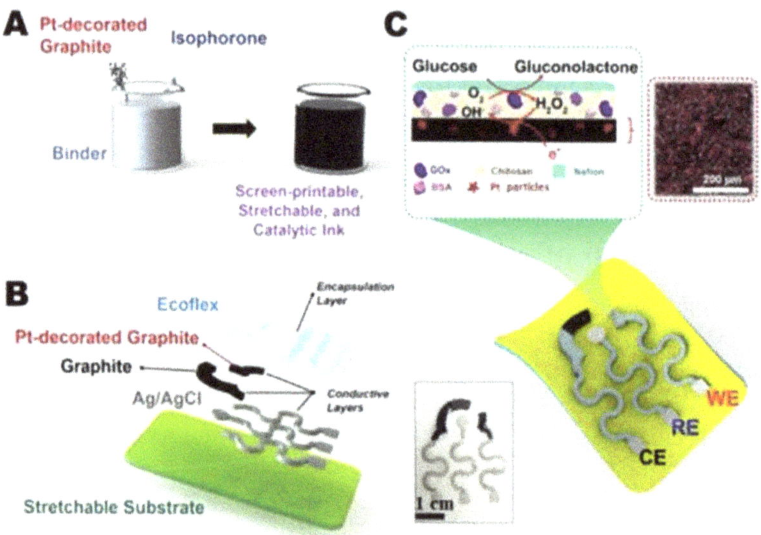

Figure 3.1. Schematic diagram of the (a) fabrication of conductive inks, (b) different layers of the sensing prototypes, and (c) working mechanism of the prototypes. Reprinted from [93] Copyright (2017), with permission from Elsevier.

Other significant works on the use of screen-printed sensors for electrochemical sensing are [94, 95].

For strain-sensing applications, screen printing was performed in [96] to develop low-cost sensors for wearable electronics. The sensors were made from silver nano-dendrite (ND) conductive inks, which have certain attributes such as rapid mass production and high performance. The Ag NDs were produced by a one-step redox reaction process of silver nitrate with copper foils. These conductive inks were mixed into a solution of polystyrene-blockpolyisoprene-block-polystyrene (SIS), xylene, and Disponer 904s. Screen printing with Ag ND inks was performed on nitrile rubber to obtain printed strain sensors. The substrates were initially cleaned with de-ionized (DI) water and ethanol. Then the printing process was carried out, followed by drying the samples at 60 °C for 2 min. Then the SIS ink was overprinted again onto the Ag ND layer, followed by drying the samples at 80 °C for 30 min. The final step was the cutting of the SIS layer and attaching copper tape on the two ends. The dimensions of the final product were a length and breadth of 60 and 12 mm, respectively. Figure 3.2 [96] shows a schematic representation of the steps of fabrication. The printing was also performed on other substrates such as PET and paper to characterize the quality of the printed inks. The sensors had high electrical conductivity and a high sensing range of 1.14×10^5 s m^{-1} and 105%, respectively.

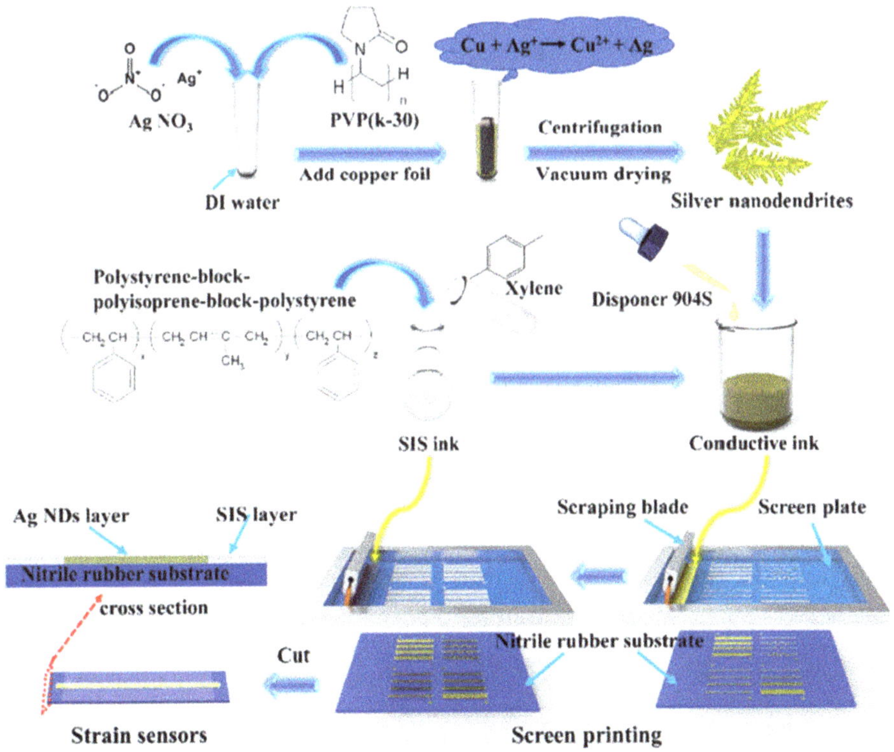

Figure 3.2. Schematic diagram of the steps of fabrication for printed strain sensors produced with Ag NDs and nitrile rubber. Reproduced from [96] with permission of The Royal Society of Chemisty.

The maximum gauge factor (GF) obtained from these sensors was 294.3. Other advantages of these sensors were a fast response time, long-term stability of the responses, and the ability to respond to different kinds of surface topographies. Other major works on the use of the screen printing technique to develop strain sensors are [97, 98].

One type of working mechanism involved in screen-printed prototypes is electrical sensors. One example can be seen in [99], where chipless and wireless temperature sensors were developed using a screen printing technique. The sensors were developed using carbon paste on flexible substrates. Resistor-inductor-capacitor circuits were formed to operate the sensors with a positive temperature coefficient. The resonant frequency of the circuit was varied by the sensing layer in accordance with the temperature values. Three different types of paste were used for the printing of the sensing prototypes. Silver paste was used to form the antennae of the sensors, whereas LOC-TITE ECI 8001 E&C paste was used to form the positive temperature coefficient. The third paste, TD-642, was used to form the dielectric in order to isolate the bridge between the inner and outer inductor ends. Initially, the silver paste was pre-heated and printed on PET substrates with a thickness of 100 μm. The screen mesh density of the printer used to carry out the process was 120 threads cm^{-1}. Then the paste were dried at 100 °C for 30 min, prior to the printing of the second type of paste. The final prototypes also consisted of a high-frequency antenna coil that was connected to a co-planar printed capacitor. The variation of the resonant frequency was less than 150 Hz for the entire range of relative humidity (RH). The operating range of the sensors was from 15 to 60 °C, obtaining sensitivities of 11.4 kHz $°C^{-1}$ and 1.8 kHz $°C^{-1}$ at 0 and 10 mm, respectively. Other significant works on the employment of the screen printing technique for electrical sensing applications are [100, 101].

3.2.2 Inkjet printing

The second most popular type of printing that has been used to fabricate printed flexible sensors is inkjet printing. This technique has been used since the 1950s, but was popularized in the 1980s [102, 103] for research purposes. One of the interesting research projects regarding the use of inkjet printing for flexible sensors is showcased in [104], where electrochemical gas sensors were produced using porous nanoparticles. The advantages of these thin-film sensors include their low cost, high electrical conductivity, excellent flexibility, and high integration. Gold nanoparticles were patterned on cellulose substrates to form the conducting layers of the prototypes. The electrodes were formed in arrays on paper-based substrates for conducting electrochemical sensing with ionic liquid (IL) electrolytes. Figure 3.3 [104] shows a schematic diagram of the fabrication of the prototypes. Initially, a mixed cellulose ester membrane was attached to the CD tray with adhesive tape for printing purposes. After the printing of the gold nanoparticles was completed, the patterns were dried at 60 °C for an hour. This was followed by placing pieces of the nanoparticle–cellulose composites onto plating solution for seeded growth. This was achieved by placing the composites onto the plating solution for 15 min, followed by

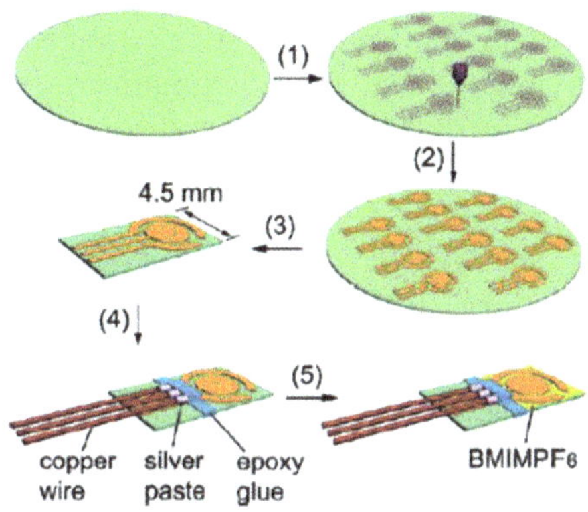

Figure 3.3. Schematic diagram of the steps of fabrication using inkjet printing to develop electrochemical sensors based on gold nanoparticle-based sensor arrays. Reprinted with permission from [104]. Copyright (2012) American Chemical Society.

gentle shaking and thorough rinsing with water. After the growth process was complete, the samples were finally dried at 60 °C under a vacuum for half an hour. Then individual gold arrays were stripped off and three thin copper wires were attached to the conducting areas of the electrode arrays using silver paste. The size of the gold nanoparticle–mixed cellulose composite was around 4.5 mm × 6.0 mm. 1-butyl-3-methylimidazolium hexafluorophosphate was used as the IL, which was dropped on the back of the substrates and allowed to reach equilibrium at room temperature. The porosity of the gold arrays assisted in the addition of electrolytes from the cellulose membranes to the top of the sensing area to achieve three-phase electrolyte–electrode–gas interfaces. The sensors achieved high sensitivity, along with a linear range from 0.054 to 0.177 v/v%. The LOD and response time of the sensors were 0.0075% and 10 s, respectively. Some other significant works on the use of inkjet printing for electrochemical sensing applications are [105, 106].

The work related to the use of inkjet printing for strain-sensing applications can be highlighted by the developed 2D-crystal based strain gauges on paper [107]. The advantages of these sensors were their biocompatibility and high sensitivity. The sensors were formed using water-based leverage technologies of biocompatible inks on flexible substrates such as paper. The graphene ink was synthesized using an ultrasonic-assisted liquid phase exfoliation technique. The graphite flakes were mixed with 1-pyrensulfonic acid sodium salt and DI water. This mixture was sonicated, followed by the removal of unexfoliated graphite via centrifugation. The centrifugation process was carried out at 4000 rpm for 20 min. This produced supernatant containing graphene and PS1 in water, which was again centrifuged at 15 000 rpm for an hour.

The final precipitate was re-dispersed in the printing solvent and was subsequently printed on flexible substrates. The graphene lines were printed with a length of 10 mm and a width of 0.5 mm. The graphene pads of the sensors were attached with copper foils, which in turn were connected with the crocodile clips for testing purposes. The performance of the sensors was dependent on certain printing factors, such as drop-spacing and the number of printing passes. When strain was applied, the thicker small-drop-spacing films produced micro-cracks with higher density and higher crack openings. A high GF of 125 was obtained when small strains such as 0.3% were applied to the prototypes. Other significant works on the use of inkjet printing techniques for the fabrication and implementation of strain sensors are [108, 109].

For electrical sensing, one interesting work is presented in [110], where low-cost, flexible printed organic sensors were developed for temperature-sensing applications. The sensors were formed using carbon nanoparticle ink along with a mixture of PEDOT:PSS and dimethyl sulfoxide (DMSO). Positive and negative thermal coefficient materials were used to form the sensors to achieve good linearity and low hysteresis in the responses of the sensors. Piezoelectric drop-on-demand (DOD) inkjet printers were used for printing purposes. Three different kinds of carbon nanoparticle inks were used for developing the prototypes. The amount of DMSO mixed in the composites was varied from 0.5 to 40 wt.% to optimize the values of the temperature coefficient of resistance and reliability of the processed materials. The sensors operated on the principle of the Wheatstone bridge, where a change in resistance was observed with the corresponding change in the input temperature values. The connections in the Wheatstone bridge were formed with alcohol-based silver nanoparticles on PEN flexible substrates. All the printing steps were carried out at a temperature of 28 °C. The voltage and frequency of the printers were 30 V and 2 kHz, respectively. Then the samples were sintered at 150 °C for 60 min and subsequently encapsulated using a barrier foil. Finally, the printing of the composites was performed, which were then cured at 150 °C for 30 min The sensors achieved a sensitivity of ~4 mV $°C^{-1}$ for an operating temperature ranging between 20 °C and 70 °C.

3.2.3 3D printing

This technique has been one of the most popular techniques in recent times. The use of 3D printed sensors has become the state-of-the-art in almost every sector. Some of the advantages these sensors possess are high durability and material strength, quick fabrication, better quality, reduced electronic waste, and a high range of customization [71]. This technique was popularized in the early 2000s to develop sensors for different kinds of applications [111, 112].

One work related to the electrochemical sensing application of these sensors is [113], where 3D printed molds were formed to generate flexible devices for monitoring hydrogen peroxide (H_2O_2) and glucose. The sensors were produced using a multi-step approach to generated multilayer graphene-based electrodes. These sensors had the additional advantage of the avoidance of any kind of requirement for inks. The processed conductive materials consisted of graphite and nano-platelet graphene, which were chemically modified using a modified Hummer's method. The samples were dried and expanded thermally to be used on 3D printed molds. Laminated electrodes were formed on different kinds of substrates to form the devices for electrochemical sensing applications. Some of the substrates used to form the prototypes were PET, PI, chitosan films, and PCL/agar composites. The substrates were cleaned initially using ethanol, followed by applying a stencil to reduce the excess adhesive residue. The deposition of the conductive material was done carefully through proper alignment of the 3D printed molds. These stamps formed electrodes through the application of manual pressure. The sensing area was kept constant by applying a resistive layer of nail polish on top of the electrodes. The surface modifications were performed using the electro-deposition process with the help of a platinizing solution. Finally, the sensors were treated with platinum nanoparticles and oxygen plasma to emphasize their use as biosensors. The sensors were tested using CV to achieve high sensitivity toward H_2O_2 and glucose. The graphite-based sensors with added platinum achieved an LOD and sensitivity of 1.91 μM and 64 nA μM^{-1} cm^{-2}. The testing of the graphene sensors for the detection of glucose was carried out between 3.7 and 9.9 μM. The treatment of nano-platelet graphene-based sensors with platinum achieved an LOD of 1.98 μM and a sensitivity of 16.5 nA μM^{-1} cm^{-2}. The glucose testing was performed via immobilization of the enzyme glucose oxidase in order to increase the specificity of the prototypes. Other research works using 3D printing techniques for developing electrochemical sensors are [114, 115].

The use of 3D printing techniques for developing strain sensors is exemplified in [116], where low-cost capacitive prototypes were produced for low-force sensing applications. The sensors were produced using graphite and PDMS as the processing materials. One of the primary advantages of these sensors includes their high sensitivity toward very low forces. 3D printed molds were formed using laboratory environment with fixed temperature and humidity conditions. Figure 3.4 [116] shows

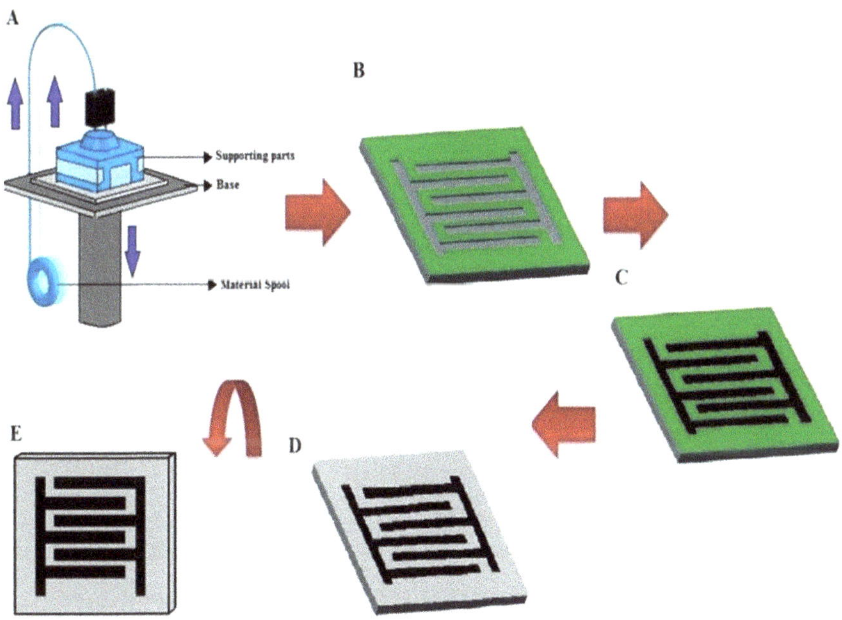

Figure 3.4. Schematic diagram depicting the fabrication steps for forming low-cost capacitive sensors using 3D printed molds. Reprinted from [116] Copyright (2018), with permission from Elsevier.

a schematic representation of the steps of fabrication of the sensors. Acrylonitrile butadiene styrene (ABS) was used as the printing filament due to its high stability towards variation in temperature. The diameter of the printing filament was 1.75 mm. The design of the electrodes was created using the CREO parameter, with each mold being generated in a total time of 25 min. The depth of the trenches in the molds was optimized to a value of 500 μm. This optimization was performed on the basis of a trade-off between the flexibility and conductivity of the formed electrodes. Graphite was initially cast on the trenches, followed by removing the redundant powders from the molds. This was followed by casting PDMS on top of the molds to form the substrates of the prototypes. The height of the PDMS was adjusted to 1000 μm using a casting knife. Finally, the samples were cured at 70 °C for 2 h to solidify the substrates. The sensors were tested for forces ranging between 3.5 and 17.5 mN, with the responses being monitored in terms of capacitive values. The prototypes obtained a sensitivity of 0.2542 pF m^{-1} N^{-1}, along with a signal-to-noise ratio (SNR) of 10.86. Some other significant works in this area are [117, 118].

The electrical sensors developed with this technique can be exemplified by [119], where nanocomposite-based sensors were developed using a 3D printing technique. The sensors were formed with graphene and PDMS as the conductive material and polymer, respectively. The advantages of these devices were their high stretchability and sensitivity toward the tested temperature range. Figure 3.5 [119] depicts the fabrication steps of the 3D printed graphene–PDMS-based sensors. The inks were prepared by including graphene nano-platelets (GNPs) with an average size of 46 μm. Exfoliated GNP–ethyl acetate dispersions were formed by sonicating a mixture of

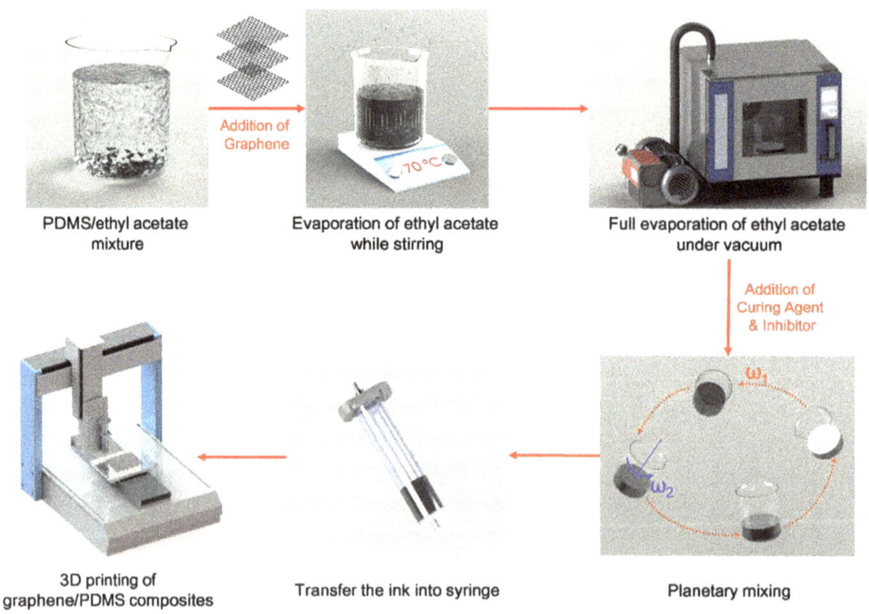

Figure 3.5. Fabrication steps followed to develop 3D printed graphene–PDMS-based temperature sensors. Reprinted with permission from [119]. Copyright (2018) American Chemical Society.

GNPs with ethyl acetate. Then PDMS was mixed with dispersions with vigorous stirring, followed by heating the samples at 70 °C to evaporate ethyl acetate. Finally, these samples were added with the current agent and inhibitor to form homogeneous graphene–PDMS inks. The nanocomposites consisted of micro-structures, which assisted in producing long-range and controlled cellular structure. The cellular structures consisted of grid, triangular, and hexagonal porous structures, all of which were sensitive and had stable performance toward external strains. The responses of the sensors were measured in terms of relative resistance with respect to changes in temperature values. The operating range was varied between 25 °C and 75 °C for the fabricated sensors. All three types of grids displayed a linear response toward the experimental values. The sensitivities of the solid and cellular composites were obtained to be 0.008 $°C^{-1}$. The ranges of the reaction and recovery times of the sensors were 1.32–3.91 s and 1.01–4.58 s, respectively. The results were much more stable compared to the solid composites. High repeatability in the responses was obtained even after on hundred cycles of heating and cooling. Other major research in this area can be exemplified by [120, 121].

3.2.4 Laser ablation

The laser-induction or laser ablation technique has been very popular in the last few years. It has provided an excellent way to generate conductive materials with high electrical conductivity in a single or a couple of steps. It also provides other advantages such as high-yield, operation at room temperature, and high customizability in the design of the electrodes. The authors of [122] were the first research group to introduce the production of laser-induced graphene from commercial polymers. This encouraged scientists to utilize this technique to form induced materials on different substrates [123, 124] for a range of applications [125, 126].

One of the works related to the use of laser-induced sensors for electrochemical sensing can be seen in [127], where graphene-based sensors were developed for the detection of nitrogen in soil. Laser-induced graphene (LIG) was developed on polyimide substrates via a one-step laser writing fabrication technique. Low-cost UV lasers were employed at various pulse widths, including 10, 20, 30, 40, and 50 ms. The best results were obtained at 20 ms, where sp^2 hybridized carbon was formed during the induction process. LIG electrodes were obtained from high-temperature Kapton tapes that were adhered to Epson printing paper. Computer-aided designs were produced on the substrates with an overall length of 25.8 mm and a circular area with a diameter of 5.3 mm. Following the fabrication process, the laser-induced electrodes were passivated with an insulating layer. This layer assisted in preventing short-circuiting and nonspecific binding during electrochemical reactions. The sensors were functionalized with ionospheres that were specific towards ammonium and nitrate ions. This ionization for ammonium ions was performed using nonactin, whereas tri-dodecyl methylammonium nitrate was used for nitrate ions. The Nernstian sensitivities and LOD for ammonium ions were 5.17 ± 7.8 mV dec^{-1} and 28.2 ± 25.0 μM, respectively, whereas for nitrate ions they were −54.8 ± 2.5 V dec^{-1} and 20.6 ± 14.8 μM, respectively. The linear range for both ions was from 10^{-5} to 10^{-1}, in addition to the recovery percentages of 96% and 95% for ammonium and nitrate ions, respectively. A few of the other research works in this area are [128, 129].

For strain-sensing applications, the employment of the laser-induced process was used to form highly sensitive graphene-based strain sensors [130]. The patterning was performed using a 355 nm pulsed laser. Polyimide films were used as the processing materials for laser-induced graphene. Around 90% adsorption of the laser took place on the films at a wavelength of 355 nm. This caused a local increase in the surface temperature, leading to ablation via the photothermal phenomenon. Figure 3.6(a) [130] shows a schematic diagram of the fabrication steps of the laser-induced strain sensors. PDMS was used as a template for the sensing prototypes in order to obtain a uniform thickness of the substrates. It was spin-coated at 100 rpm for 5 min to obtain a uniform thickness of 400 μm. Then PI films were placed on top of the PDMS substrates for the laser-induction process. After the generation of graphene, the remaining parts of the PI were removed, followed by

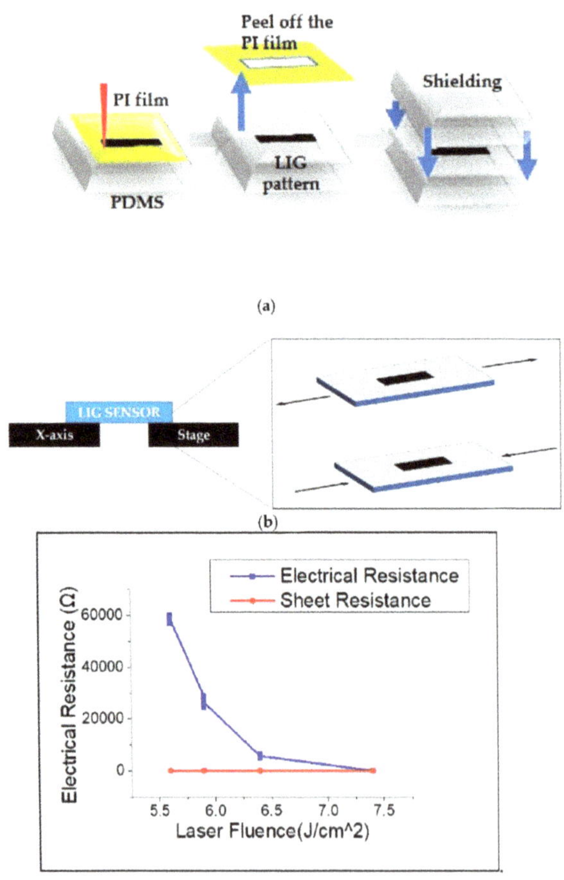

Figure 3.6. (a) Schematic diagram of the fabrication steps of laser-induced graphene sensors. (b) The sensors were bent to (c) determine the changes in the electrical resistance with respect to laser fluence. Reproduced from [130] CC BY 4.0.

covering the conductive material with an additional layer of PDMS to protect the patterns. The increase in the laser fluence, as shown in figure 3.6(c) [130], increases the intensity of photothermal ablation, which subsequently increases the electrical properties and sensitivity of the developed prototypes. The depth of the induced graphene patterns also increased with a corresponding increase in laser fluence, as a result of which the durability of the patterns and sensors increased. The sensors were highly sensitive toward small deformations, with a high GF of 160 even at very small strains. The sensors acquired a mechanical fracture at 30% tensile strain. The resistance values also increased linearly with a corresponding increase in bending values. The sensors had high durability along with high sensitivity toward the monitoring of finger motion. Similar works can be seen in [131, 132].

The electrical sensing applications of laser-induced sensors include the fabrication and implementation of flexible temperature-sensing devices [133], which showed highly linear and stable responses. These sensors were formed using the carbonization of polyimide substrates, which were subsequently used as wearable prototypes for determining and transmitting the responses from human skin wirelessly. PI substrates with a thickness of 125 μm were cleaned with IPA via ultra-sonication and DI water. The overall laser-induction process took around a few minutes, depending on the size of the prototypes. A maskless laser-induced carbonization process was carried out to form the sensors. The laser parameters consisted of a laser beam size, frequency, and scanning speed of 20 μm, 100 kHz, and 500 mm min^{-1}, respectively.

This was followed by the deposition of gold nanoparticles with a thickness of 100 nm. The final encapsulation was performed using sticky PET in order to minimize the interaction of the prototypes with the environment. Finally, the sensors were mounted on the FPCB using the anisotropic conductive film bonding method. The embedded condition circuit consists of a signal processing integrated circuit (IC), a readout integrated circuit consisting of a conventional amplifier low-pass filter, a digital inter-integrated circuit (I2C) interface, and a microcontroller. The sensors were able to distinguish the differences in temperature between when they were attached to the body compared to the open-state. The sensors were tested over a range of −10 °C to 60 °C for 10 min. A negative temperature coefficient of resistance of 0.001 42 °C^{-1} along with a high linearity of $R^2 = 0.999$ was obtained during the process. The responses were highly stable with an accuracy of ±0.2 °C. Other related research in this sector can be found in [134, 135].

3.2.5 Gravure printing

The next most favored technique that has been used to fabricate printed flexible sensors has been gravure printing. This technique has been favored due to its high speed and large volume production. This technique has been used for the last few years [136] to increase the production of sensors on an industrial scale. In terms of electrochemical sensing, flexible sensors have been developed using thin-film PET and silver nanoparticle ink [137]. The architecture of these sensors has been based on a three-electrode system where circular electrodes were formed for operation with the electrochemical impedance spectroscopy (EIS) technique. The thickness of the

PET substrates was 175 μm, along with silver nanoparticles having an average particle size ranging from 20 to 30 nm. The diameter of the working electrode was 1700 μm, while the inner and outer diameters of the counter electrode were 2900 and 3900 μm, respectively. The sensors were tested with different analytes such as mercury sulfide, lead sulfide, D-proline, and sarcosine at very low concentrations. The tested electrochemical solutions were formed using PBS solutions, with a pH of 7.4. Different concentrations of 1 pM, 100 pM, 100 nM, and 100 μM were dissolved in DI water and stored at −20 °C before their use for experimental purposes. The EIS technique was carried out with an input voltage of 100 mV, along with a frequency range from 1 Hz to 1 MHz. The results were analyzed using built-in LabVIEW software. The sensors showed significant changes in the impedance values for all the target analytes. The highest increases in impedance values for HgS and PBS were 85% and 88%, respectively. The optimum frequency at which the biggest changes were obtained was at 20 Hz. The reports in [138, 139] present similar research performed on the utilization of gravure printing for electrochemical sensing applications.

A strain-sensing application using this printing technique is shown in [140], where roll-to-roll production was performed to develop flexible strain sensors. The sensors were printed with a size of 10×17 mm^2 with a high-temperature sintering process. PI films and silver nanoparticles were processed using gravure printing to form the devices. The thickness and width of the thin-films were 300 mm and 25 μm, respectively. The average particle size, viscosity, solvent, and boiling point of the silver nanoparticles were 1–10 μm, 16 570 mPa s, $C_6H_{12}O_3$, and 156 °C, respectively. The gravure roller used for the printing process was fabricated using the laser and etching process. The input conditions during the fabrication process included an operating velocity, operating tension, nip pressure, doctor pressure, and drying temperature of 2 m min^{-1}, 10 N, 0.5 MPa, 0.5 MPa, and 150 °C, respectively. The sensors were fabricated with serpentine electrodes with a length of 9 mm, a width of 360 μm, and a gap of 250 μm. After the formation of the prototypes, a drying time of 2.5 min was used on an effective area of 300×5000 mm^2. The sensors were tested with three different types of loading, namely bending, tensile, and periodic forces. The GF ranged between 1.9 to 2.24, while the range of measurable strain was from -3.86×10^{-4} to 16.272×10^{-3}. Further strain-sensing devices developed with gravure printing are presented in [141, 142].

The study in [143] exemplifies the utilization of gravure printing for fabricating sensors for electrical applications. Fully printed flexible sensors were developed which operated on interdigitated capacitors. Gravure printing was used to process the raw materials, which were subsequently used as humidity sensors. PET and silver nanoparticle ink were used to form the substrate and electrodes, respectively. The average particle size of the silver nanoparticles ranged between 20 and 30 nm. The selectivity of the fabricated sensors was further increased using a sensitive polymer poly (2-hydroxyethyl methacrylate) (pHEMA). The pHEMA was dissolved in crystalline form in ethanol at a value of 15 wt.% to form the solution. The interdigital electrodes were developed with eight pairs of electrodes, with each electrode having a length, width, and interdigital spacing of 8600, 200, and 200 μm, respectively. The sensitive layer was deposited on top of the sensing area with thicknesses of 1 and 500 μm. The measurements were performed in a range from 30% RH to 80% RH, in order to determine the changes in capacitive values with corresponding changes in the RH. The output of the devices was analyzed using an LCR meter, which was connected to the sensors using alligator clips. The impedance analyzer, in turn, was connected to the computer with a USB cable to record the response of the printed prototypes. The biggest changes in the output of the sensors were obtained at 80% RH in comparison to the capacitance measured at 30% RH. The responses were also stable, with a maximum error of ±0.6% at 40% RH and ±0.8% at 60% RH. The studies in [144, 145] are some other works in this area.

3.3 Challenges of current sensors

Although a substantial amount of work has been performed on the fabrication of printed flexible sensors using different methodologies, there are still some issues that need to be rectified in the current framework of microelectronics. Even though researchers have favored the five techniques described above to develop printed flexible sensors with enhanced characteristics, there are still some limitations for each of them. One of the issues related to screening printing is the dynamicity of the substrates which can be used for printing purposes. This can be addressed through the use of functionalized substrates, which would increase the bonding capability of the printed conductive inks. The surface tension of the printed inks is low due to the use of templates in this technique. Also, the presence of masks causes a shadowing effect of the printed inks, which degrades the quality of the prototypes. Other major disadvantages of the screen printing technique in comparison to its counterparts are the high cost of set-up, wastage of materials, the longer time required for printing, poor ink mileage, and excessive drying time [146]. One of the possible remedies for this could be the use of adjustable or movable masks so that the printed conductive inks become properly adhered to the substrates. The issue with drying time can be optimized to obtain a trade-off with the quality control of the fabricated prototypes.

One of the major issues related to inkjet printing is the limitation in the range of inks that can be used to print the sensors. This disadvatange saturates the quality of sensors to a certain limit. The spectrum of conductive inks can be increased by introducing nanocomposite inks. The nanocomposites would not only increase the dynamicity but also increase the durability and stability of the prototypes. Similar to screen printing, the printing speed and volume are low for inkjet printing. One techniques to increase the roll-to-roll production with this process is using multiple cartridges during the printing process. This can be done by customizing the printing machines according to the need. The problems related to 3D printing are high consumption of energy, harmful emission of minute particles, which causes pulmonary diseases, and limited use of processed materials [147]. 3D printers can be customized to increase the spectrum of materials that can be used for forming the prototypes. The composites formed with this technique can be improved by printing the composites directly, instead of mixing them over the 3D printed molds [148]. The emission of harmful gases and particles can be addressed by altering the printing filament that is used to form the printed molds. Instead of using polyester-based thermoplastic, ABS, and polylactic acid (PLA), copolymer thermoplastic and polyurethane thermoplastic polymers can be used which would minimize the detrimental effects during the printing process.

The drawbacks associated with laser-induced sensors that need to be addressed are the extreme limitation of processing materials to form the conductive material and the requirement for high-temperature conditions during the printing process. One of the ways to address both these limitations is by increasing the types of commercial polymers on which laser-induction is performed to form the conductive material. Apart from the commonly used PI films that are laser-induced to form micro-supercapacitors and energy-harvesting devices, scientists are also trying to work on other polymers such as PET [149] and polyvinyl chloride (PVC) [150] as an alternative to increasing the dynamicity in the spectrum of processed materials. For gravure printing, the issues can be associated with the high initial set-up cost and high cost of sensors fabricated with this technique compared to other techniques. The way to deal with this problem can be the production of a large volume of sensors in a single roll-to-roll production in order to decrease the per unit cost. This technique can also be linked with other printing techniques such as screen printing to form the prototypes to obtain sensors with enhanced quality [151]. If these shortcomings are addressed properly with further modifications of the processed materials, the quality and range of applications of sensors can be increased to a great extent.

3.4 Conclusion

This chapter presents an overview of some of the techniques with which printed flexible sensors have been produced and implemented in different sectors. These techniques have become popular due to the distinct advantages in terms of electrical, mechanical, and thermal attributes over their counterparts. The deployment of these sensors has been performed on a large scale for industrial, biomedical, and environmental applications. Each of these methodologies described in this chapter has been elucidated with certain examples based on their operating mechanism. The chapter also presents some of the challenges that exist in the current printed flexible sensors, along with some of the possible solutions. Each of these solutions should be addressed to improve the quality of the sensors in terms of efficiency. With the improvement of the current techniques, the selectivity, sensitivity, and robustness of these sensors can be increased to a great extent, thus extending the dynamic range of these sensors in the real world.

Acknowledgement

Funded by the German Research Foundation (DFG, Deutsche Forschungsgemeinschaft) as part of Germany's Excellence Strategy—EXC 2050/1—Project ID 390696704—Cluster of Excellence 'Centre for Tactile Internet with Human-in-the-Loop' (CeTI) of Technische Universität Dresden.

References

[1] Wei Z 1970 A study on column-parallel ADCs using DMOS capacitors *J. Solid-State Circuits* **5** 250–54

[2] Pandey A, Yadav D, Singh R and Nath V 1970 Design of ultra low power CMOS temperature sensor for space applications *Int. J. Adv. Res. Electr. Electron. Instrum. Energy* **2** 4117–25

[3] Sze S M 1994 *Semiconductor Sensors* (New York: Wiley)

[4] Jaaniso R and Tan O K 2013 *Semiconductor Gas Sensors* (Amsterdam: Elsevier)

[5] Xu Y *et al* 2019 Silicon-based sensors for biomedical applications: a review *Sensors* **19** 2908

[6] 2020 Advantages and disadvantages of silicon detectors *Nuclear Power for Everyone* https://nuclear-power.net/nuclear-engineering/radiation-detection/semiconductor-detectors/silicon-based-semiconductor-detectors/advantages-and-disadvantages-of-silicon-detectors/ (Accessed: 17 Aug. 2020)

[7] Pramanik C, Saha H and Gangopadhyay U 2006 Design optimization of a high performance silicon MEMS piezoresistive pressure sensor for biomedical applications *J. Micromech. Microeng.* **16** 2060

[8] Errachid A, Ivorra A, Aguilo J, Villa R, Zine N and Bausells J 2001 New technology for multi-sensor silicon needles for biomedical applications *Sens. Actuators* B **78** 279–84

[9] Gabriel G *et al* 2007 Manufacturing and full characterization of silicon carbide-based multi-sensor micro-probes for biomedical applications *Microelectron. J.* **38** 406–15

[10] Nag A, Zia A I, Li X, Mukhopadhyay S C and Kosel J 2015 Novel sensing approach for LPG leakage detection: part I—operating mechanism and preliminary results *IEEE Sensors J.* **16** 996–1003

[11] Zia A I *et al* 2013 Technique for rapid detection of phthalates in water and beverages *J. Food Eng.* **116** 515–23

[12] Zia A I, Mukhopadhyay S C, Yu P-L, Al-Bahadly I H, Gooneratne C P and Kosel J 2015 Rapid and molecular selective electrochemical sensing of phthalates in aqueous solution *Biosens. Bioelectron.* **67** 342–49

[13] Alahi M E E, Xie L, Mukhopadhyay S and Burkitt L 2017 A temperature compensated smart nitrate-sensor for agricultural industry *IEEE Trans. Ind. Electron.* **64** 7333–41

[14] Alahi M E E, Mukhopadhyay S C and Burkitt L 2018 Imprinted polymer coated impedimetric nitrate sensor for real-time water quality monitoring *Sens. Actuators* B **259** 753–61

[15] Barhoumi L *et al* 2017 Silicon nitride capacitive chemical sensor for phosphate ion detection based on copper phthalocyanine–acrylate-polymer *Electroanalysis* **29** 1586–95

[16] Nag A, Zia A I, Mukhopadhyay S and Kosel J 2015 Performance enhancement of electronic sensor through mask-less lithography *2015 9th Int. Conf. on Sensing Technology (ICST)* (Piscataway, NJ: IEEE) pp 374–9

[17] Grayson A C R *et al* 2004 A BioMEMS review: MEMS technology for physiologically integrated devices *Proc. IEEE* **92** 6–21

[18] Bryzek J 1996 Impact of MEMS technology on society *Sens. Actuators* A **56** 1–9

[19] Khumpuang S, Maekawa H and Hara S 2013 Photolithography for minimal fab system *IEEJ Trans. Sens. Micromach.* **133** 272–77

[20] Altomare L, Gadegaard N, Visai L, Tanzi M C and Fare S 2010 Biodegradable micro-grooved polymeric surfaces obtained by photolithography for skeletal muscle cell orientation and myotube development *Acta Biomater.* **6** 1948–57

[21] Rossi V A and Carpenter M A 2020 Reusable polystyrene wafer coating as an antiadhesive layer for PDMS thin film production *Mater. Lett.* **262** 127045

[22] Ogi T, Modesto-Lopez L B, Iskandar F and Okuyama K 2007 Fabrication of a large area monolayer of silica particles on a sapphire substrate by a spin coating method *Colloids Surf.* A **297** 71–8

[23] Gao C *et al* 2010 22 nm silicon nanowire gas sensor fabricated by trilayer nanoimprint and wet etching *Microelectron. Eng.* **87** 927–30

[24] Rangelow I, Skocki S and Dumania P 1994 Plasma etching for micromechanical sensor applications *Microelectron. Eng.* **23** 365–68

[25] Al-Salman H S and Abdullah M 2015 Preparation of ZnO nanostructures by RF-magnetron sputtering on thermally oxidized porous silicon substrate for VOC sensing application *Measurement* **59** 248–57

[26] Rydosz A, Szkudlarek A, Ziabka M, Domanski K, Maziarz W and Pisarkiewicz T 2015 Performance of Si-doped WO_3 thin films for acetone sensing prepared by glancing angle DC magnetron sputtering *IEEE Sensors J.* **16** 1004–12

[27] Nag A, Mukhopadhyay S C and Kosel J 2016 Flexible carbon nanotube nanocomposite sensor for multiple physiological parameter monitoring *Sens. Actuators* A **251** 148–55

[28] Nag A, Menzies B and Mukhopadhyay S C 2018 Performance analysis of flexible printed sensors for robotic arm applications *Sens. Actuators* A **276** 226–36

[29] Lei K F, Lee K-F and Lee M-Y 2012 Development of a flexible PDMS capacitive pressure sensor for plantar pressure measurement *Microelectron. Eng.* **99** 1–5

[30] Nag A, Mukhopadhyay S C and Kosel J 2016 Tactile sensing from laser-ablated metallized PET films *IEEE Sensors J.* **17** 7–13

[31] Yaqoob U, Phan D-T, Uddin A I and Chung G-S 2015 Highly flexible room temperature NO_2 sensor based on MWCNTs-WO_3 nanoparticles hybrid on a PET substrate *Sens. Actuators* B **221** 760–68

[32] Li S *et al* 2018 The room temperature gas sensor based on Polyaniline@ flower-like WO_3 nanocomposites and flexible PET substrate for NH_3 detection *Sens. Actuators* B **259** 505–13

[33] Nag A, Mukhopadhyay S C and Kosel J 2017 Sensing system for salinity testing using laser-induced graphene sensors *Sens. Actuators* A **264** 107–16

[34] Han T *et al* 2019 Gold/polyimide-based resistive strain sensors *Electronics* **8** 565

[35] Qin Y *et al* 2015 Lightweight, superelastic, and mechanically flexible graphene/polyimide nanocomposite foam for strain sensor application *ACS Nano* **9** 8933–41

[36] Fang P, Qiu X, Wirges W, Gerhard R and Zirkel L 2010 Polyethylene-naphthalate (PEN) ferroelectrets: cellular structure, piezoelectricity and thermal stability *IEEE Trans. Dielectr. Electr. Insul.* **17** 1079–87

[37] Lamanna L *et al* 2019 Flexible and transparent aluminum-nitride-based surface-acoustic-wave device on polymeric polyethylene naphthalate *Adv. Electron. Mater.* **5** 1900095

[38] Sekitani T *et al* 2009 Organic nonvolatile memory transistors for flexible sensor arrays *Science* **326** 1516–19

[39] Latessa G, Brunetti F, Reale A, Saggio G and Di Carlo A 2009 Piezoresistive behaviour of flexible PEDOT:PSS based sensors *Sens. Actuators* B **139** 304–9

[40] Seekaew Y, Lokavee S, Phokharatkul D, Wisitsoraat A, Kerdcharoen T and Wongchoosuk C 2014 Low-cost and flexible printed graphene–PEDOT:PSS gas sensor for ammonia detection *Org. Electron.* **15** 2971–81

[41] Wang Z *et al* 2014 Facile preparation of highly water-stable and flexible PEDOT:PSS organic/inorganic composite materials and their application in electrochemical sensors *Sens. Actuators* B **196** 357–69
[42] Han T, Nag A, Mukhopadhyay S C and Xu Y 2019 Carbon nanotubes and its gas-sensing applications: a review *Sens. Actuators* A **291** 107–43
[43] Jeong H Y *et al* 2010 Flexible room-temperature NO_2 gas sensors based on carbon nanotubes/reduced graphene hybrid films *Appl. Phys. Lett.* **96** 213105
[44] Tadakaluru S, Thongsuwan W and Singjai P 2014 Stretchable and flexible high-strain sensors made using carbon nanotubes and graphite films on natural rubber *Sensors* **14** 868–76
[45] Nag A, Mitra A and Mukhopadhyay S C 2018 Graphene and its sensor-based applications: a review *Sens. Actuators* A **270** 177–94
[46] Shao Y, Wang J, Wu H, Liu J, Aksay I A and Lin Y 2010 Graphene based electrochemical sensors and biosensors: a review *Electroanalysis* **22** 1027–36
[47] Young R J, Kinloch I A, Gong L and Novoselov K S 2012 The mechanics of graphene nanocomposites: a review *Compos. Sci. Technol.* **72** 1459–76
[48] Cai W, Lai T, Du H and Ye J 2014 Electrochemical determination of ascorbic acid, dopamine and uric acid based on an exfoliated graphite paper electrode: a high performance flexible sensor *Sens. Actuators* B **193** 492–500
[49] Bessonov A, Kirikova M, Haque S, Gartseev I and Bailey M J 2014 Highly reproducible printable graphite strain gauges for flexible devices *Sens. Actuators* A **206** 75–80
[50] Shih W-P *et al* 2010 Flexible temperature sensor array based on a graphite-polydimethylsiloxane composite *Sensors* **10** 3597–610
[51] Gong S *et al* 2014 A wearable and highly sensitive pressure sensor with ultrathin gold nanowires *Nat. Commun.* **5** 1–8
[52] Zhu B *et al* 2019 Hierarchically structured vertical gold nanowire array-based wearable pressure sensors for wireless health monitoring *ACS Appl. Mater. Interfaces* **11** 29014–21
[53] Yao S and Zhu Y 2014 Wearable multifunctional sensors using printed stretchable conductors made of silver nanowires *Nanoscale* **6** 2345–52
[54] Nag A, Afasrimanesh N, Feng S and Mukhopadhyay S C 2018 Strain induced graphite/PDMS sensors for biomedical applications *Sens. Actuators* A **271** 257–69
[55] Nag A *et al* 2018 A transparent strain sensor based on PDMS-embedded conductive fabric for wearable sensing applications *IEEE Access* **6** 71020–27
[56] Rahman M S A, Mukhopadhyay S C and Yu P-L 2014 Novel planar interdigital sensors *Novel Sensors for Food Inspection: Modelling, Fabrication and Experimentation* (Berlin: Springer) pp 11–35
[57] Luo N, Ding J, Zhao N, Leung B H and Poon C C 2014 Mobile health: design of flexible and stretchable electrophysiological sensors for wearable healthcare systems *2014 11th Int. Conf. on Wearable and Implantable Body Sensor Networks* (Piscataway, NJ: IEEE) pp 87–91
[58] Kim D H *et al* 2012 Thin, flexible sensors and actuators as 'instrumented' surgical sutures for targeted wound monitoring and therapy *Small* **8** 3263–68
[59] Culha U, Wani U, Nurzaman S G, Clemens F and Iida F 2014 Motion pattern discrimination for soft robots with morphologically flexible sensors *2014 IEEE/RSJ Int. Conf. on Intelligent Robots and Systems* (Piscataway, NJ: IEEE) pp 567–72

[60] Bakytbekov A, Nguyen T Q, Huynh C, Salama K N and Shamim A 2018 Fully printed 3D cube-shaped multiband fractal rectenna for ambient RF energy harvesting *Nano Energy* **53** 587–95

[61] Hota M K, Jiang Q, Mashraei Y, Salama K N and Alshareef H N 2017 Fractal electrochemical microsupercapacitors *Adv. Electron. Mater.* **3** 1700185

[62] Mashraei Y, Sivashankar S, Buttner U and Salama K N 2016 Integration of fractal biosensor in a digital microfluidic platform *IEEE Sensors J.* **16** 8775–83

[63] Nag A, Mukhopadhyay S C and Kosel J 2017 Wearable flexible sensors: a review *IEEE Sensors J.* **17** 3949–60

[64] Chung M, Fortunato G and Radacsi N 2019 Wearable flexible sweat sensors for healthcare monitoring: a review *J. R. Soc. Interface* **16** 20190217

[65] Chang W-Y, Fang T-H, Lin H-J, Shen Y-T and Lin Y-C 2009 A large area flexible array sensors using screen printing technology *J. Disp. Technol.* **5** 178–83

[66] Moorthi A, Narakathu B, Reddy A, Eshkeiti A, Bohra H and Atashbar M 2012 A novel flexible strain gauge sensor fabricated using screen printing *2012 Sixth Int. Conf. on Sensing Technology (ICST)* (Piscataway, NJ: IEEE) pp 765–8

[67] Pyo S *et al* 2014 Development of a flexible three-axis tactile sensor based on screen-printed carbon nanotube-polymer composite *J. Micromech. Microeng.* **24** 075012

[68] Huang L, Huang Y, Liang J, Wan X and Chen Y 2011 Graphene-based conducting inks for direct inkjet printing of flexible conductive patterns and their applications in electric circuits and chemical sensors *Nano Res.* **4** 675–84

[69] Dankoco M, Tesfay G, Bènevent E and Bendahan M 2016 Temperature sensor realized by inkjet printing process on flexible substrate *Mater. Sci. Eng.:* B **205** 1–5

[70] Wang C-T, Huang K-Y, Lin D T, Liao W-C, Lin H-W and Hu Y-C 2010 A flexible proximity sensor fully fabricated by inkjet printing *Sensors* **10** 5054–62

[71] Han T, Kundu S, Nag A and Xu Y 2019 3D printed sensors for biomedical applications: a review *Sensors* **19** 1706

[72] Guo S Z, Qiu K, Meng F, Park S H and McAlpine M C 2017 3D printed stretchable tactile sensors *Adv. Mater.* **29** 1701218

[73] Liu C *et al* 2018 3D printing technologies for flexible tactile sensors toward wearable electronics and electronic skin *Polymers* **10** 629

[74] Han T, Nag A, Afsarimanesh N, Mukhopadhyay S C, Kundu S and Xu Y 2019 Laser-assisted printed flexible sensors: a review *Sensors* **19** 1462

[75] Fonseca M A, Allen M G, Kroh J and White J 2006 Flexible wireless passive pressure sensors for biomedical applications *Tech. Dig. Solid-State Sensor, Actuator, and Microsystems Workshop (Hilton Head 2006)* pp 37–42

[76] Alahi M E E, Nag A, Mukhopadhyay S C and Burkitt L 2018 A temperature-compensated graphene sensor for nitrate monitoring in real-time application *Sens. Actuators,* A **269** 79–90

[77] Wang H *et al* 2019 Stamp-assisted gravure printing of micro-supercapacitors with general flexible substrates *2019 IEEE 32nd Int. Conf. on Micro Electro Mechanical Systems (MEMS)* (Piscataway, NJ: IEEE) pp 950–3

[78] Pavinatto F J, Paschoal C W and Arias A C 2015 Printed and flexible biosensor for antioxidants using interdigitated ink-jetted electrodes and gravure-deposited active layer *Biosens. Bioelectron.* **67** 553–59

[79] Bariya M *et al* 2018 Roll-to-roll gravure printed electrochemical sensors for wearable and medical devices *ACS Nano* **12** 6978–87
[80] Briand D, Molina-Lopez F, Quintero A V, Ataman C, Courbat J and de Rooij N F 2011 Why going towards plastic and flexible sensors? *Procedia Eng.* **25** 8–15
[81] Aleksandrova D 2020 Five benefits of flexible electronics for displays and sensors *FlexEnable* https://flexenable.com/blog/five-benefits-of-flexible-electronics-for-displays-and-sensors/ (Accessed: 17 Aug. 2020)
[82] Global flexible sensors market—industry trends and forecast to 2026 https://databridgemarketresearch.com/reports/global-flexible-sensors-market (Accessed: 17 Aug. 2020)
[83] Ngo H-D *et al* 2018 A novel low cost wireless incontinence sensor system (screen-printed flexible sensor system) for wireless urine detection in incontinence materials *Proceedings* **2** 716
[84] Barmpakos D, Segkos A, Tsamis C and Kaltsas G 2017 A disposable flexible humidity sensor directly printed on paper for medical applications *J. Phys.: Conf. Ser.* **931** 012003
[85] Zirkl M *et al* 2010 Fully printed, flexible, large area organic optothermal sensors for human–machine-interfaces *Procedia Eng.* **5** 725–29
[86] Chlaihawi A *et al* 2016 Novel screen printed flexible magnetoelectric thin film sensor *Procedia Eng.* **168** 684–87
[87] Transparency Market Research 2019 Printed and flexible sensors market—global industry analysis, size, share, growth, trends and forecast 2019–2029 *Report* TMRGL270
[88] Allied Market Research Printed and flexible sensors market 2020 https://alliedmarketresearch.com/printed-and-flexible-sensors-market (Accessed: 17 Aug. 2020)
[89] Trending news: printed and flexible sensors market growth, size, analysis and trends to 2025: Canatu Oy, Fujifilm Holding Corporation, ISORG 2020 https://3wnews.org/uncategorised/2110872/trending-news-printed-flexible-sensors-market-growth-size-analysis-and-trends-to-2025-canatu-oy-fujifilm-holding-corporation-isorg/
[90] Visual Arts Cork History of screen printing *Encyclopedia of Art Education* http://visual-arts-cork.com/printmaking/screen-printing.htm#:~:text=History%20of%20Screen%20Printing, began%20to%20pick%20up%20popularity (Accessed: 17 Aug. 2020)
[91] Metters J P, Kadara R O and Banks C E 2011 New directions in screen printed electroanalytical sensors: an overview of recent developments *Analyst* **136** 1067–76
[92] Liu X, Yao Y, Ying Y and Ping J 2019 Recent advances in nanomaterial-enabled screen-printed electrochemical sensors for heavy metal detection *TrAC, Trends Anal. Chem.* **115** 187–202
[93] Abellán-Llobregat A *et al* 2017 A stretchable and screen-printed electrochemical sensor for glucose determination in human perspiration *Biosens. Bioelectron.* **91** 885–91
[94] Khan S, Ali S and Bermak A 2019 Recent developments in printing flexible and wearable sensing electronics for healthcare applications *Sensors* **19** 1230
[95] Honeychurch K C 2012 Screen-printed electrochemical sensors and biosensors for monitoring metal pollutants *Insci. J.* **2** 1–51
[96] Tian B *et al* 2019 All-printed, low-cost, tunable sensing range strain sensors based on Ag nanodendrite conductive inks for wearable electronics *J. Mater. Chem.* C **7** 809–18
[97] Bose A *et al* 2019 Highly sensitive screen printed strain gauge for micro-strain detection *2019 IEEE Int. Conf. on Flexible and Printable Sensors and Systems (FLEPS)* (Piscataway, NJ: IEEE) pp 1–3

[98] Lee C G *et al* 2018 Stretchable strain sensors fabricated by screen printing of silver paste on the surface modified transparent elastomeric polyurethane films *Mater. Sci. Appl.* **9** 1008
[99] Albrecht A, Salmeron J F, Becherer M, Lugli P and Rivadeneyra A 2019 Screen-printed chipless wireless temperature sensor *IEEE Sensors J.* **19** 12011–15
[100] Aliane A *et al* 2014 Enhanced printed temperature sensors on flexible substrate *Microelectron. J.* **45** 1621–26
[101] Knoll M, Offenzeller C, Mayrhofer B, Jakoby B and Hilber W 2018 A screen printed thermocouple-array on a flexible substrate for condition monitoring *Proceedings* **2** 803
[102] A very brief history of industrial inkjet printing https://screenweb.com/content/a-very-brief-history-industrial-inkjet-printing (Accessed: 17 Aug. 2020)
[103] DuraFast A brief history of inkjet printer technology *DuraFast* https://blog.durafastlabel.com/a-brief-history-of-inkjet-printer-technology/ (Accessed: 17 Aug. 2020)
[104] Hu C, Bai X, Wang Y, Jin W, Zhang X and Hu S 2012 Inkjet printing of nanoporous gold electrode arrays on cellulose membranes for high-sensitive paper-like electrochemical oxygen sensors using ionic liquid electrolytes *Anal. Chem.* **84** 3745–50
[105] Pu Z, Wang R, Wu J, Yu H, Xu K and Li D 2016 A flexible electrochemical glucose sensor with composite nanostructured surface of the working electrode *Sens. Actuators* B **230** 801–9
[106] Moya A, Gabriel G, Villa R and del Campo F J 2017 Inkjet-printed electrochemical sensors *Curr. Opin. Electrochem.* **3** 29–39
[107] Casiraghi C *et al* 2018 Inkjet printed 2D-crystal based strain gauges on paper *Carbon* **129** 462–67
[108] Eshkeiti A *et al* 2014 A novel self-supported printed flexible strain sensor for monitoring body movement and temperature *SENSORS, 2014 IEEE* (Piscataway, NJ: IEEE) pp 1615–18
[109] Chang X *et al* 2018 ZnO nanorods/carbon black-based flexible strain sensor for detecting human motions *J. Alloys Compd.* **738** 111–17
[110] Bali C, Brandlmaier A, Ganster A, Raab O, Zapf J and Hübler A 2016 Fully inkjet-printed flexible temperature sensors based on carbon and PEDOT: PSS *Mater. Today Proc.* **3** 739–45
[111] Birol H, Maeder T, Jacq C and Ryser P 2004 3-D structuration of LTCC for sensor micro-fluidic applications *European Microelectronics and Packaging Symp.* pp 366–71
[112] Kanbach H, Wilde J, Kriebel F and Meusel E 2000 3D Si-on-Si stack packaging *Solder. Surf. Mount Technol.* **12** 35–9 .
[113] Stromberg L R *et al* 2019 Stamped multilayer graphene laminates for disposable in-field electrodes: application to electrochemical sensing of hydrogen peroxide and glucose *Microchim. Acta* **186** 533
[114] Demuru S, Haque R, Joho M O, Bionaz A, van der Wal P and Briand D 2019 3D-integration of printed electrochemical sensors in pet microfluidics for biochemical sensing *20th Int. Conf. on Solid-State Sensors, Actuators and Microsystems and Eurosensors 33* (Piscataway, NJ: IEEE) pp 2464–7
[115] Tan C, Nasir M Z M, Ambrosi A and Pumera M 2017 3D printed electrodes for detection of nitroaromatic explosives and nerve agents *Anal. Chem.* **89** 8995–9001
[116] Nag A, Feng S, Mukhopadhyay S, Kosel J and Inglis D 2018 3D printed mould-based graphite/PDMS sensor for low-force applications *Sens. Actuators,* A **280** 525–34
[117] Gul J Z, Sajid M and Choi K H 2020 Retraction: 3D printed highly flexible strain sensor based on TPU–graphene composite for feedback from high speed robotic applications *J. Mater. Chem.* C **8** 2597

[118] Le T *et al* 2015 A novel strain sensor based on 3D printing technology and 3D antenna design *2015 IEEE 65th Electronic Components and Technology Conf. (ECTC)* (Piscataway, NJ: IEEE) pp 981–6

[119] Wang Z *et al* 2018 3D-printed graphene/polydimethylsiloxane composites for stretchable and strain-insensitive temperature sensors *ACS Appl. Mater. Interfaces* **11** 1344–52

[120] Ali S, Hassan A, Hassan G, Bae J and Lee C H 2016 All-printed humidity sensor based on graphene/methyl-red composite with high sensitivity *Carbon* **105** 23–32

[121] Kim K, Park J, Suh J-H, Kim M, Jeong Y and Park I 2017 3D printing of multiaxial force sensors using carbon nanotube (CNT)/thermoplastic polyurethane (TPU) filaments *Sens. Actuators* A **263** 493–500

[122] Lin J *et al* 2014 Laser-induced porous graphene films from commercial polymers *Nat. Commun.* **5** 1–8

[123] Ye R *et al* 2017 Laser-induced graphene formation on wood *Adv. Mater.* **29** 1702211

[124] Luong D X *et al* 2019 Laser-induced graphene composites as multifunctional surfaces *ACS Nano* **13** 2579–86

[125] You R, Liu Y Q, Hao Y L, Han D D, Zhang Y L and You Z 2020 Laser fabrication of graphene-based flexible electronics *Adv. Mater.* **32** 1901981

[126] Ye R, James D K and Tour J M 2019 Laser-induced graphene: from discovery to translation *Adv. Mater.* **31** 1803621

[127] Garland N T *et al* 2018 Flexible laser-induced graphene for nitrogen sensing in soil *ACS Appl. Mater. Interfaces* **10** 39124–33

[128] Aparicio-Martínez E, Ibarra A, Estrada-Moreno I A, Osuna V and Dominguez R B 2019 Flexible electrochemical sensor based on laser scribed graphene/Ag nanoparticles for non-enzymatic hydrogen peroxide detection *Sens. Actuators* B **301** 127101

[129] Zhang Y *et al* 2020 A flexible non-enzymatic glucose sensor based on copper nanoparticles anchored on laser-induced graphene *Carbon* **156** 506–13

[130] Jeong S-Y, Ma Y-W, Lee J-U, Je G-J and Shin B-S 2019 Flexible and highly sensitive strain sensor based on laser-induced graphene pattern fabricated by 355 nm pulsed laser *Sensors* **19** 4867

[131] Tian H *et al* 2014 Scalable fabrication of high-performance and flexible graphene strain sensors *Nanoscale* **6** 699–705

[132] Tao L-Q *et al* 2017 An intelligent artificial throat with sound-sensing ability based on laser induced graphene *Nat. Commun.* **8** 1–8

[133] Gandla S *et al* 2020 Highly linear and stable flexible temperature sensors based on laser-induced carbonization of polyimide substrates for personal mobile monitoring *Adv. Mater. Technol.* **5** 2000014

[134] Bobinger M R *et al* 2019 Flexible and robust laser-induced graphene heaters photo-thermally scribed on bare polyimide substrates *Carbon* **144** 116–26

[135] Tao L-Q *et al* 2016 A flexible 360-degree thermal sound source based on laser induced graphene *Nanomaterials* **6** 112

[136] Huang Q and Zhu Y 2019 Printing conductive nanomaterials for flexible and stretchable electronics: a review of materials, processes, and applications *Adv. Mater. Technol.* **4** 1800546

[137] Narakathu B, Reddy S, Atashbar M, Rebrosova E, Rebros M and Joyce M 2011 A novel gravure printed impedance based flexible electrochemical sensor *SENSORS, 2011 IEEE* (Piscataway, NJ: IEEE) pp 577–80

[138] Maddipatla D, Narakathu B, Turkani V, Bazuin B and Atashbar M 2018 A gravure printed flexible electrochemical sensor for the detection of heavy metal compounds *Proceedings* **2** 950

[139] Reddy A, Narakathu B, Atashbar M, Rebros M, Rebrosova E and Joyce M 2011 Gravure printed electrochemical biosensor *Procedia Eng.* **25** 956–59

[140] Park J, Nam D, Park S and Lee D 2018 Fabrication of flexible strain sensors via roll-to-roll gravure printing of silver ink *Smart Mater. Struct.* **27** 085014

[141] Maddipatla D 2016 Development of fully printed and flexible strain, pressure and electrochemical sensors *MSEng Thesis* Western Michigan University, Kalmazoo, MI

[142] Parameswaran C and Gupta D 2019 Large area flexible pressure/strain sensors and arrays using nanomaterials and printing techniques *Nano Converg.* **6** 1–23

[143] Reddy A, Narakathu B, Atashbar M, Rebros M, Rebrosova E and Joyce M 2011 Fully printed flexible humidity sensor *Procedia Eng.* **25** 120–23

[144] Turkani V S *et al* 2019 A highly sensitive printed humidity sensor based on a functionalized MWCNT/HEC composite for flexible electronics application *Nanoscale Adv.* **1** 2311–22

[145] Zhang X *et al* 2019 Novel printed carbon nanotubes based resistive humidity sensors *2019 IEEE Int. Conf. on Flexible and Printable Sensors and Systems (FLEPS)* (Piscataway, NJ: IEEE) pp 1–3

[146] Andò B, Baglio S, Bulsara A R, Emery T, Marletta V and Pistorio A 2017 Low-cost inkjet printing technology for the rapid prototyping of transducers *Sensors* **17** 748

[147] Pearson A 2018 Disadvantages of 3D printing *3DInsider* https://3dinsider.com/3d-printing-disadvantages/ (Accessed: 17 Aug. 2020)

[148] Campbell T A and Ivanova O S 2013 3D printing of multifunctional nanocomposites *Nano Today* **8** 119–20

[149] Sato Y, Tsukamoto M, Shinonaga T and Kawa T 2016 Femtosecond laser-induced periodic nanostructure creation on PET surface for controlling of cell spreading *Appl. Phys.* A **122** 184

[150] Lum R 1976 Laser-induced decomposition of PVC *J. Appl. Polym. Sci.* **20** 1635–49

[151] Narakathu B *et al* 2012 A novel fully printed and flexible capacitive pressure sensor *SENSORS, 2012 IEEE* (Piscataway, NJ: IEEE) pp 1–4

IOP Publishing

Printed and Flexible Sensor Technology
Fabrication and applications
Subhas Chandra Mukhopadhyay and Anindya Nag

Chapter 4

Advances in printable devices for biomedical applications

Soma Zandi, Mahroo Baharfar, Parisa Moazzam, Amir Razmjou, Amin Beheshti and Mohsen Asadnia

This chapter presents an overview of recent advances in the development of printable devices for biomedical applications. These devices have advantages such as low-cost production and high accuracy, which are desirable features in mass production and disposable biosensor applications. We discuss recent progress in the development of printable biosensors and their applications, and move on to the fabrication process of such devices. Screen printing, three-dimensional (3D) printing, inkjet printing, and drop-on-demand (DOD) printing technologies, as common methods for the fabrication of printable devices for biomedical applications, are introduced and critically discussed. Through specific examples we elaborate on the potential of 3D wax printing technology to create sophisticated biomedical devices. We also touch on the application of printable devices for point-of-care testing and various biosensing purposes. Finally, this review details some recommendations and future directions for improving sensing accuracy and robustness.

4.1 Introduction

Sensors are defined as devices capable of discovering, displaying, and responding intelligently to changes in the surrounding environment [1]. Often these changes are visualised through parameters such as a change in resistance, voltage, ampere, resonance, optical properties, etc [2]. Recently, sensors have become widespread in daily life owing to their diverse applications, ranging from monitoring changes in various physical parameters to (bio)chemical detection [1, 3]. Among the present uses of sensors, sensing devices which are based on the detection of biologically critical species have received considerable attention due to their remarkable potential in the realm of healthcare monitoring [4, 5].

doi:10.1088/978-0-7503-3439-6ch4

© IOP Publishing Ltd 2021

Biosensors are particular types of integrated devices in which a biological interaction is converted to an analytical signal to provide quantitative or semi-quantitative information [6]. The difference between biosensors and conventional bio-analytical devices lies primarily in the additional processing steps (such as reagent addition) required in a bio-analytical system. A biosensor consists of two main integrated components—the bio-recognition element(s) and the transducer or sensing interface [7, 8]. A bio-receptor or bio-recognition element is responsible for recognising the desired analyte in a complex biological matrix. In contrast, the transducer translates the interaction of the analyte with the bio-receptor into a relevant analytical signal [9].

Since the initiation of biosensors in 1962, the necessity to improve the applicability of such devices in terms of fabrication cost, sensitivity, availability, and the possibility of on-site measurement has inspired intensive research around the world [10]. Recent advancements in the area of biosensors are primarily dominated by multi-disciplinary endeavours to amalgamate biosensing devices with emerging technologies such as bioelectronics and advanced printing techniques. The growing field of printable biosensors has shaped the way that biosensors are fabricated and manufactured in the both laboratory and at larger scales [11]. Printable biosensors are biosensors in which some type of printing technology is involved in the fabrication of the biosensing interface or sensing platform. The utilisation of printing technology is expected to increase the potential of biosensors for mass production at low cost and minimum inter-device standard deviation [12, 13].

Printable biosensor production is gaining popularity due to its valuable characteristics, such as high resolution, rapid prototyping, low cost, and facile fabrication procedure. Interestingly, current printing technologies are compatible with a wide range of substrate materials ranging from paper to textiles and polymers, which makes the fabrication of various types of biosensors quite feasible and cost-effective. Additionally, printable biosensors are not limited to two-dimensional (2D) planar platforms [13]. 3D printing and additive manufacturing have been used widely in the development of more complex and 3D biosensing platforms [14]. 3D printed biosensors hold great promise in the fabrication of multi-functional biosensors, enabling integrated sample collection, separation, sample delivery to the detection zone, and reagent addition [15].

A wide variety of printing techniques have been used for fabrication of printable biosensing platforms including stencil/screen printing [16, 17], inkjet printing [18], laser printing [19], wax printing [16, 20, 21], and various versions of 3D printing [22–26]. Each of these printing methods has its advantages and limitations for the fabrication of different types of biosensing devices. Microfabrication and additive manufacturing [21, 27] have been used widely in the development of various physical and biomedical sensors [28, 29]. Accordingly, the best-suited printing technique should be selected considering the fabrication cost, the detection method of interest, the desired substrate to establish the biosensing platform, the required functions, and the final application.

This chapter provides a comprehensive overview of printable biosensors in terms of detection approaches, signal processing, substrate materials, and current printing technologies. The related advantages and disadvantages of each printing technique are discussed, and future opportunities in the emerging field of printable biosensors are highlighted. This chapter is focused on the principles and applications of printable biosensors and other types of printable sensors which are based on changes in physical signals such as strain sensors (pressure and temperature sensors are excluded).

4.2 Printable biosensors

Biosensors are categorised based on the type of bio-recognition element or the transducer. Considering the types of bio-recognition features, biosensors can be classified into two broad categories: catalytic biosensors and affinity biosensors. In catalytic biosensors, enzymes are typically used as the bio-receptor, while in affinity-based biosensors various receptors with specific bio-affinity to the desired analyte, such as antibodies, aptamers, and DNA, are utilised [2].

Based on the transducing surface, there are various types of biosensing platforms including electrochemical biosensors [30], optical biosensors (including the different transducing modules of fluorescence, surface plasmon resonance, colorimetry, etc), acoustic biosensors, and thermal biosensors [8]. Herein, the diverse detection approaches and transducing surfaces utilised in printable biosensors are presented.

The detection process in printable biosensors is dependent on the type of transducing surface. Electrochemical, optical, piezoelectric, and field-effect transistor transducers are among the most common detection modules in printable biosensors.

Electrochemical transducers are one of the most popular transducing surfaces in biosensors thanks to their accuracy, simplicity, portability, miniaturised set-up, and sensitivity [31, 32]. Briefly, electrochemical transducing surfaces operate based on measuring the change in the current, voltage, or surface resistance caused by molecular interactions between the analyte and the recognition element. Often, electrochemical biosensors are composed of a counter electrode (CE), a reference electrode (RE), and a working electrode (WE) [33], demonstrated in figure 4.1. A reference electrode (usually silver/silver chloride) provides a stable potential without being influenced by the medium. The proposed interaction between the immobilised bio-recognition element and the analyte occurs at the surface of the working electrode, while the counter electron allows complete current flow in the equivalent circuit to the electrochemical cell. There have been many attempts to develop electrochemical transducers without a reference electrode to simplify the experimental set-up [34, 35].

To study the specific interaction between the sensing surface and the analytes, some kind of electrical input (voltage or current) is usually applied on the working electrode, and changes in current or voltage are recorded as the analytical signal. Based on the applied electrical input or the final analytical signal, there are various types of electrochemical sensors, including amperometric sensors (measuring current as an analytical signal), coulometric sensors (measuring the amount of transferred charge), voltammetric sensors (applying various voltage inputs such as cyclic voltammetry,

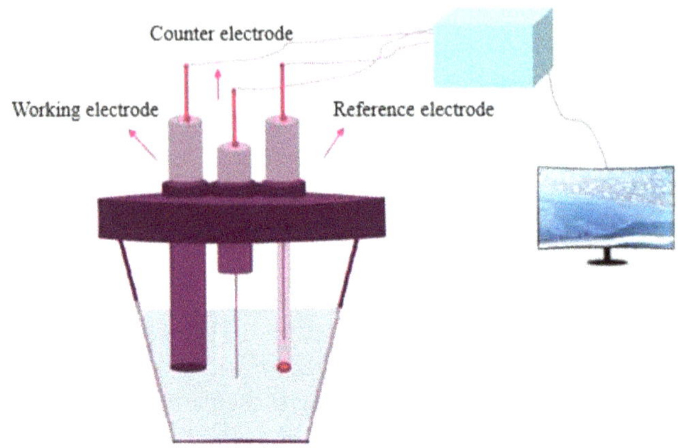

Figure 4.1. Schematic of an electrochemical cell. (Reproduced with permission from [33]. Copyright 2017 Elsevier.)

square wave voltammetry, differential pulse voltammetry, and stripping voltammetric methods), and impedance sensors (measuring changes in surface resistance) [9].

Optical transducers are used widely in biosensor technologies. These transducers are based on the detection of the change in the physical quantity of optical response (such as intensity, light scattering, refraction index, absorption, and fluorescence) resulting from biomolecular interactions between an analyte with a bio-recognition element [36]. The transducer converts the change into an electrical signal proportional to the concentration of the analyte. Optical transducers have several advantages, such as high sensitivity, requiring a low amount of analyte, rapid analysis and detection, and affordability [10]. It is worth noting that there are other types of optical signal processing based on colorimetric modules. These colorimetric sensors use the change in three colour components (red–green–blue) as a concentration-dependant analytical signal [11].

Piezoelectric biosensors use piezoelectric materials, known as anisotropic crystals, which exhibit mechanical oscillations when an external alternating voltage is applied [29, 37]. The obtained resonance frequency in piezoelectric transducers is directly proportional to the mass bound on the surface of crystal and the resonant frequency decreases with increasing mass on the crystal. In other words, the interaction between the bio-recognition element and the target analyte produces extra mass on the surface of the crystal. Thus the concentration of the analyte can be determined by measuring the change in resonance frequency. This type of biosensor is used widely for assays based on the hybridisation reaction, binding of the antigen–antibody, and protein and nucleic acid interactions [12].

Field-effect transistor (FET)-based transducers are not as common as the above mentioned transducing surfaces. In FET-based transducers, a surface that is very sensitive to the interfacial potential, such as metal oxide semiconductors, is exploited. In these transducers any change in interfacial potential resulting from the interaction of the analyte with the sensing surface will result in a shift of conductance proportional to the analyte concentration [13].

4.3 The fabrication process of printable sensors

Printing technologies have been utilised for different purposes in the fabrication of biosensors, including the fabrication of entire biosensing devices, transducing surfaces, and bio-sensitive surfaces (the deposition of bio-receptors). The following sections present the printing methods for the fabrication of biosensors along with their general principles and potential applications.

4.3.1 Screen printing

Screen printing is one of the most commonly used methods in the fabrication of transducing surfaces for biosensing platforms. This method is the most prominent method for printing conductive tracks as the transducing surface on diverse substrates [14]. In this method, a customised screen or stencil with the desired pattern is used to transfer a paste through a screen mesh. The screen masks can be prepared easily using photolithography, laser cutting, and craft cutting [15]. A schematic of this method is shown in the figure 4.2.

While the screen printing method is a reasonably new technology, it has been utilised widely in biosensors, in particular for point-of-care disposable electrochemical devices [38, 39]. Some of the advantages of electrodes fabricated using the screen printed method are their economy, reproducibility, versatility, and suitability for use in portable systems [38, 40].

The screen printing and inkjet printing methods are the most popular methods for producing paper-based electrochemical sensors [41]. In 2009 Dungchai *et al* [42]

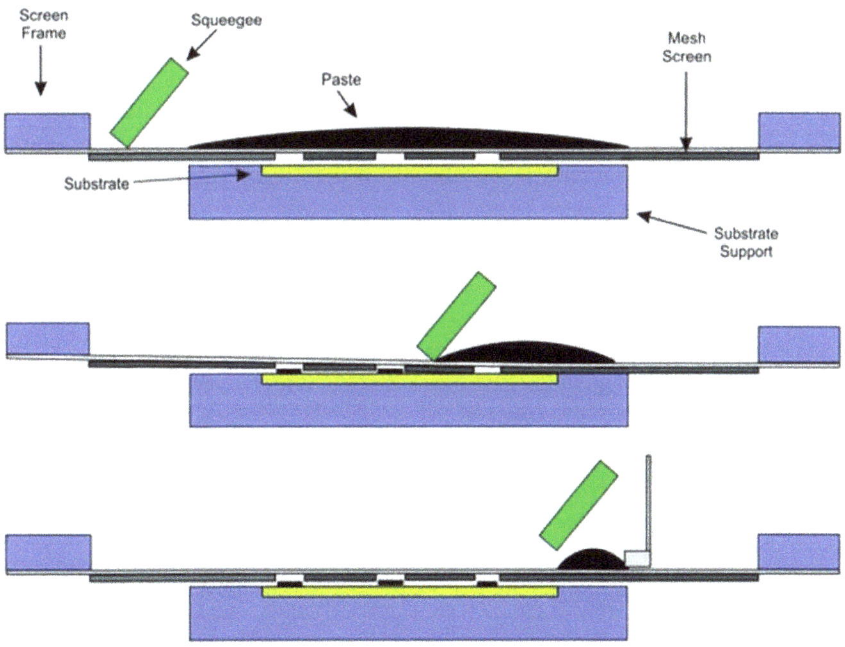

Figure 4.2. An illustration of the screen printing method.

reported, for the first time, electrochemical detection using paper-based microfluidic devices. First, microfluidic channels and test zones were patterned on filter paper using photolithography. Then the CE and WE were screen printed using carbon ink containing Prussian blue (Pb), which modifies the selectivity for H_2O_2 detection [42]. Also, the RE and conductive pads were defined by applying silver/silver chloride (Ag/AgCl) ink through screen printing [42]. The basic design of these electrochemical sensors is illustrated in figure 4.3(A) [42]. The fabricated device (shown in figure 4.3(B)) was used for simultaneously measuring the glucose, lactate, and uric acid levels in real biological samples. This study showed that a paper microfluidic device could be suitable for low-cost point-of care monitoring in medical diagnosis as the obtained level of glucose, lactate, and uric acid were within the error range of the concentration measured for the same samples using traditional tests [42].

Furthermore, Yang *et al* [43] developed a paper–fluidic electrochemical biosensor for glucose detection by taking advantage of the porous structure of paper for enzyme immobilisation [43]. In their work, platinum nanoparticles (PtNPs) were used for modification of the screen printed electrode (SPE) through electrochemically coating the WE with PtNPs as well as using the dip-coating approach to make an enzyme modified paper pad (EPP) [43]. Then the modified paper pad and SPEs were placed between two ABS plastic holders to decrease evaporation, holding the SPEs and EPP in alignment to provide stable signal sensing [43]. In this method the glucose level was determined by measuring the H_2O_2 produced from an enzymatic reaction and sensing them through the enzymeless PtNP–SPE. Using enzyme modified paper leads to improvement in the mechanical properties and hydrophilicity as well as high stability and reproducibility, and suitability for real-time sensing of several samples in one run, which make this platform suitable for creating a wide range of electrochemical biosensors based on paper-fluidic electrochemical sensors [43].

However, some disadvantages of this method include: (i) the large amount of material required and subsequently wasting some material during the fabrication process, and (ii) the limitation of the maximum resolution due to using a stencil for patterning the electrodes [41].

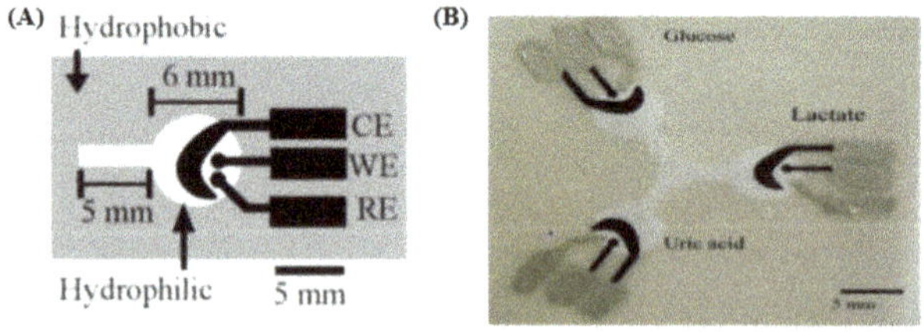

Figure 4.3. (A) The basic design of paper-based electrodes for electrochemical detection in microfluidic devices. (B) Schematic of the final fabricated paper-based microfluidic devices with three electrodes for multianalyte determination. (Reproduced with permission from [42]. Copyright 2009 American Chemical Society.)

4.3.2 3D printing

3D printing technology has attracted increasing attention in biosensor development due to its unique features compared to conventional printing methods. It offers advantages such as a less time-consuming procedure, lower fabrication cost, shorter processing time, etc. This technology has been used for the fabrication of microfluidic devices, biosensor platforms, and components such as sensing areas and transduction surfaces. 3D printed biosensors often have high sensitivity, resolution, and selectivity, and rapid measurement time [44, 45]. This method has been used widely to fabricate cell-based sensors, biomolecule-based sensors (enzyme sensors, DNA sensors, and immunosensors), cell-based sensors, etc [44, 46].

The most common types of 3D printing methods used to create prototypes in the biomedical applications include stereolithography (SLA), fused deposition modelling (FDM), and photopolymer jetting (Polyjet) (see figure 4.4). In the following, the principles of these methods and their advantages and disadvantage are described in detail.

In 1988 Charles Hull introduced the SLA printing method, which was based on photo-curing, and it is known as the first type of 3D printing technology [47]. In this method a laser with an ultraviolet beam was used to provide the required energy for the polymerisation reaction of photopolymer liquid resin and solidify the resin

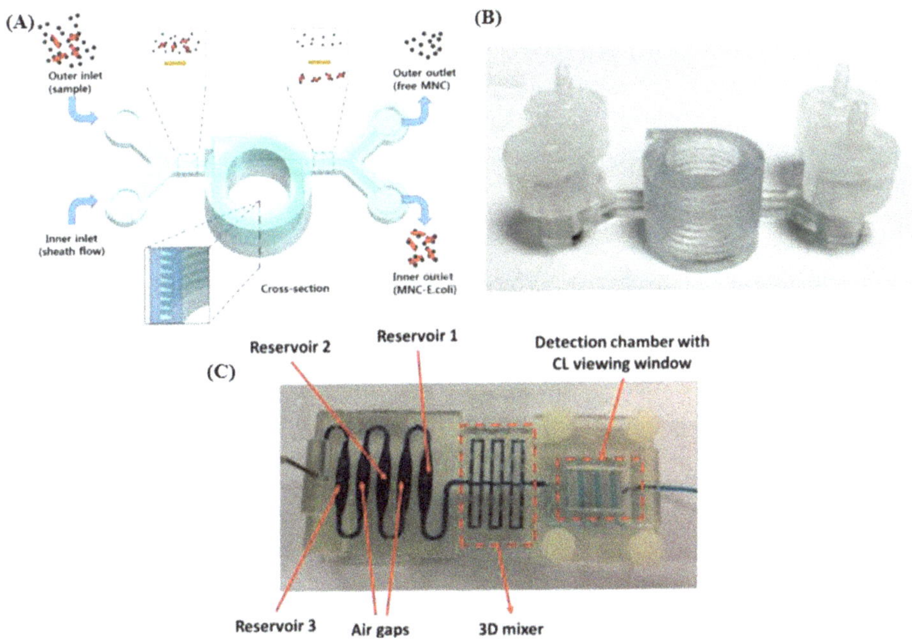

Figure 4.4. (A) Schematic demonstrating the layout of a 3D printed microfluidic device and the separation of the captured bacteria via inertial focusing. (B) Photograph of a fabricated 3D printed microfluidic device. (Reproduced with permission from [59]. Copyright 2015 Springer.) (C) Photograph of a 3D printed CL immunoarray device with a unibody design. (Reproduced with permission from [60]. Copyright 2017 Royal Society of Chemistry.)

layer-by layer to form a 3D structure on the surface of the resin via photochemical solidification. This method has some unique features, such as a smooth surface finish, high resolution, and high accuracy, such that the smallest details can be as small as 10–20 μm, whereas in most other fabrication methods this range is 50–200 μm [48, 49]. These features also make it a suitable fabrication method to produce sensors for biomedical applications (in particular complex structures) based on various materials [48]. However, one of the main limitations facing this technology is the shortage of suitable photo-sensitive resins, such as being a sufficiently low-viscosity liquid [50]. Also, other problems associated with this method are its inability to apply multiple printing materials, as well as the fact that moisture and heat can reduce its durability [44].

In 1992 the FDM printing method was introduced, for the first time, by Crump *et al* [51]. In this method, a thermoplastic filament is extruded through a nozzle, which is heated to above the melting temperature of thermoplastic materials. Then the melted filament is deposited layer-by-layer onto a platform and solidifies [52]. The most suitable types of thermoplastic materials for this method are polyamide (PA), acrylonitrile butadiene styrene (ABS), polylactic acid (PLA), polyamide (PA), and polycaprolactone (PCL), because of its high thermal stability and low glass transition temperature (T_g) [50]. The significant advantage of this technology is the ability to use multi-material printing by applying several nozzles and filaments [44]. This method can be employed to create a complex structure with good mechanical properties, such as 3D-scaffolds embedded with living cells for tissue engineering [53, 54] and it is a low-cost printing method. However, compared to other printing methods it has a lower printing resolution and slower printing speed [44].

In 2001 the first Polyjet printer was produced by Gothait [55], based on a photo-curing 3D printing technique, like the SLA method [56]. In this technology small droplets of photopolymer are deposited on a platform through multiple nozzles. Once the deposited resin has been cured using ultraviolet light, the platform is lowered to deposit the next layer on previous one, and this process is repeated until the entire prototype is produced [56]. In this method supporting material is needed during the fabrication, which is removed after completing the process [56]. In this technology different materials with different properties, such as colour, stiffness, etc, can be printed simultaneously, thanks to the multiple nozzles [44, 56]. In addition to this, the printed structures made using the Polyjet printing method have high resolution (16 μm) and accuracy [44, 49], this it is suitable for producing prototypes that require high precision. However, it is a costly technology, and parts fabricated using this method may suffer from poor mechanical properties and are vulnerable to heat and moisture [49, 56], thus the features produced are likely to lose their strength over time.

In 2015 Dirkzwager *et al* proposed prototypes of sensing devices based on aptamer-tethered enzyme capture (APTEC) using stereolithography-based 3D printing technology for point-of-care diagnosis of malaria [57]. This research demonstrated that magnetic bead-based wells have better performance, in particular in comparison to a paper-based syringe test. In 2014 Lee *et al* developed a microfluid device made using the SLA printing method to detect pathogenic bacteria, which contribute to many health issues. The results of this study illustrated that this 3D immunomagnetic flow assay is one of the fastest detection methods with high sensitivity and selectivity [58]. They also extended their research to fabricate a helical microchannel with the same printing approach for rapid and straightforward sensing of *Escherichia coli* bacteria in milk. Figures 4.4(A) and 4.4(B) show a schematic and a photograph of the microfluidic device fabricated using the SLA printing method in this study [59].

The importance of measuring the protein level of blood in the early diagnosis of cancer promoted researchers to develop and prototype a microfluidic immunoarray for quantification of cancer biomarker proteins [60]. To this end, in 2011 a research group fabricated a 3D printed unibody immunoarray using the SLA printing method to simultaneously measure two cancer biomarker proteins through chemiluminescence (CL) detection (figure 4.4(C)) [60]. This assay offers a faster result with lower cost in comparison to other technologies used for the measurement of biomarker proteins [60]. Another example of using the SLA printing method to detect protein level expression was presented by Chan *et al* [61]. For the first time, in this research movable components of a microfluidic chip were fabricated, including a torque-actuated pump, rotary valve, and pushing valve, using the SLA 3D printing method. In particular, in this study the designed chip was used for measuring the concentration of urinary protein [61]. The results of this research proved that an integrated microchip could be a promising technology for general colourimetric assays, due to its lower consumption of reagent, manufacturing method, and simple operation protocol [61].

In 2016 paper-based enzymatic reactors (PERs) were constructed using the FDM printing method for the first time, and were used for glucose detection in a batch injection analysis (BIA) cell coupled with an amperometric detection method [62]. In detail, H_2O_2 was injected via an electronic micropipette onto the SPE (containing the WE), which was placed in the 3D printed BIA cell, and was collected and analysed for the purpose of determining the glucose level [62]. The obtained results show that the glucose level measured with the 3D printed BIA cell coupled with SPE is in good agreement with the data achieved from the conventional method. The proposed approach also enjoys several advantages over conventional manufacturing methods, such as cost-effectiveness, simplicity, robustness, shorter manufacturing time, and favourable analytical performance [62].

In 2015 Gowers *et al* proposed a microfluidic device with integrated microdialysis probes and needle biosensors for real-time monitoring of tissue metabolism, made using 3D printing technology. Figure 4.5(A) shows the design of the assembled 3D microfluid device [63]. In detail, the Polyjet and FDM methods were employed to create needle holders and a microfluid chip (figure 4.5(B) and 4.5C), respectively. The part fabricated using the FDM method has good mechanical stability and the needle holders have a suitable tightness, as utilising the Polyjet method allows soft and rigid structures to be printed [63]. The printed microfluidic device was utilised for simultaneous determination of subcutaneous lactate and glucose levels using the needle biosensors inserted in the needle holder. The obtained results demonstrate that the fabricated device could be employed effectively for online measurement and

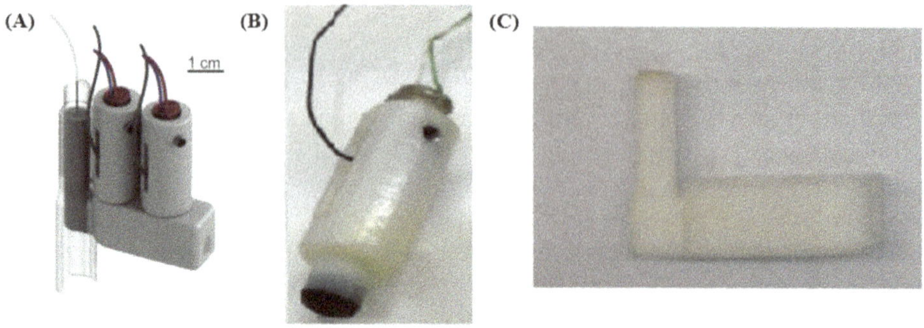

Figure 4.5. (A) Photograph of the 3D printed microfluidic device with two electrode holders containing needle biosensors, and a microdialysis outlet holder. (B) An image of the device needle holder containing a needle biosensor and (C) the microfluidic chip fabricated, using two different 3D printing technologies. (Reproduced with permission from [63]. Copyright 2015 American Chemical Society.)

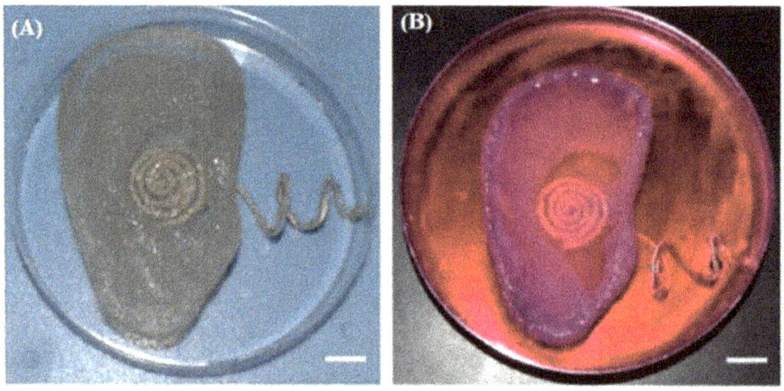

Figure 4.6. Photograph of a 3D bionic ear (A) immediately after 3D printing and (B) during *in vitro* culturing. (Reproduced with permission from [64]. Copyright 2013 American Chemical Society.)

monitoring of metabolite levels in tissue and, for instance it can be used by athletes for assessing the effectiveness of their training in real time [63].

Mannoor and co-workers [64] developed a 3D bionic ear with the anatomical geometry of the human ear (figure 4.6(A)) through the 3D printing of a cell-seeded alginate hydrogel matrix with chondrocytes, along with infused silver nanoparticles to capture electromagnetic frequencies in the radiofrequency range [64]. For this purpose, the printed bioelectronic ear was cultured *in vitro* for ten weeks (figure 4.6(B)) to grow cartilage tissue around an inductive coil antenna. The fabricated bionic ear demonstrates better functionality in detecting sounds over a radio-frequency range than the human counterpart [64].

In 2020 Cardoso *et al* developed a 3D printed electrochemical biosensor for analysing biological fluid samples using FDM printing technology, in which polylactic acid-containing graphene (G-PLA) filaments were utilised to fabricate the biosensor platform [65]. The results for the 3D printed glucose biosensor with G-PLA as the transducer for detecting the concentration of glucose in blood plasma prove its suitability for measuring glucose [65]. Furthermore, in this study, the performance of the proposed 3D printed G-PLA sensors was demonstrated for simultaneous measurement of nitrite (NIT) and uric acid (UA) in saliva and urine samples, using both voltammetric and amperometric methods [65]. The obtained response showed that surface treatment enhances the electrochemical properties of the G-PLA 3D printed sensor. Also, in comparison the analytical parameters obtained using the (batch injection analysis-multiple-pulse amperometry) BIA-MPA method proved the superior performance of this method over (differential pulse voltammetry) DPV for simultaneous determination of UA and NIT in urine and saliva in terms of sample throughput, precision, detection limits, and linear ranges [65]. Thus biosensors fabricated using the FDM printing method could be used successfully for the analysis of real biological samples [65].

In wax printing, solid wax is used as the printing material and the printing process is carried out easily by melting the wax, and subsequent wax printing with the desired pattern is printed on the target substrate [4]. This method is used widely for the fabrication of paper-based biosensing devices and the wax printed paper is treated thermally to create hydrophobic barriers to keep a liquid sample within a defined area [4]. This technique is known as one of the simplest fabrication methods with the fewest steps to print designed patterns on paper, and is also an inexpensive printing process. Thus these properties make it the most suitable method for fabricating large numbers of microfluidic paper-based analytical devices (μPADs) [41, 66]. However, the main limitation associated with this method is the low resolution caused by the melted wax diffusing in both the horizontal and vertical directions when the wax melts on the substrate. While a hydrophobic wall is created by vertical spreading across the paper, lateral spread has a destructive effect on the resolution of the created pattern since it causes widening of the barrier compared to the original design, in particular a wider patterned barrier on the front surface than the back surface [66]. To address this problem, in 2019 Carrilho *et al* [66] introduced two equations for the purpose of predicting the total width of the hydrophobic barrier and to increase our understanding for designing features with higher resolution. In addition, fabricated wax printed prototypes have weak durability at high temperature caused by the low melting temperature of solid wax.

4.3.3 Inkjet printing

This method is based on Rayleigh's instability theory, in which the reason behind the natural break up of a stream of liquid into small droplets under gravity is described [67]. This method is a very inexpensive and rapid way to develop flexible and high-resolution point-of-care biosensors at a large production scale. Typically, it is classified into two categories: continuous inkjet (CIJ) printing and drop-on-demand (DOD) printing [18]. The CIJ printer was invented by Sweet *et al* [18] in 1972 and three years later Zoltan developed the first type of DOD printer [68].

In CIJ printing, the liquid ink flows continuously by applying a constant pressure to the chamber [69, 70]. Once the liquid ink is ejected from the nozzle it breaks into droplets because of Rayleigh scattering and a piezoelectronic crystal controls the size and speed of the flowing droplets [70]. Then, to print a particular pattern, the droplets are charged by a charge electrode. While the charged droplets are directed to specific positions on the substrate as they pass between deflection electrodes, a gutter captures uncharged droplets and returns them to the ink bottle [70].

Unlike CIJ printers, DOD printers have several nozzles to expel ink droplets and involves ejecting the ink droplets only when needed via an actuator, which produces the pulse pressure [18, 69]. Compared to CIJ printing, the DOD printing method has achieved greater popularity as this method can generate droplets with a smaller diameter ($\leqslant 20\ \mu m$) and better placement accuracy [71]. The thermal inkjet method and piezoelectric inkjet method are the two main approaches for creating pulses.

In thermal inkjet printers the ink is heated up to vaporisation temperature by electrical heater elements [72]. Subsequently, a pulse pressure, which is caused by the bubbles created at the surface of the heaters, pushes the droplets out from the nozzle [18, 69]. As the bubble of vapour collapses, the droplets are ejected through the nozzle.

In contrast, in DOD printers the pressure pulse is generated by deformation of a piezoceramic plate due to applying electric pulse, causing ink droplets to be ejected through the nozzle [72]. Once the electric pulse is removed the piezoceramic plate moves back to its normal shape. The piezoelectric method can use a broader range of liquids. In contrast, the liquids used in the thermal inkjet method are limited to volatile liquids. The piezoelectric printheads have longer lifetimes since they function at a low temperature [18]. However, the thermal inkjet method is more inexpensive and simpler [18, 69].

In general, the inkjet printing method can be used to deposit the biosensor's components, including the bio-recognition element, and patterning the reaction surface and transducing molecule in a fast, straightforward, high precision, and low-cost way [18]. The critical parameters of the printing method are the compatibility of the ink formulation and the ink's rheological properties with the selected printhead, as incompatibilities have detrimental effects on the precision of deposition and subsequently on the resolution.

In 2005 Set *et al* [73] created a prototype of an amperometric glucose biosensor by applying electronic and biological inks through a thermal inkjet printer (figure 4.7) [73]. Electronic ink of the conductive polymer blend PEDOT/PSS was deposited onto an ITO-coated glass. Then biological ink, containing the enzyme GOD, was ejected onto the polymeric film (figure 4.7(A)) [73]. The working electrode formed by printed PEDOT/PSS provided a suitable reaction surface for the redox reaction between the glucose and the GOD with no significant enzyme activity loss, and with outstanding sensitivity [73]. Two years later they utilised the same method to fabricate an (Horseradish peroxidase) HRP-based amperometric biosensor using a thermal inkjet printer for the detection of hydrogen peroxide [74]. The results of this research were in line with their previous studies, but with better printing resolution. This printing method is not only used for surface patterning but also direct and indirect patterning of reaction channels or zones.

It has been demonstrated that the inkjet printing method is the leading technique to fabricate paper-based electrode sensors. Määttänen *et al* [75] proposed one of the first electrochemical paper-based biosensors made using an inkjet printing method. As illustrated in figure 4.7(B), through the piezoelectric printing method nano- gold

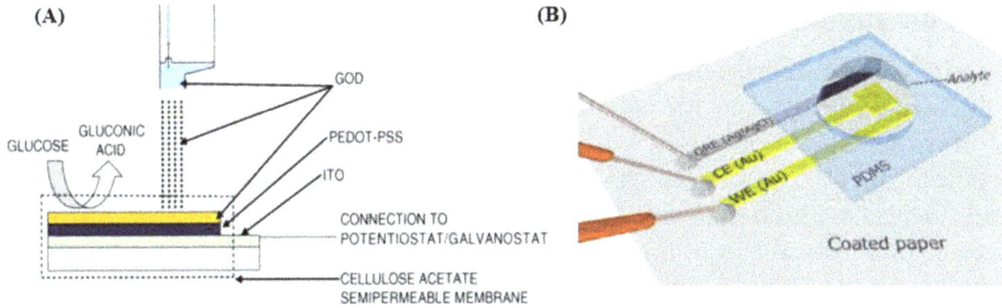

Figure 4.7. (A) The structure of the completed prototype of an amperometric GOD electrode fabricated using thermal inkjet printing. (Reproduced with permission from [73]. Copyright 2003 Elsevier.) (B) Schematic diagram of paper-based electrochemical sensors fabricated using a piezoelectric printer. (Reproduced with permission from [75]. Copyright 2005 Elsevier.)

ink was used to pattern a working and counter electrode on a paper substrate [75]. Also, the reference electrode was formed with an electrochemical decomposition of AgCl on printed silver strips to obtain Ag/AgCl, and the reaction well was created using PDMS for the electrolyte/analyte solution [75]. In this study, various modification approaches were utilised to enhance the performance of this electrochemical printed device [75]. Such devices were used for various sensing applications. For example, polyaniline film was electropolymerised on the WE surface using cyclic voltammetry for potentiometric pH measurement [75]. Additionally, on the WE surface, PEDOT–GOx was electropolymerised for sensing the glucose concentration in blood [75]. This method provided a successful tool for measuring the glucose level and detecting PH sensing even after a number of weeks [75]. This study proved that this disposable paper-based sensor has the potential for use in a wide variety of electrochemical experiments as it is more convenient and inexpensive than conventional sensors and has the same performance [75].

Moreover, in 2012 Li *et al* developed a paper-based blood typing device with the ability to diagnose all blood types and display the results in a written format without requiring interpretation of the results by technicians [76]. The proposed paper text-reporting was to some extent based on the haemagglutination reactions between antigens in the blood sample and antibodies in a text pattern. For this purpose, hydrophilic patterns of letters and symbols were formed by inkjet printing of an alkenyl ketene dimer (AKD) on the paper sample, which was surrounded by unprinted hydrophobic regions, and antibodies were ejected into corresponding patterns and also solution of saline used to ensure removal of excessive blood. The blood type introduced into the patterns is typed based on the reaction between antibodies and antigens existed on the surface of red blood cells (RBC), and also the saline solution is used to better clarify the occurrence of the specific haemagglutination reaction [76]. For example, if the haemagglutination reaction occurs, the agglutinated blood sample would not be removed by the saline solution [76] and thus the blood type can be reported by visually displaying a specific text pattern. It should be noted that the results of this research are in excellent agreement with pathological laboratory results.

4.4 The application of printable biosensors in biomedicine

The exploitation of printable biosensors as biomedical devices is rapidly growing in both clinical settings and for on-site end-user measurements. As was elaborated in the previous section, a variety of printable biosensors based on various detection modules, different substrates, and diverse printing technologies have been developed. Generally, such printable sensing platforms have gained considerable importance in point-of-care testing (POCT) using disposable printed sensors, as well as other biosensing applications of printed medical sensing devices such as wearable devices and even implantable devices.

4.4.1 Application as disposable biosensors for point-of-care testing

The literature is replete with different types of disposable printed biosensors for the detection of a wide variety of biologically essential analytes in human biofluids such as urine, serum, plasma, and blood [77]. Thus far, several printable biosensors have been developed for analysis of human urine to detect phosphate ions [78, 79], uric acid [80–82], morphine [83], progesterone [84], urinary albumin [85], and creatinine [86]. All these printed biosensors proposed for the analysis of urinary samples are based on electrochemical detection on screen printed electrodes, and their differences depend on the exploited enzymes and electrochemical mediators.

Human blood and other blood-derived bio-matrices such as plasma and serum contain a wealth of analytes for POCT applications via printable biosensors. During the past few decades, many printable biosensors for the detection of numerous analytes in blood (or blood-derived) samples have been proposed. The majority of these printable biosensors are based on conventional screen/stencil printing techniques. However, in some cases, the exploitation of advanced printing technologies such as inkjet printing has also been reported. Blood glucose [87–89], cholesterol [90], alcohol [91], urea [92], uric acid [93], hormones [94], lactic acid [95], disease biomarkers [96], metabolites [97], and blood electrolytes such as calcium [98] and potassium [99] are among the widely studied analytes using printable blood biosensors. In an interesting study, Wang *et al* proposed a low-cost electrochemical biosensor for detection of beta-hydroxybutyrate (BHB) as a biomarker of diabetic acidosis in whole blood. The biosensor was fabricated using wax printing to create microfluidic channels on a piece of chromatographic paper, and conventional stencil printing was utilised for fabrication of the transducing surface. They proposed a 3D multi-layered structure to integrate the sample reaction zone using pre-stored enzymatic reagents in the detection zone. The disposable platform was coupled to a personal glucometer to enable multiplex detection of glucose and BHB with minor modifications [100]. A schematic of their device is shown in figure 4.8(A).

More interestingly, the application of disposable printable biosensors for *in situ* reporting of blood type has also been reported [76].

Disposable printable biosensing devices have been developed widely for the POCT of various analytes in human saliva. Saliva is a readily available biofluid which also provides the opportunity for POCT in an easy and non-invasive manner. Thus far biogenic amines [101], lactate [102], alpha-amylase [103], glucose, urea,

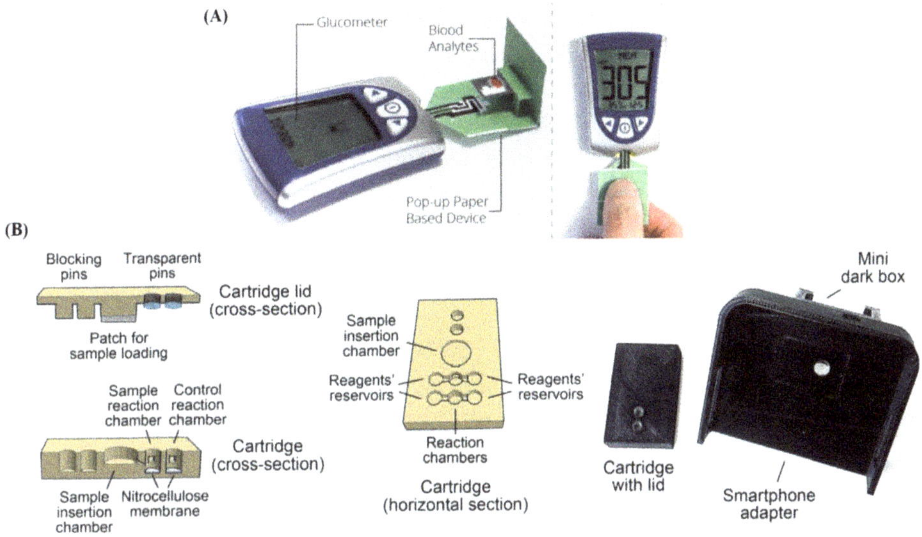

Figure 4.8. (A) An integrated printable biosensor with a 3D pop-up design for low-cost detection of BHB in whole blood using a universal personal glucometer. (Reproduced with permission from [100]. Copyright 2016 American Chemical Society.) (B) Schematic of a 3D printed disposable biosensor for lactate determination by chemiluminescence of enzyme coupled reactions, illustrating a cross-section and horizontal view of the device, device assembly and integration with a smartphone to develop a mini-luminometer. (Reproduced with permission from [107]. Copyright 2014 The Royal Society of Chemistry.)

phosphate [104], cancer biomarkers [105], and cortisol [106] have been detected in saliva using disposable printable biosensing devices. In a study by Roda *et al* a disposable 3D printed biosensor based on an optical transducer was developed for the detection of lactate in oral fluid and sweat. The signal acquisition was carried out by the detection of chemiluminescence emission generated by enzyme coupled reactions (lactate oxidase coupled to horseradish peroxidase) using a smartphone and subsequent red–green–blue (RGB) analysis. The exploitation of 3D printing enabled production of a disposable minicartridge as a portable luminometer coupled to a smartphone [107]. Figure 4.8(B) shows a schematic of 3D printed biosensors showing different compartments and overall device assembly.

4.4.2 Application as wearable and implantable sensing devices

One of the fascinating applications of printable biosensors is the exploitation of such devices as wearable biosensing platforms or implantable biosensors.

Wearable biosensors are capable of rapid and non-invasive monitoring of diverse analytes in real time. Human sweat, saliva, tears, and breath are among the popular and readily available matrices which can be monitored easily using on-body printable biosensing devices. Interestingly, wearable printable (bio)chemical sensors have also been used for the detection of hazardous substances in the surrounding environment [108]. The field of wearable printable biosensors is dominated mostly by electrochemical detection using screen printed electrodes fabricated on flexible substrates such as gloves, textiles, contact lenses, silicon, and various flexible polymers [11]. In some other cases, inkjet printing [75] and laser printing [109] have been used to create the conductive tracks of the transducing surface on flexible substrates. Printed wearable biosensing platforms have been established for the real-time detection of various target compounds such as electrolytes, cytokines, hormones, antibodies, glucose, lactate, cortisol, drugs, metabolites, and uric acid in human sweat [110].

One of the most successful modulations of printable wearable biosensors is temporary tattoo-based skin-worn devices. To integrate detection and data acquisition in these wearable biosensors, a transceiver for wireless data transmission can be easily incorporated into the sensing platform. Figure 4.9 shows typical steps involved in the fabrication of printable tattoo-based biosensors and a standard flexible wearable device in the form of a temporary tattoo with screen printed electrodes as the transducing surface [111].

Tai *et al* [112] proposed a wearable printable biosensor for real-time detection of caffeine in human sweat using electrochemical detection module overprinted electrodes as a transducing surface. The whole printable wearable biosensor containing the printed transducing surface and printed circuit board (PCB) for applying electrochemical input was packed into an integrated wristband. The surface of the working

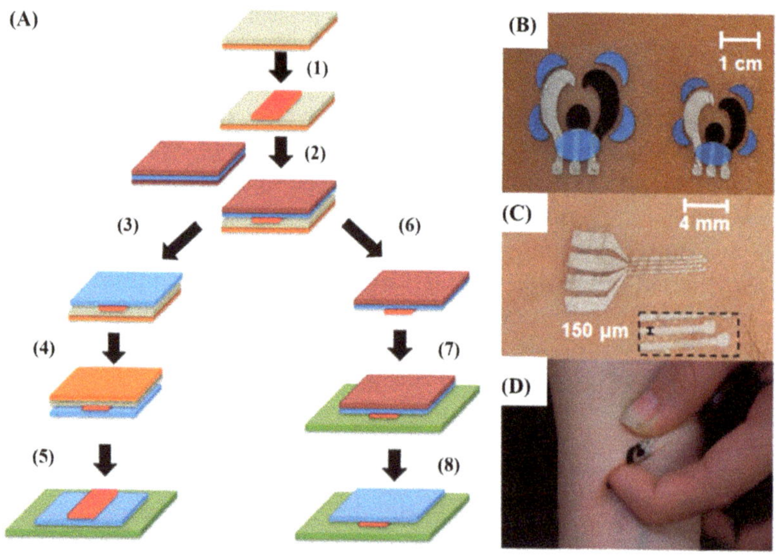

Figure 4.9. (A) A schematic of the successive steps involved in the fabrication of a tattoo-based printable biosensor as a wearable device for both physiological and external environment monitoring: (1) screen printing the electrodes onto a surface coated with a release agent, (2) the adhesive sheet is stuck to the printed electrodes, (3) the transparent protective layer is removed and the sensor is flipped, (4) the paper is coated with a release agent, and (5) the sensor is applied on the body for monitoring the external environment, or (6) the temporary paper is removed, (7) the sensor is attached to the skin, and (8) the protective layer is removed. (B)–(E) show different temporary transfer tattoo-based wearable electrochemical biosensors. (Reproduced with permission from [111]. Copyright 2012 The Royal Society of Chemistry.)

electrode was modified with carbon nanotubes and Nafion for anti-fouling protection and improvement of the sensitivity. The printed electrodes were produced using a roll-to-roll printing process [112].

Bariya *et al* [113] developed a cost-effective method for the fabrication of flexible substrates through roll-to-roll gravure printing. Using this method, mechanically and electrochemically stable electrodes with uniform redox kintetics could be printed onto a roll of 150 m flexible substrate. The real applicability of the wearable sensor was demonstrated for the detection of metal ions, metabolites, and other small molecules by functionalization of the working zone. Their work paves a new way towards large-scale and low-cost fabrication of disposable wearable devices [113]. Images of the fabrication method, wearable device, and the operation process are provided in figure 4.10.

Although on-body printable biosensors hold great promise in non-invasive healthcare monitoring, such devices are limited to analytes and physiochemical parameters that are readily available on the human body. Therefore, the application of printable biosensing devices has also been extended to in-body (implantable) biosensors. For instance, Voskerician *et al* proposed a universal printable enzymatic

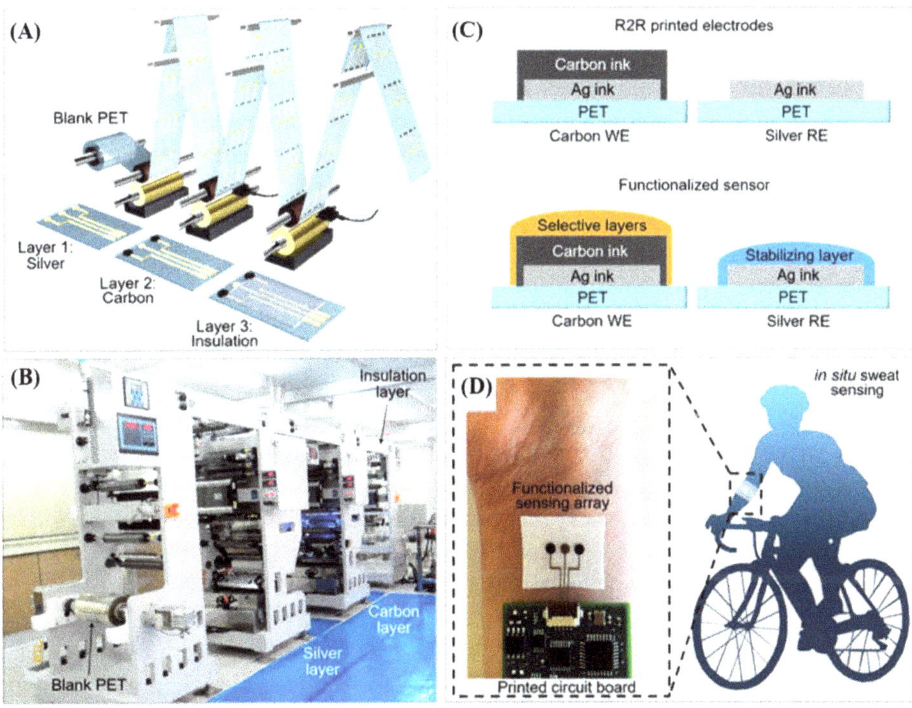

Figure 4.10. (A) Roll-to-roll printing of electrodes on flexible poly(ethylene terephthalate) (PET) substrate, (B) photograph of roll-to-roll gravure printing machines with modular units for each layer, (C) cross-section view of multiple layers gravure printed on PET substrate, and (D) application of the gravure printed substrate for *in situ* and real-time monitoring of analytes in human sweat. (Reproduced with permission from [113]. Copyright 2018 American Chemical Society.)

biosensor for *in vivo* measurement of various bio-species such as glucose using a thick-film screening method [114]. To this end, two carbon electrodes as the working and counter electrodes and one silver/silver chloride electrode were printed on polyester substrate and the performance of the sensors was evaluated through subcutaneous implantation in rats. It was shown that the sensor functionality was not affected by the *in vivo* environment [114].

4.5 Parameters of printable biosensing devices

In this section we intend to provide a critical discussion of the parameters of printable biosensors in terms of their analytical characteristics (i.e. sensitivity, selectivity, and standard deviation), multiplexing, assay time, and the final cost. For this purpose, the relevant parameters for the performance of printable biosensors and related challenges as well as possible strategies will be discussed.

4.5.1 Analytical characteristics of printable biosensors

Sensitivity (the limit of detection (LOD)) is one of the most critical analytical parameters of printable biosensing devices. This issue has inspired intensive research to enhance the sensitivity of printable biosensors through innovation in the formulation of the inks (or pastes) utilised for printing or post-treatment of printed tracks by incorporation of modified materials. Many classes of modifiers have been incorporated into printable biosensors thus far, including metallic nanomaterials (i.e. nanoparticles, nano-flowers, and nano-rods) [54], molecularly imprinted polymers [115], conductive polymers [116], carbonaceous materials (i.e. graphitic compounds, carbon nanotubes, and carbon nitride) [117], ionic liquids [117], etc.

Additionally, diverse signal amplification strategies such as the application of nano-carriers [118], nano-tracers [18], and nano-enzymes [119] have been used widely in printable biosensors to improve the sensitivity of such devices. Fluid flow manipulation [120], hyphenation of analytical procedures (i.e. hyphenation of separation with detection) [18], variation in device design [121], type of exploited substrate [120], and utilisation of arrays of transducing surfaces [120] (in the case on electrochemical detection) are among the other parameters which have been proven to be effective for increased sensitivity and linear dynamic range of printable biosensors.

In a study by Walgama *et al* a combination of different printing methods was used to fabricate a single biosensor, capable of hyphenated magnetic separation and electrochemical detection. Their device consisted of a paper-based electrochemical detection zone and a 3D printed minicartridge for facile transportation of micro-scale magnetic beads. The detection zone was fabricated by printing hydrophobic wax barriers and stencil printing of electrodes on a piece of chromatography paper which was subsequently coupled to the 3D printed microfluidic channel. Using this innovative design and combination of isolation and electrochemical detection, a detection limit down to 12 pM could be obtained for a silver–DNA-biotin (bDNA) conjugate [122]. Their device is shown in figure 4.11.

Selectivity in printable biosensors is achieved through the incorporation of bio-recognition elements such as aptamers, antibodies, enzymes, proteins, nucleic acids,

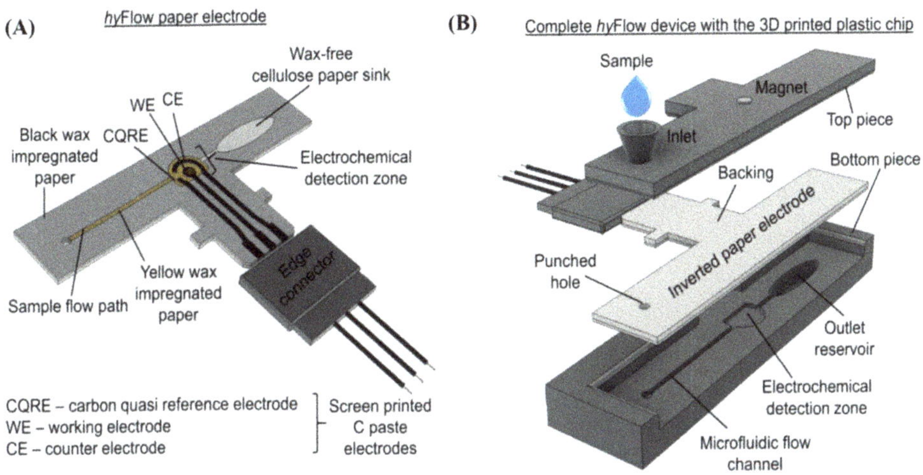

Figure 4.11. A schematic of (A) the design and (B) the assembly of a device fabricated using three different printing fabrication methods, 3D printing, stencil printing, and wax printing, for versatile combination of magnetic isolation and ultrasensitive paper-based electrochemical detection over a single biosensing platform. (Reproduced with permission from [122]. Copyright 2020 The Royal Society of Chemistry.)

and even cells or micro-organisms on the sensing interface [18]. Typically, the interaction between the sensing surface and target analytes are specific, which guarantees the selectivity of the biosensing platform. Although post-treatment of printed biosensors is normally used to create the selective sensing surface, printing methods—in particular inkjet printing—have also been used for the deposition of bio-recognition elements. The utilisation of printing technology for the fabrication of the sensing surface comes with the inherent advantages of improved reproducibility (sensor-to-sensor standard deviations), rapid fabrication, higher special resolution, the deposition of multiple bio-recognition elements (for multiplex measurements), and the possibility of gradient deposition. However, the quantitative loss of bio-recognition elements due to non-specific adsorption on the printing chamber and maintaining the activity and stability of biomolecules during the printing process are significant challenges for printing biologically active tracks [18].

4.5.2 Other parameters of printable biosensors

Fabrication procedure, integration, assay time, and the final device cost are among the other parameters of printable biosensors. Fabrication procedure and the final cost are primarily dependent on the printing method of interest. Table 4.1 is provided to make a comparison between current printing technologies in terms of their cost, fabrication time, potential application in the fabrication of biosensing platforms, and advantages and disadvantages.

There are many important parameters for both the final assay time as well as the system integration, such as type of assay, analyte of interest, and desired sample matrix or biofluid, and the exploitation of emerging technologies such as microfluidics in printable biosensors hold great promise in enhancing the performance of

Table 4.1. A comparison between different printing methods and their potential application for the fabrication of printable biosensors.

Printing technique	Application in fabrication of biosensors	Advantages	Disadvantages	Reference
Screen printing	• Fabrication of biosensing device • Fabrication of transducing surface	• Low cost • Facile fabrication • Well-suited for mass production • Wide availability of commercial pastes	• High paste consumption • High waste generation • Imprecision • Nonuniformity of the final tracks of the transducing surface	[123]
Wax printing	• Fabrication of biosensing device (paper-based devices)	• Simple fabrication • Low cost • Suitable for mass production • Rapid fabrication	• Poor resolution • Requires post-treatment	[16]
Inkjet printing	• Fabrication of biosensing device • Fabrication of transducing surface • Fabrication of sensing surface	• Wide availability • Availability of a wide range of commercial inks • Compatible with various types of inks • High resolution • High precision • Suitable for mass production	• Challenges in ink formulation • Maintaining the activity of biomolecules during the printing process	[18, 124, 125]
3D printing	• Fabrication of biosensing device • Fabrication of transducing surface • Fabrication of sensing surface	• Ease of use • Availability of instrument • Low cost	• Maintaining the activity of biomolecules during the printing process • Time consuming	[126]

such devices. The combination of microfluidics with printable biosensing devices not only reduces the assay time thanks to the rapid fluid delivery and reagent mixing, but also enables the integration of multiple steps associated with a single assay. On top of that, multiplexing is more feasible in microfluidic-enabled printable biosensors, since several recognition elements can be accommodated over a small footprint and sample flow can be divided and directed to several detection zones at the micro-scale. One of the most well-known and successful examples of such devices are the various versions of microfluidic paper-based analytical devices (μPADs) in which microfluidics are combined with printing methods (wax printing, inkjet printing, and screen printing) to fabricate different types of biosensors [4].

4.6 Summary and future outlook

Printable biosensing devices have been adopted widely in recent years as an easy-to-fabricate yet efficient platform in both POCT and healthcare monitoring through medical devices such as wearables and implantable biosensors. The exploitation of printing technologies in the fabrication of biosensing platforms provides several advantages, including reducing the fabrication cost and time, enabling mass production, and improving the reproducibility of fabricated biosensors, the feasibility of both 2D and 3D fabrication, and the possibility of fabricating biosensors on diverse substrates. Thus far various printing techniques have been used to fabricate biosensing platforms ranging from simple printing techniques such as wax printing and screen/stencil printing to more advanced ones such as inkjet printing and different modules of 3D printing.

Recent trends in the area of printable biosensors are focused on the exploitation of novel ink materials well-suited for direct printing of the desired track without any need for post-treatment with modifiers. New formulation of inks (or pastes) based on advanced materials is promising for developing motifs for enhancing the sensitivity and efficiency of printable biosensors, in particular for electrochemical measurement on printed electrodes. The development of various bio-inks containing target bio-recognition elements and suitable additives to maintain the stability of biomolecules during the printing process is a promising prospect to achieve fully printable biosensors. This development will also pave a new way towards multiplex measurement on printable biosensors through the simultaneous and high-resolution deposition of several bio-probes on the sensing surface.

Additionally, the combination of printable biosensors with other emerging technologies such as microfluidics and flexible electronics is a fascinating approach to integrate several functionalities on a single platform, make multiplexing more feasible, and establish cost-effective wearable biosensing devices.

References

[1] Ejeian F *et al* 2018 Biosensors for wastewater monitoring: a review *Biosens. Bioelectron.* **118** 66–79

[2] Bhalla N, Jolly P, Formisano N and Estrela P 2016 Introduction to biosensors *Essays Biochem.* **60** 1–8

[3] Nag A, Mukhopadhyay S C and Kosel J 2019 *Printed Flexible Sensors: Fabrication, Characterization and Implementation* vol 33 (Berlin: Springer)
[4] Huerta-Nuñez L F E *et al* 2019 A biosensor capable of identifying low quantities of breast cancer cells by electrical impedance spectroscopy *Sci. Rep.* **9** 6419
[5] Khan H, Razmjou A, Ebrahimi Warkiani M, Kottapalli A and Asadnia M 2018 Sensitive and flexible polymeric strain sensor for accurate human motion monitoring *Sensors* **18** 418
[6] Grieshaber D, MacKenzie R, Vörös J and Reimhult E 2008 Electrochemical biosensors—sensor principles and architectures *Sensors* **8** 1400–58
[7] Ranjan R, Esimbekova E N and Kratasyuk V A 2017 Rapid biosensing tools for cancer biomarkers *Biosens. Bioelectron.* **87** 918–30
[8] Ejeian F *et al* 2019 Design and applications of MEMS flow sensors: a review *Sens. Actuators* A **295** 483–502
[9] Mehrotra P 2016 Biosensors and their applications–a review *J. Oral Biol, Craniofac. Res.* **6** 153–59
[10] Wang J 2006 Electrochemical biosensors: towards point-of-care cancer diagnostics *Biosens. Bioelectron.* **21** 1887–92
[11] Yu Y, Nyein H Y Y, Gao W and Javey A 2020 Flexible electrochemical bioelectronics: the rise of *in situ* bioanalysis *Adv. Mater.* **32** 1902083
[12] Windmiller J R and Wang J 2013 Wearable electrochemical sensors and biosensors: a review *Electroanalysis* **25** 29–46
[13] Kivimäki L, Hakalahti L, Mäkelä I, Ylikunnari M, Keränen M and Takkinen K 2009 Printable biosensor surface *Report* VTT Technical Research Centre of Finland
[14] Sajad A M *et al* 2020 Development of an ultra-sensitive and flexible piezoresistive flow sensor using vertical graphene nanosheets *Nano-Micro Letters* **12** 109
[15] Patel B A, Anastassiou C A and O'Hare D 2006 Biosensor design and interfacing *Body Sensor Networks* (Berlin: Springer) pp 41–87
[16] Akyazi T, Basabe-Desmonts L and Benito-Lopez F 2018 Review on microfluidic paper-based analytical devices towards commercialisation *Anal. Chim. Acta* **1001** 1–17
[17] Izzah Binti Mohammad A R M, Liang K, Asadnia M and Chen V 2019 MOF-based enzymatic microfluidic biosensor via surface patterning and biomineralization *ACS Appl. Mater. Interfaces* **11** 1807–20
[18] Li J, Rossignol F and Macdonald J 2015 Inkjet printing for biosensor fabrication: combining chemistry and technology for advanced manufacturing *Lab Chip.* **15** 2538–58
[19] Duocastella M, Fernández-Pradas J, Morenza J, Zafra D and Serra P 2010 Novel laser printing technique for miniaturized biosensors preparation *Sens. Actuators* B **145** 596–600
[20] Raoufi M A *et al* 2020 Fabrication of unconventional inertial microfluidic channels using wax 3D printing *Soft Matter* **16** 2448–59
[21] Kottapalli A G P, Asadnia M, Miao J and Triantafyllou M 2015 Soft polymer membrane micro-sensor arrays inspired by the mechanosensory lateral line on the blind cavefish *J. Intell. Mater. Syst. Struct.* **26** 38–46
[22] Palenzuela C L M and Pumera M 2018 (Bio)analytical chemistry enabled by 3D printing: sensors and biosensors *TrAC, Trends Anal. Chem.* **103** 110–18
[23] Bazaz S R *et al* 2020 3D printing of inertial microfluidic devices *Sci. Rep.* **10** 1–14
[24] Razavi Bazaz S *et al* 2019 Microfluidics: rapid softlithography using 3D-printed molds *Adv. Mater. Technol.* **4** 1970056

[25] Syed M S, Rafeie M, Henderson R, Vandamme D, Asadnia M and Warkiani M E 2017 A 3D-printed mini-hydrocyclone for high throughput particle separation: application to primary harvesting of microalgae *Lab Chip* **17** 2459–69
[26] Hagihghi R, Razmjou A, Orooji Y, Warkiani M E and Asadnia M 2020 A miniaturized piezoresistive flow sensor for real-time monitoring of intravenous infusion *J. Biomed. Mater. Res.* B **108** 568–76
[27] Syed M S *et al* 2018 Selective separation of microalgae cells using inertial microfluidics *Bioresour. Technol.* **252** 91–9
[28] Asadnia M, Kottapalli A G P, Miao J, Warkiani M E and Triantafyllou M S 2015 Artificial fish skin of self-powered micro-electromechanical systems hair cells for sensing hydrodynamic flow phenomena *J. R. Soc. Interface* **12** 20150322
[29] Asadnia M *et al* 2013 High temperature characterization of PZT (0.52/0.48) thin-film pressure sensors *J. Micromech. Microeng.* **24** 015017
[30] Amir Razmjou M A, Hosseini E, Habibnejad Korayem A and Chen V 2019 Design principles of ion selective nanostructured membranes for the extraction of lithium ions *Nat. Commun.* **10** 5793
[31] Wang Y, Xu H, Zhang J and Li G 2008 Electrochemical sensors for clinic analysis *Sensors* **8** 2043–81
[32] Asadnia M *et al* 2016 Mercury (II) selective sensors based on AlGaN/GaN transistors *Anal. Chim. Acta* **943** 1–7
[33] Anik Ü 2017 Electrochemical medical biosensors for POC applications *Medical Biosensors for Point of Care (POC) Applications* (Amsterdam: Elsevier) pp 275–92
[34] Asadnia M *et al* 2017 Ca^{2+} detection utilising AlGaN/GaN transistors with ion-selective polymer membranes *Anal. Chim. Acta* **987** 105–10
[35] Mahmud M *et al* 2020 Recent progress in sensing nitrate, nitrite, phosphate, and ammonium in aquatic environment *Chemosphere* **259** 127492
[36] Borisov S M and Wolfbeis O S 2008 Optical biosensors *Chem. Rev.* **108** 423–61
[37] Asadnia M, Kottapalli A G P, Shen Z, Miao J and Triantafyllou M 2013 Flexible and surface-mountable piezoelectric sensor arrays for underwater sensing in marine vehicles *IEEE Sensors J.* **13** 3918–25
[38] Yamanaka K, Vestergaard M. d. C. and Tamiya E 2016 Printable electrochemical biosensors: a focus on screen-printed electrodes and their application *Sensors* **16** 1761
[39] Hayat A, Haider W, Rolland M and Marty J-L 2013 Electrochemical grafting of long spacer arms of hexamethyldiamine on a screen printed carbon electrode surface: application in target induced ochratoxin A electrochemical aptasensor *Analyst* **138** 2951–57
[40] Tudorache M and Bala C 2007 Biosensors based on screen-printing technology, and their applications in environmental and food analysis *Anal. Bioanal.Chem.* **388** 565–78
[41] Tortorich R P, Shamkhalichenar H and Choi J-W 2018 Inkjet-printed and paper-based electrochemical sensors *Appl. Sci.* **8** 288
[42] Dungchai W, Chailapakul O and Henry C S 2009 Electrochemical detection for paper-based microfluidics *Anal. Chem.* **81** 5821–26
[43] Yang J *et al* 2014 fluidic electrochemical biosensing platform with enzyme paper and enzymeless electrodes *Sens. Actuators* B **203** 44–53
[44] Razmjou A *et al* 2017 Preparation of iridescent 2D photonic crystals by using a mussel-inspired spatial patterning of ZIF-8 with potential applications in optical switch and chemical sensor *ACS Appl. Mater. Interfaces* **9** 38076–80

[45] Liu C *et al* 2018 3D printing technologies for flexible tactile sensors toward wearable electronics and electronic skin *Polymers* **10** 629
[46] Khosravani M R and Reinicke T 2020 3D-printed sensors: current progress and future challenges *Sens. Actuators* A **305** 111916
[47] Hull C W and Lewis C W 1991 Methods and apparatus for production of three-dimensional objects by stereolithography *US patent* US4999143A
[48] Melchels F P, Feijen J and Grijpma D W 2010 A review on stereolithography and its applications in biomedical engineering *Biomaterials* **31** 6121–30
[49] Han T, Kundu S, Nag A and Xu Y 2019 3D printed sensors for biomedical applications: a review *Sensors* **19** 1706
[50] Vanderburgh J, Sterling J A and Guelcher S A 2017 3D printing of tissue engineered constructs for *in vitro* modeling of disease progression and drug screening *Ann. Biomed. Eng.* **45** 164–79
[51] Crump S S 1989 Apparatus and method for creating three-dimensional objects *US patent* US5121329A
[52] Zein I, Hutmacher D W, Tan K C and Teoh S H 2002 Fused deposition modeling of novel scaffold architectures for tissue engineering applications *Biomaterials* **23** 1169–85
[53] Khalil S, Nam J and Sun W 2005 Multi-nozzle deposition for construction of 3D biopolymer tissue scaffolds *Rapid Prototyp. J.* **11** 9–17
[54] Hutmacher D W, Schantz T, Zein I, Ng K W, Teoh S H and Tan K C 2001 Mechanical properties and cell cultural response of polycaprolactone scaffolds designed and fabricated via fused deposition modeling *J. Biomed. Mater. Res.* **55** 203–16
[55] Gothait H 2001 Apparatus and method for three dimensional model printing *US patent* US6259962B1
[56] Quan H, Zhang T, Xu H, Luo S, Nie J and Zhu X 2020 Photo-curing 3D printing technique and its challenges *Bioact. Mater.* **5** 110–15
[57] Dirkzwager R M, Liang S and Tanner J A 2016 Development of aptamer-based point-of-care diagnostic devices for malaria using three-dimensional printing rapid prototyping *ACS Sens.* **1** 420–26
[58] Lee W *et al* 2014 Ultrarapid detection of pathogenic bacteria using a 3D immunomagnetic flow assay *Anal. Chem.* **86** 6683–88
[59] Lee W *et al* 2015 3D-printed microfluidic device for the detection of pathogenic bacteria using size-based separation in helical channel with trapezoid cross-section *Sci. Rep.* **5** 7717
[60] Tang C, Vaze A and Rusling J 2017 Automated 3D-printed unibody immunoarray for chemiluminescence detection of cancer biomarker proteins *Lab Chip* **17** 484–89
[61] Chan H N *et al* 2016 Simple, cost-effective 3D printed microfluidic components for disposable, point-of-care colorimetric analysis *ACS Sens.* **1** 227–34
[62] Dias A A *et al* 2016 Based enzymatic reactors for batch injection analysis of glucose on 3D printed cell coupled with amperometric detection *Sens. Actuators* B **226** 196–203
[63] Gowers S A *et al* 2015 3D printed microfluidic device with integrated biosensors for online analysis of subcutaneous human microdialysate *Anal. Chem.* **87** 7763–70
[64] Mannoor M S *et al* 2013 3D printed bionic ears *Nano Lett.* **13** 2634–39
[65] Cardoso R M *et al* 2020 3D-printed graphene/polylactic acid electrode for bioanalysis: biosensing of glucose and simultaneous determination of uric acid and nitrite in biological fluids *Sens. Actuators* B **307** 127621

[66] Carrilho E, Martinez A W and Whitesides G M 2009 Understanding wax printing: a simple micropatterning process for paper-based microfluidics *Anal. Chem.* **81** 7091–95

[67] Rayleigh L 1878 On the instability of jets *Proc. Lond. Math. Soc.* **1** 4–13

[68] Zoltan S 1974 Pulsed droplet ejecting system *US patent* US3683212A

[69] Daly R, Harrington T S, Martin G D and Hutchings I M 2015 Inkjet printing for pharmaceutics—a review of research and manufacturing *Int. J. Pharm.* **494** 554–67

[70] Martin G D, Hoath S D and Hutchings I M 2008 Inkjet printing-the physics of manipulating liquid jets and drops *J. Phys.: Conf. Ser.* **105** 012001

[71] De Gans B J, Duineveld P C and Schubert U S 2004 Inkjet printing of polymers: state of the art and future developments *Adv. Mater.* **16** 203–13

[72] Kipphan H 2001 *Handbook of Print Media: Technologies and Production Methods* (Berlin: Springer)

[73] Setti L, Fraleoni-Morgera A, Ballarin B, Filippini A, Frascaro D and Piana C 2005 An amperometric glucose biosensor prototype fabricated by thermal inkjet printing *Biosens. Bioelectron.* **20** 2019–26

[74] Setti L, Fraleoni-Morgera A, Mencarelli I, Filippini A, Ballarin B and Di Biase M 2007 An HRP-based amperometric biosensor fabricated by thermal inkjet printing *Sens. Actuators* B **126** 252–57

[75] Määttänen A *et al* 2013 A low-cost paper-based inkjet-printed platform for electrochemical analyses *Sens. Actuators* B **177** 153–62

[76] Li M, Tian J, Al-Tamimi M and Shen W 2012 Paper-based blood typing device that reports patient's blood type in writing *Angew. Chem.* **124** 5593–97

[77] Taleat Z, Khoshroo A and Mazloum-Ardakani M 2014 Screen-printed electrodes for biosensing: a review (2008–2013) *Microchim. Acta* **181** 865–91

[78] Gilbert L, Browning S, Jenkins A T and Hart J P 2010 Studies towards an amperometric phosphate ion biosensor for urine and water analysis *Microchim. Acta* **170** 331–36

[79] Gilbert L, Jenkins A T A, Browning S and Hart J P 2011 Development of an amperometric, screen-printed, single-enzyme phosphate ion biosensor and its application to the analysis of biomedical and environmental samples *Sens. Actuators* B **160** 1322–27

[80] Ballesta-Claver J, Ortega I D, Valencia-Mirón M and Capitán-Vallvey L 2011 Disposable luminol copolymer-based biosensor for uric acid in urine *Anal. Chim. Acta* **702** 254–61

[81] Markas A 1994 Novel, reagentless, amperometric biosensor for uric acid based on a chemically modified screen-printed carbon electrode coated with cellulose acetate and uricase *Analyst* **119** 833–40

[82] Kanyong P, Pemberton R, Jackson S and Hart J P 2012 Development of a sandwich format, amperometric screen-printed uric acid biosensor for urine analysis *Anal. Biochem.* **428** 39–43

[83] Ahmar H, Tabani H, Koruni M H, Davarani S S H and Fakhari A R 2014 A new platform for sensing urinary morphine based on carrier assisted electromembrane extraction followed by adsorptive stripping voltammetric detection on screen-printed electrode *Biosens. Bioelectron.* **54** 189–94

[84] Ditcham W, Palmer C, Warren K, Monaghan C, Kappelle W and Matson P 2009 Development of an immunoassay to measure progesterone using printed biosensors, and its application to the assessment of ovarian function in the numbat (*Myrmecobius fasciatus*) *J. Immunoas. Immunochem.* **30** 232–43

[85] Omidfar K, Dehdast A, Zarei H, Sourkohi B K and Larijani B 2011 Development of urinary albumin immunosensor based on colloidal AuNP and PVA *Biosens. Bioelectron.* **26** 4177–83
[86] Kumar V *et al* 2017 Creatinine–iron complex and its use in electrochemical measurement of urine creatinine *IEEE Sensors J.* **18** 830–36
[87] Wang T, Cook C and Derby B 2009 Fabrication of a glucose biosensor by piezoelectric inkjet printing *2009 Third Int. Conf. on Sensor Technologies and Applications* pp 82–5
[88] Crouch E, Cowell D C, Hoskins S, Pittson R W and Hart J P 2005 Amperometric, screen-printed, glucose biosensor for analysis of human plasma samples using a biocomposite water-based carbon ink incorporating glucose oxidase *Anal. Biochem.* **347** 17–23
[89] Zhang F *et al* 2017 Over 20% PCE perovskite solar cells with superior stability achieved by novel and low-cost hole-transporting materials *Nano Energy* **41** 469–75
[90] Li G, Liao J, Hu G, Ma N and Wu P 2005 Study of carbon nanotube modified biosensor for monitoring total cholesterol in blood *Biosens. Bioelectron.* **20** 2140–44
[91] Luo P, Xie G, Liu Y, Xu H, Deng S and Song F 2008 Electrochemical detection of blood alcohol concentration using a disposable biosensor based on screen-printed electrode modified with Nafion and gold nanoparticles *Clin. Chem. Lab. Med.* **46** 1641–47
[92] O'Reilly E, Kelly M, Morrin A, Smyth M R and Killard A J 2011 Chronocoulometric determination of urea in human serum using an inkjet printed biosensor *Anal. Chim. Acta* **697** 98–102
[93] Piermarini S *et al* 2013 Uricase biosensor based on a screen-printed electrode modified with Prussian blue for detection of uric acid in human blood serum *Sens. Actuators* B **179** 170–74
[94] Li N, Chow A M, Ganesh H V, Ratnam M, Brown I R and Kerman K 2019 Diazonium-modified screen-printed electrodes for immunosensing growth hormone in blood samples *Biosensors* **9** 88
[95] Sprules S D, Hart J P, Wring S A and Pittson R 1995 A reagentless, disposable biosensor for lactic acid based on a screen-printed carbon electrode containing Meldola's blue and coated with lactate dehydrogenase, NAD+ and cellulose acetate *Anal. Chim. Acta* **304** 17–24
[96] Lin P-Y *et al* 2012 Detection of alpha-methylacyl-CoA racemase (AMACR), a biomarker of prostate cancer, in patient blood samples using a nanoparticle electrochemical biosensor *Biosensors* **2** 377–87
[97] Schaffar B P 2002 Thick film biosensors for metabolites in undiluted whole blood and plasma samples *Anal. Bioanal.Chem.* **372** 254–60
[98] Wang Y, Xu H, Yang X, Luo Z, Zhang J and Li G 2012 All-solid-state blood calcium sensors based on screen-printed poly (3, 4-ethylenedioxythiophene) as the solid contact *Sens. Actuators* B **173** 630–35
[99] Xu H, Yang X, Wang Y, Zheng J, Luo Z and Li G 2010 Disposable blood potassium sensors based on screen-printed thick film electrodes *Meas. Sci. Technol.* **21** 055802
[100] Wang C-C *et al* 2016 A paper-based 'pop-up' electrochemical device for analysis of beta-hydroxybutyrate *Anal. Chem.* **88** 6326–33
[101] Piermarini S, Volpe G, Federico R, Moscone D and Palleschi G 2010 Detection of biogenic amines in human saliva using a screen-printed biosensor *Anal. Lett.* **43** 1310–16
[102] Claver J B, Mirón M V and Capitán-Vallvey L 2009 Disposable electrochemiluminescent biosensor for lactate determination in saliva *Analyst* **134** 1423–32

[103] Mahosenaho M, Caprio F, Micheli L, Sesay A M, Palleschi G and Virtanen V 2010 A disposable biosensor for the determination of alpha-amylase in human saliva *Microchim. Acta* **170** 243–49

[104] Kwan R C, Leung H, Hon P Y, Cheung H C, Hirota K and Renneberg R 2005 Amperometric biosensor for determining human salivary phosphate *Anal. Biochem.* **343** 263–67

[105] Viswanathan S, Rani C, Anand A V and Ho J.-A. 2009 Disposable electrochemical immunosensor for carcinoembryonic antigen using ferrocene liposomes and MWCNT screen-printed electrode *Biosens. Bioelectron.* **24** 1984–89

[106] Zangheri M *et al* 2015 A simple and compact smartphone accessory for quantitative chemiluminescence-based lateral flow immunoassay for salivary cortisol detection *Biosens. Bioelectron.* **64** 63–8

[107] Roda A, Guardigli M, Calabria D, Calabretta M M, Cevenini L and Michelini E 2014 A 3D-printed device for a smartphone-based chemiluminescence biosensor for lactate in oral fluid and sweat *Analyst* **139** 6494–501

[108] Colozza N *et al* 2019 A wearable origami-like paper-based electrochemical biosensor for sulfur mustard detection *Biosens. Bioelectron.* **129** 15–23

[109] Huang G-W, Feng Q-P, Xiao H-M, Li N and Fu S-Y 2016 Rapid laser printing of paper-based multilayer circuits *ACS Nano* **10** 8895–903

[110] Ray T R *et al* 2019 Bio-integrated wearable systems: a comprehensive review *Chem. Rev.* **119** 5461–533

[111] Windmiller J R, Bandodkar A J, Valdés-Ramírez G, Parkhomovsky S, Martinez A G and Wang J 2012 Electrochemical sensing based on printable temporary transfer tattoos *Chem. Commun.* **48** 6794–96

[112] Tai L C *et al* 2018 Methylxanthine drug monitoring with wearable sweat sensors *Adv. Mater.* **30** 1707442

[113] Bariya M *et al* 2018 Roll-to-roll gravure printed electrochemical sensors for wearable and medical devices *ACS Nano* **12** 6978–87

[114] Voskerician G, Liu C-C and Anderson J M 2005 Electrochemical characterization and *in vivo* biocompatibility of a thick-film printed sensor for continuous *in vivo* monitoring *IEEE Sensors J.* **5** 1147–58

[115] Wang P, Sun G, Ge L, Ge S, Yu J and Yan M 2013 Photoelectrochemical lab-on-paper device based on molecularly imprinted polymer and porous Au–paper electrode *Analyst* **138** 4802–11

[116] Chen Z, Wright C, Dincel O, Chi T-Y and Kameoka J 2020 A low-cost paper glucose sensor with molecularly imprinted polyaniline electrode *Sensors* **20** 1098

[117] Rana A *et al* 2020 Graphitic carbon nitride as an amplification platform on an electrochemical paper-based device for the detection of norovirus-specific DNA *Sensors* **20** 2070

[118] Wang X, Deng W, Shen L, Yan M and Yu J 2016 A 3D electrochemical immunodevice based on an Au paper electrode and using Au nanoflowers for amplification *New J. Chem.* **40** 2835–42

[119] Chaiyo S *et al* 2018 Non-enzymatic electrochemical detection of glucose with a disposable paper-based sensor using a cobalt phthalocyanine–ionic liquid–graphene composite *Biosens. Bioelectron.* **102** 113–20

[120] Adkins J A, Noviana E and Henry C S 2016 Development of a quasi-steady flow electrochemical paper-based analytical device *Anal. Chem.* **88** 10639–47

[121] Punjiya M, Moon C H, Matharu Z, Nejad H R and Sonkusale S 2018 A three-dimensional electrochemical paper-based analytical device for low-cost diagnostics *Analyst* **143** 1059–64

[122] Walgama C, Nguyen M P, Boatner L M, Richards I and Crooks R M 2020 Hybrid paper and 3D-printed microfluidic device for electrochemical detection of Ag nanoparticle labels *Lab Chip* **20** 1648–57

[123] Paschoalino W J *et al* 2019 Emerging considerations for the future development of electrochemical paper-based analytical devices *ChemElectroChem* **6** 10–30

[124] Yamada K, Henares T G, Suzuki K and Citterio D 2015 Paper-based inkjet-printed microfluidic analytical devices *Angew. Chem. Int. Ed.* **54** 5294–310

[125] Singh M, Haverinen H M, Dhagat P and Jabbour G E 2010 Inkjet printing—process and its applications *Adv. Mater.* **22** 673–85

[126] Sharafeldin M, Jones A and Rusling J F 2018 3D-printed biosensor arrays for medical diagnostics *Micromachines* **9** 394

Chapter 5

Laser induced graphene: advances in electro-biochemical sensing and energy applications

Khairunnisa Amreen, Avinash Kothuru and Sanket Goel

Laser induced graphene (LIG) is a porous material fabricated directly through laser ablation on various carbon precursors. Since its discovery, LIG has attracted the interest of researchers owing to its excellent physical, mechanical, electrical, and chemical properties. LIG has proven to be a cost-effective and scalable approach for producing graphene. The literature reports the generation of LIG on a variety of carbon precursors, such as cloth, coconut, potato, bread, cork, etc, in addition to polymers such as polyimide and polyetherimide. Depending upon the carbon source, various lasers, such as CO_2, UV, IR, and femtosecond lasers, etc, are utilized for ablation. Fine tuning of the preparation technique enables surface modification of LIG with desired functional groups as well as doping with required atoms. Therefore LIG is potentially an excellent candidate for diversified applications. Such applications range from electrochemical sensing, biosensing, energy storage, energy harvesting, energy generation, immunosensing, gas sensing, liquid sensing, artificial intelligence, circuits, electronics and communications, wearable electronics, etc. This chapter summarizes in brief some of the LIG preparation methods and their broad spectrum applications, particularly in electrochemical and biosensing. It also focuses on future outlooks and research gaps in the reported literature.

5.1 Introduction

For several decades, different carbon materials, such as carbon nanotubes, carbon nanofibres, carbon black, mesoporous carbon, etc, have been explored for their mechanical, chemical, and physical properties in various applications. Carbon presents exceptional behavior and a variety of structures ranging from 3D graphites, diamonds, 1D nanotubes, and fullerenes. The unique 2D structure of carbon allotrope is called graphene, which was discovered quite recently [1]. Typically, it is a hexagonal, planar assembly of carbon atoms. It is made up of sp^2-hybridized

doi:10.1088/978-0-7503-3439-6ch5 © IOP Publishing Ltd 2021

carbon and has aroused a substantial amount of interest among researchers owing to its valuable properties. From the time when graphene was first prepared by Novoselov *et al* in 2004 [2], there has been an abundance of publications regarding the synthesis and applications of this attractive material. Owing to its structural make-up, wherein 2p orbitals delocalize onto carbon sheets, graphene is highly rigid and has good thermal conductivity [3]. Due to its unique and cross-functional properties, graphene research and its application in miscellaneous fields has progressed at a fast pace for over 40 years. Due to its distinctive electronic and crystallographic nature [4], graphene has a large theoretical specific surface area (2630 $m^2\ g^{-1}$), high intrinsic mobility (200,000 $cm^2\ v^{-1}\ s^{-1}$), a high Young's modulus (~ 1.0 TPa), and high thermal conductivity (~ 5000 $Wm^{-1}\ K^{-1}$) with exceptionally high optical transmittance (~ 97.7%), and excellent electrical conductivity [5].

5.2 Properties of graphene

(i) *Stiffness*: Arguably, graphene is considered to be the strongest material. This exceptional property is due to the hexagonal lattice and strong stable sp^2 bonds. Several reports offer different stiffnesses and breaking forces, however, Nicholl *et al* reported a unique procedure using an interferometric profilometer and a non-contact approach to measure the stiffness and it was discovered that at room temperature the stiffness is 20–100 N m^{-1} [6].

(ii) *Strength*: A monolayer of graphene with zero defects is an extremely strong material. Hone *et al* measured the strength of pristine single layer graphene as 42 Nm^{-1}, which leads to a strength of 130 GPa [7]. It was found that the introduction of sp^3 defects in the material lowered the breaking strength by 14%, however, the stiffness was maintained even with the defects [8].

(iii) *Toughness*: Zhang *et al* [9] produced a micromechanical device for measuring the fracture toughness of graphene prepared using a chemical vapor deposition (CVD) approach. Using an iron beam, a central crack was introduced on the graphene surface followed by applying load. The stress of the fracture decreased with an increase in the length of the crack. A 15.9 J m^{-2} the release rate of critical strain energy was calculated. The toughness (stress intensity factor) was measured as 4.0 ± 0.6 MPa [9].

(iv) *Thermal conductivity*: Graphene has a high thermal conductivity of about 2500–5000 W mK^{-1} at room temperature, which is greater than that of diamond. Hence, it is highly useful in Joule heating applications [10].

(v) *Electrical resistivity*: The electrical resistivity of graphene is 35% lower than that of copper. At room temperature it is about $1 \times 10^{-8}\ \Omega \times m$, which is the lowest of any known material [11].

(vi) *Impermeability*: Graphene is highly impermeable. The smallest atoms of helium cannot pass through a graphene sheet. Thus it is a useful material in many electronic and mechanical sensors [12].

(vii) *Electron mobility*: Amongst all the other electronic materials, graphene has the highest electron mobility with a theoretical limit of 200 000 cm^2 $(V \times s)^{-1}$, i.e. more than one hundred times greater than that of silicon [11].

(viii) *Optical property*: Graphene shows 97% transmittance and absorbs only 2.3% of reflecting light, therefore it leads to better performance than that of ITO [13].

(ix) *Stretchability*: Graphene has a high Young's modulus and high fracture strength, and hence is strong, flexible, and stretchable. It can be stretched upto 25%, and is therefore extremely valuable in flexible electronics [7].

(x) *High surface area*: Graphene has a high surface area of 2630 m^2 g^{-1}. An entire soccer field can be covered with less than 3 gm of graphene [14].

(xi) *Atomic thickness*: A single layer of graphene is only one atom thick, hence it is called a 2D material. It is about 0.335 nanometers thick.

Such advantages have led to its use as a transparent conductive electrode, which has opened the door for several other ancillary applications after suitable integration and comprehension. With a current market size of ~$80 million, the market for graphene is expected to grow to >$500 million by 2025 [15]. The increasing demand for graphene and the exceptional growth of its market are plausibly due to the explosion of graphene usage in varied applications such as microelectronics, energy storage (supercapacitors), flexible display (LEDs), transparent conductive electrodes, composite (electrochemical), biological (sensors), and biomimetic devices, energy harvesting (fuel cells), etc [4]. The fabrication of graphene using a method with broad commercial potential is needed. The current commercial level synthesis is carried out by either thermal or extensive chemical processes [16]. The use of extreme temperatures and microwaves, and the involvement of processes such as annealing, pyrolysis, hydrothermal treatment, inert gas environments, etc, make the process economically non-viable [17, 18]. Similarly, the chemical reduction method uses hydrazine, a strong reducing agent, a carcinogen, and an environmental pollutant which is highly toxic in nature [19–21], for the bulk synthesis. The CVD method where graphene is grown on an etched substrate at elevated temperatures has also been explored for graphene synthesis. A hydrothermal process in which graphene oxide forms in hot water is also used. However, these approaches have certain limitations, such as a lack of control on shape and high cost [22].

5.3 The commercial synthesis of graphene

In order to meet the demand for graphene is various sectors, several routes for the large scale production of graphene have been adopted. Some of the commonly used procedures are described in the following.

5.3.1 Bottom up approach

Just like any other nanomaterial synthesis, a bottom up approach is also employed to produce graphene. Graphene is grown chemically through the formation of molecules and clusters via an atom-by-atom arrangement. Generally, elevated temperatures and starting precursors are required for this method. This approach produces high quality graphene due to automatic and precise control [23].

5.3.2 Chemical vapor deposition

CVD is the most commonly applied technique. Various hydrocarbon precursors, such as hexane, pentane, methane, ethylene, acetylene, etc, are used for exposing large metal substrates at extreme temperatures. CVD comprises approaches such as plasma based, thermal based, hot/cold wall based, etc. The hydrocarbons decompose over the metal surface and produce the growth of carbon atoms [24].

5.3.3 Mechanical exfoliation

Mechanical exfoliation is a simple technique, wherein a graphite material is taken and tape is used to exfoliate layers and transfer them to a substrate. The number of layers can be manipulated as desired and can further be characterized by physicochemical techniques such as Raman spectroscopy, atomic force microscopy, scanning tunneling microscopy, etc [25].

5.3.4 Liquid-phase exfoliation

This technique is a three step procedure: (i) dispersion, (ii) exfoliation, and (iii) purification. Graphite pieces are dispersed or dissolved in a suitable solvent. The exfoliation takes place through ultrasonication. Purification is carried out to remove non-exfoliated particles and the solvent is removed from the obtained graphene [26].

5.3.5 Electrochemical exfoliation

This method incorporates the use of an electrolyte (liquid) and an applied potential that is utilized to consume a graphite electrode. It is an anodic/cathodic reaction. The anodic reaction is the most commonly used as it gives graphene oxide, however, the cathodic reaction is more efficient in terms of obtaining few layers of graphene with good quality [27].

Although these techniques are used widely on a commercial level, the need for a simple, scalable, and economically viable process is acute. Thus various methods for the preparation of functionalized graphene at low temperatures and using chemical free processes are required [28]. Despite ample research, the development of high quality graphene in a cost-effective way while eradicating the use of precursors remains challenging [29]. However, a recent emerging advancement is a laser-based processing technique [4]. The laser is employed to convert carbon precursors, of both natural as well as synthetic substrates, into a graphitic material. Based on the power and speed of the laser, the derived graphitic material forms graphene/graphene oxide. Hence, the thus obtained graphene is called laser induced graphene (LIG) [30]. Some of the commonly used laser ablations are CW lasers (which involves a CO_2 infrared laser), pulsed lasers such as femtosecond pulsed laser, ultrashort pulsed lasers, etc [4, 28].

5.4 Laser induced graphene (LIG) fabrication

LIG fabrication is influenced greatly by the interaction of the material and the laser. Several methods such as ablation, localized heating, thermal reaction, photo-chemical reaction, laser induced exfoliation, writing, and CVD, etc, have been developed. Various substrates, such as carbon, silicon, paper, glass, plastic, etc, are being utilized for patterning graphene and are referred as 'on chip' electrodes. In the last decade, several research groups have been working to develop LIG. For example, in 2011 Kurra *et al* [31] used a pulsed laser technique to obtain graphene nanoribbons using a pyrolytic graphite substrate. The fabricated region had 5–6 layers of graphene ribbons [31]. Likewise, Wakaya *et al* [32] used a pulsed UV laser to exfoliate the graphene from SiO_2 and Si substrates. The thickness of the graphene determined the power density of the laser [32]. For fabricating graphene over graphite surfaces, pulsed laser techniques such as femtosecond, nanosecond, CO_2, etc, were also utilized. These lasers were also used efficiently in reducing graphene to graphene oxide and reduced graphene oxide. In 2019 Kurra *et al* summarized some of the various lasers employed for preparing LIG. Table 5.1 lists various lasers used in literature for fabricating LIG and their potential applications [4, 28].

5.4.1 Procedure

Recently, our group worked on the fabrication of LIG on a polyimide sheet using a CO_2 laser for various resistive, capacitive, and electrochemical sensing applications. A simple design was produced using AutoCAD Fusion 360 software. The design file was saved in the .dxf format. Then the .dxf file was transferred to CorelDraw X7. The outline and the design width were optimized as per the CO_2 laser preferences. Figure 5.1 gives the schematic representation of the fabrication procedure of LIG on the polyimide sheet [48]. By changing the speed and power of the CO_2 laser various LIG samples were formed over polyimide.

A 5.4 W laser power produced 91.3% carbon, 7.7% oxygen, and 1.0% nitrogen. A comparative study was carried out where the 30 W power of the CO_2 laser was varied from 1.35 W to 1.95 W and different LIG samples were obtained. Table 5.2 gives a comparison of various samples. More than 95% carbon was obtained in all the samples. It was also discovered that the thickness of LIG can be varied by scanning the power of the laser, hence enhancing the conductivity. Physico-chemical characterization of the fabricated LIG was performed using SEM, EDX, XRD, XPS, and Raman spectroscopy.

5.4.2 An LIG based microfluidic device

Our research group has also worked on developing microfluidic devices using LIG. The design was created in CAD software and saved as a .dxf file. As per the settings of the CO_2 laser machine, the file was opened in CorelDraw and aligned accordingly. For the design layout an acrylic sheet was used. The dimensions were 76 mm × 26 mm. Polyimide of the same dimensions was taken and bonded with the acrylic using adhesive tape. Various speeds and powers of laser were employed to obtained graphene. The acquired graphene was then peeled off via the sticky tape leaving behind the depth. The laser ablation was repeated over the same space and the

Table 5.1. The various types of lasers used for fabricating LIG and their potential applications.

S. No.	Carbon precursor	Laser	Application	Reference
1	Graphene oxide	Light scribe DVD burner	LED, earphone, pressure sensor, photodetector	[33]
2	Graphene oxide	650 nm laser	Heater	[34]
3	Graphene oxide	450 nm laser	Strain sensor	[35]
4	Graphene oxide	248 nm UV laser	Sound sensor	[36]
5	Graphene oxide	788 nm CW laser		[36]
6	Graphene oxide	800 nm femtosecond laser		[37]
7	Graphene oxide	650 nm laser	Photodetector	[37]
8	Graphene oxide	532 nm diode pump solid-state laser		[38]
9	Polyimide	450 nm laser	Intelligent throat	[39]
10	Polyimide, polyetherimde	10.6 m CO_2 pulsed laser	Super capacitor	[16]
11	Polyimide,	CO_2 pulsed laser	Strain sensor	[40]
12	Sulfonated poly(ether ether ketone)	Ar–Kr laser, 514.5 nm	Super capacitor	[41]
13	Paper, coconut, bread, potato, etc	10.6 m CO_2 pulsed laser		[42]
14	Teflon	9.3 m CO_2 pulsed laser		[43]
15	Methane	532 nm optically pumped, solid-state laser		[44]
16	Wood	10.6 m CO_2 pulsed laser	Electrocatalyst	[45]
17	Single layer graphene	CO_2 laser	Thermal effect	[46]
18	Graphene oxide	Ultra short pulsed laser		[36]
19	Single layer graphene	Nd:YAG/5 ns		[47]
20	Single layer graphene	CW laser	Fabrication of graphene patterns	[44]

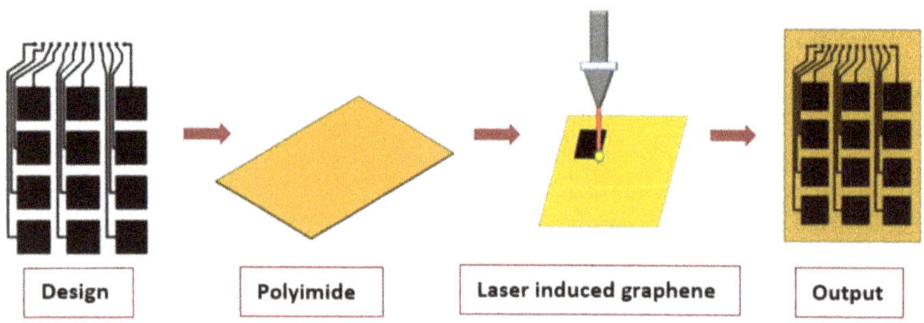

Figure 5.1. Schematic representation of LIG fabrication on polyimide using CO_2 laser. Reproduced with permission from [48]. Copyright 2020 IEEE.

Table 5.2. Comparison of the various prepared samples.

	Carbon (%)	Oxygen (%)	Laser power (W)
Bare PI	79.97	20.03	
Sample 1	96.25	3.75	1.35
Sample 2	96.20	3.8	1.5
Sample 3	96.22	3.78	1.65
Sample 4	96.50	3.5	1.8
Sample 5	96.55	3.45	1.95

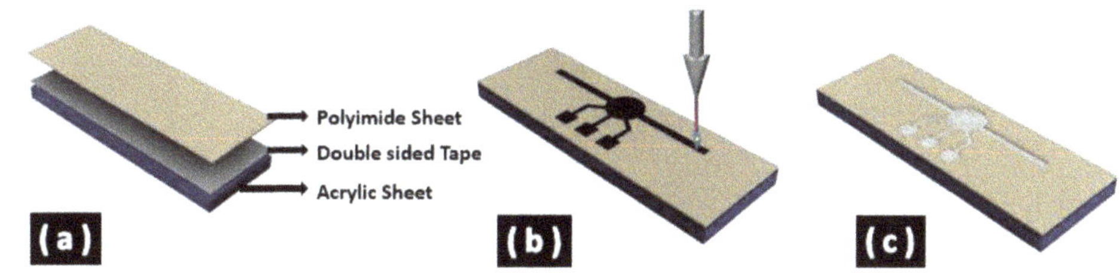

Figure 5.2. Schematic representation of the LIG based microfluidic device fabrication for electrochemical sensing application.

obtained graphene was peeled off again until an optimal depth representing the microchannel was obtained. A microchannel with a depth of around 84 μm was obtained. The second part dealt with the fabrication of LIG electrodes through laser ablation. PDMS was used to cover the microfluidic device via plasma treatment. Figure 5.2 shows a complete schematic representation of the procedure.

5.4.3 Chemical modification of LIG (composites)

The fabricated LIG can be chemically modified either through *in situ* or *ex situ* approaches. The modification of the surface using various functional groups improves the conductivity and properties of the LIG greatly. Several research groups have reported various methods of surface modification. For example, the *in situ* approach of doping of heteroatoms on LIG has been proven to be an effective approach for enhancing the performance and applications of LIG. Owing to the high surface area the formation of composites is easier. By varying the substrate composition, laser speed, and power, heteroatoms such as C, N, and O were doped successfully [16]. Peng *et al* prepared LIG with boron doped via an *in situ* method. Lasing a PI film with boric acid resulted in boron doped LIG [49]. Likewise, Ye *et al* fabricated nanoparticles incorporated into LIG, such a Co_3O_4, Fe_3O_4, and MoO_2, by lasing a PI sheet containing a complex metal solution [50]. Li *et al* designed a chamber wherein the atmosphere of the laser ablation can be modified for inducing chemical modifications on LIG. The chamber was saturated with Ar, O_2, H_2, or SF_6, which changed the hydrophilic and hydrophobic properties of the LIG [51]. The *ex situ* approach involves the modification of the surface after LIG fabrication.

For example, the electrodeposition of polymers, metal oxides, nanomaterials, etc, was reported on previously formed LIG [52]. Zhang *et al* developed another method where the LIG was treated with metal salts and then re-ablated with the laser [53].

5.4.4 Different carbon sources for LIG

In addition to polyimide and polyetherimide as substrates for LIG, sulfonated polyether ketone, polyethersulfone, and polysulfone are also potential polymer candidates that can be used for LIG fabrication [54]. Yieu *et al* discovered that when a focused laser is ablated multiple times over substrates such as cross linking polymers, phenolic resins, and epoxy, LIG can be formed [42]. The same research group also discovered that LIG can be obtained by laser ablation of lignin containing substrates. They were successful in producing LIG over substrates such as coconut, potato, fabric, and bread, see figure 5.3 [42].

5.5 Electrochemical and biosensing applications of LIG

With the emerging preparation techniques, the variety in the applications of LIG is also increasing. Figure 5.4 depicts some generalized applications of LIG [31].

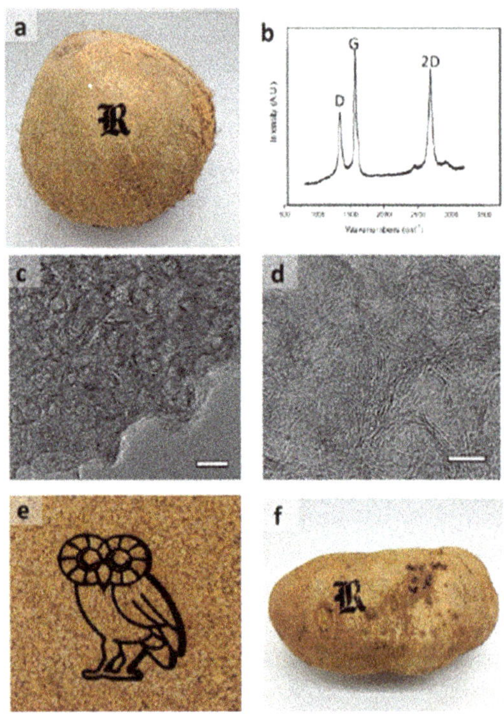

Figure 5.3. LIG scribed on different substrates. (a) LIG patterned as "R" on a coconut. (b) Raman spectrum of coconut-derived LIG. (c) Low-resolution TEM of coconut LIG (d) High-resolution TEM coconut LIG portraying the characteristic 0.34 nm d-spacing of graphene (5nm). (e) LIG on cork as shape of an owl. (f) LIG on Potato scribed with the laser. Reproduced with permission from [42]. Copyright 2018 American Chemical Society.

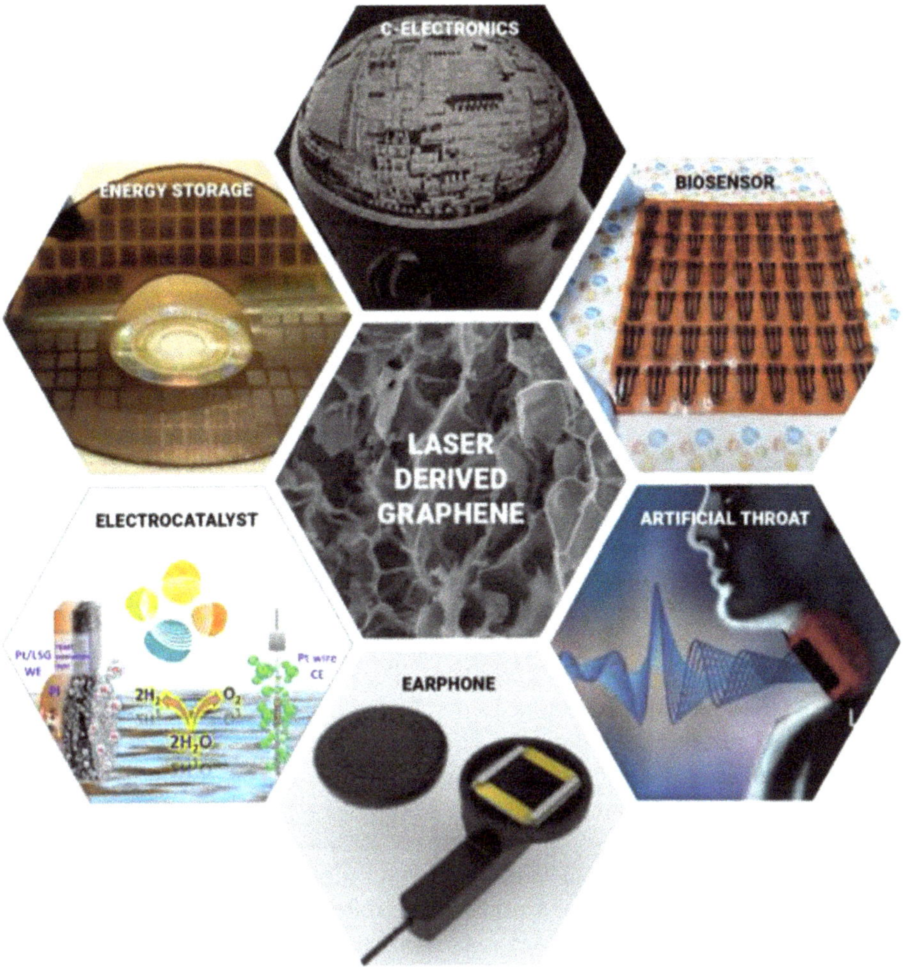

Figure 5.4. Applications of LIG in electronics and electrochemical sensing. Reproduced with permission from [28]. Copyright (2019) Elsevier.

However, this chapter focuses mainly on the emerging applications of LIG in electrochemical and biosensing applications.

5.5.1 LIG in electrochemical, biosensor, and immunosensor applications

Owing to its distinctive properties, LIG has received extensive interest in electrochemical and biosensor applications, such as the sensing of chemicals, heavy metals, biochemicals, proteins, DNA, RNA, etc [55]. Nguyen *et al* established that graphene and neural cells interface, and can be used for biosensing [56]. Usage of bare LIG and surface chemically modified LIG as working electrodes for electrochemical sensing has been attempted by various research groups. For example, Sarda *et al* attempted sensing of hydrazine, an environmental pollutant, with a bare LIG as the working electrode versus

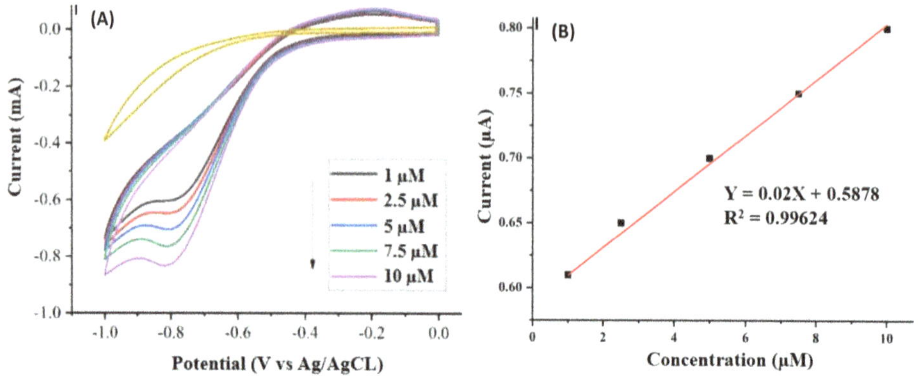

Figure 5.5. (A) Cyclic Voltammetry response in various H_2O_2 concentration at 10mV s^{-1} in pH 7 phosphate buffer electrolyte. (B) Corresponding calibration plot. Reproduced with permission from [48]. Copyright 2020 IEEE.

Ag/AgCl as the reference electrode in a conventional three-electrode system, achieving a substantial limit of detection of 70 μM [57]. Similarly, laser induced bare porous graphene working electrodes were used by Chao *et al* for electrochemical detection of *trans*-resveratrol, a therapeutic chemical known for its anti-inflammatory and anti-carcinogenic properties, present in grape wine. The prepared electrode had an LOD of 0.16 μmol L [58]. Araujo *et al* developed a pyrolyzing technique for scribing graphene on paper. Three-electrode based miniature devices were prepared and tested for detection of ascorbic acid and caffeic acid in food samples and picric acid in clinical and forensic analysis [59]. Our group has reported a bare LIG based electrochemical sensor for non-enzymatic hydrogen peroxide electrochemical sensing. Figure 5.5 [48] shows the performance of the electrode in various concentrations of H_2O_2. A three-electrode based system wherein bare LIG as a working electrode, platinum as a counter electrode, and Ag/AgCl as a reference electrode were used in pH phosphate buffer supporting electrolyte. Cyclic voltammetry curves showed the electrode was selective towards H_2O_2 detection in a linear range of 1–10 μM and gave LOD as 0.3 μM.

To further enhance the electron transfer mechanism and conductivity, the bare LIG can be chemically modified using various conducting nanomaterials, polymers, metals, enzymes, etc. This not only improves the sensitivity and selectivity, but also increases the stability of the fabricated electrodes. In addition, the major advantage is that the modifications help to perform enzymeless sensing. For example, several non-enzymatic glucose sensors based on LIG have been reported. Yue *et al* reported a non-enzymatic glucose sensor using copper nanoparticle tethered laser induced graphene. The fabricated Cu–NP–LIG sensor showed a remarkable sensitivity towards glucose of 495 μA mM^{-1} cm^{-2} with LOD as 0.39 μM. In addition, the sensor was highly stable, reproducible, and selective [53]. Similarly, an electrodeposited copper nanoparticle based laser scribed electrochemical sensing platform was designed by Lin *et al* [60]. A remarkably selective, interference mitigated response towards glucose oxidation in the linear range of 1 μM to 4.54 mM with limit of detection of 0.35 μM was obtained [60]. A unique disposable enzymeless glucose sensing strip was fabricated via LIG modified with copper nanocubes. The prepared sensor had exceptional selectivity towards glucose in the presence of other interfering sugars such as lactose, fructose, etc, and gave a linear response in the range 25 μM–4 mM and limit of detection as low as 250 nM [61]. Zhang *et al* demonstrated a novel approach of synthesizing LIG from a phenolic resin precursor using a visible light laser. Further, they used this for biosensing of glucose in the linear range 0.2 to 10 mM [62]. Eider and co-workers developed flexible silver nanoparticle (AgNP) decorated LIG for non-enzymatic hydrogen peroxide sensing. An electrodeposition technique in 1 mM silver-nitrate solution was employed for AgNP modification. The thus the fabricated electrode gave a limit of detection of H_2O_2 as 7.9 μM, furthermore, real milk samples were subjected to analysis [63]. Likewise, Yuhan *et al* reported platinum nanoparticle deposited LIG using a magnetron sputtering method for selective H_2O_2 non-enzymatic sensing. They could obtain a further low detection limit of 0.1 μM [64]. Hyosang *et al* presented a laser induced graphene stamp and transferred it to a gold (Au) plate for electrochemical sensing applications [65].

Owing to the large surface area of LIG, immobilizing various physiological enzymes for studying their electrochemical behavior is plausible. For example, Chang *et al* utilized a UV laser for LIG formation over a glass substrate. Further, it was employed for biomolecule sensing, where glucose oxidase was immobilized over LIG for the fabrication of a glucose sensing device [66]. Hyosang *et al* reported a Pt nanoparticle tethered glucose oxidase (GOX) LIG biosensor for the sensitive detection of glucose in sweat. They obtained an exceptional LOD of 300 nm [65]. Zhengfen *et al* fabricated an excellent biosensor for femtomolar micro-RNA detection. They fabricated nitrogen doped LIG which was used as a working electrode in combination with a three-electrode-based system [67]. Vanegas *et al* reported a laser scribed graphene biosensor for detection of an bioamine, i.e. histamine. The fabricated graphene was chemically modified with copper micro-particles and diamineoxidase enzyme. The sensor was highly selective with a detection limit of 11.6 μM in real food samples [68]. Mamleyev *et al* developed LIG on polyimide using an IR laser. Further, chitosan anchored urease enzyme was immobilized to detect urea concentrations as low as 10^{-4} M [15]. A few reports are also available where a composite of more of than one metal nanoparticle was used for surface modification of LIG. For example, Xue *et al* made a flexible electrochemical sensor using a combination of platinum (Pt) and gold (Au) nanoparticle decorated LIG. The fabricated electrode was mounted on a polydimethylsiloxane (PDMS) substrate making it highly flexible. Further, the electrode was applied to detection of the biochemical dopamine, a neurotransmitter. An excellent limit of the detection of 75 nM was obtained with high selectivity without interference from co-existing biochemicals, such as ascorbic acid and uric acid in the physiological system [69]. Nayak *et al* demonstrated an efficient sensor for simultaneous detection of ascorbic acid, dopamine, and uric acid, via Pt nanoparticle modified LIG [70].

Another exceptional example of metal nanoparticle modification of LIG was reported by Lei *et al* [71]. They fabricated TiO_2 anchored LIG over an ITO electrode utilizing spin coating followed by a laser ablation technique. Further, the prepared electrode was used for detection of acetylcholinesterase inhibitor enzyme. A remarkable LOD was obtained of 5.4 pg mL^{-1} [71]. In addition to metal nanoparticles, various carbon nanomaterials, or even a combination of metal and carbon nanoparticles is being used for the surface modification of LIG. For example, Salam *et al* developed a unique LIG decorated with a composite of multiwalled carbon nanotubes and gold nanoparticles (AuNP/MWCNT). The formed electrode demonstrated an impressive electro-catalytic detection of nitrite with an LOD of 7.45 μM [72]. Various other impressive modifications of LIG using polymers and ionic liquid have also being attempted. Zhiwei *et al* reported ionic liquid/poly-L-cysteine modified LIG for the detection of lead, a heavy metal ion. They obtained a remarkable LOD of 0.17 $\mu g\ L^{-1}$ and further tested the electrode in real water samples for practical applications [55]. Further, the LIG was modified with a conductive polymer—polyaniline. The fabricated LIG/polyaniline composite acted as an excellent pH sensor. The experiment was carried out with a unit change of pH value from 4 to 10 [73]. Poly(3,4-ethylenedioxythiophene) (PEDOT) conductive polymer modified LIG was reported by Xu *et al*, where the fabricated electrode gave a highly selective response to dopamine (a neurotransmitter) with other co-existing interfering biochemicals such as uric acid and ascorbic acid. They achieved a low limit of detection of 0.33 μM [74]. Fenzl *et al* prepared a 1-pyrenebutyric acid/amino-functionalized bioreceptor modified LIG electrode for the biosensing of thrombin. They were able to achieve a low detection limit of 1 pM and also tested the sensor in a real serum sample with appreciable recovery values [75]. Garland *et al* reported LIG prepared using a UV laser and further modified with an ion selective membrane to develop selective nitrogen sensing in plant soil. They achieved distinctive sensing of ammonia and nitrate ions. A detection limit of 28.2 μM for ammonium and 20.6 μM for nitrate was obtained [76].

In addition to this, LIG is proving to be an efficient material for developing immunosensors. Zhiheng *et al* recently reported noble metal (Au, Ag, Pt) decorated LIG for immunosensing application. A layer of metal solution was spin coated over a polyimide sheet and subjected to laser irradiation. This resulted in the formation of a uniform, metal nanoparticle decorated LIG. Further, an antigen–antibody approach was carried out for the detection of *E. coli* [77]. Soares *et al* demonstrated an LIG based immunosensor for *Salmonella* detection. LIG was chemically modified with a metal catalyst via CVD followed by immobilization of *Salmonella* antibodies. The designed sensor was successfully applied for real sample analysis, where live *Salmonella* in chicken broth was detected. The linear range was 25–105 CFU mL^{-1} and the LOD was 13 CFU mL^{-1} [78]. Currently, bare LIG and surface modified LIG are being used extensively for electrochemical and biosensing applications in various fields.

5.5.2 Application of LIG as a liquid, gas, and pressure sensor

Usage of LIG for multiplexed applications, such as smart liquid sensors, gas sensors, and pressure sensors, has gained much interest recently. The development of flexible and embeddable gas sensors plays an important role in air quality management, leakage analysis, flue gas analysis, etc. Recently, Michael *et al* developed LIG based sensitive gas sensors for the determination of various CO_2 and N_2 mixtures in the presence of argon and helium. A potential was applied across the LIG gas sensor which gave heating of the LIG channel. Thus the fabricated device could measure the thermal conductivity of an array of gases. Further, this was embedded in a cement substrate and checked for practical applications [79]. Similarly, Manan *et al* designed a flexible methane sensor. An interdigitated LIG electrode was fabricated and chemically modified with palladium nanoparticles. Amperometric studies, in accordance with a changing concentration of methane gas in an ionic liquid/polyvinylidene fluoride electrolyte, were examined [80]. LIG as liquid sensors has also managed to gain attention recently. Yanan *et al* reported LIG paper based liquid sensors with customized shapes, performance, and structure. Unlike conventional sensors, the prepared sensor was free standing and avoids the binder dissolving mechanism. It gave a faster response, and greater reproducibility and accuracy in micro-volume detection. The obtained results showed that the sensor can be used in real-time applications [81].

Pressure sensors based on flexible electronics are emerging to be significant for applications such as electronic skin, physiological signal detection, etc [82]. A remarkable pressure sensor based on a bean pod structure has been reported very recently by Qiong *et al.* Polystyrene beads sandwiched between two layers of LIG were developed. A highly sensitive, self-healing, wide range sensing (up to 100 kPa) pressure sensor was obtained. As a real-time application, the sensor was subjected to human arterial monitoring and gave an appreciable result [82]. Ashok *et al* developed a piezoresistive strain sensor, where MoS_2 modified LIG was used. A layer of MoS_2 was coated on a polyimide sheet and subjected to laser ablation, resulting in MoS_2 decorated LIG. A highly sensitive, reproducible, and ultra-low detection limit sensor was obtained [83]. In another report, Alexander *et al* developed a LIG formation on a thin polyurethane substrate and further tested them as comfortable, skin-friendly sensors [84]. LIG fabricated on PDMS with a pulsed laser has been developed by Sung *et al.* The thus-developed strain sensor was sensitive and had a high gauge factor of ~160. In addition, it was manufactured by a simple process and was effective for human finger motion monitoring [85].

Recently, our group has also reported a liquid level sensor based on bare LIG and also developed a capacitive touch pad with the same. Figure 5.6 [48] shows a picture of a bare LIG as capacitive touch pad and liquid level sensor. Liquid level sensors

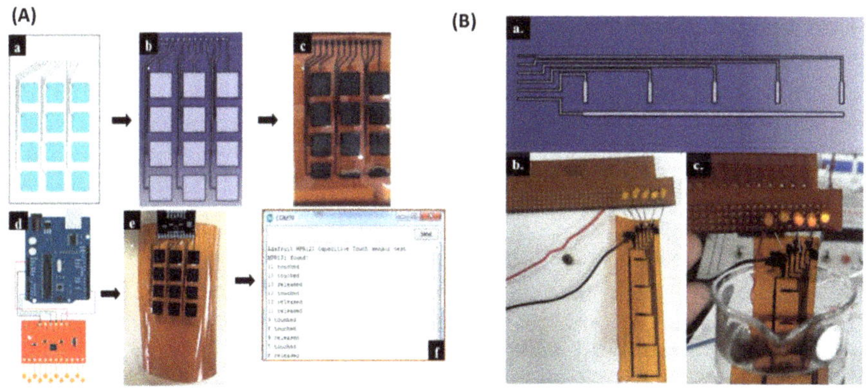

Figure 5.6. (A) Capacitive touch pad. (B) Liquid level sensor. Reproduced with permission from [48]. Copyright 2020 IEEE.

are employed to determine the liquid levels and interfaces in liquids such as oil and water. They portray whether the level of fluid is above or below the point of sensation. LED lights are used to mark the level. These sensors can detect the level in a container and therefore has ample scope in real-life applications. The capacitive touchpads were designed in various ways such as resistive and capacitive. The values of the capacitance and resistance were measured using an LCR meter.

5.6 LIG in energy applications

5.6.1 Application of LIG as a supercapacitor and microsupercapacitor

One of the emerging applications of LIG is in energy storage devices. These devices are of great significance as sensors, wearable electronics, smart electronics, supercapacitors, microsupercapacitors, etc [86]. Amongst various energy storage devices, batteries are most commonly used. Although the batteries have exceptional energy density, they have a short life span [87]. In contrast, supercapacitors stand out due to several advantages, such as good cycling stability, high power density, and a faster charging/discharging rate [86]. However, since supercapacitors are made up of symmetrical electrodes, they also face issues relating to low power density during real-time applications [87]. To overcome this different materials have been employed over the base electrodes to enhance their power density. LIG with chemical modification and hybrid formation as a supercapacitor electrode has gained a lot of attention recently [86].

Various approaches such as the use of fiber supercapacitors are being adapted. For example, Chao *et al* developed a unique laser induced metal sulfide–graphene nanoribbon composite fiber supercapacitor [88]. The hybrid supercapacitor gave an excellent specific capacitance of 58.3 mF cm^{-2} at 50 μA cm^{-2}, an energy density of 7.0 μWh cm^{-2} at 50 μA cm^{-2}, and a power density of 49.9 μW cm^{-2} at 50 μA cm^{-2}. In addition, a high cycling stability (93.6%, 10 000 cycles) was also achieved [88]. Furthermore, functional group modification of LIG has also been attempted to improve the performance. In 2019 Kyung *et al* came up with nitrogen doped LIG for supercapacitor applications using double laser pyrolysis [89]. A remarkable specific capacitance of 49.0 mF cm^{-2} at 0.2 mA cm^{-2} was obtained [89]. Likewise, Wentao *et al* developed an LIG surface morphology modified with Co_3O_4. In comparison to the performance of pristine LIG, they discovered that the chemically modified LIG gave a higher areal specific capacitance and high cycle stability even after 10 000 cycles [90]. Seung *et al* fabricated a hybrid supercapacitor using LIG modified with $LiNi_{1/3}Mn_{1/3}Co_{1/3}O_2$. The prepared electrode gave a high capacitance of 141.5 F g^{-1}, an energy density of ~123.5 Wh kg^{-1}, a power density of ~14 074.8 W kg^{-1}, and a cycle stability of 94.6% after 20 000 cycles [87]. Yeong *et al* designed a flexible microsupercapacitor-like sticker utilizing highly swollen reduced graphene oxide electrodes. An ultrashort-pulse laser was deployed for the formation of these electrodes. The obtained microsupercapacitor gave an energy density of 1.08 mWh cm^3 and a 13 times higher power density of 83.5 mW cm^{-3} in comparison to traditional lithium batteries [91]. Huilong *et al* synthesized a KOH-activated 3D LIG for microsupercapacitor applications. They compared the performance of LIG as a supercapacitor based on KOH concentration. This method provided doping of nitrogen and oxygen atoms on the LIG, therefore improving the performance tremendously. An areal capacitance of 32.00 mF cm^{-2} (4.27 μWh cm^{-2}) was achieved that was ten times higher than that without KOH activation [92].

Kumar *et al* prepared laser induced reduced graphene oxide over graphene oxide film for a solid state porous supercapacitor. Significantly, the laser treated graphene oxide film electrodes showed a good areal capacitance of 149.7 mF cm^{-2} at 3 mV s^{-1} with a maximum energy density of 7.5 mWh cm^{-3}. In addition, these electrodes provided an exceptional cycling stability of more than 85% after 5000 charge–discharge cycles [93]. Recently, Gillian *et al* fabricated a biomass derived polymer, a poly furfuryl alcohol/graphene oxide doped film LIG, for supercapacitor applications. This resulted in an extraordinary specific areal capacitances of 16.0 mF cm^{-2} at 0.05 mA cm^{-2}, and gave 97% of its capacitance over 10 000 cycles [94]. Er-Chieh reported a simple PEDOT (poly (3,4-ethylenedioxythiophene)) modified LIG as an efficient supercapacitor. It provided reversible capacities of 115.2, 97.0, and 78.4 F g^{-1} at rates of 0.5, 2, and 6 A g^{-1}, respectively. They also exhibited less loss in capacitance after 4000 cycles [95]. Kandy *et al* developed reduced graphene oxide film using an IR laser. They obtained high electrical conductivity films (17.4 s cm^{-1}). They further used it as an electrochemical capacitor providing a power density of 20 W cm^{-3}, which was 20 times greater than that of commonly used activated carbon supercapacitors [33]. Peng *et al* reported a vertical stacked graphene based supercapacitor where they transformed both sides of a polyimide sheet to LIG [96].

A pioneering LIG microsupercapacitor (MSC) was designed on a polyimide film using a CO_2 laser in an optimized atmosphere, providing an interdigitated electrode pattern. An electrochemical double layer was the source of energy storage. The specific areal capacitance (C_A) was maintained above 1 mF cm^{-2}, hence this sensor is applicable for high power electronic applications. The MSC provided a C_A of 3.9 mF cm^{-2} and 1.3 mF cm^{-2} at discharge current densities of 0.2 and 25 mA cm^{-2}, respectively [22]. LIG doped with boron developed by Peng *et al* provided a C_A of 16.5 mF cm^{-2}, i.e. three times higher than bare LIG. Furthermore, the energy density was improved by five to ten times [49]. Cai *et al* reported a plasma treatment method for LIG which enhanced wettability by increasing the oxygen content and in turn increased the capacitance by 11.6% [97]. By modifying the LIG with metal oxides such as MnO_2 and FeOOH, along with polyaniline (PANI), Li *et al* fabricated an MSC. The performance of this MSC was found to be better than that of commercially available storage devices [52]. Overall, LIG with chemical modifications has proven to be a potential candidate for energy storage, specifically as high performance super capacitors. Table 5.3, provided by Kurra *et al*, gives a detailed comparison between different LIG based super capacitors reported in the literature [31].

5.6.2 Application of LIG in fuel cells and nanogenerators (energy harvesting)

The continuous pressure on non-renewable sources of energy is an immediate environmental concern. Thus significant research to develop renewable sources, such as fuel cells, microbial fuel cells, batteries, water splitting, etc, are being looked at as

Table 5.3. LIG based super capacitors reported in the literature.

Type	Voltage (V)	C_A (mF cm^{-2})	C_V (F cm^{-3})	E_V (mWh cm^{-3})	P_V (W cm^{-3})	Reference
GO→rGO (22 μm)	1	1.2		1	100	[98]
GO→LSG (68.2 μm)	3	4.82		0.4	10	[99]
GO→LSG (7.6 μm)	1	2.3	2.35		200	[99]
PI→LIG (25 μm)	3.5	2	0.8	1	100	[100]
PI/H_3BO_3→B doped LIG (25 μm)	1	16.5	6	0.3	1	[49]
GO→LSG + ED (MnO_2)	0.9	400	250	30	1	[101]
GO/RuCl3→LSG/RuO_2 (25 μm)	1	70	158	1	1	[102]
PI→LIG (25 μm)	1	9		0.2	1	[96]
GO/HAuCl4→rGO/ AuNPs (13.7 μm)	1	0.77	17.2			[103]
PI→LIG + ED (FeOOH (41 μm)//MnO2 (101 μm))	1.8	21.9	5.4	3.2	2.89	[52]
GO→rGO + ED (CoNi2S_4)	1.7		122.4	49	10	[104]
PI→LIG (80 μm)	1	22.4	2.8			[105]
PI→LIG	1	10				[106]
Lignin→LIG	1	25.1	6.27	1	2	[107]

GO: graphene oxide; PVA: polyvinyl alcohol; LSG: laser scribed graphene; LIG: laser induced graphene; ED: electrodeposition [28].

alternatives. The major reaction behind these renewable energy alternatives is either an oxygen reduction reaction or oxygen evolution reaction. In order to carry out this reaction, catalysts such as noble metals, carbides, oxides, etc, are used. In this context LIG plays a key role. However, the use of LIG in fuel cells is still quite recent. A couple of reports have been published in the literature recently. For example, Ye *et al* fabricated LIG with metal oxide decoration and demonstrated an oxygen reduction reaction (ORR). They discovered that doping of LIG with metal oxides can enhance the efficiency of the ORR [50]. Zhang *et al* exposed the LIG surface to plasma treatment to achieve oxidization. This led to an increase in the adsorption of oxygen evolution reaction (OER) intermediates and therefore it is an efficient candidate for ORR and OER. The same electrode was also proven effective for water splitting [108]. Tiliakos *et al* demonstrated the fabrication of LIG with hydrophobicity and catalyst coated on the LIG over Nafion membranes as a microporous layer for proton exchange membrane fuel cells. A peak power of 266 mW cm^{-2} and current density of 800 mA cm^{-2} were obtained. This was 20% more efficient than carbon black based fuel cells [109]. Recently, our group developed a PDMS-based microfluidic glucose biofuel cell integrated with laser induced graphene bioelectrodes. Figure 5.7 gives a detailed schematic representation of the fabrication [5].

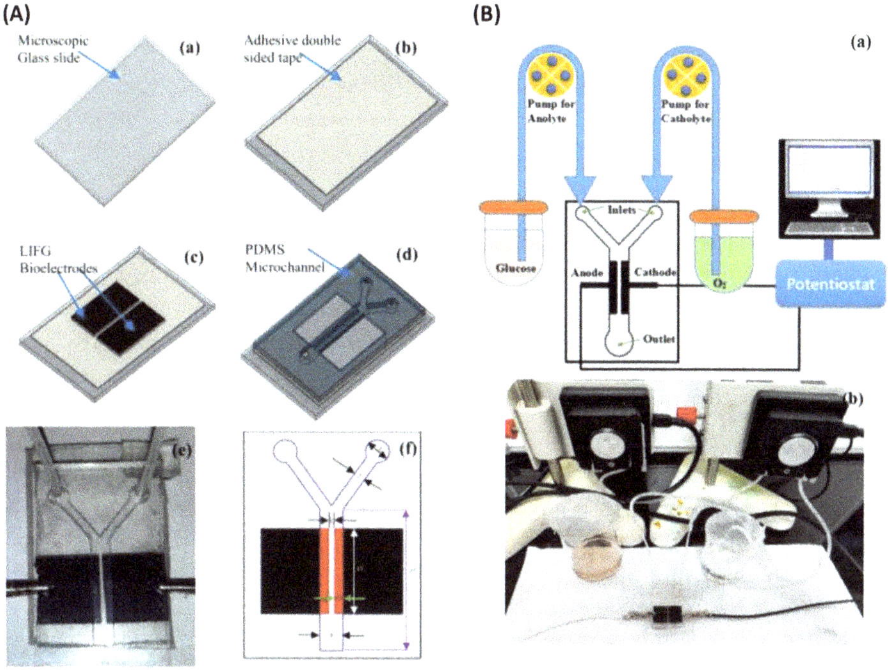

Figure 5.7. (A) Schematic representation of the complete fabrication process of PDMS based LIG bioelectrodes device for bio fuel cell application (B) Experimental set-up. Reproduced with permission from [5]. Copyright 2020 IEEE.

The fabrication of LIG electrodes was achieved by a CO_2 infrared laser over a polyimide sheet. A PDMS–silicone based microfluidic device with a Y channel was fabricated using soft lithography. The prepared LIG was incorporated into the microfluidic device through bonding. The fabricated device showed excellent redox activity in the presence of oxygen and glucose. It produced a power density of 13 μW cm^{-2} at 0.425 V at room temperature.

Apart from being employed in a fuel cell, LIG is also being explored as a nanogenerator. Triboelectric nanogenerators (TENGs) exhibit exceptional performance for converting wasted mechanical energy into electrical energy. In this regard, a couple of papers reported LIG as a nanogenerator. For example, Jiang *et al* developed a flexible, multifunctional triboelectric nanogenerator with laser induced graphene integrated with MXene/PDMS-enabled porous film. This proved to be a promising tool in harvesting mechanical energy from the agricultural field and various human activities. Furthermore, it can be applied in writing or tactile sensing for pads, robotics, and the field of human–machine interactions [30]. Likewise, Stanford *et al* reported an LIG/polyimide sheet composite to fabricate TENGs. Two pieces of polyimide sheet were used, of which one was ablated to give the LIG/polyimide sheet composite. TENGs relying on conductor-to-dielectric and metal-less dielectric-to-dielectric device geometry led to an open circuit voltage >3.5 kV and a peak power >8 mW. A single sheet of polyimide was transformed to metal free, flexible TENG.

It was further embedded with PDMS to provide LIG/PDMS composite TENG [79]. Overall, the use of LIG for energy harvesting as fuel cells and nanogenerators is not yet much explored, but nevertheless shows exceptional potential.

5.6.3 Future outlook and conclusion

The current chapter deals with the discovery, fabrication, characterization, and applications of LIG. Owing to the extraordinary electronic, mechanical, physical, and chemical properties of graphene, the demand for graphene has been growing on daily basis. Compared to the conventional micromachining technique, laser induction and laser scribing of graphene on different carbon precursors has proven to be one the most effective approach for commercial synthesis. Various types of lasers are being used, such as CO_2 lasers, pulsed lasers, ultrashort pulsed lasers, femtosecond lasers, etc, for synthesis of LIG. Numerous applications and uses of graphene in commercial products are also being reported. With fine tuning and controlled porosity, composition, and conductivity, graphene has proven to be a versatile candidate for many applications. A broad variety of applications of LIG in the electrochemical sensing of various bioanalytes, biochemicals, environmental pollutants, DNA, RNA, amino acids, etc, have been reported. The biosensing of analytes, both by enzymatic and non-enzymatic approaches, has also shown significant growth in recent times. LIG based immunosensing is in its early stages, with only a few reports present in literature. However, due to its high surface area, LIG has shown excellent immobilization towards antibodies and enzymes, thus in near future numerous immunosensors with an LIG matrix can be expected. LIG has also proven itself as a worthy matrix for supercapacitor and microsupercapacitor devices. Numerous reports in the literature are available, dealing with a variety of modifications for LIG to enhance the supercapacitor performance. However, the integration of LIG in microfluidic devices has just begun. LIG as nanogenerators and fuel cells has also not been much explored and very few reports are available. Hence, in the future modification of the LIG surface could enhance its performance in energy harvesting microdevices. Currently there are a number of research gaps and studies that need to be addressed. For example, although the preparation of LIG is repeatable, the structure and properties still need to achieve uniformity. LIG based devices are yet to be used at the industrial level. In particular, devices such as supercapacitors, batteries, nanogenerators, solar cells, wearable electronics, etc, are still at the laboratory level. Nevertheless, foreseeing the growing applications and use of LIG, it is anticipated that in the near future there will be tremendous scope for use of LIG in various academic, laboratory, and industry sectors.

References

[1] Novoselov K S *et al* 2007 Electronic properties of graphene *Phys. Status Solidi Basic Res.* **244** 4106–11

[2] Marengo M, Marinaro G and Kosel J 2017 Flexible temperature and flow sensor from laser-induced graphene *Proc. IEEE Sensors* (Piscataway, NJ: IEEE) pp 1–3

[3] Papageorgiou D G, Kinloch I A and Young R J 75 Mechanical properties of graphene and graphene-based nanocomposites *Prog. Mater. Sci.* **90** 75–127, 2017
[4] Wang F *et al* 2018 Laser-induced graphene: preparation, functionalization and applications *Mater. Technol.* **33** 340–56
[5] Rewatkar P, Kothuru A and Goel S 2020 PDMS-based microfluidic glucose biofuel cell integrated with optimized laser-induced flexible graphene bioelectrodes *IEEE Trans. Electron Devices* **67** 1832–8
[6] Nicholl R J T *et al* 2015 The effect of intrinsic crumpling on the mechanics of free-standing graphene *Nat. Commun.* **6** 1–7
[7] Lee C, Wei X, Kysar J W and Hone J 2008 Measurement of the elastic properties and intrinsic strength of monolayer graphene *Science* **321** 385–88
[8] Zandiatashbar A *et al* 2014 Effect of defects on the intrinsic strength and stiffness of graphene *Nat. Commun.* **5** 1–9
[9] Zhang P *et al* 2014 Fracture toughness of graphene *Nat. Commun.* **5** 1–7
[10] Alemour B, Yaacob M H, Lim H N and Hassan M R 2018 Review of electrical properties of graphene conductive composites *Int. J. Nanoelectron. Mater.* **11** 371–98
[11] Chen J H, Jang C, Xiao S, Ishigami M and Fuhrer M S 2008 Intrinsic and extrinsic performance limits of graphene devices on SiO_2 *Nat. Nanotechnol.* **3** 206–9
[12] Berry V 2013 Impermeability of graphene and its applications *Carbon* **62** 1–10
[13] Sheehy D E and Schmalian J 2009 Optical transparency of graphene as determined by the fine-structure constant *Phys. Rev.* B **80** 2–5
[14] Peigney A, Laurent C, Flahaut E, Bacsa R R and Rousset A 2001 Specific surface area of carbon nanotubes and bundles of carbon nanotubes *Carbon* **39** 507–14
[15] Mamleyev E R *et al* 2019 Laser-induced hierarchical carbon patterns on polyimide substrates for flexible urea sensors *NPJ Flex. Electron.* **3** 2
[16] Lin J *et al* 2014 Laser-induced porous graphene films from commercial polymers *Nat. Commun.* **5** 1–8
[17] Larciprete R, Fabris S, Sun T, Lacovig P, Baraldi A and Lizzit S 2011 Dual path mechanism in the thermal reduction of graphene oxide *J. Am. Chem. Soc.* **133** 17315–21
[18] Paquin F, Rivnay J, Salleo A, Stingelin N and Silva C 2015 Multi-phase semicrystalline microstructures drive exciton dissociation in neat plastic semiconductors *J. Mater. Chem.* C **3** 10715–22
[19] Moon I K, Lee J and Lee H 2011 Highly qualified reduced graphene oxides: the best chemical reduction *Chem. Commun.* **47** 9681–3
[20] Gao X, Jang J and Nagase S 2010 Hydrazine and thermal reduction of graphene oxide: reaction mechanisms, product structures, and reaction design *J. Phys. Chem.* C **114** 832–42
[21] Park S, An J, Potts J R, Velamakanni A, Murali S and Ruoff R S 2011 Hydrazine-reduction of graphite- and graphene oxide *Carbon* **49** 3019–23
[22] Ye R, James D K and Tour J M 2019 Laser-induced graphene: from discovery to translation *Adv. Mater.* **31** 1–15
[23] Tour J M 2014 Top-down versus bottom-up fabrication of graphene-based electronics *Chem. Mater.* **26** 163–71
[24] Li X *et al* 2009 Large-area synthesis of high-quality and uniform graphene films on copper foils *Science* **324** 1312–4
[25] Geim A K 2011 Nobel lecture: random walk to graphene *Rev. Mod. Phys.* **83** 851–62

[26] Niu L, Coleman J N, Zhang H, Shin H, Chhowalla M and Zheng Z 2016 Production of two-dimensional nanomaterials via liquid-based direct exfoliation *Small* **12** 272–93
[27] Abdelkader A M, Cooper A J, Dryfe R A W and Kinloch I A 2015 How to get between the sheets: a review of recent works on the electrochemical exfoliation of graphene materials from bulk graphite *Nanoscale* **7** 6944–56
[28] Kurra N, Jiang Q, Nayak P and Alshareef H N 2019 Laser-derived graphene: a three-dimensional printed graphene electrode and its emerging applications *Nano Today* **24** 81–102
[29] Xu M, Fujita D, Sagisaka K, Watanabe E and Hanagata N 2011 Production of extended single-layer graphene *ACS Nano* **5** 1522–8
[30] Jiang C *et al* 2019 A multifunctional and highly flexible triboelectric nanogenerator based on MXene-enabled porous film integrated with laser-induced graphene electrode *Nano Energy* **66** 104121
[31] Kurra N, Sagade A A and Kulkarni G U 2011 Ultrafast direct ablative patterning of HOPG by single laser pulses to produce graphene ribbons *Adv. Funct. Mater.* **21** 3836–42
[32] Wakaya F, Kurihara T, Yurugi N, Abo S, Abe M and Takai M 2015 Maskless laser processing of graphene *Microelectron. Eng.* **141** 203–6
[33] El-Kady M F, Strong V, Dubin S and Kaner R B 2012 Laser scribing of high-performance and flexible graphene-based electrochemical capacitors *Science* **335** 1326–30
[34] Lin S Y *et al* 2017 High-performance graphene-based flexible heater for wearable applications *RSC Adv.* **7** 27001–6
[35] Wang D Y *et al* 2016 High performance flexible strain sensor based on self-locked overlapping graphene sheets *Nanoscale* **8** 20090–5
[36] Arul R, Oosterbeek R N, Robertson J, Xu G, Jin J and Simpson M C 2016 The mechanism of direct laser writing of graphene features into graphene oxide films involves photo-reduction and thermally assisted structural rearrangement *Carbon* **99** 423–31
[37] Mohammad M A *et al* 2016 Tunable graphene oxide reduction and graphene patterning at room temperature on arbitrary substrates *Carbon* **109** 173–81
[38] Fatt Teoh H, Tao Y, Soon Tok E, Wei Ho G and Haur Sow C 2012 Direct laser-enabled graphene oxide-reduced graphene oxide layered structures with micropatterning *J. Appl. Phys.* **112** 064309
[39] Tao L Q *et al* 2017 An intelligent artificial throat with sound-sensing ability based on laser induced graphene *Nat. Commun.* **8** 1–8
[40] Rahimi R, Ochoa M, Yu W and Ziaie B 2015 Highly stretchable and sensitive unidirectional strain sensor via laser carbonization *ACS Appl. Mater. Interfaces* **7** 4463–70
[41] Lamberti A *et al* 2017 All-SPEEK flexible supercapacitor exploiting laser-induced graphenization *2D Mater.* **4** 035012
[42] Chyan Y, Ye R, Li Y, Singh S P, Arnusch C J and Tour J M 2018 Laser-induced graphene by multiple lasing: toward electronics on cloth, paper, and food *ACS Nano* **12** 2176–83
[43] Ye R *et al* 2018 Laser-induced conversion of Teflon into fluorinated nanodiamonds or fluorinated graphene *ACS Nano* **12** 1083–8
[44] Park J B *et al* 2011 Fast growth of graphene patterns by laser direct writing *Appl. Phys. Lett.* **98** 16–9
[45] Ye R *et al* 2017 Laser-induced graphene formation on wood *Adv. Mater.* **29** 1–7
[46] Huang T *et al* 2013 The effects of low power density CO_2 laser irradiation on graphene properties *Appl. Surf. Sci.* **273** 502–6

[47] Kiisk V, Kahro T, Kozlova J, Matisen L and Alles H 2013 Nanosecond laser treatment of graphene *Appl. Surf. Sci.* **276** 133–7

[48] Kothuru A, Hanumanth Rao C, Puneeth S B, Salve M, Amreen K and Goel S 2020 Laser-induced flexible electronics (LIFE) for resistive, capacitive and electrochemical sensing applications *IEEE Sens. J.* **20** 7392–9

[49] Peng Z *et al* 2015 Flexible boron-doped laser-induced graphene microsupercapacitors *ACS Nano* **9** 5868–75

[50] Ye R *et al* 2015 *In situ* formation of metal oxide nanocrystals embedded in laser-induced graphene *ACS Nano* **9** 9244–51

[51] Li Y *et al* 2017 Laser-induced graphene in controlled atmospheres: from superhydrophilic to superhydrophobic surfaces *Adv. Mater.* **29** 1–8

[52] Li L *et al* 2016 High-performance pseudocapacitive microsupercapacitors from laser-induced graphene *Adv. Mater.* **28** 838–45

[53] Zhang J, Ren M, Li Y and Tour J M 2018 *In situ* synthesis of efficient water oxidation catalysts in laser-induced graphene *ACS Energy Lett.* **3** 677–83

[54] Singh S P, Li Y, Zhang J, Tour J M and Arnusch C J 2018 Sulfur-doped laser-induced porous graphene derived from polysulfone-class polymers and membranes *ACS Nano* **12** 289–97

[55] Lu Z *et al* 2019 Ionic liquid/poly-L-cysteine composite deposited on flexible and hierarchical porous laser-engraved graphene electrode for high-performance electrochemical analysis of lead ion *Electrochim. Acta* **295** 514–23

[56] Nguyen P and Berry V 2012 Graphene interfaced with biological cells: opportunities and challenges *J. Phys. Chem. Lett.* **3** 1024–9

[57] Sharma S, Ganeshan S K, Pattnaik P K, Kanungo S and Chappanda K N 2020 Laser induced flexible graphene electrodes for electrochemical sensing of hydrazine *Mater. Lett.* **262** 127150

[58] Zhang C, Ping J and Ying Y 2020 Evaluation of trans-resveratrol level in grape wine using laser-induced porous graphene-based electrochemical sensor *Sci. Total Environ.* **714** 136687

[59] de Araujo W R, Frasson C M R, Ameku W A, Silva J R, Angnes L and Paixão T R L C 2017 Single-step reagentless laser scribing fabrication of electrochemical paper-based analytical devices *Angew. Chemie—Int. Ed.* **56** 15113–7

[60] Lin S *et al* 2018 A flexible and highly sensitive nonenzymatic glucose sensor based on DVD-laser scribed graphene substrate *Biosens. Bioelectron.* **110** 89–96

[61] Tehrani F and Bavarian B 2016 Facile and scalable disposable sensor based on laser engraved graphene for electrochemical detection of glucose *Sci. Rep.* **6** 1–10

[62] Zhang Z, Song M, Hao J, Wu K, Li C and Hu C 2018 Visible light laser-induced graphene from phenolic resin: a new approach for directly writing graphene-based electrochemical devices on various substrates *Carbon* **127** 287–96

[63] Aparicio-Martínez E, Ibarra A, Estrada-Moreno I A, Osuna V and Dominguez R B 2019 Flexible electrochemical sensor based on laser scribed graphene/Ag nanoparticles for non-enzymatic hydrogen peroxide detection *Sens. Actuators* B **301** 127101

[64] Zhang Y *et al* 2019 Laser-induced graphene-based non-enzymatic sensor for detection of hydrogen peroxide *Electroanalysis* **31** 1334–41

[65] Yoon H, Nah J, Kim J, Xuan X and Park J 2019 Laser-induced graphene stamp for high performance electrochemical sensing applications *Proc. IEEE Int. Conf. Micro Electro Mech. Syst.* pp 537–40

[66] Chang T L and Chen Z C 2015 Surface patterning of multilayer graphene by ultraviolet laser irradiation in biomolecule sensing devices *Appl. Surf. Sci.* **359** 543–9

[67] Wan Z, Nguyen N T, Gao Y and Li Q 2020 Laser induced graphene for biosensors *Sustain. Mater. Technol.* **25** e00205

[68] Vanegas D C *et al* 2018 Laser scribed graphene biosensor for detection of biogenic amines in food samples using locally sourced materials *Biosensors* **8** 42

[69] Hui X, Xuan X, Kim J and Park J Y 2019 A highly flexible and selective dopamine sensor based on Pt–Au nanoparticle-modified laser-induced graphene *Electrochim. Acta* **328** 135066

[70] Nayak P, Kurra N, Xia C and Alshareef H N 2016 Highly efficient laser scribed graphene electrodes for on-chip electrochemical sensing applications *Adv. Electron. Mater.* **2** 1600185

[71] Ge L, Hong Q, Li H and Li F 2019 A laser-induced TiO_2-decorated graphene photoelectrode for sensitive photoelectrochemical biosensing *Chem. Commun.* **55** 4945–8

[72] Nasraoui S *et al* 2019 Investigation of laser induced graphene electrodes modified by MWNT/AuNPs for detection of nitrite *16th Int. Multi-Conf. Syst. Signals Devices* pp 615–20

[73] Rahimi R, Ochoa M, Tamayol A, Khalili S, Khademhosseini A and Ziaie B 2017 Highly stretchable potentiometric pH sensor fabricated via laser carbonization and machining of carbon–polyaniline composite *ACS Appl. Mater. Interfaces* **9** 9015–23

[74] Xu G, Jarjes Z A, Desprez V, Kilmartin P A and Travas-Sejdic J 2018 Sensitive, selective, disposable electrochemical dopamine sensor based on PEDOT-modified laser scribed graphene, *Biosens. Bioelectron.* **107** 184–91

[75] Fenzl C, Nayak P, Hirsch T, Otto S, Alshareef H N and Baeumner A J 2017 Laser-scribed graphene electrodes for aptamer-based biosensing *ACS Sens.* **2** 616–20

[76] Garland N T *et al* 2018 Flexible laser-induced graphene for nitrogen sensing in soil *ACS Appl. Mater. Interfaces* **10** 39124–33

[77] You Z *et al* 2020 Laser-induced noble metal nanoparticle–graphene composites enabled flexible biosensor for pathogen detection *Biosens. Bioelectron.* **150** 111896

[78] Soares R R A *et al* 2020 Laser-induced graphene electrochemical immunosensors for rapid and label-free monitoring of *Salmonella enterica* in chicken broth *ACS Sensors* **5** 1900–11

[79] Stanford M G, Yang K, Chyan Y, Kittrell C and Tour J M 2019 Laser-induced graphene for flexible and embeddable gas sensors *ACS Nano* **13** 3474–82

[80] Dosi M, Lau I, Zhuang Y, Simakov D S A, Fowler M W and Pope M A 2019 Ultrasensitive electrochemical methane sensors based on solid polymer electrolyte-infused laser-induced graphene *ACS Appl. Mater. Interfaces* **11** 6166–73

[81] Wang Y, Niu Z, Chen J, Zhai Y, Xu Y and Luo S 2019 Freestanding laser induced graphene paper based liquid sensors *Carbon* **153** 472–80

[82] Tian Q, Yan W, Li Y and Ho D 2020 Bean pod-inspired ultrasensitive and self-healing pressure sensor based on laser-induced graphene and polystyrene microsphere sandwiched structure *ACS Appl. Mater. Interfaces* **12** 9710–7

[83] Chhetry A, Sharifuzzaman M, Yoon H, Sharma S, Xuan X and Park J Y 2019 MoS_2-decorated laser-induced graphene for a highly sensitive, hysteresis-free, and reliable piezoresistive strain sensor *ACS Appl. Mater. Interfaces* **11** 22531–42

[84] Dallinger A, Keller K, Fitzek H and Greco F 2020 Stretchable and skin-conformable conductors based on polyurethane/laser-induced graphene *ACS Appl. Mater. Interfaces* **12** 19855–65

[85] Jeong S Y, Ma Y W, Lee J U, Je G J and Shin B S 2019 Flexible and highly sensitive strain sensor based on laser-induced graphene pattern fabricated by 355 nm pulsed laser *Sensors* **19** 1–12

[86] Fu X-Y, Chen Z-D, Han D-D, Zhang Y-L, Xia H and Sun H-B 2020 Laser fabrication of graphene-based supercapacitors *Photon. Res.* **8** 04000577

[87] Lee S H, Kim K Y and Yoon J R 2020 Binder- and conductive additive-free laser-induced graphene/$LiNi_{1/3}Mn_{1/3}Co_{1/3}O_2$ for advanced hybrid supercapacitors *NPG Asia Mater.* **12** 28

[88] Zhou C *et al* 2019 Laser-induced bi-metal sulfide/graphene nanoribbon hybrid frameworks for high-performance all-in-one fiber supercapacitors *J. Power Sources* **438** 227044

[89] Kim K Y, Choi H, Van Tran C and Bin In J 2019 Simultaneous densification and nitrogen doping of laser-induced graphene by duplicated pyrolysis for supercapacitor applications *J. Power Sources* **441** 227199

[90] Wang W *et al* 2020 Tailoring the surface morphology and nanoparticle distribution of laser-induced graphene/Co_3O_4 for high-performance flexible microsupercapacitors *Appl. Surf. Sci.* **504** 144487

[91] Lee Y A *et al* 2020 Attachable micropseudocapacitors using highly swollen laser-induced-graphene electrodes *Chem. Eng. J.* **386** 123972

[92] Liu H *et al* 2020 Laser-induced and KOH-activated 3D graphene: a flexible activated electrode fabricated via direct laser writing for in-plane micro-supercapacitors *Chem. Eng. J.* **393** 124672

[93] Kavinkumar T, Kavitha P, Naresh N, Manivannan S, Muneeswaran M and Neppolian B 2019 High performance flexible solid-state symmetric supercapacitors based on laser induced porous reduced graphene oxide-graphene oxide hybrid nanostructure devices *Appl. Surf. Sci.* **480** 671–9

[94] Hawes G F, Yilman D, Noremberg B S and Pope M A 2019 Supercapacitors fabricated via laser-induced carbonization of biomass-derived poly(furfuryl alcohol)/graphene oxide composites *ACS Appl. Nano Mater.* **2** 6312–24

[95] Cho E C *et al* 2020 PEDOT-modified laser-scribed graphene films as binder- and metallic current collector-free electrodes for large-sized supercapacitors *Appl. Surf. Sci.* **518** 146193

[96] Peng Z, Lin J, Ye R, Samuel E L G and Tour J M 2015 Flexible and stackable laser-induced graphene supercapacitors *ACS Appl. Mater. Interfaces* **7** 3414–9

[97] Cai J, Lv C and Watanabe A 2016 Cost-effective fabrication of high-performance flexible all-solid-state carbon micro-supercapacitors by blue-violet laser direct writing and further surface treatment *J. Mater. Chem.* A **4** 1671–9

[98] Gao W *et al* 2011 Direct laser writing of micro-supercapacitors on hydrated graphite oxide films *Nat. Nanotechnol.* **6** 496–500

[99] El-Kady M F and Kaner R B 2013 Scalable fabrication of high-power graphene micro-supercapacitors for flexible and on-chip energy storage *Nat. Commun.* **4** 1475

[100] Ye X, Long J, Lin Z, Zhang H, Zhu H and Zhong M 2014 Direct laser fabrication of large-area and patterned graphene at room temperature *Carbon* **68** 784–90

[101] Li M *et al* 2018 Embedding hollow Co_3O_4 nanoboxes into a three-dimensional macroporous graphene framework for high-performance energy storage devices *Nano Res.* **11** 2836–46

[102] Hwang J Y *et al* 2015 Direct preparation and processing of graphene/RuO_2 nanocomposite electrodes for high-performance capacitive energy storage *Nano Energy* **18** 57–70

[103] Li R Z *et al* 2016 High-rate in-plane micro-supercapacitors scribed onto photo paper using: *in situ* femtolaser-reduced graphene oxide/Au nanoparticle microelectrodes *Energy Environ. Sci.* **9** 1458–67

[104] Moosavifard S E, Shamsi J, Altafi M K and Moosavifard Z S 2016 All-solid state, flexible, high-energy integrated hybrid micro-supercapacitors based on 3D LSG/$CoNi_2S_4$ nano-sheets *Chem. Commun.* **52** 13140–3

[105] Wang S *et al* 2017 High-performance stacked in-plane supercapacitors and supercapacitor array fabricated by femtosecond laser 3D direct writing on polyimide sheets *Electrochim. Acta* **241** 153–61

[106] Zhang J, Zhang G, Zhou T and Sun S 2020 Recent developments of planar micro-supercapacitors: fabrication, properties, and applications *Adv. Funct. Mater.* **30** 1–21

[107] Zhang W, Lei Y, Ming F, Jiang Q, Costa P M F J and Alshareef H N 2018 Lignin laser lithography: a direct-write method for fabricating 3D graphene electrodes for micro-supercapacitors *Adv. Energy Mater.* **8** 1–9

[108] Zhang J, Ren M, Wang L, Li Y, Yakobson B I and Tour J M 2018 Oxidized laser-induced graphene for efficient oxygen electrocatalysis *Adv. Mater.* **30** 1–7

[109] Tiliakos A, Trefilov A M I, Tanasă E, Balan A and Stamatin I 2020 Laser-induced graphene as the microporous layer in proton exchange membrane fuel cells *Appl. Surf. Sci.* **504** 144096

IOP Publishing

Printed and Flexible Sensor Technology
Fabrication and applications
Subhas Chandra Mukhopadhyay and Anindya Nag

Chapter 6

Fabrication and applications of wearable microfluidic devices for point-of-care sampling, manipulation, and testing

Shilun Feng, Shaoxi Wang, Qianbin Zhao, Gaozhe Cai, Zizhen Ming, Anindya Nag and Yupan Wu

Wearable microfluidic devices have become the state-of-the-art in the field of point-of-care testing (POCT) in recent times. Wearable sensing devices can be worn or implanted within human skin or other body parts to continuously monitor an individual's state of health, enabling real-time tracking of an individual's physical activities. Moreover, researchers all over the world have combined microfluidics with wearable devices to detect the user's physiological biomarkers. This chapter gives an overview of the significant development achieved in wearable microfluidic devices in terms of their fabrication and characterization. Moreover, this chapter presents the different categories for their use in physical property and body fluid sampling, manipulation, and detection. In particular, examples of the application of such wearable microfluidic devices for analysing body fluids, including sweat, urine, and saliva, are provided. Future prospects and developments in the field of POCT are also described.

6.1 Introduction

6.1.1 Point-of-care testing (POCT)

There are growing issues regarding the efficient diagnosis and monitoring of biomedical problems. More than 95% of the around 8 billion people in the world face some form of health problem [1, 2], which suggests that only one in twenty people on average can be considered healthy, which is very alarming. It is estimated that more and more people will be affected by diseases caused by genetic disorders or infection in the near future [3]. It is thus necessary to address this issue and develop devices capable of monitoring and managing such health problems efficiently.

doi:10.1088/978-0-7503-3439-6ch6

© IOP Publishing Ltd 2021

Diagnostics are a central part of the healthcare system, and they play an important role in clinical and epidemiological decision-making. However, problems arise with existing techniques in which the diagnosis and monitoring are performed in hospitals and require specialists, with high costs and long times required to complete the overall process. The situation is even worse in developing countries [4]. A rapid test kit that is affordable, simple to operate, and accurate at diagnosis could be a lifesaver, in particular for the people in the developing world.

POCT (or bedside testing) involves rapid test kits and is defined as medical diagnostic testing at or near the point of care (POC)—that is, at the time and place of patient care [5]. Compared to standard laboratory testing, POC diagnostics are rapid, simple, and inexpensive, and are thus suitable for resource-limited settings. Therefore POC diagnostics are increasingly valuable for initiating and scaling on-site medical care for the prevention and control of infectious diseases. The WHO has set seven guidelines for the development of diagnostics in resource-poor settings, which are: (i) affordability, (ii) sensitivity, (iii) specificity, (iv) user-friendliness, (v) rapidity and robustness, (vi) equipment-free nature, and (vii) delivery to those who need it, corresponding to the acronym. Some of the advantages of using point-of-care testing (POCT) sensors are:

1. Regular monitoring can be performed on the patient to scrutinize small changes.
2. Wearable systems remove the use of visual sensors which could cause a risk to the privacy of a person.
3. The reaction time is much shorter when using wearable systems compared to non-wearable ones as the patient can immediately inform the caretaker in the case of an emergency.
4. The system is much more compact and easier to use compared to workbench systems which require the patient to visit a special place for testing purposes.

6.1.2 Wearable devices

In the last two decades or so, wearable sensors have become popular, where monitoring can be carried out on-site and in nearly real-time on body parts such as the arms and legs. This has proven helpful, in particular for elderly people for whom their reaction time can be very slow in the case of an emergency due to their low body strength and frailty. This led to the introduction of wearable sensing systems that are worn continually by the patient to monitor the problems faced by him/her and give proper assessment and a quick response.

POCT wearable devices can be worn on or implanted in human skin to continuously monitor an individual's state of health, enabling real-time tracking of an individual's physical activities [6] and detecting the user's physiological biomarkers [7, 8]. Wearable devices have also been applied for food intake sensing [9]. Different applications have been achieved using this wearable technology, and the effectiveness has been demonstrated of measuring several biophysical markers, such as heart rate (HR) [10], blood pressure [11], and so on. Compared to the conventional bulky sensors, their lightweight, non-irritating nature, and biocompatibility are striking features of the new generation of wearable sensors based on soft materials. Recently, more and more researchers are developing different techniques for wearable microfluidic devices for the monitoring of physiology biomarkers of different body fluids, such as sweat, urine, saliva, etc. The soft, skin-mounted, and wearable sensors adhere and conform to the body in a manner that routes and captures body fluids through a network of microchannels for analysis *in situ*.

6.1.3 The microfluidic lab-on-a-chip technique

The lab-on-a-chip (LOC) is the integration of one or more functional laboratory units on one microfluidic chip with the size of the components ranging in sub-micrometers. Normally, different processes can be completed on one chip. The LOC has been utilized in different fields, including analysis (e.g. chemical analysis, medical diagnostics, and environmental monitoring) and synthetic chemistry (e.g. rapid screening and microreactors for pharmaceutics). There are several advantages to using microfluidic chips, which include lower reagent consumption, quick response times, and reduced reaction times, because of the short diffusion distances and high surface-to-volume ratios. LOCs can also be fabricated in batches, thereby providing cheap portable platforms for chemical and biological studies. They can sample, handle, and react to tiny amounts of reagents for real-time detection.

Some disadvantages exist as well. The technology has not been developed fully because of its novelty, and physical and chemical effects, such as capillary forces, surface roughness, and chemical interactions of production materials during the reaction processes, which become more dominant on a small scale, need to be addressed. Therefore, the processes used in LOCs are often more complicated than those implemented using conventional lab equipment.

6.1.4 The significance of wearable microfluidics for biomedical applications

An early microfluidic platform based on glass and silicon substrates for electrophoresis was first demonstrated in 1992 [12]. This miniaturized analytical chip was rigid and planar, so was unsuited to close bonding with the human body. However, after about a decade, subsequent efforts were directed at soft lithography, in particular using elastomeric materials such as polydimethylsiloxane (PDMS), to build microfluidic systems. The ground-breaking material PDMS is easy to fabricate and is also biocompatible with human skin, and thus has unique advantages as a building block for wearable biosensing systems. Later, the more biocompatible material silk was used for further research [13]. Various wearable microfluidic prototypes for quantitative sampling, manipulation, and detection established the foundations for emerging applications capable of real-time physiological monitoring as well as instant feedback therapy based on body fluids.

For example, real-time tracking of the physiological state of athletes using soft wearable biosensors is essential to offer insights about their health and performance. Microfluidic and electrochemical sweat sensors can detect various electrolytes and sweat contains a wealth of physiologically relevant information that can reflect hydration, electrolyte balance, exercise intensity, and also renal function. Another example is a biosensor located in a wearable diaper for elderly people for on-site monitoring of urinary tract infections in nearly real-time. POCT microfluidics has the potential to be used in sampling, on-line manipulation, and on-line detection. POCT sampling devices make it possible to sample and store signals for detection with spatial, temporal, and chemical resolutions and information using a tiny amount of body fluid. POCT manipulation allows on-chip sample preparation, separation, and enrichment. POCT detection allows real-time feedback to end-users to increase the efficiency of diagnostics and therapy.

In this chapter, we will list several examples to show how the materials and microfabrication methods were chosen for wearable microfluidic devices, what theories and designs were involved, and where such POCT wearable microfluidic devices are used.

6.2 Materials and fabrication of wearable microfluidic devices

To develop a wearable microfluidic sensor, different kinds of flexible materials, differing in their Young's modulus (E), electrical conductivity, mechanical properties, and properties to form fluidic channels, are used to develop the substrate or platform. These materials are further combined with several sensing materials, which depend on the sensing method. The above two components use different processes and fabrication methods to produce the final design and dimensions of the wearable microfluidic device. For example, structures with dimensions in microns can be developed using photolithography techniques rather than screen printing for refined structures. After the fabrication of the device, different parameters, such as fluorescence signals and electrochemical signals, are considered for monitoring for experimental purposes. The chosen parameters depend mainly on the targeted analytes and applications. Cost, scalability, and recyclability should be considered when choosing materials and fabrication techniques. Photolithography is the most commonly utilized method for fabricating most LOCs. Silicon materials are often chosen because of their well-developed fabrication processes and techniques.

To meet other requirements, additional novel and promising fabrication methods have been developed quickly, including those involving glass, ceramic, and metal etching, deposition and bonding, polydimethylsiloxane (PDMS) processing (e.g. soft lithography), stereolithography, and fast replication methods using electroplating, injection molding, and embossing, in particular on paper and textiles.

In this section, information on substrate and sensing materials, fabrication techniques, and the characteristics of the devices will be provided.

6.2.1 Substrate materials and sensing materials

The dynamicity, durability, sensitivity, biocompatibility, etc, of the sensing system are some of the attributes in which flexible sensing systems outperform their counterparts. With wearable sensors gaining popularity for commercial applications, work on the development of wearable flexible systems has been performed on a large scale to integrate the advantages of flexible sensors with wearable devices.

Flexible sensors have prominent advantages which make them a significantly better choice in most applications. Some of the advantages of flexible sensors are:

1. They are thinner and have high flexibility and bendability, which reduces the effects of inflexibility caused by their counterparts.
2. The impact resistance of flexible sensors is lower than that of non-flexible sensors.
3. They are lower in cost which is highly advantageous for their use on a large scale.
4. The substrates used to fabricate flexible sensors are organic which reduces the chances of thermal injury.

Due to the above-mentioned advantages, flexible sensors are preferable for used in wearable sensing systems for different applications [14, 15]. The raw materials are processed using different fabrication techniques to form sensors which are characterized and embedded with signal conditioning circuits to form sensors.

The structure of a sensor can be classified broadly into two categories: the substrate and sensing parts. Substrate materials are defined as materials used on the base of the sensor which are developed as a platform for the sensing materials which are used for diagnostic purposes. Flexible sensors are designed depending on the use of the sensor.

6.2.1.1 Platform materials

The flexible platforms can be classified into polymer-based, paper-based, and textile-based platforms.

Polymer-based platforms

Polymers are large molecules that are formed by repetitive covalent bonding of single functional monomers. The occurrence of polymers can be divided into two different types: naturally occurring and synthetic (those developed in the laboratory). These two types can be differentiated into three categories:

1. *Organic polymers*: These types of polymers contain carbon. They can be both naturally occurring and laboratory-based. Examples are polyethylene naphthalate (PEN), polyethylene terephthalate (PET), silk, paper, and textiles.
2. *Inorganic polymers*: These types of polymers are the opposite of the organic ones. They are devoid of carbon. Examples are PDMS and polyimide.
3. *Hybrid polymers*: These polymers are a mixture of the above two categories. They are developed in the laboratory.

The above three categories can be fabricated with different structures and materials. They can be divided into two categories based on their structure and occurrence:

1. *Elastomers*: These polymers have high viscosity and elasticity leading to a lower Young's modulus compared to other types of polymers. An example is epichlorohydrin rubber (ECO) [16].

2. *Thermoplastics*: These polymers are developed with plastics, which are initially molded and then solidified at a specific temperature. The intermolecular forces between these monomers are higher compared those of elastomers, thus they have a higher molecular weight. An example is Teflon [17].

Some of the common polymers used to develop flexible sensors are PDMS [18], polyethylene terephthalate (PET) [19], PEN [20], butyl rubber [21], etc. The differences in these polymers lies in their dielectric constant, Young's modulus, permittivity, and permeability. Any polymer used to develop the substrate depends on the application that the sensor would be used for. Other than the generally used non-conducting polymers, there are some electrically conducting polymers, such as PEDOT:PSS [22] and polyaniline [23], which are mixed with flexible nanoparticles to form a nanocomposite to make electrodes. The thickness of the sensor substrates plays a crucial role in determining the application of the sensor. For example, the monitoring of physiological parameters, such as ECG [24], respiration [25], limb movement [26], etc, would require the sensors to be very thin for high sensitivity. The thickness of the substrates is varied using the different techniques used for fabrication. Figure 6.1(a) shows a mix of graphite and PDMS as the sensing material and PDMS as the supporting substrate material [27]. Figure 6.1(b) shows another example of a flexible temperature sensor array developed with an organic polymer [28]. The device used graphite/PDMS as the sensing material and Teflon was used as the substrate material.

Paper-based platforms

Most microfluidics platforms use materials such as glass, silicon, and polymers, such as PDMS and polymethylmethacrylate (PMMA), which require complicated fabrication and external equipment to connect with. Because of these two drawbacks, these platforms are not available for development in resource-limited countries. Therefore, paper-based LOCs will be more popular because they can avoid the above-mentioned drawbacks of a complex process and the need to be combined with external equipment for operation. Moreover, it is easy to develop portable and cheap paper-based LOCs. In the past two decades, biomedical microfluidics platforms have developed from a cleanroom product to a commercial product and have been demonstrated to be suitable for POC diagnostics. For these reasons, paper-based LOC biomedical microfluidics platforms have potential in developing countries. 'Paper' refers to the ubiquitous material made of cellulose fibers. Such sensors provide simple, inexpensive, and portable diagnostics in less-industrialized countries [30].

Microfluidic paper-based analytical devices (μPADs) were first proposed by Whitesides and co-workers in 2007, who utilized chromatography paper with photoresist patterning for colorimetric detection of glucose and protein in an artificial urine sample [31]. Since then there has been a growing development of μPADs in the field of analyte diagnostics.

In the work reported by Mu *et al*, a paper-based skin patch was demonstrated for the diagnostic screening of cystic fibrosis (CF) [32]. This flexible and stretchable

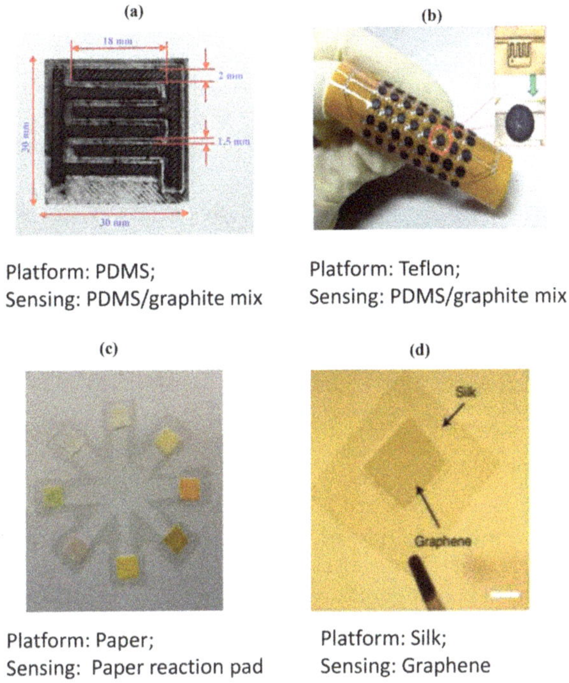

Figure 6.1. (a) A PDMS platform with a PDMS/graphite mix for stress–strain sensing. (Reproduced with permission from [27]. Copyright 2018 Elsevier.) (b) A Teflon platform and PDMS/graphite mix for temperature sensing. (Reproduced with permission from [28]. Copyright 2010 MDPI.) (c) A hydrophobic wax barrier formed on the paper to create a channel to the reaction paper pad. (Reproduced with permission from [29]. Copyright 2015 S Feng.) (d) A silk platform and graphene sensing components that can be located on a tooth to monitor saliva contents. (Reproduced with permission from [13]. Copyright 2012 Springer Nature.)

'Band-Aid' type device enables conformal contact with the subject's skin. The level of anions in secreted sweat was detected by colorimetric assays using a combination of anion exchange paper and pH test paper.

In the field, using PMMA slices and a PSA channel to encapsulate the paper component is an effective approach to prevent contamination of the collected body fluid with open-air dust and dirt. It can isolate immediate contacts between chemical reagents coated on the test area and the human body, as shown in figure 6.1(c) [29].

Textile-based platforms

Textiles are a basic material in the daily lives of human. It is the standard material of the apparel industry, and is also adopted widely to dress wounds and tissues, and in organ repair. In recent decades, textile materials have been developed for body fluid biosensing systems taking advantage of their good biocompatibility and hydrophilicity, and the ability to integrate them into our clothes [33]. A surface-treated hydrophilic thread or textile can provide fluid flow from the sampling site to the sensing electrodes by capillary force. Commercial products can benefit from its low cost compared to a polymeric material, such as PDMS. For example, a tooth platform for monitoring of breath and saliva was made of a silk substrate and

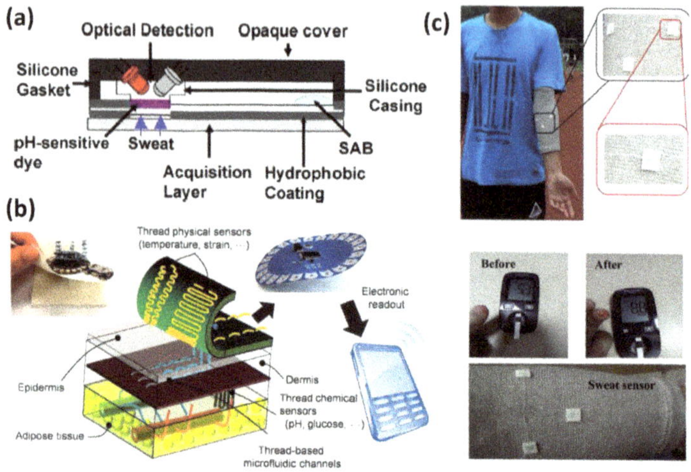

Figure 6.2. (a) Schematic of a pH sensing and optical detection platform based on textile fluid transport. (Reproduced with permission from [35]. Copyright 2009 Elsevier.) (b) Schematics of the structure and use of textile-based pH and strain sensors. The thread-based diagnostic device (TDD) is realized for transdermal health monitoring. (Reproduced with permission from [36]. Copyright 2016 Springer.) (c) Implementation of sweat pH and glucose of volunteer testing before and after exercise. Reprinted from [27], copyright (2017), with permission from Elsevier

graphene sensing materials, as shown in figure 6.1(d) [13]. Roy *et al* developed PDMS–textile wearable devices with robust performance for antenna research [34].

Textile or thread materials can deliver a fluid through their intrinsic wicking force. A textile-based sweat pH testing system was proposed by Morris *et al*, and they successfully observed a pH increase phenomenon in exercise [35]. The secreted sweat was collected on human skin and transported by the hydrophilic textile in a defined way down a channel coupled with emitter-detector LEDs for pH screening, shown as figure 6.2(a). Furthermore, researchers investigated an autonomous real-time wearable sweat pH screening system using a microfluidic device. In the proposed textile-based platform, a smart textile microfluidic chip was able to draw and deliver secreted sweat to a colorimetric sensing module in a downstream chamber with LEDs. The sample collection efficiency, detection time, and dead sample volume are improved significantly by the wicking force offering continuous flow.

Recently, Mostafalu *et al* presented a toolkit system combining textile-based microfluidics to detect the strain in different parts of the human body, as shown in figure 6.2(b) [36]. A functional textile with different physical, chemical, and biological characteristics was produced to implement the microfluidic channel (hydrophilic textile), electronic wires (conductive textile), and sensors (textile electrode) in a TDD platform. An external smartphone or computer was connected to the TDD by a wireless connection or a conductive textile for the data analysis and readout. Moreover, the textile can be coupled with a paper material system due to its flexibility for further development of a wearable platform. Xiao *et al* introduced a non-invasive textile/paper-based microfluidic wearable device for the *in situ* screening of human sweat glucose using a smartphone (figure 6.2(c)) [37].

6.2.1.2 Common sensing materials

The sensing materials use depend on the detection method. For electrochemical detection methods the electrodes are the main sensing materials, which include normal conductive materials and nanocomposites. For optical detection methods, detection components consisting of functional surfaces as well as microchambers containing reaction reagents could work.

Electrodes are the conductive part of the sensor which interact with the material under consideration to determine its properties. The electrodes in flexible sensors are developed from different kinds of nanoparticles [38]. The size of these nanoparticles varies from a nanometer to a few micrometers depending on the type of nanomaterial. Nanoparticles can be categorized into different types, such as nanotubes [39], nanoribbon [40], nano-beads [41], nanowires [42], etc. These types differ in terms of their surface area, mechanical and thermal properties, and electrical conductivity. Some of the common elements used to develop flexible sensors are carbon [43], gold [44], silver [45], silicon [46], iron [47], nickel [48], cobalt [49], and copper [50]. This step is performed after the fabrication of the sensor and before it is used for any application. Different elements such as gold [51], carbon [52–54], aluminum [55], and iron oxide [56] have been functionalized for biomedical applications. For example, carbon was functionalized [57] to increase its application for bio-imaging purposes.

These electrodes were developed mainly with single nanoparticles or in the form of nanocomposites. Nanocomposites can be defined as the mixture of a nanoparticle with a polymer or other elements to amalgamate the properties of the individual components. A flexible electrode developed with assembled graphene and PET displaying outstanding flexibility [58], with graphite/PDMS mixtures as the conductive electrodes. Some of the common functional groups attached with gold NPs include the amine group [59], polyethylene glycol (PEG) group [60, 61], the carboxyl group [62], DNA [63, 64], and the methacrylate group [65].

Microchambers can be made to trap the liquids and reactions can happen during this. Droplets are common microchambers. Other microchambers are made from microfabricated structures. Figure 6.1(c) is one example of using a paper pad to keep the reaction reagents in a specific location for colorimetric reactions.

6.2.2 Fabrication techniques

After the raw materials are chosen for a particular sensing device, the fabrication technique processes these materials to produce the sensors with a specific shape and structure. A particular fabrication technique is chosen based on the following attributes:

1. *Raw materials*: The types of raw materials that are used to develop the sensor determines which technique has best ability to process them with the best efficiency.
2. *Dimensions of the sensors*: The dimensions of the sensors can be varied greatly by the techniques used for fabrication. For example, in the case of

screen printing the dimensions will go down to a few micrometers, while using photolithography will bring the dimensions down to a few nanometers.

3. *Application*: The use of the sensors can also determine the technique chosen for their fabrication. Some applications such as monitoring of body movements or physiological parameters would require much smaller and more sensitive sensors compared to applications such as shoe sensing or daily life activities [108].

The different techniques used for developing the sensors depend primarily on the applications for which the sensing system will be used.

6.2.2.1 Photolithography

Photolithography is the most standard technique for fabricating structures for a wearable microfluidic sensor [66, 67]. This is achieved by using (ultra-violet) light to apply a pattern on a silicon or quartz substrate. The light transfers the developed design onto the photo-mask using masked lithography [68] or it is uploaded on the system using maskless lithography [69]. There are two types of photolithography based on a template, removing the exposed patterns for a positive photoresist [70], such as AZ 1514 and AZ 9260, and keeping the exposed patterns for a negative photoresist [71], such as SU-8 photoresists. Also, in wet etching [72] the exposed substrate is treated with chemicals to remove the soluble resist and in dry etching [73] anisotropic patterns are required on the exposed substrate to achieve high aspect ratio structures.

6.2.2.2 Screen printing

The screen printing technique [76, 77] has been used widely in commercial applications since the early nineteenth century. In the last two decades this technique has been applied in flexible electronics. In this technique a mesh is used to transfer inks from one side to the other, except for the regions blocked by the mesh. The flow of the ink from the mesh onto the substrate is important as it controls the thickness of the substrate. This develops a pattern on the substrate that is designed on the mesh. Screen printing where the design developed on the coated screen is transferred on the printed shirt with the ink poured over the screen [78].

There are many commercially available conductive inks such as silver [79] and graphene [80], which are used to develop electrodes via screen printing on the substrate. The greatest advantage of using this technique is the versatility in the type of materials used as a substrate for the ink, such as polymers [81], fabrics [82], circuit boards [83], etc. Even though this technique has been used for a long time for manufacturing different devices, it is not as effective as other techniques as the finished products are not very precise.

Laser printing

Although lasers have been used for a long time, the introduction of laser printing has only been perfected in the last two decades [26, 84–86]. This technique is advantageous compared to others for the following reasons:

a. Unlike photolithography, no prior sample preparation is required.
b. Very thin and flexible materials can be produced using this technique, which cannot be achieved by techniques such as screen printing.
c. The finished products have relatively smooth edges that are perpendicular to the surface. Thus no extra step is required to polish the end product.

In laser printing the substrate is placed under the laser to scan or bend part of the substrate to form a design. The proportion of samples removed from the sample depends on the material and the settings of the particular laser system used for that sample. Depending on the maximum output power produced by the laser nozzle, different types of laser machines can be used. Other important properties that can change the thickness of curved samples are:

a. *z-axis*: This axis determines the height of the platform below the laser nozzle.
b. *Speed*: This determines the speed at which the nozzle moves over the sample. In the case of laser printing there has to be a trade-off between the input power and the speed of the nozzle for obtaining optimum features.
c. *Frequency*: This determines the frequency of the laser nozzle moving over the sample.

Ink-jet printing

The ink-jet printing technique for printing electronics was developed from the ink-jet printers used to print documents [87, 88]. The idea of using ink-jet printers developed from replacing the ordinary ink with any conductive ink to develop the sensing part on a sensing substrate [89]. The common types of conductive inks used in ink-jet printing are silver [90], carbon [91], and copper [92]. An example of a temperature sensor was developed by ink-jet printing on a PET substrate [93]. An commercial 3D printer (Formslab form2, figure 6.3(a)) was used. Some other popular and high resolution 3D printers are from Nanoscribe and Boston Micro Fabrication (BMF). Compared to other technologies, using ink-jet printing technology for manufacturing has the following advantages:

a. The fabricating device, a printer with the use of different conductive inks for fabrication purposes, is cheaper.

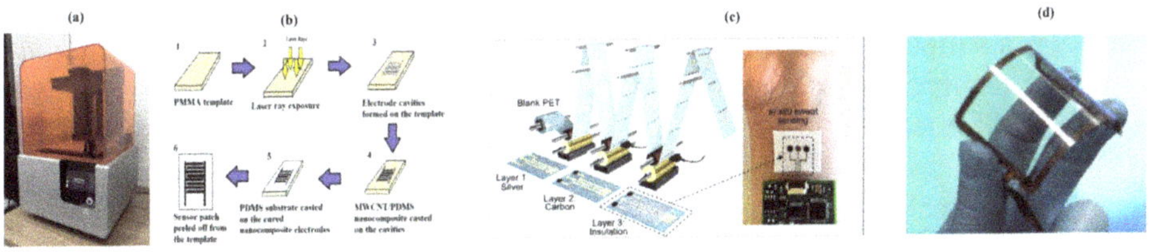

Figure 6.3. (a) The 3D printing Formslab form2 device. (b) A schematic diagram for the individual steps followed in the casting technique to develop a strain-sensitive sensor. (Reproduced with permission from [74]. Copyright 2016 IEEE.) (c) The roll-to-roll process for a tattoo-based biosensor. (Reproduced with permission from [75]. Copyright 2018 American Chemical Society.) (d) Another example of the roll-to-roll process for graphene on PTE. (Reproduced with permission from [58]. Copyright 2010 Nature.)

b. High quality of images or sensing devices can be achieved with this technique.
c. The time required to complete a process is very short. This is particularly beneficial in the case of the fabrication of large-scale devices.
d. The devices are easier to use and can be operated by any non-specialist.

Casting
Like screen printing, this technique has also been used for many years for commercial applications [74, 94]. It was not until the last decade or so that this technique was also used for the fabrication of flexible sensors. This process is carried out with a template forming a mold of the desired shape. A liquid or semi-liquid material is then be poured in the mold which is then be solidified to take the shape of that mold. After the material has solidified it is cast off the mold and used as a separate device. In the case of flexible sensors, the liquid or semi-liquid material that is poured is the conductive material that then solidifies and forms the electrodes. On top of the solidified mold, a layer of different material, such as a polymer, is poured and solidified, and thus cast to form the substrate for the sensor. A schematic diagram of the explained steps is shown in figure 6.3(b) [74]. This method was followed to produce a flexible sensor with PDMS as the substrate and a nano-composite of multi-walled carbon nanotubes (MWCNTs) and PDMS as the electrode material. Figure 6.1(a) shows one structure made using this method to form PDMS as the substrate and graphite/PDMS as the electrodes.

Even though the casting technique has seen significant use lately [95, 96], some limitations curb its use to a large extent. The limitations are concerned mainly with the sensitivity, which is not a high in the electrodes as those developed with printing or laser cutting techniques. This is because the second material (preferably a polymer) poured over the first solidified material decreases the conductivity of the electrodes.

6.2.2.3 Tattoo-based wearable electrochemical devices
Roll-to-Roll Gravure Printed Electrochemical Sensors were developed for Wearable and Medical Devices [75], as shown in figure 6.3(c). A Tattoo-Based Noninvasive Glucose Monitoring was also developed [97]. It was shown that 30-inch graphene films for transparent electrodes can be attached to PET using the roll-to-roll method [58], as in figure 6.3(d). There has also been an attempt to use tattoo ink to detect cancer cells [98].

There is a similar technique called weaving [99, 100]. Compared to the previously mentioned techniques, this is a comparatively recent approach to develop flexible electronics. This technique has been used mainly to develop fabric embedded sensors [101, 102]. This process is carried out using handlooms or any weaving device. This is because this technique is comparatively new and the weaving machines designed for developing fabric embedded flexible sensors are still high in cost and need to be operated by specialists.

Some of the advantages of using this technique are:

a. The person using the sensor does not have to wear anything separate as the sensor is be embedded in the textile.

b. The sensor can be produced in a single step. There are no individual steps to produce the electrodes and substrates separately.
c. The sensors can be produced with any substrate without much effect on their characteristics. This is because the conductive material is woven into the garments [103].

6.2.3 Characteristics of the sensor

To achieve both high temporal and spatial resolution, imaging technologies are used to examine the sensors. The fundamental trade-off in imaging technologies is between acquisition speed, sensitivity, and the amount of information that can be acquired. Multi-pixelated imaging devices (camera-based), such as charge-coupled devices (CCDs), and complementary metal-oxide-semiconductors (CMOS) [104], scanning electron microscopy (SEM), and transmission electron microscopy (TEM) are frequently used to check the surface features of sensors.

Stylus profilometers and optical profilers are commonly used to measure the x–y–z microstructures on the surface of sensors. A stylus profilometer uses a physical probe to move along the surface. The size dimensions of x, y, and z can be obtained with a feedback loop that monitors the force from the sample pushing up against the probe as it scans along the surface. Optical profilers (figure 6.3(c)) are interference microscopes measuring the microstructures' z dimension on the surface using the wavelength of light as the ruler. A comparison between the optical path difference between a reference surface and a test surface is used to obtain the dimension information based on the wave properties of light. A normal stress/strain machine can measure the strength of the device. An LCR (inductance, capacitance, and resistance) machine, for example a HIOKI IM 3536 High Precision Tester, can be used to measure the basic resistance, capacity, and so on.

6.3 Theories and designs

6.3.1 Wearable microfluidic devices for physical properties

To achieve physiological measurements for practical applications outside of labs or clinical environments, it is essential to use robust interfaces with the skin without chemical and physical irritation and develop integrated systems with weights, sizes, and shapes that do not generate discomfort. By developing the next generation of wearable electronic technologies, it could be possible to detect physical characteristics such as stiffness, strain, motion, electric impedance, conductivity, bio-potential, and other relevant parameters. This use of wearable devices for stress/strain sensing and pressure measurement will be described here.

6.3.1.1 Bending stress/strain measurements

Nag and Feng *et al* used a 3D-printed mold-based graphite/PDMS sensor for low-force sensing applications and characterized the sensor patch for the change in responses with stress using the effective capacitance [105]. The electrodes were shaped in an interdigitated manner and the sensor exhibited a capacitive principle. They further explored more devices for other stress/strain applications [27, 106–108].

The capacitance of the sensor can be expressed by the following equation:

$$C = \frac{\varepsilon_0 \varepsilon_r A}{d}. \tag{6.1}$$

The relative change in capacitance of a flexible sensor due to an external load can be defined as

$$\frac{\Delta C}{C} = \left(\frac{\Delta A}{A} - \frac{\Delta d}{d} \right). \tag{6.2}$$

Since the total area (A) is related to the width (W) and length (L) of the electrodes, the capacitance change can be rewritten as

$$\frac{\Delta C}{C} = \left(\frac{\Delta L}{L} + \frac{\Delta W}{W} - \frac{\Delta d}{d} \right), \tag{6.3}$$

where A is the effective area of the sensor patch, d denotes the interdigital distance between the two consecutive electrodes, ε_0 is the dielectric constant of a vacuum, ε_r is the relative permittivity of the material, C is the capacitance, $\frac{\Delta C}{C}$ is the relative capacitance, $\frac{\Delta A}{A}$ is the relative change in the area, and $\frac{\Delta d}{d}$ is the relative distance between the interdigitated electrodes.

6.3.1.2 Pressure measurements

Wang *et al* [109] developed a transparent and flexible triboelectric sensing array for touch security applications and characterized the sensing device electrically using an open-circuit voltage under sliding and tapping modes. In the contact-separation mode, the object is controlled by an external force to separate and contact the fluorinated ethylene propylene polymer film, generating a reciprocating current between the top indium tin oxide (ITO) electrode and ground thanks to the electrostatic induction effect. The open-circuit voltage can be calculated qualitatively as

$$U_{OC} = U_{top} - U_{ground} = \frac{-\sigma d}{2\varepsilon_0 \varepsilon_r}, \tag{6.4}$$

where U_{top} is the potential of the top ITO electrode, U_{ground} is zero, d denotes the thickness of the PET substrate, which is the distance between the parallel electrodes, ε_0 is the absolute permittivity, and ε_r is the relative dielectric constant of PET.

6.3.2 Wearable microfluidic devices for body fluids

Wearable microfluidic devices for body fluids are characterized using POC sampling, POC manipulation, and POCT and signal processing. Through this section we describe how body fluids can be sampled, manipulated (stored, mixed with reagents, transported), detected (using optical as well as electrochemical methods), and the signal processed using POCT wearable devices, which construct the basis for research and product development.

6.3.2.1 POC sampling and manipulation

Modules were attached for detection using wearable devices. Using hydrophilic working components, a liquid such as urine, sweat, or saliva can be sampled easily.

Figure 6.4(a) shows a sampling module for sweat [110]. Figure 6.4(b) shows an example of microneedle sampling. Figure 6.4(c) shows a sampling probe for tiny biosamples using a microfabricated needle with good chemical, temporal, and spatial resolutions [113]. The details of the microfabrication process can be found in [114]. A collection of such works is described in the review [115], including a cheap plastic microfluidic chip with a hydrophilic membrane at the sampling tip [116].

After the fluid was sampled, it was broken into droplets for storage, mixing with reagents, separation, enrichment, and transport to the detection areas. The fluids can be sampled, stored, and mixed with reaction reagents using the microfabricated probe [110]. The microfluidic chip can perform the separation or enrichment of the cells [117, 118]. Lateral flow or a pressure gradient can transport the detection analytes to the detection area.

6.3.2.2 POCT and signal processing

Wearable sensors enable continuous tracking of body movements, temperature, cardiac activity, blood pressure, and others biosignals, providing invaluable insights into physiological status, but are limited in obtaining direct information on dynamic biochemical and metabolic processes. By analyzing biofluids, such as saliva, sweat, and interstitial fluids, they can track the changes in metabolic processes for monitoring a patient's health status. Wearable sweat sensors can be exploited for

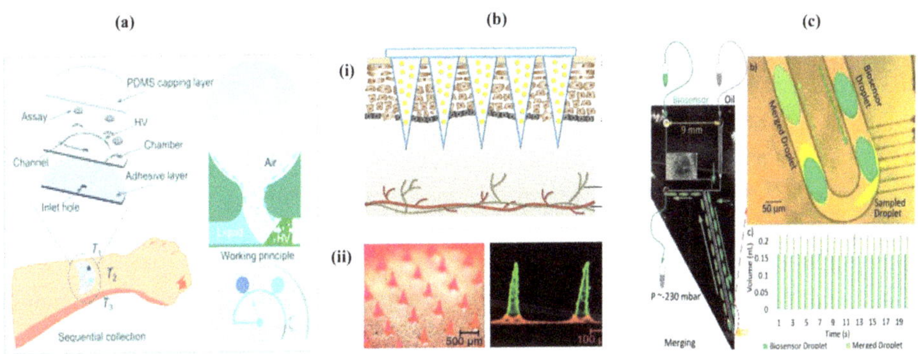

Figure 6.4. (a) Sampling module for sweat. (Reproduced with permission from [110]. Copyright 2020 The Royal Society of Chemistry.) (b) Working principles, classification, and representative applications of microneedles. (i) Common working principles of microneedles for drug delivery and four typical microneedles. (Reproduced with permission [111]. Copyright 2017 The Royal Society of Chemistry.) (ii) Multilayered pyramidal dissolving mironeedles, composed of silk fibroin tips supported on flexible pedestals, for improved drug delivery. (Reproduced with permission from [112]. Copyright 2017 Elsevier.) (c) Sampling probe for an injury. (Reproduced with permission from [113]. Copyright 2019 The Royal Society of Chemistry.)

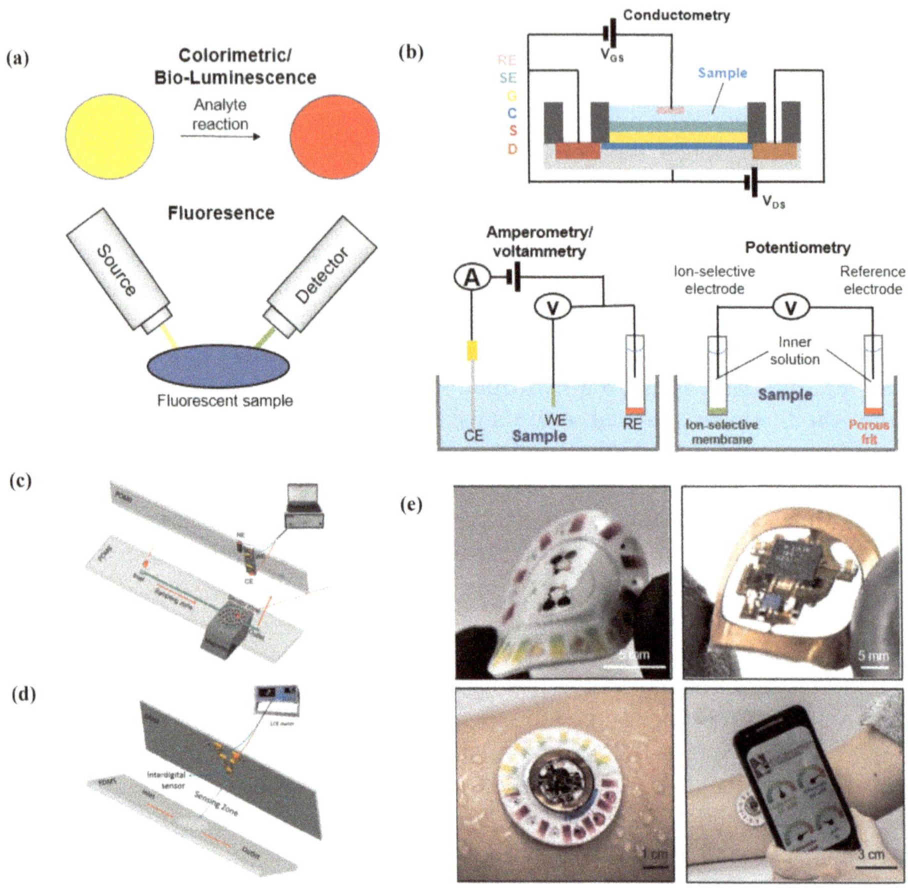

Figure 6.5. (a) Schematic showing different sensing modalities for optical analysis methods. (b) Schematic showing different sensing modalities for electrochemical analysis methods. (c) Designed detection module for impedance change. (Reproduced with permission from [121]. Copyright 2018 MDPI.) (d) Designed modules for electrochemical property changes. (Reproduced with permission from [122]. Copyright 2019 Nature.) (e) Hybrid devices for both electrochemical and optical sensing. (Reproduced with permission from [123]. Copyright 2019 AAAS.)

real-time and on-skin analysis based on electrical approaches and optical technologies. Electrical approaches such as potentiometry, voltammetry, amperometry, and conductometry provide sensing capabilities in a variety of cases.

This section will focus on the recent progress of wearable sensor detection methods, which are mainly for optical and electrochemical analysis [119]. A schematic is shown in figure 6.5(a) for optical methods, including colorimetric, bioluminescence, and fluorescence detections. A schematic is shown in figure 6.5(b) for electrochemical analysis, including conductometry, amperometry/voltammetry, and potentiometry analysis. Feng has developed impedance sensing [110] and electrochemical sensing [111] in microfluidic devices, as shown in figures 6.5(c) and (d), respectively. For the impedance sensing, Nag *et al* have explored phosphate detection using microfluidic devices [120].

Skin-mounted sensors have received more and more attention in recent years, in particular those for non-invasive detection of epidermal biofluids using electrochemical or colorimetric monitoring methodologies. Electrical approaches such as potentiometry, voltammetry, amperometry, and conductometry provide sensing capabilities in a variety of cases. The methodologies show remarkable prospects for real-time tracking of many electrolytes (e.g. sodium, potassium, calcium, and chloride), heavy metals (e.g. zinc, hydrargyrum, and copper), and metabolites (e.g. glucose and lactate). By analyzing these sweat biomarkers, wearable electrochemical sensors can retrieve physiologically rich information, such as exercise intensity, which can be determined by measuring the concentration of lactate in sweat. To show their enormous promise in sports monitoring (i.e. their suitability for field studies, remote monitoring, and continuous assessment), representative attempts to establish wearable electrochemical sweat sensors will be reviewed in this section. Recent advances in electrochemical sensors have paved the way for simultaneous multianalyte biosensing in real-time monitoring due to their flexibility and conformality, as well as remarkable capacities for *in situ* data processing and wireless data transmission.

In contrast to electrochemical sensing, colorimetry is a visualized sensing method without the requirement for intricate electronic parts and components. Colorimetric detection has been introduced to monitor the concentration of target analytes with the assistance of color reagents. With the aid of different indicator dyes, a normal smartphone can be used for the analysis. Toward a compact on-body wearable system, utilizing colorimetric detection methods can be independent of the added integrated electronic element for analyte electrochemical sensing functionalities, as well as the corresponding power supply system, which contributes to reducing the size, weight, and complexity of wearable systems, further enabling a more comfortable wearing experience. An epifluidic device collects freshly secreted sweat directly without any contamination and evaporation through the enclosed microfluidic channel.

Wearable sensors need to be able to simultaneously monitor multiple parameters to efficiently analyze an athlete's health status and sports performance at the molecular levels. The previously reported epidermal sensor can only monitor a single analyte or a few analytes at a time, which makes it difficult to yield a comprehensive evaluation of the user's physiological status. Therefore, with the increasing demand for simultaneous monitoring of multiple parameters, the emerging hybrid sensors have promise. These sensors integrate the capabilities of simultaneously performing electrochemical, colorimetric, or volumetric analysis, and *in situ* data processing as well as wireless communication. Moreover, several hybrid attempts merging both electronic electrochemical sensors and colorimetric assays on a comprehensive epifluidic wearable device have been presented [123, 124].

Hybrid sensors will attract considerable research interest and gradually become the research focus for both health monitoring and sports physiological assessments, because the integration of soft electronics, electrochemical sensors, and soft microfluidics would allow for the emergence of miniaturized, fully integrated platforms, which give a full picture of users' health and physiological status, as shown in figure 6.5(d), where optical and electrochemical detection are combined for sweat sensing.

6.4 Applications

To achieve physiological measurements for practical applications outside of laboratories or clinical environments, it is essential to exploit robust interfaces to the skin without chemical or physical irritation and develop integrated systems with weights, sizes, and shapes that do not generate discomfort. By developing the next generation of wearable electronic technologies, it could not only be possible to detect physical characteristics such as stiffness, strain, motion, electric impedance, conductivity, bio-potential, and other relevant parameters, but also enable the measurement of complementary information through the capture and biochemical analysis of biofluids (for example, sweat).

6.4.1 Wearable microfluidic devices for sweat

Fully integrated microfluidic sweat analytic platforms have usually been employed to receive the weak naturally induced secreting pressure from the sweat glands and deliver sweat through microchannels collecting microreservoirs for subsequent sensing and analysis.

Sweat analysis is currently utilized for various applications including disease diagnosis, athletic performance optimization, and drug abuse detection. Collection, manipulation, and chemical analysis of sweat can noninvasively monitor physiologically relevant electrolytes, molecules, and metabolites. Monitoring of glucose levels in sweat can be introduced to qualitatively track the blood glucose level, secreted sweat chloride, and monitor the glucose loss in exercise. The pH of sweat can be related to the average sweat secreting rate and offer significant insights into electrolyte balance and hydration status. Moreover, excessive loss of sodium and potassium in sweat could lead to hyponatremia, hypokalemia, and muscle cramps. In addition, sweat lactate can serve as a sensitive biomarker of pressure. Also, usually but importantly, skin temperature is informative for monitoring many diseases and skin injuries such as pressure ulcers. These parameters can thus help to identify certain specific conditions of the human body [125].

Conventional and typical approaches for sweat analysis usually involve multistep processes of sweat collection, transport, storage, and analysis using benchtop instruments, and the accuracy is limited by potential contamination, degradation, and loss of samples during the preparation process, which are also unable to achieve real-time sweat detection.

Javey *et al* [129] proposed a mechanically flexible and integrated wearable sensor for multiplexed *in situ* perspiration analysis by simultaneously detecting sweat metabolites (glucose and lactate), electrolytes (potassium and sodium ions), and skin temperature.

Amperometric glucose and lactate sensors have been developed using glucose oxidase and lactate oxidase immobilized in a permeable film of polysaccharide chitosan. An Ag/AgCl electrode was set as a shared reference electrode and counter electrode for both sensors, and Prussian blue dye was used as a mediator. The enzymatic sensors produced current signals proportional to the concentration of the relevant metabolites between the Ag/AgCl electrode and the working electrode. The analysis of sodium and potassium ions was conducted using ion-selective electrodes (ISEs), combined with a polyvinyl butyral (PVB)-coated reference electrode. To achieve long term analysis without significant voltage drift, poly(3,4-ethylenedioxythiophene) polystyrene sulfonate (PEDOT:PSS) was used as an ion-to-electron transducer in the ISEs and carbon nanotubes in the PVB membrane. For temperature detection, a resistance-based temperature sensor was exploited based on Cr/Au metal microwires and parylene served as an insulating layer to avoid electrical contact of the metal lines with sweat and skin.

Wearable sensors not only depend on electronics for electrochemical detection, but also rely on colorimetric methods for a visual readout. Rogers *et al* [123, 126] proposed battery-free, skin-interfaced microfluidic and electronic systems for electrochemical and colorimetric analysis of secreted sweat by monitoring the rate and loss of pH, lactate, glucose, and chloride simultaneously, as shown in figure 6.6(a) [109].

Choi *et al* proposed a customizable epifluidic device to quantitatively estimate the secretory fluidic pressures generated at the surface of skin using a collection of modified capillary bursting valves (CBVs) [127] (figure 6.6(b)(i)). For this skin-mounted system, each integrated CBV was independent with a private inlet, and the characteristic burst pressure (BP) was gradually increased with the designed structure. Moreover, Wang developed a microfluidic sweat collection and operation sensor on the skin (figure 6.6(b)(ii)).

Colomertric detections are illustrated in figure 6.6(c). It shows a device with integrated microchannels and colorimetric detection zones for biochemical, electrolyte, and metabolite analysis [126].

For colorimetric assays, the colorimetric sensing method has been developed to quantitatively evaluate the rate and volume of sweat loss, and the concentration of lactate, chloride, pH, glucose, and creatinine in sweat. Rogers *et al* provided a wearable microfluidic device for detecting multiple markers in sweat using a colorimetric method. In an orbicular serpentine channel, cobalt(II) chloride in the coating polyhydroxyethylmethacrylate (pHEMA) of the hydrogel matrix was used as an indicator. When sweat entered the microchannel, the anhydrous cobalt (II) chloride chelated with water producing hexahydrate cobalt chloride, causing a color change from deep blue to pale purple. Thus, quantitative analysis on the sweat rate and volume can be achieved based on the position of the leading edge of the change in color.

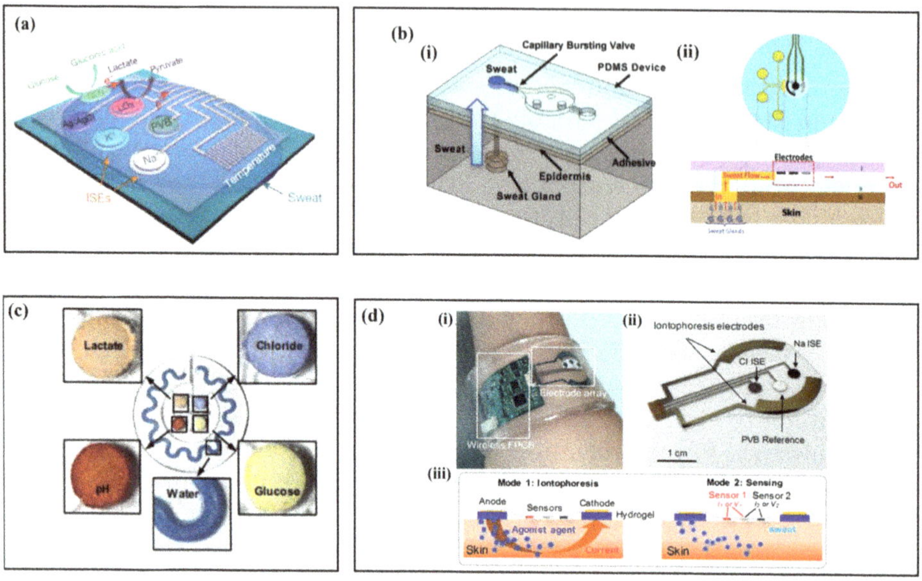

Figure 6.6. (a) Schematic illustration of an array of electrochemical sensors for analyzing the concentration of glucose, lactate, potassium, and sodium in sweat. GOx = glucose oxidase; LOx = lactate oxidase; PVB = polyvinyl butyral. (Reproduced with permission from [126]. Copyright 2018 AAAS.) (b) Image of an autonomous sweat extraction and sensing platform. (i) Example 1. (Reproduced with permission from [127]. Copyright 2017 Royal Society of Chemistry.) (ii) Example 2. (Reproduced with permission from [128]. Copyright 2017 American Chemical Society.) (c) Colorimetric methods. (Reproduced with permission from [126]. Copyright 2018 AAAS.) (d) Schematic drawing of strategies for stimulating the release of sweat by iontophoresis and for electrochemically analyzing the chemistry of this sweat. (Reproduced with permission from [126]. Copyright 2018 AAAS.)

The concentration of lactate in sweat can be detected based on the enzymatic reactions between lactate and cofactor NAD+ (nicotinamide adenine dinucleotide) by lactate dehydrogenase and diaphorase, which generates a color change through the chromogenic reagent. The color change in the targeted reservoir would be proportional to the concentration of lactate.

Concerning the colorimetric detection of glucose concentration, glucose oxidase coated in a cellulose matrix generates hydrogen peroxide based on the oxidation of glucose and the reduction of oxygen, and iodide then oxidizes to iodine by peroxidase, inducing a change in color from yellow to brown. As a result, the glucose concentration can be determined by the color change. Creatinine, a sensitive marker for monitoring hydration status and renal function, was detected by a mixture of enzymes (creatininase, creatinase, and peroxidase) with the assistance of a corresponding responsive dye (4-aminophenazone) at the same time.

A common pH indicator consists of bromothymol blue, methyl red, and phenolphthalein to generate colorimetric changes within a corresponding range. The detection of chloride can be conducted by the competitive binding between Hg^{2+} and Fe^{2+} with

2,4,6-tris(2-pyridyl)-s-triazine (TPTZ). Iron ions bind with TPTZ in the presence of chloride ions and Hg^{2+} participates as $HgCl_2$, further leading to a change in color.

On the other hand, electrochemical sensors have been developed in a variety of wearable devices and the sensing electrodes can be modified differently in terms of specific applications, as shown in figure 6.6(d). Davis *et al* [130] proposed a fully integrated wearable platform to extract and analyze sweat autonomously for cystic fibrosis and glucose monitoring. They used electrochemical sensors for sweat chloride, sodium, and glucose analysis. Ag/AgCl electrodes were selected for chloride ion measurement while the detection of sodium ions was conducted based on sodium ionophore X select phone-based ion-selective electrodes. A polyvinyl butyral (PVB) coated electrode including saturated chloride ions was set as the control electrode owing to its stable potentials in various analyte solutions. The potential differences between the PVB-coated electrode and the ion-selective electrodes were detected using a differential amplifier. The concentration of glucose solutions was analyzed based on the chronoamperometry response from the glucose sensor [131].

In the optical analysis, the top layers of the sensor must be transparent for optical detection. Most colorimetric assays generate irreversible color changes that limit such devices to single time-point analysis.

There is a need for high-throughput and low-cost fabrication of wearable sensing platforms with limited performance variation in POC health monitoring and diagnosis that are capable of delivering consistent detecting performance when multiplexed and complex detections are conducted on the same device.

Javey *et al* [75] developed a high-throughput and low-cost method for producing roll-to-roll gravure printed electrodes with controllable size and density which are designed for various electrochemical sensing applications, such as detecting ions, metabolites, heavy metals, and other small molecules, demonstrating a key translational step in enabling large-scale and cost-effective production of wearable sensors for personalized health and fitness monitoring.

The intrinsically simple, low-cost, and wearable devices show great potential in offering warning signs associated with physical activities that cause abnormal responses. Gao *et al* [132] reported an entirely laser engraved sensor for sweat sampling and sensing which was easy to fabricate at a large scale.

Briefly, soft microfluidic sensors are broadly applicable for ambulatory modes of physiological monitoring of athletes, which technically bypasses the requirement for trained personnel in laboratories or hospitals and the requirements for standard laboratory equipment. The flexible and stretchable attributes of microfluidic devices promise better athlete compliance than conventional planar wearable sensors. Most importantly, these sensors enable imperceptible sweat collection and real-time, multianalyte monitoring of sweat biochemistry. They provide valuable information about the physiological status, such as dehydration, electrolyte balance, fatigue, renal function, etc. The great potential of the applications of soft microfluidics in sports monitoring should not be neglected by sports scientists and biomedical engineers.

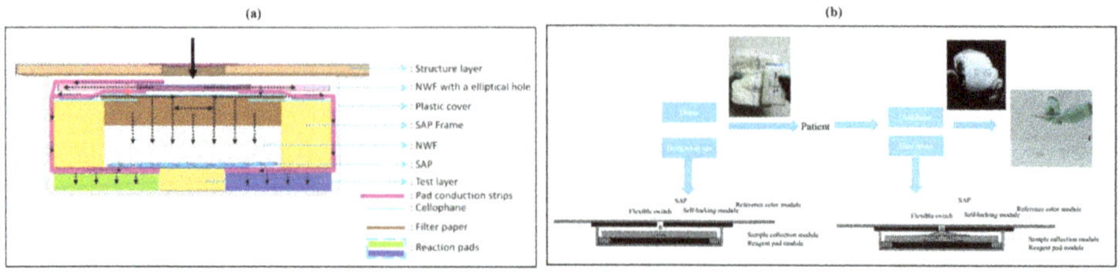

Figure 6.7. (a) A urine sampling module located on a diaper can be used to extract the urine directly. (Reproduced with permission from [140]. Copyright 2018 Springer.) (b) On-diaper colorimetric detections can be achieved. (Reproduced with permission from [141]. Copyright 2018 The Royal Society of Chemistry.)

6.4.2 Wearable microfluidic devices for urine

In one approach for analysing urine, the sample is obtained from the superabsorbent materials inside a diaper by adding calcium salt, which allows the urine to be released quickly [29, 133]. Continuous ATP bioluminescence [134, 135], electrochemical analysis [133, 136], and biomarker detection using paper-based methods have also been explored [137, 138]. Different methods have also tried to use extraction methods on the diaper directly [139] and one example is shown in figure 6.7(a) [140]. On-diaper colorimetric detection using paper-based methods has also been explored [141], as shown in figure 6.7(b).

Several patents have focused on developing these kinds of POCT devices on diapers. The patent EP 0560099 A2, with the title ‘Device for determining chemical and/or biological conditions in liquid, especially aqueous, media, e.g. urine, using at least one indicator’, relates to a device for detecting and/or measuring, in particular, chemical and/or biological conditions with the aid of at least one indicator in a liquid medium. One attempt is ongoing to use a credit card sized card indicator, which can be applied to the clear plastic sheet of a compatible urine collection device. The patent US 5707818A describes a device and related methods for simultaneously performing a plurality of immunoassays to detect the presence of the respective analytes in a sample. EP 1 150 609 B1describes a device for receiving and controlling voided body fluids. US 2008/0269707 A1 is a patent for a lateral flow device to be attached to an absorbent article for body fluid detections.

POCT urine cytology has also been created for capturing and detecting urothelial carcinoma of the bladder (UCB) cancer cells in urine, it also has the ability to process entire samples and the presence of background cells [142].

6.4.3 Wearable microfluidic devices for saliva

Similar to sweat, saliva is another well-accepted diagnostic biological fluid which can be collected easily and noninvasively, and it also contains abundant potential target analytes for disease diagnostics (ions, lipids, proteins, etc) [143–145]. Tooth platform monitoring of breath and saliva was made using a silk platform and graphene sensing materials [13], as shown in figure 6.8(a). A wearable salivary uric acid mouthguard biosensor with integrated wireless electronics is reported in [146] and figure 6.8(b) panel (i) shows a photograph of the the mouthguard biosensor

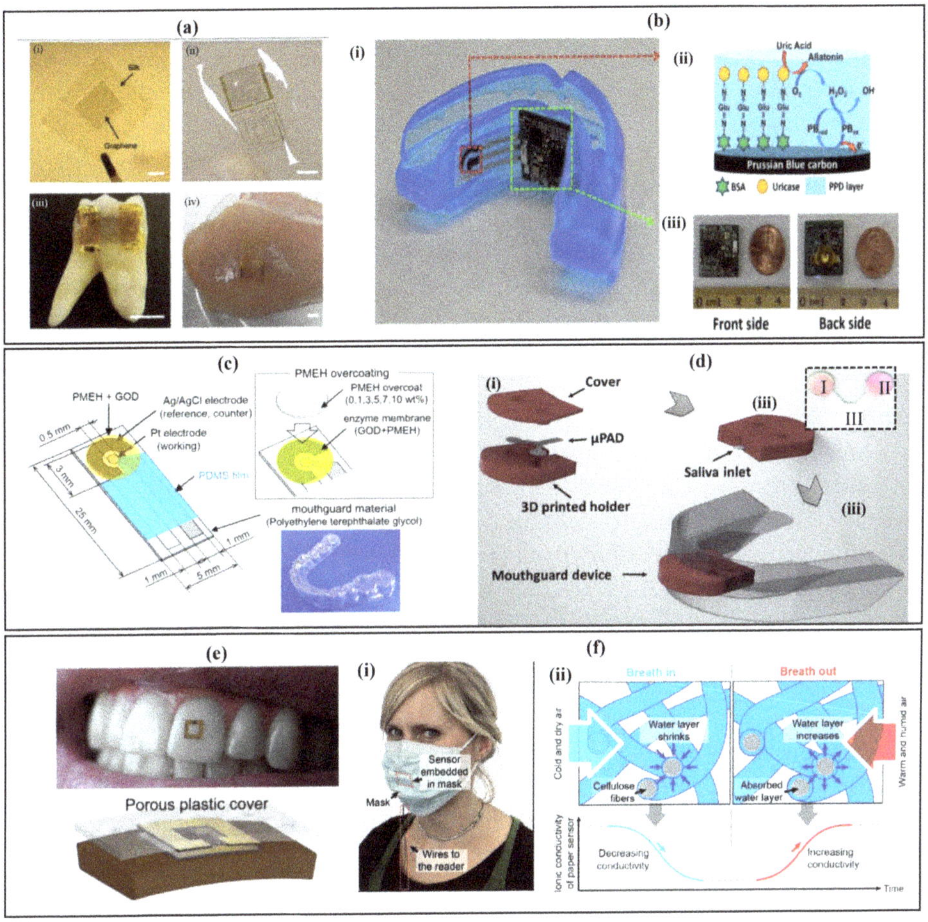

Figure 6.8. (a) A silk–graphene platform for monitoring of breath and saliva: (i) the graphere structure on silk; (ii) the sensing part with the signal transmission part; (iii) the structure on the tooth; (iv) the sturcutre on the hand for the bending force measurements. (Reproduced with permission from [13]. Copyright 2012 Nature.) (b) Images of a mouthguard biosensor: (i) photograph of the mouthguard biosensor integrated with a wireless amperometric circuit board, (ii) reagent layer of the chemically modified printed Prussian blue carbon working electrode containing uricase for the salivary uric acid (SUA) biosensor, (iii) photograph of the wireless amperometric circuit board: front (left) and back (right). (Reproduced with permission from [146]. Copyright 2015 Elsevier.) (c) Electrochemical sensing involving a mouthguard biosensor with a telemetry system for monitoring of saliva glucose. (Reproduced with permission from [147]. Copyright 2016 Elsevier.) (d) Schematic diagram of the wearable salivary glucose monitoring mouthguard assembly process. The μPAD is mounted within a 3D-printed polymeric holder with a saliva inlet (i) to isolate the contact between the water-soluble colorimetric reagents and the human mouth. Afterward this sensing module containing μPAD (ii) is installed into the silicone mouthguard, and (iii) the inset in the upper right corner shows an optical image of the salivary diagnostic μPAD, which is Y-shaped and composed of a sampling zone (III) and two detection zones for the colorimetric assays of glucose (I) and nitrite (II). (Reproduced with permission from [151]. Copyright 2019 Springer.) (e) The designed functional, RF-trilayer sensors for tooth-mounted, wireless monitoring of the oral cavity and food consumption. (Reproduced with permission from [149]. Copyright 2018 Wiley.) (f) A designed paper-based electrical respiration sensor on a mask: (i) one wearing mask; (ii) the principle showing how breath in and out can change the conductivities. (Reproduced with permission from [150]. Copyright 2016 Wiley.)

integrated with wireless amperometric circuit board, (ii) shows a schematic of the reagent layer of the chemically modified printed Prussian blue carbon working electrode containing uricase for the SUA biosensor, and (iii) shows photographs of the wireless amperometric circuit board with the front shown on the left and that back shown on the right. A mouthguard biosensor with telemetry was used to monitor saliva glucose [147] and is shown in figure 6.8(c).

Recently, Castro *et al* proposed a craft cutter printing colorimetric microfluidic paper-based device (μPAD) for the salivary diagnostics of periodontitis and diabetes mellitus through detection of the concentration of glucose and nitrite in human saliva [148]. They further describe a linear correlation relation observed between the measured colorimetric intensity and the concentration of glucose and nitrite in an effective analytical range. Additionally, for the development of a wearable glucose monitoring device, the μPAD was sealed within a 3D-printed holder and integrated into a silicone mouthguard (figure 6.8(d)), and then the assembled wearable device was tested through being used to monitor the glucose concentration in subject's saliva after ingesting some chocolate. The color intensity recorded on the μPAD declined constantly over time from 10 to 40 min due to the consumption of chocolate. The cost of this overall wearable platform is affordable for mass production, costing \$0.01 for the disposable μPAD and \$10 for the silicone mouthguard. Functional RF-trilayer sensors were developed for tooth-mounted, wireless monitoring of the oral cavity and food consumption [149], as shown in figure 6.8(e).

In addition, many proof-of-concept wearable integrated paper-based microfluidic technologies have been demonstrated. Compared to microfluidic devices constituting a PDMS architecture, the implementation of μPADs can be improved and broadened by exploiting the material behavior, such as the hygroscopicity of cellulose paper. Güder *et al* reported a paper-based electrical respiration wearable facemask to monitor individual breathing patterns [150], as shown in figure 6.8(f). A printed paper-based moisture sensing module was integrated inside a common medical mask and was able to convert the changes of humidity induced by the breathing inhalation and exhalation into an amplified electrical signal. The collected respiration data were be transmitted to a smartphone or personal computer through wireless connectivity and subsequently uploaded to the cloud network for professional healthcare monitoring.

6.4.4 Wearable microfluidic devices for drug delivery

Recently, Jiang *et al* presented a smart epidermal drug delivery system using a flexible passive pH-regulated pump to inhibit and heal wound infection [152]. As shown in figure 6.9(a), the pH-sensitive hydrogel (poly mAA-co-AAm gel) was inserted into the PDMS chamber. When the proliferating bacterial colonies on the wound bed caused a more alkaline pH (normally from 7.3–10) on the epidermis [153], the pH-sensitive gel would swell and compress the overhead drug reservoir across a flexible elastic membrane to slowly release antibiotic drug to the infected wound bed at a constant flow rate, <0.1 μl/min. For each single cell, the passive

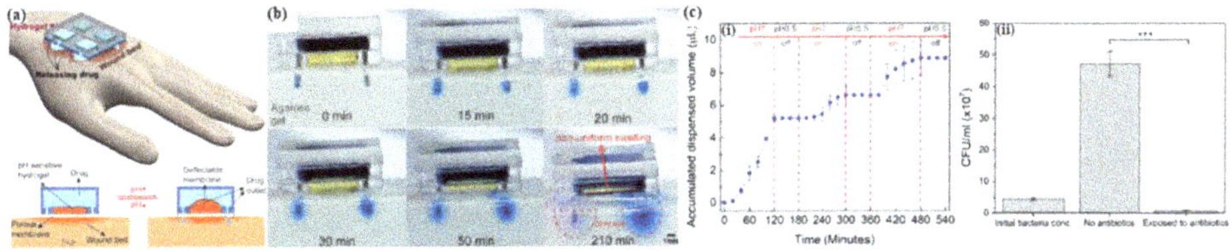

Figure 6.9. (a) Schematic of a smart pH-regulated drug delivery dermal system. The multilayer structure device is composed of two PDMS chambers separated by a flexible elastic PDMS membrane. The top chamber contains the antibiotic drug serving as the drug reservoir and the bottom chamber contains the pH-sensitive gel serving as the passive pump. (b) Photograph of the implementation of the pH-regulated dressing patch on a mimicked infected wound bed. The non-uniform pH distribution induces non-uniform gel swelling and each outlet of the patch can cover a 20 mm^2 wound area after 210 min. (c) (i) The 'on–off' characteristics of the pH-regulated passive pump. After three pH change cycles (from 7 to 5.5) over 540 min, the pump delivers 8.9 μl of the drug, and it is noted that the normal pH (5.5) cannot actuate the proposed drug delivery system. (ii) The investigation of the bacteria inhibitory effects of the dermal dressing patch. The initial concentration of *P. aeruginosa* is 4.5 × 107 CFU/ml. Through CFU assays of bacteria inhibition after 24 h, the sharp decline in the concentration of bacteria with the treatment by antibiotics is significant compared with the control group, which reveals the effective bacteria eliminating effects of the pH-regulated drug delivery system. (Reproduced with permission from [152]. Copyright 2019 The Royal Society of Chemistry.)

hydrogel actuator could offer an anticipated drug delivery performance with a 40 mm^2 wound coverage area (figure 6.9(b)). The layer-by-layer fabrication method enabled the dressing patch to fit the size and site of chronic wounds for a contour attachment. Simultaneously, the pH-sensitive pump promises 'on–off' characteristics for the real-time response of the drug delivery system to the pH fluctuation of infected wounds (figure 6.9(c) (i)). Figure 6.9(c) (ii) further shows that for *in vitro* simulating experiments, the concentration of bacteria without treatment of the dermal dressing patch was increased to over 4.7×10^8 CFU/ml, while with the patch including the antibiotics released by the proposed passive pump could significantly reduce the concentration of bacteria to 0.8×10^7 CFU/ml, a nearly 58 times decrease compared to the former. Bagherifard *et al* demonstrated a thermo-stimulating dermal patch for the application of sustained topical drug delivery [154]. To control the release rate, the drug was encapsulated within a thermo-responsive material (poly N-isopropylacrylamide (PNIPAM)) microparticle which was fabricated using droplet-based microfluidic technology in a high-throughput manner. Such an active wearable dermal patch has the potential to serve as an effective healing approach while avoiding inappropriate drug administration.

6.5 Conclusions and outlook

The chapter describes the fabrication, categories, and functionalization of wearable microfluidic sensors, starting with the explanation of the materials, fabrication, and characteristics considered for sensor devices. Then the different theories, designs, and categories were described. Finally, this chapter presented a few examples of wearable microfluidic sensors along with the different sensing parameters considered for monitoring purposes using body fluids. These show the perfect combination of the wearable field and microfluidic field, which the authors hope can be used broadly and help the development of both fields.

Acknowledgments

The authors would like to thank Dr Anindya Nag and Professor Subhas Mukhopadhyay for their long-term cooperation in developing wearable flexible devices. Professor Shilun Feng drafted and organized all of the materials. Professor Shaoxi Wang, Professor Yupan Wu, and Dr Anindya Nag provided the materials and helped build the structures. Dr Zhao, Mr Gaozhe Cai, and Dr Zizhen Ming provided support on the field of microfluidics and helped with the draft. Funded by the German Research Foundation (DFG, Deutsche Forschungsgemeinschaft) as part of Germany's Excellence Strategy—EXC 2050/1—Project ID 390696704—Cluster of Excellence 'Centre for Tactile Internet with Human-in-the-Loop' (CeTI) of Technische Universität Dresden.

References

[1] Del Ben F *et al* 2016 A method for detecting circulating tumor cells based on the measurement of single-cell metabolism in droplet-based microfluidics *Angew. Chem. Int. Ed. Engl.* **55** 8581–4

[2] Zilionis R *et al* 2017 Single-cell barcoding and sequencing using droplet microfluidics *Nat. Protoc.* **12** 44–73

[3] Haandbæk N, Bürgel S C, Rudolf F, Heer F and Hierlemann A 2016 Characterization of single yeast cell phenotypes using microfluidic impedance cytometry and optical imaging *ACS Sens.* **1** 1020–7

[4] Gawad S, Schild L and Renaud P H 2001 Micromachined impedance spectroscopy flow cytometer for cell analysis and particle sizing *Lab Chip* **1** 76–82

[5] Kost G J, Tran N K and Louie R F 2008 Point-of-care testing: principles, practice, and critical-emergency-disaster medicine *Encyclopedia of Analytical Chemistry: Applications, Theory and Instrumentation* (New York: Wiley)

[6] Chortos A, Liu J and Bao Z 2016 Pursuing prosthetic electronic skin *Nat. Mater.* **15** 937–50

[7] Lipomi D J *et al* 2011 Skin-like pressure and strain sensors based on transparent elastic films of carbon nanotubes *Nat. Nanotechnol.* **6** 788–92

[8] Kim D H *et al* 2011 Epidermal electronics *Science* **333** 838–43

[9] He S *et al* 2020 A comprehensive review of the use of sensors for food intake detection *Sens. Actuators,* A 112318

[10] Georgiou K *et al* 2018 Can wearable devices accurately measure heart rate variability? A systematic review *Folia Med.* **60** 7–20

[11] Zheng Y-L, Yan B P, Zhang Y-T and Poon C C 2014 An armband wearable device for overnight and cuff-less blood pressure measurement *IEEE Trans. Biomed. Eng.* **61** 2179–86

[12] Harrison D J, Manz A, Fan Z, Luedi H and Widmer H M 1992 Capillary electrophoresis and sample injection systems integrated on a planar glass chip *Anal. Chem.* **64** 1926–32

[13] Mannoor M S *et al* 2012 Graphene-based wireless bacteria detection on tooth enamel *Nat. Commun.* **3** 763

[14] Windmiller J R and Wang J 2013 Wearable electrochemical sensors and biosensors: a review *Electroanalysis* **25** 29–46

[15] Stoppa M and Chiolerio A 2014 Wearable electronics and smart textiles: a critical review *Sensors* **14** 11957–92
[16] Liu H, Li Q, Wang L, Mao Y and Wu C 2016 Effect of SnO_2 and Sb-doped SnO_2 on the structure and electrical conductivity of epichlorohydrin rubber *Polym. Compos.* **8** 2411–6
[17] Farhat O *et al* 2015 Growth of vertically aligned ZnO nanorods on Teflon as a novel substrate for low-power flexible light sensors *Appl. Phys.* A **119** 1197–201
[18] Comina G, Suska A and Filippini D 2014 PDMS lab-on-a-chip fabrication using 3D printed templates *Lab Chip* **14** 424–30
[19] El-Saftawy A, Elfalaky A, Ragheb M and Zakhary S 2014 Electron beam induced surface modifications of PET film *Radiat. Phys. Chem.* **102** 96–102
[20] Huang X, Pu Z, Feng M, Tong L and Liu X 2013 $BaTiO_3$@MWCNTs core/shell nanotubes embedded PEN nanocomposite films with high thermal stability and high permittivity *Mater. Lett.* **96** 139–42
[21] Chameswary J and Sebastian M T 2014 Development of butyl rubber–rutile composites for flexible microwave substrate applications *Ceram. Int.* **40** 7439–48
[22] Kim N *et al* 2014 Highly conductive PEDOT:PSS nanofibrils induced by solution-processed crystallization *Adv. Mater.* **26** 2268–72
[23] Cong H-P, Ren X-C, Wang P and Yu S-H 2013 Flexible graphene–polyaniline composite paper for high-performance supercapacitor *Energy Environ. Sci.* **6** 1185–91
[24] Lou C *et al* 2016 Flexible graphene electrodes for prolonged dynamic ECG monitoring *Sensors* **16** 1833
[25] Jung S, Ji T and Varadan V K 2006 Point-of-care temperature and respiration monitoring sensors for smart fabric applications *Smart Mater. Struct.* **15** 1872
[26] Nag A, Mukhopadhyay S C and Kosel J 2016 Flexible carbon nanotube nanocomposite sensor for multiple physiological parameter monitoring *Sens. Actuators* A **251** 148–55
[27] Nag A, Afasrimanesh N, Feng S and Mukhopadhyay S C 2018 Strain induced graphite/ PDMS sensors for biomedical applications *Sens. Actuators* A **271** 257–69
[28] Shih W-P *et al* 2010 Flexible temperature sensor array based on a graphite–polydimethylsiloxane composite *Sensors* **10** 3597–610
[29] Feng S 2015 Investigations of paper-based lab-on-chips for on-diaper point-of-care screening of urinary tract infections *Master's thesis* Chinese Academy of Sciences
[30] Yager P *et al* 2006 Microfluidic diagnostic technologies for global public health *Nature* **442** 412
[31] Martinez A W, Phillips S T, Butte M J and Whitesides G M 2007 Patterned paper as a platform for inexpensive, low-volume, portable bioassays *Angew. Chem. Int. Ed.* **46** 1318–20
[32] Mu X *et al* 2015 A paper-based skin patch for the diagnostic screening of cystic fibrosis *Chem. Commun.* **51** 6365–8
[33] Curto V F *et al* 2012 Concept and development of an autonomous wearable micro-fluidic platform for real time pH sweat analysis *Sens. Actuators* B **175** 263–70
[34] Simorangkir R B, Feng S, Sayem A S, Esselle K P and Yang Y 2018 PDMS-embedded conductive fabric: a simple solution for fabricating PDMS-based wearable antennas with robust performance 2018 12th Int. Symp. on Medical Information and Communication Technology (ISMICT) (Piscataway, NJ: IEEE) pp 1–3
[35] Morris D *et al* 2009 Bio-sensing textile based patch with integrated optical detection system for sweat monitoring *Sens. Actuators* B **139** 231–6
[36] Mostafalu P *et al* 2016 A toolkit of thread-based microfluidics, sensors, and electronics for 3D tissue embedding for medical diagnostics *Microsyst. Nanoeng.* **2** 16039

[37] Xiao G *et al* 2019 A wearable, cotton thread/paper-based microfluidic device coupled with smartphone for sweat glucose sensing *Cellulose* **26** 4553–62
[38] Fedlheim D L and Foss C A 2001 *Metal Nanoparticles: Synthesis, Characterization, and Applications* (Boca Raton, FL: CRC Press)
[39] Endo M, Iijima S and Dresselhaus M S 2013 *Carbon Nanotubes* (Amsterdam: Elsevier)
[40] Cai J *et al* 2014 Graphene nanoribbon heterojunctions *Nat. Nanotechnol.* **9** 896–900
[41] Sharon M *et al* 1998 Spongy carbon nanobeads—a new material *Carbon* **36** 507–11
[42] Wang Z L 2013 *Metal and Semiconductor Nanowires: Nanowires and Nanobelts: Materials, Properties and Devices* vol 1 (Berlin: Springer)
[43] Saetia K *et al* 2014 Spray-layer-by-layer carbon nanotube/electrospun fiber electrodes for flexible chemiresistive sensor applications *Adv. Funct. Mater.* **24** 492–502
[44] Zhou Y *et al* 2014 Selective and sensitive colorimetric sensor of mercury (II) based on gold nanoparticles and 4-mercaptophenylboronic acid *Sens. Actuators* B **196** 106–11
[45] McFarland A D and Van Duyne R P 2003 Single silver nanoparticles as real-time optical sensors with zeptomole sensitivity *Nano Lett.* **3** 1057–62
[46] Datta S S *et al* 2014 25th anniversary article: double emulsion templated solid microcapsules: mechanics and controlled release *Adv. Mater.* **26** 2205–18
[47] Taccola S *et al* 2013 Characterization of free-standing PEDOT:PSS/iron oxide nanoparticle composite thin films and application as conformable humidity sensors *ACS Appl. Mater. Interfaces* **5** 6324–32
[48] Zhang Y *et al* 2013 Electrochemical deposition of nickel nanoparticles on reduced graphene oxide film for nonenzymatic glucose sensing *Electroanalysis* **25** 959–66
[49] Popczun E J, Read C G, Roske C W, Lewis N S and Schaak R E 2014 Highly active electrocatalysis of the hydrogen evolution reaction by cobalt phosphide nanoparticles *Angew. Chem.* **126** 5531–4
[50] Asad M, Sheikhi M H, Pourfath M and Moradi M 2015 High sensitive and selective flexible H_2S gas sensors based on Cu nanoparticle decorated SWCNTs *Sens. Actuators* B **210** 1–8
[51] Tiwari P M, Vig K, Dennis V A and Singh S R 2011 Functionalized gold nanoparticles and their biomedical applications *Nanomaterials* **1** 31–63
[52] Wang X *et al* 2010 Bandgap-like strong fluorescence in functionalized carbon nanoparticles *Angew. Chem.* **122** 5438–42
[53] Mochalin V N and Gogotsi Y 2009 Wet chemistry route to hydrophobic blue fluorescent nanodiamond *JACS* **131** 4594–5
[54] Cao L *et al* 2007 Carbon dots for multiphoton bioimaging *JACS* **129** 11318–9
[55] Guo Z, Pereira T, Choi O, Wang Y and Hahn H T 2006 Surface functionalized alumina nanoparticle filled polymeric nanocomposites with enhanced mechanical properties *J. Mater. Chem.* **16** 2800–8
[56] Amstad E, Textor M and Reimhult E 2011 Stabilization and functionalization of iron oxide nanoparticles for biomedical applications *Nanoscale* **3** 2819–43
[57] Hu S-L *et al* 2009 One-step synthesis of fluorescent carbon nanoparticles by laser irradiation *J. Mater. Chem.* **19** 484–8
[58] Bae S *et al* 2010 Roll-to-roll production of 30-inch graphene films for transparent electrodes *Nat. Nanotechnol.* **5** 574
[59] Lee S H, Bae K H, Kim S H, Lee K R and Park T G 2008 Amine-functionalized gold nanoparticles as non-cytotoxic and efficient intracellular siRNA delivery carriers *Int. J. Pharm.* **364** 94–101

[60] Lipka J *et al* 2010 Biodistribution of PEG-modified gold nanoparticles following intratracheal instillation and intravenous injection *Biomaterials* **31** 6574–81
[61] Cho W-S *et al* 2010 Size-dependent tissue kinetics of PEG-coated gold nanoparticles *Toxicol. Appl. Pharmacol.* **245** 116–23
[62] Wangoo N, Bhasin K, Mehta S and Suri C R 2008 Synthesis and capping of water-dispersed gold nanoparticles by an amino acid: bioconjugation and binding studies *J. Colloid Interface Sci.* **323** 247–54
[63] Javier D J, Nitin N, Levy M, Ellington A and Richards-Kortum R 2008 Aptamer-targeted gold nanoparticles as molecular-specific contrast agents for reflectance imaging *Bioconjugate Chem.* **19** 1309–12
[64] HyeáJang H and SuáHan M 2010 A functionalized gold nanoparticles-assisted universal carrier for antisense DNA *Chem. Commun.* **46** 4151–3
[65] Hu W, Zhang S N, Niu X, Liu C and Pei Q 2014 An aluminum nanoparticle–acrylate copolymer nanocomposite as a dielectric elastomer with a high dielectric constant *J. Mater. Chem.* C **2** 1658–66
[66] Herzer N, Hoeppener S and Schubert U S 2010 Fabrication of patterned silane based self-assembled monolayers by photolithography and surface reactions on silicon-oxide substrates *Chem. Commun.* **46** 5634–52
[67] Khumpuang S, Maekawa H and Hara S 2013 Photolithography for minimal fab system *IEEJ Trans. Sens. Micromach.* **133** 272–7
[68] Harris D J, Hu H, Conrad J C and Lewis J A 2007 Patterning colloidal films via evaporative lithography *Phys. Rev. Lett.* **98** 148301
[69] Nag A, Zia A I, Li X, Mukhopadhyay S C and Kosel J 2016 Novel sensing approach for LPG leakage detection: part I—operating mechanism and preliminary results *IEEE Sensors J.* **16** 996–1003
[70] Lee H S and Yoon J-B 2005 A simple and effective lift-off with positive photoresist *J. Micromech. Microeng.* **15** 2136
[71] Sugiura S *et al* 2007 Photoresponsive polymer gel microvalves controlled by local light irradiation *Sens. Actuators* A **140** 176–84
[72] Zhuang D and Edgar J 2005 Wet etching of GaN, AlN, and SiC: a review *Mater. Sci. Eng.* R **48** 1–46
[73] Powell R A 2012 *Dry Etching for Microelectronics* (Amsterdam: Elsevier)
[74] Nag A, Mukhopadhyay S and Kosel J 2016 Transparent biocompatible sensor patches for touch sensitive prosthetic limbs *10th Int. Conf. on* Sensing Technology (ICST) (Piscataway, NJ: IEEE) pp 1–6
[75] Bariya M *et al* 2018 Roll-to-roll gravure printed electrochemical sensors for wearable and medical devices *ACS Nano* **12** 6978–87
[76] Ohta S *et al* 2013 All-solid-state lithium ion battery using garnet-type oxide and Li_3BO_3 solid electrolytes fabricated by screen-printing *J. Power Sources* **238** 53–6
[77] Khan S, Lorenzelli L and Dahiya R 2014 Bendable piezoresistive sensors by screen printing MWCNT/PDMS composites on flexible substrates *10th Conf. on* PhD Research in Microelectronics and Electronics (PRIME) (Piscataway, NJ: IEEE) pp 1–4
[78] Simorangkir R B, Yang Y, Esselle K P and Zeb B A *et al* 2017 A method to realize robust flexible electronically tunable antennas using polymer-embedded conductive fabric *IEEE Trans. Antennas Propag.* **66** 50–8

[79] Merilampi S, Laine-Ma T and Ruuskanen P 2009 The characterization of electrically conductive silver ink patterns on flexible substrates *Microelectron. Reliab.* **49** 782–90
[80] Ping J, Wu J, Wang Y and Ying Y 2012 Simultaneous determination of ascorbic acid, dopamine and uric acid using high-performance screen-printed graphene electrode *Biosens. Bioelectron.* **34** 70–6
[81] Pardo D A, Jabbour G E and Peyghambarian N 2000 Application of screen printing in the fabrication of organic light-emitting devices *Adv. Mater.* **12** 1249–52
[82] Locher I and Tröster G 2007 Screen-printed textile transmission lines *Textile Res. J.* **77** 837–42
[83] Siegel A C *et al* 2010 Foldable printed circuit boards on paper substrates *Adv. Funct. Mater.* **20** 28–35
[84] Nag A, Mukhopadhyay S C and Kosel J 2016 Tactile sensing from laser-ablated metallized PET films *IEEE Sensors J.* **17** 7–13
[85] Barron J A, Ringeisen B R, Kim H, Spargo B J and Chrisey D B 2004 Application of laser printing to mammalian cells *Thin Solid Films* **453** 383–7
[86] Birnbaum A J, Kim H, Charipar N A and Piqué A 2010 Laser printing of multi-layered polymer/metal heterostructures for electronic and MEMS devices *Appl. Phys.* A **99** 711–6
[87] Hossain S Z *et al* 2009 Development of a bioactive paper sensor for detection of neurotoxins using piezoelectric inkjet printing of sol–gel-derived bioinks *Anal. Chem.* **81** 5474–83
[88] Homenick C M *et al* 2016 Fully printed and encapsulated SWCNT-based thin film transistors via a combination of R2R gravure and inkjet printing *ACS Appl. Mater. Interfaces* **8** 27900–10
[89] Singh M, Haverinen H M, Dhagat P and Jabbour G E 2010 Inkjet printing—process and its applications *Adv. Mater.* **22** 673–85
[90] Shen W, Zhang X, Huang Q, Xu Q and Song W 2014 Preparation of solid silver nanoparticles for inkjet printed flexible electronics with high conductivity *Nanoscale* **6** 1622–8
[91] Wang S *et al* 2015 Inkjet printing of conductive patterns and supercapacitors using a multi-walled carbon nanotube/Ag nanoparticle based ink *J. Mater. Chem.* A **3** 2407–13
[92] Farraj Y, Grouchko M and Magdassi S 2015 Self-reduction of a copper complex MOD ink for inkjet printing conductive patterns on plastics *Chem. Commun.* **51** 1587–90
[93] Zhu H, Lin X, Su Y, Dong H and Wu J 2015 Screen-printed microfluidic dielectrophoresis chip for cell separation *Biosens. Bioelectron.* **63** 371–8
[94] Eswaraiah V, Balasubramaniam K and Ramaprabhu S 2011 Functionalized graphene reinforced thermoplastic nanocomposites as strain sensors in structural health monitoring *J. Mater. Chem.* **21** 12626–8
[95] Lin L *et al* 2013 Transparent flexible nanogenerator as self-powered sensor for transportation monitoring *Nano Energy* **2** 75–81
[96] Latessa G, Brunetti F, Reale A, Saggio G and Di Carlo A 2009 Piezoresistive behaviour of flexible PEDOT:PSS based sensors *Sens. Actuators* B **139** 304–9
[97] Bandodkar A J *et al* 2015 Tattoo-based noninvasive glucose monitoring: a proof-of-concept study *Anal. Chem.* **87** 394–8
[98] Salinas H R *et al* 2020 A colorful approach towards developing new nano-based imaging contrast agents for improved cancer detection *Biomater. Sci.* **9** 482–95

[99] Takamatsu S *et al* 2011 Flexible fabric keyboard with conductive polymer-coated fibers *IEEE Sensors* (Piscataway, NJ: IEEE) pp 659–62

[100] Chen N, Engel J, Pandya S and Liu C 2006 Flexible skin with two-axis bending capability made using weaving-by-lithography fabrication method *19th IEEE Int. Conf. on* Micro Electro Mechanical Systems *(Istanbul)* (Piscataway, NJ: IEEE) pp 330–3

[101] Lee J B and Subramanian V 2005 Weave patterned organic transistors on fiber for e-textiles *IEEE Trans. Electron Devices* **52** 269–75

[102] Cherenack K, Zysset C, Kinkeldei T, Münzenrieder N and Tröster G 2010 Woven electronic fibers with sensing and display functions for smart textiles *Adv. Mater.* **22** 5178–82

[103] Briedis U, Valisevskis A and Grecka M 2017 Development of a smart garment prototype with enuresis alarm using an embroidery-machine-based technique for the integration of electronic components *Procedia Comput. Sci.* **104** 369–74

[104] Hain R, Kähler C J and Tropea C 2007 Comparison of CCD, CMOS and intensified cameras *Exp. Fluids* **42** 403–11

[105] Nag A, Feng S, Mukhopadhyay S, Kosel J and Inglis D 2018 3D printed mould-based graphite/PDMS sensor for low-force applications *Sens. Actuators* A **280** 525–34

[106] He S *et al* 2020 Recent progress in 3D printed mold-based sensors *Sensors* **20** 703

[107] Nag A, Feng S, Afsarimanesh N, Mukhopadhyay S and Kosel J 2018 Development of novel Gold/PDMS sensors for medical applications 12th Int. Symp. on Medical Information and Communication Technology (ISMICT) (Piscataway, NJ: IEEE) pp 1–5

[108] Nag A, Feng S, Mukhopadhyay S and Kosel J 2018 Development of printed sensors for shoe sensing applications 12th Int. Symp. on Medical Information and Communication Technology (ISMICT) (Piscataway, NJ: IEEE) pp 1–6

[109] Yuan Z *et al* 2017 Transparent and flexible triboelectric sensing array for touch security applications *ACS Nano.* **11** 8364–9

[110] Zhang Y *et al* 2020 Skin-interfaced microfluidic devices with one-opening chambers and hydrophobic valves for sweat collection and analysis *Lab Chip* **20** 2635–45

[111] Wang M, Hu L and Xu C 2017 Recent advances in the design of polymeric microneedles for transdermal drug delivery and biosensing *Lab Chip* **17** 1373–87

[112] Lau S, Fei J, Liu H, Chen W and Liu R 2017 Multilayered pyramidal dissolving microneedle patches with flexible pedestals for improving effective drug delivery *J. Control. Release* **265** 113–9

[113] Feng S, Clement S, Zhu Y, Goldys E M and Inglis D W 2019 Microfabricated needle for hydrogen peroxide detection *RSC Adv.* **9** 18176–81

[114] Feng S *et al* 2017 A microfluidic needle for sampling and delivery of chemical signals by segmented flows *Appl. Phys. Lett.* **111** 183702

[115] Feng S, Shirani E and Inglis D W 2019 Droplets for sampling and transport of chemical signals in biosensing: a review *Biosensors* **9** 80

[116] Feng S, Nguyen M N and Inglis D W 2017 Microfluidic droplet extraction by hydrophilic membrane *Micromachines* **8** 331

[117] Feng S, Skelley A M, Anwer A G, Liu G and Inglis D W 2017 Maximizing particle concentration in deterministic lateral displacement arrays *Biomicrofluidics* **11** 024121

[118] Inglis D, Vernekar R, Krüger T and Feng S 2020 The fluidic resistance of an array of obstacles and a method for improving boundaries in deterministic lateral displacement arrays *Microfluid. Nanofluid.* **24** 1–8

[119] Bandodkar A J, Jeang W J, Ghaffari R and Rogers J A 2019 Wearable sensors for biochemical sweat analysis *Annu. Rev. Anal. Chem.* **12** 1–22

[120] Nag A, Alahi M E E, Feng S and Mukhopadhyay S C 2019 IoT-based sensing system for phosphate detection using graphite/PDMS sensors *Sens. Actuators,* A **286** 43–50

[121] Yuan Y *et al* 2018 Development of an internet of things based electrochemical microfluidic system for free calcium detection *Appl. Sci.* **8** 1357

[122] Liu G, Cao C, Ni S, Feng S and Wei H 2019 On-chip structure-switching aptamer-modified magnetic nanobeads for the continuous monitoring of interferon-gamma *ex vivo Microsyst. Nanoeng.* **5** 1–11

[123] Bandodkar A J *et al* 2019 Battery-free, skin-interfaced microfluidic/electronic systems for simultaneous electrochemical, colorimetric, and volumetric analysis of sweat *Sci. Adv.* **5** eaav3294

[124] Bandodkar A J *et al* 2019 Soft, skin-interfaced microfluidic systems with passive galvanic stopwatches for precise chronometric sampling of sweat *Adv. Mater.* 1902109

[125] Heikenfeld J 2016 Bioanalytical devices: technological leap for sweat sensing *Nature* **529** 475–6

[126] Choi J, Ghaffari R, Baker L B and Rogers J A 2018 Skin-interfaced systems for sweat collection and analytics *Sci. Adv.* **4** eaar3921

[127] Choi J *et al* 2017 Soft, skin-mounted microfluidic systems for measuring secretory fluidic pressures generated at the surface of the skin by eccrine sweat glands *Lab Chip* **17** 2572–80

[128] Martín A *et al* 2017 Epidermal microfluidic electrochemical detection system: enhanced sweat sampling and metabolite detection *ACS Sens.* **2** 1860–8

[129] Gao W *et al* 2016 Fully integrated wearable sensor arrays for multiplexed *in situ* perspiration analysis *Nature* **529** 509–14

[130] Emaminejad S *et al* 2017 Autonomous sweat extraction and analysis applied to cystic fibrosis and glucose monitoring using a fully integrated wearable platform *Proc. Natl Acad. Sci. USA* **114** 4625–30

[131] Kim J *et al* 2016 Noninvasive alcohol monitoring using a wearable tattoo-based iontophoretic-biosensing system *ACS Sens.* **1** 1011–9

[132] Yang Y *et al* 2019 A laser-engraved wearable sensor for sensitive detection of uric acid and tyrosine in sweat *Nat. Biotechnol.* **38** 217–24

[133] Feng S L and Dong T 2014 Applied technology in diaper-based UTI testing for elder people by using nitrite ion selective electrode *Appl. Mech. Mater.* **662** 225–28

[134] Feng S and Dong T 2014 Design and characterization of a lab-on-chip for continuous bioluminescent measurements of ATP *IEEE Int. Symp. on Medical Measurements and Applications* (Piscataway, NJ: IEEE) pp 1–4

[135] Feng S, Dong T and Yang Z 2014 Detection of urinary tract infections on lab-on-chip device by measuring photons emitted from ATP bioluminescence 36th Annual Int. Conf. of the IEEE Engineering in Medicine and Biology Society (Piscataway, NJ: IEEE) pp 3114–7

[136] Feng S, Roseng L E and Dong T 2015 Quantitative detection of *Escherichia coli* and measurement of urinary tract infection diagnosis possibility by use of a portable, handheld sensor IEEE Int. Symp. on Medical Measurements and Applications (Piscataway, NJ: IEEE) pp 586–9

[137] Chen C and Dong T 2015 Microfluidic paper-based analytical devices for colorimetric detection of urinary tract infection biomarkers on adult diapers 37th Annual Int. Conf. of the IEEE Engineering in Medicine and Biology Society (EMBC) (Piscataway, NJ: IEEE) pp 5892–5

[138] Chen C, Wu Y and Dong T 2014 Dipsticks integrated on smart diapers for colorimetric analysis of urinary tract infections in the field Proc. of the 16th Int. Conf. on Mechatronics (Piscataway, NJ: IEEE) pp 423–7

[139] Song X 2008 Lateral flow device for attachment to an absorbent article *Australian patent* AU2008243894B2

[140] Karlsen H, Dong T and Suo Z 2018 A diaper pad for diaper-based urine collection and colorimetric screening of urinary biomarkers *Ann. Biomed. Eng.* **46** 717–25

[141] Zhou J and Dong T 2018 Design of a wearable device for real-time screening of urinary tract infection and kidney disease based on smartphone *Analyst* **143** 2812–8

[142] Birkhahn M *et al* 2013 A novel precision-engineered microfiltration device for capture and characterisation of bladder cancer cells in urine *Eur. J. Cancer* **49** 3159–68

[143] Liu J and Duan Y 2012 Saliva: a potential media for disease diagnostics and monitoring *Oral Oncol.* **48** 569–77

[144] Bhakta S A, Borba R, Taba M Jr, Garcia C D and Carrilho E 2014 Determination of nitrite in saliva using microfluidic paper-based analytical devices *Anal. Chim. Acta* **809** 117–22

[145] Kim J *et al* 2014 Non-invasive mouthguard biosensor for continuous salivary monitoring of metabolites *Analyst* **139** 1632–6

[146] Kim J *et al* 2015 Wearable salivary uric acid mouthguard biosensor with integrated wireless electronics *Biosens. Bioelectron.* **74** 1061–8

[147] Arakawa T *et al* 2016 Mouthguard biosensor with telemetry system for monitoring of saliva glucose: a novel cavitas sensor *Biosens. Bioelectron.* **84** 106–11

[148] de Castro L F *et al* 2019 Salivary diagnostics on paper microfluidic devices and their use as wearable sensors for glucose monitoring *Anal. Bioanal.Chem.* **411** 4919–28

[149] Tseng P, Napier B, Garbarini L, Kaplan D L and Omenetto F G 2018 Functional, RF-trilayer sensors for tooth-mounted, wireless monitoring of the oral cavity and food consumption *Adv. Mater.* **30** 1703257

[150] Güder F *et al* 2016 Paper-based electrical respiration sensor *Angew. Chem. Int. Ed.* **55** 5727–32

[151] de Castro L F *et al* 2019 Salivary diagnostics on paper microfluidic devices and their use as wearable sensors for glucose monitoring *Anal. Bioanal.Chem.* **411** 4919–28

[152] Jiang H, Ochoa M, Waimin J F, Rahimi R and Ziaie B 2019 A pH-regulated drug delivery dermal patch for targeting infected regions in chronic wounds *Lab Chip* **19** 2265–74

[153] Schneider L A, Korber A, Grabbe S and Dissemond J 2007 Influence of pH on wound-healing: a new perspective for wound-therapy? *Arch. Dermatol. Res.* **298** 413–20

[154] Bagherifard S *et al* 2016 Dermal patch with integrated flexible heater for on demand drug delivery *Adv. Healthcare Mater.* **5** 175–84

Chapter 7

Single-walled carbon nanotubes for flexible and printed electronics

S Bhadra

Flexible and printed electronic devices are of interest in many application areas such as displays, solar cells, sensors, circuits, and RFID tags. Single-walled carbon nanotubes (SWNTs) are an emerging and one of the most promising classes of material in the field of flexible and printed electronics because of their high intrinsic carrier mobility, conductivity, and mechanical flexibility. Random networks or thin films of SWNTs demonstrate electrical, mechanical, and optical properties very close to the properties of a single SWNT. Since thin films or random networks are suitable for integration into devices in a scalable manner, unlike single SWNTs, they have been used as channel materials or conductive layers for flexible and printed devices. This chapter reviews the recent developments in assembly techniques, including printing processes, for forming such films or networks of SWNTs. The applications of SWNT thin films or networks as channel semiconductors in flexible and printed thin film transistors and integrated circuits and as conductors in sensors, solar cells, and supercapacitors are discussed. There remain are a number of challenges to solve to improve the processing and performance of SWNT based flexible and printed devices. These challenges along with potential research work that can be performed with SWNTs are also presented.

7.1 Introduction

Flexible electronic devices can radically change people's perspectives on electronics [1]. They are of research interest and under great demand in novel applications such as flexible circuits, conformable RFID tags, flexible displays, electronic paper, wearable health monitors, and implantable medical devices [2–9]. The development of flexible and mechanically durable materials and substrates, novel processing and manufacturing methods, deformable electrode and circuits, and system integration are the crucial steps toward advancement of flexible devices. One common strategy

doi:10.1088/978-0-7503-3439-6ch7 © IOP Publishing Ltd 2021

to achieve flexible devices is using naturally flexible polymer substrates in combination with emerging nanostructured and molecular materials deposited by different techniques. The development of large-area flexible electronics relies on the improvements in manufacturing techniques. The microfabrication techniques used in the conventional electronics industry require high temperature processes, and have high operation cost, restrictions in substrate material, limitations in sample size, and wastage of material. Therefore many of these techniques are not compatible with flexible electronics. Printing is an alternative promising fabrication method for flexible devices. It has the advantages of low cost, low annealing temperature, compatibility of flexible electronics, minimal material waste, and better printability of polymer materials. Substrate of any material and size can be used in printed electronics. Due to low material waste this technique offers cost-effective manufacturing [10, 11].

The development of printed and flexible electronics relies heavily on advancements in materials. Among the different materials carbon nanotubes (CNTs) have attracted widespread interest for high-performance flexible electronics due to their high carrier mobility, high electrical and thermal conductivities, mechanical flexibility, high aspect ratio, high melting point, and stability [12]. They are also one of the lightest and strongest known materials in the world. SWNTs are a special form of CNTs which only contain one layer. Different to those CNTs with multiple layers which are semi-conductive, SWNTs can be either semiconductors or metallic depending on its structure. After years of research, the detailed molecular structure of SWNTs is now clear. A pure SWNT can be viewed as a rolled-up tubular shell of graphene sheet which is made up of benzene-type hexagonal rings of carbon atoms [13], as shown in figure 7.1. The schematic representation of the formation of an SWNT after unfolding the hollow tubular shell and laying it flat is displayed in figure 7.2. The SWNT is defined by a chiral vector, $C = n\mathbf{a}_1 + m\mathbf{a}_2$, which is an OA vector represented in the unit vectors ($\mathbf{a}_1$ and $\mathbf{a}_2$) of the graphene lattice [14]. Depending on the magnitude of the unit vector (integers n and m), there are three types SWNTs. If either m or n is equal to 0, the corresponding CNT is called a zigzag nanotube. If m is equal to n, the CNT is called an armchair nanotube. All other configurations are called chiral nanotubes. Figure 7.2 also presents schematic representations of the three types of CNTs according to their related chiral vectors. The magnitude of the translational vector T depends on the chirality of the CNT and is equal to the length of a unit cell of the CNT. The vector T is also orthogonal to the

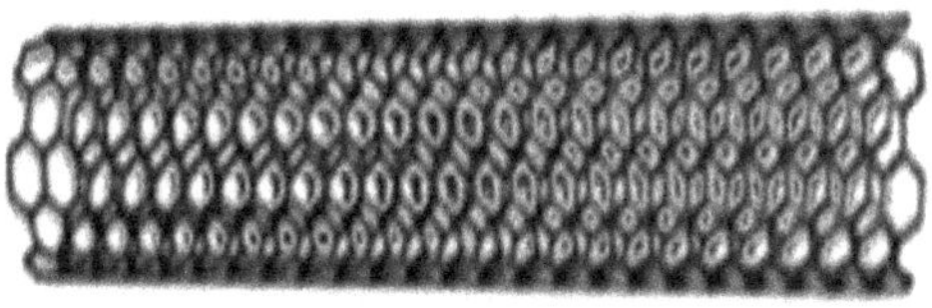

Figure 7.1. Molecular representation of an SWNT. (Reproduced from [13]. Copyright 2006 Oxford University Press.)

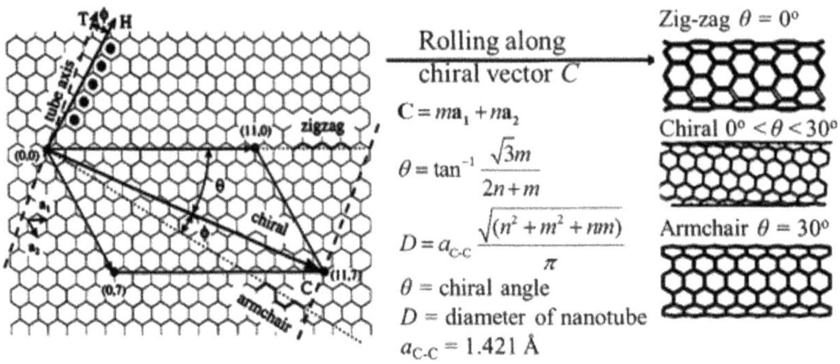

Figure 7.2. Schematic representation of the formation of SWNTs and three types of CNTs. (Reproduced with permission from [14]. Copyright 2014 Springer.)

chiral vector C. Depending on the indices, SWNTs can have different electrical attributes. The SWNT index (n, m) with 3 multiples of $n - m$ are metallic, and others are semiconducting SWNTs. [15].

Many groups have researched flexible and printed devices and circuits using solution-processed SWNTs [16–23]. Many of them have demonstrated that individual SWNTs show excellent electronic properties. For example, the current carrying capability of m-SWNTs can be up to 10^9 A cm^{-2} (higher than for copper or aluminium) and the mobility of sc-SWNTs can be up to 10^4 cm^2 V^{-1} s^{-1} [24–26]. Although devices composed of individual SWNTs have shown excellent performance, the deposition of individual SWNTs at a desired location is very difficult and has proved to be impractical for commercial deposition processes. Additionally, devices fabricated with individual SWNTs show poor uniformity and reproducibility as it is difficult to synthesis individual SWNTs with homogeneous properties reliably [12, 27]. Random networks and thin film based SWNT devices are the most practical approaches for commercial purposes as they offer feasible processing and fabrication, and reproducible performance [12, 28, 29]. The lateral deformation of the curvy and entangled SWNTs can withstand large strain and it is easy to make ink with them. Therefore SWNT networks are appropriate for flexible and printed electronics. SWNT networks or thin films have been used as channel materials and/or conducting electrodes in various types of flexible and printed devices such as thin film transistors (TFTs) and logic circuits [30], sensors [31–33], organic light-emitting diodes (OLEDs) [34, 35], touchscreens [36], supercapacitors [37, 38], and so on.

In this chapter, we review the preparation of SWNT networks and thin films, and their potential applications for flexible and printed electronics. We present the methods to produce networks or thin films of SWNT using both dry and solution-based processes and strategies to align and sort SWNTs. Among the different dry and solution-based processes we place the most emphasis on chemical vapour deposition (CVD) growth and printing processes. In most recent flexible and printed electronic devices, SWNT networks and films are used as either electrodes or channel materials. SWNT films and networks as channel materials have been reviewed extensively in the literature [12, 39, 40]. We focus on the application of

SWNT networks and films as channel materials as well as electrode materials in different devices. Finally, we conclude the chapter with current remaining challenges and opportunities for future work with SWNTs.

7.2 The preparation of SWNT networks and thin films

SWNT films and networks as semiconductor materials in TFTs or conductors should have the necessary electrical performance and their fabrication process should be simple, reliable, and scalable to large areas. Only a few techniques are currently available for preparing SWNT networks and thin films, as presented figure 7.3 [40]. In general, the fabrication of thin films or networks of SWNTs is classified in two categories: dry processes and solution processes [10, 11]. Among the different dry processes, direct CVD growth and dry drawing from vertically aligned CNT arrays are commonly used [39]. On the other hand, several methods such as

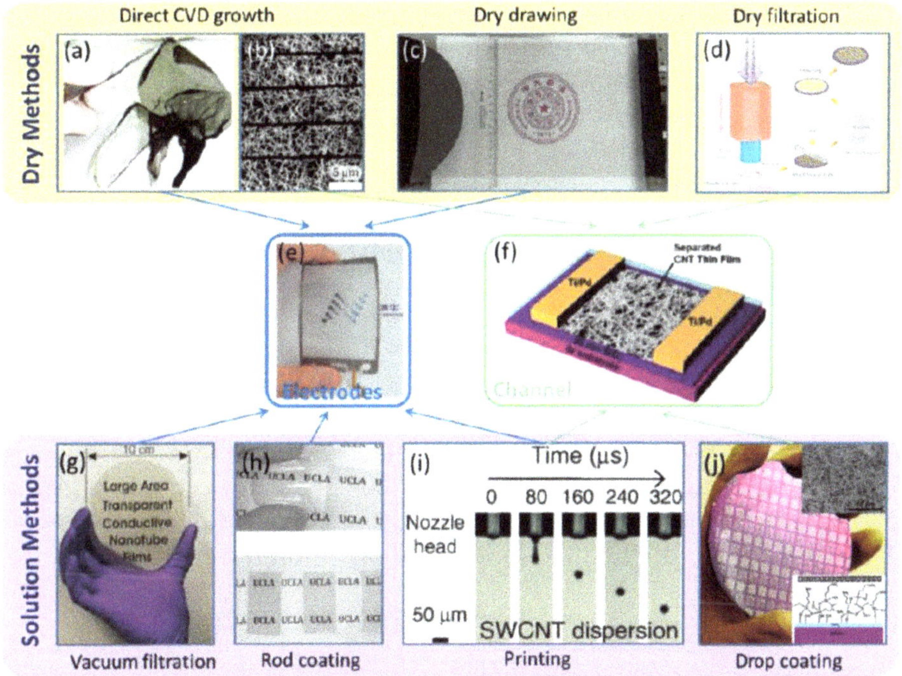

Figure 7.3. Different methods for preparing carbon nanotube networks and thin films and demonstrations of their electronic applications. (a)–(d) Dry processes: (a) Photograph of CVD directly grown freestanding SWNT thin film. (b) SEM image of a CVD-grown SWNT network patterned into strips for the active layer of a flexible TFT. (c) Photograph of of multi-walled carbon nanotube (MWNT) thin films obtained using a dry drawing process. (d) Schematic of the dry filtration process for collecting SWNT networks for flexible TFTs. (e), (f) The use of SWNT networks for flexible electrodes and active layers for flexible TFTs, respectively. (g)–(j) Solution processes: (g) Photograph of SWNT thin film fabricated using vacuum filtration. (h) Photograph of SWNT films with different transmittance fabricated by rod coating. (i) Time-dependent snapshot of an SWNT ink droplet during inkjet printing. (j) Photograph of wafer-scale SWNT TFTs fabricated using drop-coated SWNT networks. Reproduced from [40] CC BY 4.0..

rod coating [41], drop coating [42], and printing [16–18] have been reported as solution process methods. Among the different dry and solution process methods we will discuss the CVD growth and the printing process methods in detail.

7.2.1 Growth, alignment, and purification of CVD-grown SWNTs

1. *Growth*: For CVD growth, a substrate with catalyst particles is placed in a furnace (generally at temperatures >800 °C). Gaseous or volatile compounds of carbon are supplied and decomposed in the chamber, and catalyzed by catalyst particles on the substrate. The catalyst particles serve as nucleation sites for the initiation of SWNT growth. As a result, the SWNTs are grown directly on the surface of the substrate [43]. The SWNTs grown in the CVD process can be made as long as hundreds of microns and do not contain bundles. Therefore, CVD-grown SWNTs have better electrical properties than SWNTs grown by other techniques. Figure 7.4(a) shows an SEM image of a CVD-grown SWNT film on a quartz substrate. The film achieved a mobility in the range of 16–22 $cm^2\ V^{-1}\ s^{-1}$ and the on–off ratio ranged from 10° to 10^4 depending on the channel length. The density of SWNTs in the film was ~5 to 7 SWNTs per micrometer [39].

   Two methods are followed to obtain the catalyst on a substrate to grow SWNTs with the CVD technique. A catalyst solution consisting of ferritin or metal oxides is spin coated on the substrate. Then the solution is reduced at elevated temperatures under hydrogen gas and a random network of SWNT film is produced. The density of the film depends on the type of catalyst solution, the concentration of the solution, and the reduction period. In the dry process, the catalyst is deposited by evaporating 0.2–0.5 nm of metal using e-beam evaporation directly on the surface [44–46].
2. *Alignment*: CVD-grown SWNTs can be aligned easily via directed growth methods using laser ablation [47], gas flow [48], an electric field [48], and surface-oriented growth [49]. The alignment of SWNTs produces denser film

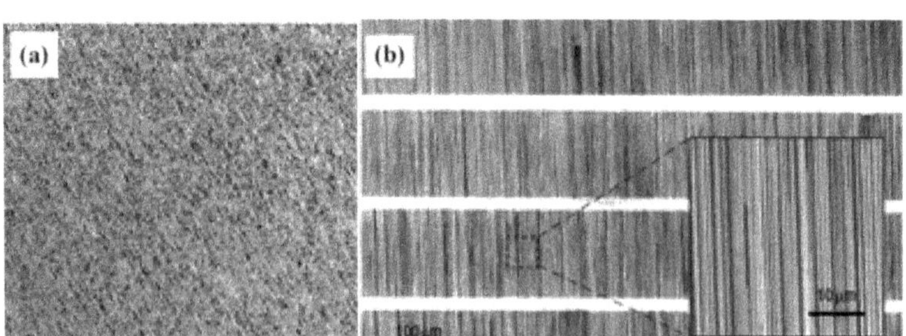

Figure 7.4. (a) A random network of CVD-grown SWNT film. It was grown on quartz substrate with e-beam evaporated Fe particles used as catalysts. (b) Aligned SWNT film fabricated by surface-oriented growth. (Reproduced with permission from [39]. Copyright 2013 The Royal Society of Chemistry.)

than a random network of CNTs and enhances the electrical properties of the SWNT thin films. Generally, aligned SWNT film devices have a mobility of about 10–40 times higher than random network SWNT film devices [39]. Figure 7.4(b) shows the SEM image of an aligned SWNT film on a quartz substrate. The aligned film was formed using surface-oriented growth [39]. The SWNTs were well aligned with a density of approximately five SWNTs per micrometer and the film achieved a mobility of 1100 $cm^2\ V^{-1}\ s^{-1}$ (the channel width set as the width of the electrodes), which is much larger than what can be achieved using random network SWNT film in figure 7.4(a). The general densities of surface-oriented grown SWNTs are between one and ten SWNTs per micrometers. One group achieved a density of 22 SWNTs per micrometers [50] and another group achieved a density of 45 SWNTs per micrometer [51] by using successive growth.

3. *Purification and film transfer*: The biggest disadvantage of CVD-grown SWNTs is the lack of homogeneity in the films as the films contain mixtures of SWNTs with different chirality. Assuming that all chiralities grow with equal probability, one third of the grown SWNTs are m-SWNTs and two thirds are semiconducting SWNTs (sc-SWNT) in a random network. Such impure films cannot be used as active materials for transistors, as a percolation path of m-SWNTs can be created from source to drain electrodes in the presence of m-SWNTs. As a result, the device will be shorted. Therefore, m-SWNTs should be eliminated by special processes such as electrical breakdown, dry filtration, electrophoresis, dielectrophoresis, or stripe patterning before using CVD-grown SWNTs as the active layer of transistors [39]. Dielectrophersis is one of the most popular approaches for electronic sorting of SWNTs and is based on an AC electric field. The SWNTs become sorted according to the differences of their dielectric constants. Figure 7.5 shows an experimental set-up for the dielectrophoresis of SWNT solution using a microelectrode array. In this set-up, a drop of SWNT solution is placed on a microelectrode array. When an AC electric field is applied, the m-SWNTs are polarized much faster than sc-SWNTs. This results in preferential polarization in m-SWNTs and the polarized SWNTs are selectively deposited between the microelectrodes. By increasing the frequency of the AC electric field the sorting is accelerated. Sorting with microelectrode arrays has limited throughput. The throughput of dielectrophoresis can be improved by using larger electrodes or dielectrophretic field-flow fractionation [52].

As the CVD process requires high temperature processing, SWNT films cannot be directly grown on the flexible substrates using CVD process. Normally SWNT films are grown on quartz and then transferred to the flexible substrate. There are several methods to transfer CVD-grown SWNTs onto a flexible substrate. Figure 7.6 shows a diagram of a common method [53]. In this method a metal layer such as gold is deposited on the SWNT film grown on quartz. Then a transfer substrate such as a PDMS stamp or heat

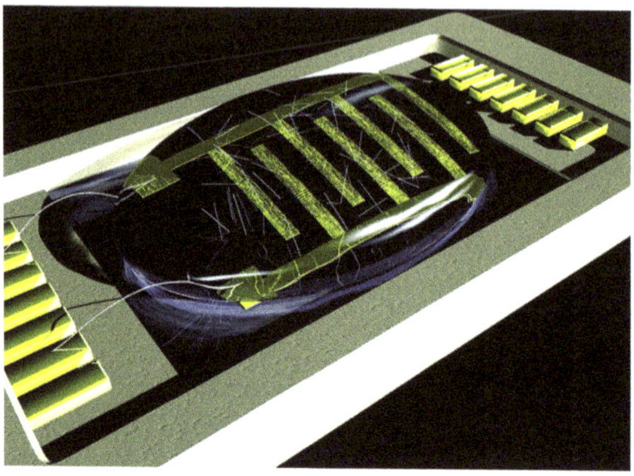

Figure 7.5. Schematic of the experimental set-up for dielectrophoretic sorting of SWNTs. (Reproduced with permission from [52]. Copyright 2003 AAAS.)

release tape is placed on the metal layer. When the PDMS or the release film is peeled off it holds the gold and SWNT layer. The PDMS stamp or the release film is then pressed on the receiving flexible substrate and the gold and SWNT layers are transferred onto the flexible substrate. On the flexible substrate the metal layer is removed using chemical etchants. CVD is believed to be the most suitable CNT synthesis method in terms of product purity and large-scale production. It has the advantages of mild operation conditions, low cost, and controllable processes [22].

7.2.2 Deposition, alignment, and purification of solution-processed SWNTs

Processes such as arc discharge, high pressure carbon monoxide, and laser ablation synthesize SWNTs in powder form. These powders can be purified and dissolved in solution to deposit on flexible substrates with well-established solution deposition techniques such as rod coating [41], drop coating [42], and inkjet/aerosol printing [16–18]. It is important to sort the SWNTs by electric type during or before the deposition, in particular removing m-SWNTs from sc-SWNTs to deposit transistor channel material. The solution processing required successful dissolution of SWNTs in aqueous solution or in organic solvents. Most of the time that is achieved with the assistance of certain types of surfactants [54, 55]. SWNT thin films obtained by rod or drop coating have been used as flexible, stretchable, and transparent electrodes [56]. On the other hand, printing SWNTs has been shown to deposit electrode and channel materials for TFTs [16]. The main advantage of solution-based processes is low temperature requirements and hence compatibility with various types of widely used flexible substrates. Therefore solution-based processes enable low-cost and large-area deposition on various flexible substrates [10, 11]. There has been tremendous progress in SWNT separation and purification after the growth. Therefore it is possible to obtain SWNTs with the same electronic type or chirality in large quantities and fabricate sc-SWNT TFTs with uniform performance [16].

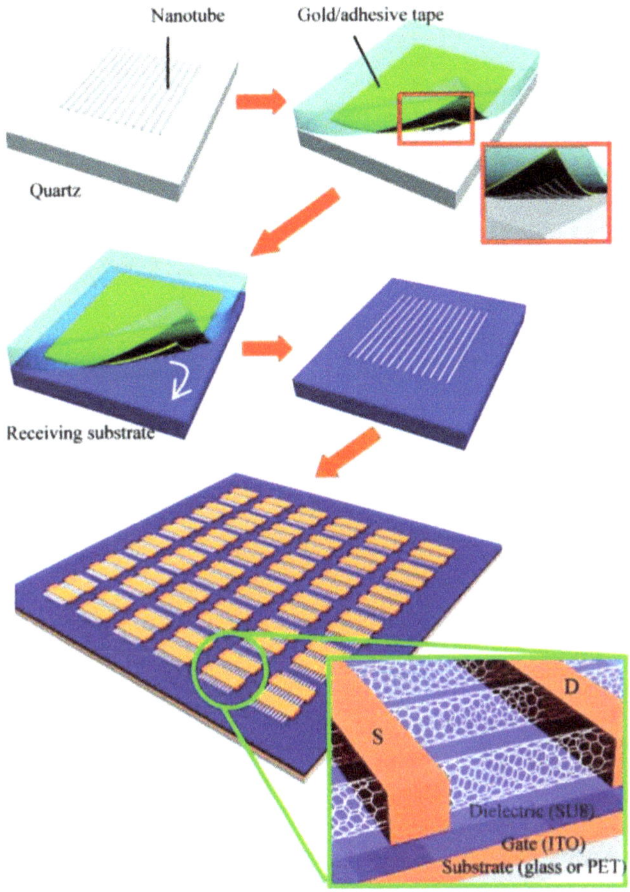

Figure 7.6. Schematic of a film transfer method. In this method CVD-grown SWNT film has been transferred from quartz substrate to glass or PET. (Reproduced with permission from [53]. Copyright 2008 The American Chemical Society.)

7.2.2.1 Inkjet and aerosol printing of SWNTs

Several research groups have demonstrated the potential to deposit SWNTs with the use of inkjet and aerosol printing [17, 18, 57, 58]. In [59] researchers patterned SWNTs on flexible paper and polymer substrate using inkjet printing. Another group inkjet printed sc-SWNTs onto pre-patterned source–drain electrodes to demonstrate SWNT transistors [19, 60]. However, these works used unsorted SWNTs and the density of m-SWNTs was high in the films. As a result the devices obtained low on–off ratios. However, by using sorted sc-SWNT ink different groups achieved high on–off ratios with high mobility. For example, the authors of [61] demonstrated an inkjet printed SWNT active layer based TFT which achieved a mobility of up to 6.6 cm^2 V^{-1} s^{-1} and on–off ratios of up to 10^5. In [16] the researchers demonstrated TFTs with inkjet printed SWNTs and achieved a mobility of 1.6 to 4.2 cm^2 V^{-1} s^{-1} and an on–off ratio of four to five digits by controlling both the densities and the electrical properties of the SWNT networks. Figure 7.7 shows the fabrication of the inkjet

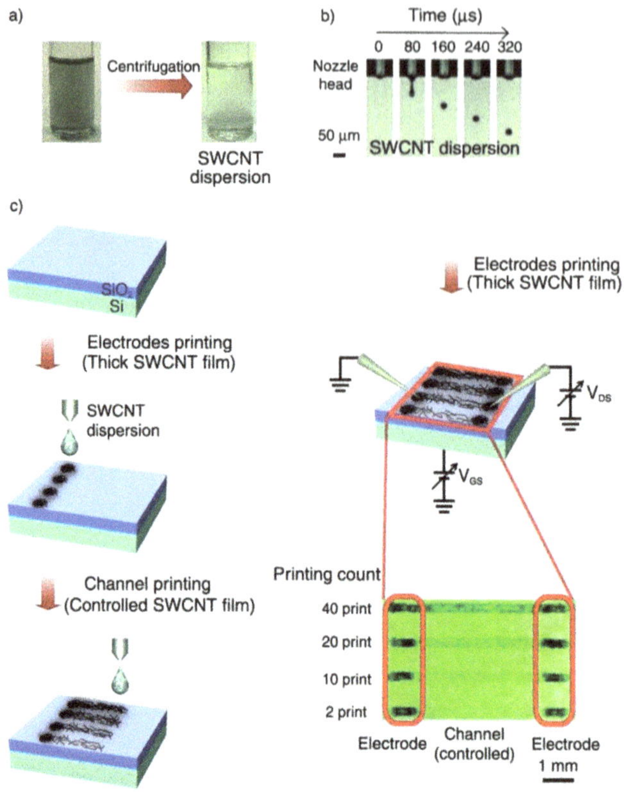

Figure 7.7. Schematic of inkjet printing of SWNTs with variable density. Variable density was achieved by dropping a different number of droplets. (Reproduced with permission from [16]. Copyright 2010 Wiley.)

printed SWNT TFTs of this work. Here different drops of SWNT solution were inkjet printed in a controlled way to obtain a low density of film. Highly density SWNT films were deposited to form the source and drain, and low density SWNT films were deposited as channels and connected the source and drain.

Ha *et al* aerosol-jet printed sc-SWNT onto pre-patterned Cr/Au electrodes on polyimide. Later they printed ion gel as the dielectric and PEDOT:PSS as the gate material by aerosol printing [17]. These TFTs achieved mobility in the range of 20–50 $cm^2 V^{-1} s^{-1}$ and on–off ratios in the range of 10^3 to 10^5. A combination of inkjet printing and roll-to-roll gravure printing was used to fabricate an SWNT active layer based TFT in [62]. This device was built on PET. In this case inkjet printing was used to print the SWNT channel, whereas the source, drain, gate electrodes, and dielectric layer were printed using gravure printing. Figure 7.8 shows a photograph of mass-produced devices along with SEM images of the channel region.

7.2.2.2 Alignment of solution-based SWNTs

Solution-processed SWNT film generates a random network of SWNTs. The mobility of solution-processed SWNT film can potentially be increased by aligning

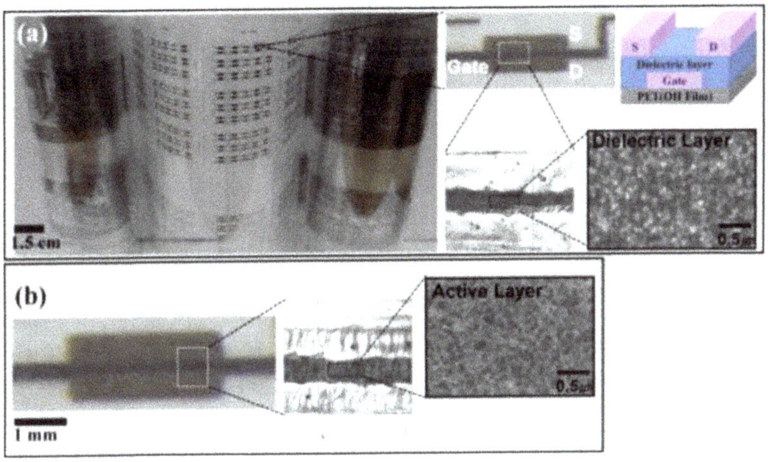

Figure 7.8. Mass-produced printed TFTs on a flexible substrate along with SEM images of the channel region. (Reproduced with permission from [62]. Copyright 2012 Elsevier.)

the SWNTs. The charge transport through the film is limited by the number of tube-to-tube junctions. This tube-to-tube junction can be reduced by aligning the SWNTs in the film. The alignment of SWNTs can be achieved by applying an external force such as an electric field (ac-dielectrophoresis) [63], liquid flow [64], or centrifugal force (spin-coating) [65]. Achieving a self-assembled layer by convective flow [66] and the Langmuir–Schaefer assembly technique [49] is an alternative to applying an external force. Among the different alignment techniques for solution-processed SWNT films, ac-dielectrophoresis (DEP) is one of the most popular ones. In this technique, an AC voltage is applied between two electrodes to align the SWNTs. This method produces highly aligned SWNTs. The density of the SWNTs can be controlled by varying the solution concentration or the AC voltage frequency [63]. Although the DEP technique is very effective for aligning SWNTs, DEP aligns m-SWNTs more than sc-SWNTs. This alignment of m-SWNTs in the channel of TFTs can result in low on–off ratios even with sc-enriched solutions.

7.3 Applications of sc-SWNTs

TFTs play an important role in the field of electronics. sc-SWNTs require low temperature processing, have optical transparency, mechanical compliance, and high mobility. Therefore sc-SWNTs are suitable to be used as an active layer material [67]. Gate dielectrics are important parts of SWNT based TFTs. Gate dielectric materials with high capacitances are required for hysteresis-free operation. Moreover, high-capacitance gate dielectrics are needed to operate TFTs at low voltage. This results in power efficiency of the TFT operation. Additionally, the fabrication method of the dielectric layer has to be compatible with a flexible substrate. One approach is to utilize inorganic oxides such as ITO (2–5 nm) formed by atomic layer deposition (ALD). Since the ALD process is not compatible with flexible substrates, crosslinked epoxies of around 10 nm thickness are spin coated on top. The epoxy layer acts as an adhesive layer for transfer printing [68]. Such dielectrics offer high capacitance (up to 330 nF cm^{-2}) and low leakage current density, interface charge density, interface state density, and dissipation factors. Another attractive approach is to use some 3D-crosslinked organic multilayers (~16 nm) as the dielectric layer. These organic multilayers are formed by self-assembly processes at room-temperature. These organic multilayers produce large capacitances (~170 nF cm^{-2}), small leakage current (densities less than 10^{-9} A cm^{-2}), and smooth surface morphologies [69]. These high-capacitance dielectrics reduce the operating voltage of the SWNT based TFTs. Therefore the hysteresis for SWNT devices with bilayer nano-dielectrics or organic multilayers is smaller than that of devices with thick oxide or polymer dielectrics [69–71]. In [68] the researchers used high-capacitance bilayer dielectrics based on atomic-layer-deposited HfO_2 and spin-cast epoxy with SWNT networks to enable low-voltage, hysteresis-free, and high-performance TFTs on silicon and flexible plastic substrates. These HfO_2–epoxy dielectrics demonstrated mechanical flexibility, large capacitance, and low leakage current. The dielectric layers were deposited at low temperature (around 150 °C) which makes the fabrication process compatible with a range of plastic substrates. Figure 7.9(a) shows the schematic of the TFT. Figure 7.9(b) shows the transfer curves of the TFTs supported by a plastic substrate and with channel lengths, from top to bottom, of 50, 75, and 100 μm. The channel width in each case is 250 μm and V_{DS} is –0.2 V. The inset shows I_{ON} versus the reciprocal of channel length ($1/L$). In this work the researchers first used pristine SWNTs which typically lead to p-channel operation in the TFTs, as seen in the transfer curve in figure 7.9(b). Later they achieved n-channel operation by charge-transfer doping with polyethyleneimine (PEI) coatings. In the n-channel mode with similar channel lengths of p-channel devices, the devices exhibited very low hysteresis. Figure 7.9(c) shows the transfer curves of the TFT in the n-channel mode with channel lengths, from top to bottom, of 50, 75, and 100 μm. The channel width is 250 μm and V_{DS} is 0.2 V in this case.

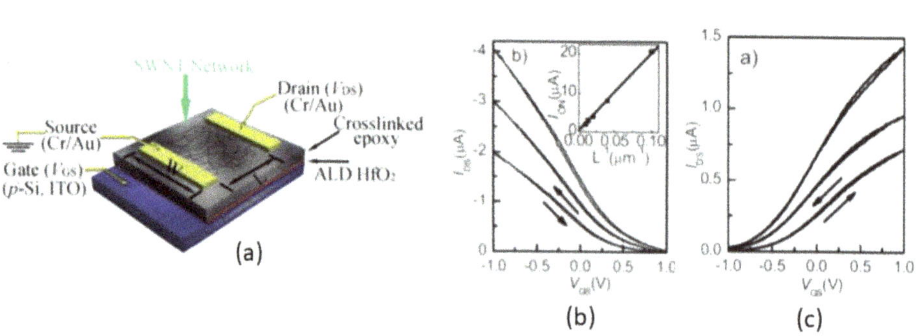

Figure 7.9. (a) Schematic of the TFT. (b) Transfer curves of the TFTs supported by a plastic substrate, and with channel lengths, from top to bottom, of 50, 75, and 100 μm. The channel width in each case is 250 μm and V_{DS} is –0.2 V. The inset shows I_{ON} versus the reciprocal of channel length ($1/L$). (c) Transfer curves of the TFTs in n-channel mode (after PEI coatings of the SWNTs) with channel lengths, from top to bottom, of 50, 75, and 100 μm The channel width is 250 μm and V_{DS} is 0.2 V in this case. (Reprinted with permission from [68]. Copyright 2006 Wiley.)

TFTs with such gate dielectrics are generally fabricated in a bottom-gate structure. In this structure, a continuous conductive film acting as gate electrode is deposited on flexible substrate [68, 72]. This structure is easy to fabricate and is useful for determining the electronic properties of the devices. The disadvantage of this structure is that the nanotubes are exposed to oxygen from the air resulting in instability of the TFTs. Therefore, this structure cannot be used to build circuits. However, as the electronic properties of exposed SWNTs can be changed with chemical modification, this structure is good for making chemical sensors. There is one approach to avoid the limitations of the bottom-gate structure. In this approach the gate dielectric is deposited on top of SWNTs for a top-gate device structure, as shown in figure 7.10(a) [12]. In this structure high k dielectrics can be deposited by ALD, for example, on CVD-grown SWNTs transferred to a flexible substrate. Such designs with a random network of SWNTs can offer good device properties and stability. In [73] researchers reported top-gate structure SWNT TFTs built on PI substrate and fabricated logic circuits with those TFTs. Figures 7.10(b) and (c) show the cross-sectional diagram of their PMOS inverter and a photograph of a collection of their SWNT transistors and circuits on the PI substrate, respectively [73]. Their transistors demonstrated good device properties such as a mobility of ~70 $cm^2 V^{-1} s^{-1}$, subthreshold slope ~200 $mVdec^{-1}$, operating voltages less than 4 V, transconductances as high as 0.12 $\mu S \mu m^{-1}$ and on–off ratios $>10^3$ [73]. Additionally, because high-capacitance gate dielectrics reduce the relative contribution of voltage across the dielectric to the threshold voltage (V_T), V_T can be controlled using gate metals with different work functions. In [73] researchers replaced Au with Al as the gate metal which shifted V_T by −(0.6–0.8) V resulting in a device functioning from depletion mode to enhancement mode. Figure 7.10(d) shows the transfer characteristics of the TFT with a high k HfO_2 dielectric and high work function (Au, blue) and low work function (Al, black) gate electrodes. Measured and simulated data are

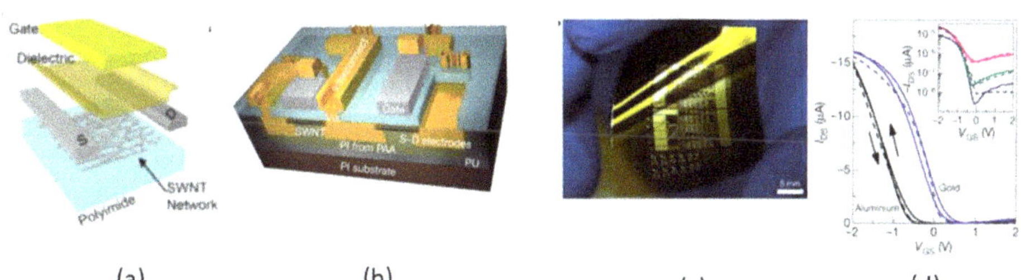

Figure 7.10. (a) A top-gate device structure for an SWNT TFT. (Reproduced with permission from [12]. Copyright 2009 Wiley.) (b) Cross-sectional diagram of an SWNT PMOS inverter on a polyimide (PI) substrate. Here PU is polyurethane, PAA is polyamic acid, V_{dd} is the common power supply voltage, V_{out} is the output voltage, V_{in} is the input voltage, and GND is the common ground. (c) A collection of SWNT transistors and circuits on a flexible plastic sheet. (d) Transfer characteristics of the TFT with high k HfO_2 dielectric and high work function (Au, blue) and low work function (Al, black) gate electrodes. Measured and simulated data are represented by solid and dashed lines, respectively. Inset: transfer characteristics of the SWNT with the Al gate plotted on a logarithmic scale, with different V_{DS} =−0.5 V (navy), −2 V (green), and −5 V (magenta). (Reproduced with permission from [73]. Copyright 2008 Springer.)

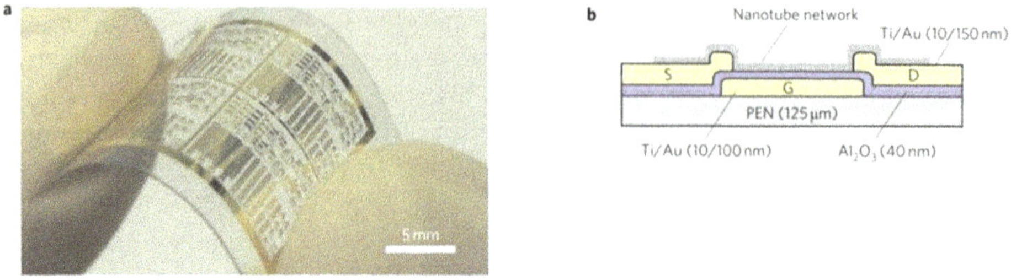

Figure 7.11. (a) Photograph of different integrated circuits such as inverters, ring oscillators (3, 11 and 21 stages), NOR and NAND gates, reset–set flip-flops, and delay flip-flops on a flexible and transparent PEN substrate. (b) Schematic cross-section of the bottom-gate TFT used to build the circuits in (a). (Reproduced with permission from [74]. Copyright 2011 Springer.)

represented by solid and dashed lines, respectively. The inset shows the transfer characteristics of the SWNT with an Al gate plotted on a logarithmic scale, with different V_{DS}=−0.5 V (navy), −2 V (green), and −5 V (magenta). The group fabricated different logic gates with these TFTs. They also fabricated medium-scale integrated circuits based on the SWNT network TFTs, on a thin plastic substrate [73]. A disadvantage of the top-gate TFT structure is that SWNTs are passivated and are not affected by external dopants, such as oxygen or polymers. Therefore, these transistors often demonstrate some level of ambipolar behavior.

Sun *et al* fabricated TFTs and integrated circuits on a flexible substrate using random networks of SWNTs. The SWNTs were prepared using a dry filtration method [74]. The TFTs demonstrated mobility of 35 $cm^2\ V^{-1}\ s^{-1}$ and I_{on}/I_{off} of 10^6. They used the TFTs to build different integrated circuits such as inverters, ring oscillators (3, 11 and 21 stages), NOR and NAND gates, reset–set flip-flops, and delay flip-flops on a polyethylene naphthalate (PEN) substrate, as seen in figure 7.11(a). Figure 7.11(b) shows a schematic cross-section of their bottom-gate carbon nanotube TFT on a flexible substrate. The length and width of the TFTs in the circuits were 100 mm each. The TFTs achieved high electrical performance from the long nanotube lengths (~10 μm) and Y-type inter-nanotube junctions. These have a large contact area and thus low contact resistance. In addition to integrated circuits, flexible SWNT TFTs have also been used for various functional systems, such as backplanes for driving organic light-emitting diodes [75], displays [76, 77], and sensors [78, 79].

7.4 Applications of m-SWNTs

SWNTs have a high aspect ratio and they are naturally highly curved and entangled in their assemblies. It enables them to flex and stretch. Therefore, metallic SWNTs are ideal for flexible and stretchable interconnects and electrodes [31, 35, 80]. In addition, thin film of SWNT is optically transparent. This property makes them suitable for application as electrodes in optoelectronics devices. In this section we will discuss the application of m-SWNTs in different electronic devices.

7.4.1 Pressure and strain sensors

m-SWNT electrodes have been used widely in pressure and strain sensors [31, 33, 82] due to their intrinsic flexibility and outstanding piezoresistive properties. These sensors are made of individual SWNTs, SWNT thin films, or SWNT/polymer composites [81]. The SWNT/polymer composite based strain sensors are popular and have demonstrated a higher gauge factor than conventional foil-type gauges. However, the major drawback of them is that the strain sensitivity is strongly dependent on many factors, such as the aspect ratio, content, and dispersion state of the SWNTs, as well as the properties of the host polymer. Some groups reported strain sensors with SWNT thin films [82–87]. The SWNT thin film based sensors showed higher strain sensitivity than composite sensors. However, under tension they showed nonlinear behaviour. In [81] the researchers developed a simple and low-cost microfabrication process to fabricate SWNT film strain gauges. The SWNT thin films used in this work were spray-deposited on a flexible polyimide base. Figure 7.12 shows the microfabrication process of the SWNT based strain sensor. As can be seen, conventional photolithography was used to pattern the gauges followed by SWNT etching in an O_2 plasma. The group has controlled the gauge sensitivity of the spray-coated SWNT film by experimentally varying design parameters such as a strain-sensitive CNT grid and the film thickness. In [31] the

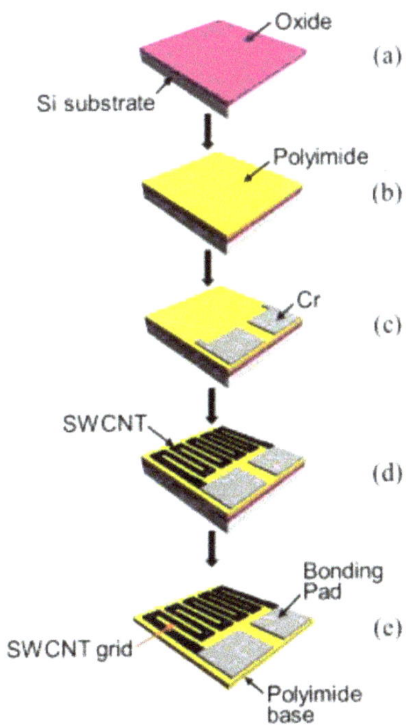

Figure 7.12. Fabrication steps of an SWNT film strain gauge: (a) oxide deposition, (b) polyimide film coating, (c) Cr deposition and patterning, (d) spray coating and patterning of SWNT, and (e) dipping into BOE and releasing flexible strain gauges from the substrate. © IOP Publishing. Reproduced with permission. All rights reserved.

researchers fabricated capacitive pressure sensors comprised of two strips of PDMS bearing stretchable nanotube films. The strips were laminated together, face-to-face, with Ecoflex silicone elastomer. SWNT based pressure and strain sensors can be used in touch screen panels and displays [88, 89]. In [90] solution-processed SWNT films were patterned on PET and used as touch electrodes in a resistive touch panel. These electrodes demonstrated better flexibility and mechanical stability than ITO electrodes. The touch panel worked with good performance even after 3 million actuations.

7.4.2 Biological and chemical sensors

The electrical properties of m-SWNTs depend exclusively on surface atoms. These surface atoms are very sensitive to adsorbents [91, 92]. This sensitivity can be measured electrically in resistor structures. As a result, it is possible to use SWNT films as sensing elements for various absorbents such as chemical vapours, bio-macromolecules [93–95]. SWNT thin films have a greater number of SWNTs exposed to the analyte of interest than an individual nanotube. As a result, a thin film of SWNT based sensors shows a better signal-to-noise ratio and detection limit than individual SWNT based sensors.

Conventional gas sensors respond to the analyte concentration, whereas SWNT based gas sensors respond to surface coverage of the analytes. Therefore the SWNT gas sensor offers very high sensitivity and detection limits for low vapor pressure analytes [96, 97]. The simplest form of an m-SWNT based gas sensor is a chemiresistor. In a chemiresistor, electrical contacts are made at two ends of an SWNT thin film. This is presented in the flow cell chemiresistor shown in figure 7.13 [98]. Gas analytes absorb onto the surfaces of the SWNTs, in particular at defect

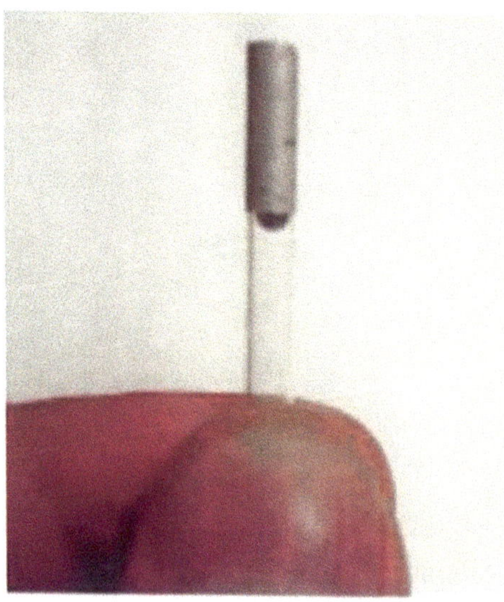

Figure 7.13. (a) Optical micrograph of an SWNT flow cell chemiresistor. (Reproduced with permission from [98]. Copyright 2013 AIP Publishing.)

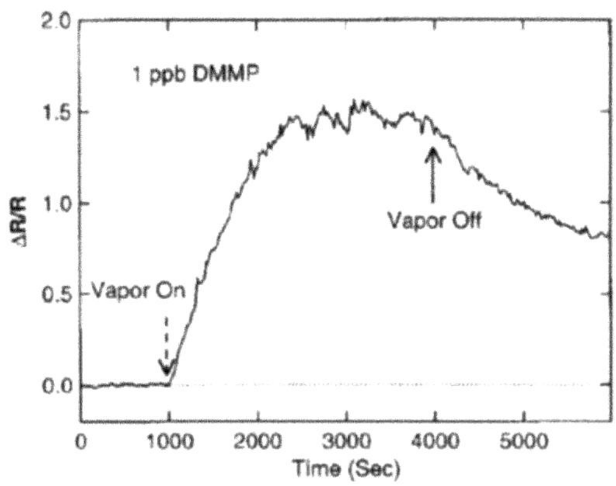

Figure 7.14. Relative change in resistance ($\Delta R/R$) over time for an SWNT chemiresistor loaded in a flow cell when exposed to 1 ppb DMMP. (Reproduced with permission from [98]. Copyright 2013 AIP Publishing.)

points. Dimethyl methylphosphonate (DMMP) can be detected with this sensor at ppb levels due to its high electron-donating properties (figure 7.14) [98]. As seen from the response curve of figure 7.14, the recovery time of the sensor is slow. In [99] Li *et al* used vertically grown CVD into a PDMS matrix as an ethanol gas sensing layer. Other groups have also demonstrated flexible gas sensors using m-SWNTs for the detection of NH_3 [100, 101], NO_2 [102], H_2 [103], and other gases. SWNT gas sensors lack specificity. By functionalizing SWNTs with receptors for targeted gas this problem can be solved. For example, in [104] the researchers decorated SWNTs with palladium nanoparticles to form an SWNT chemresistor specific to hydrogen detection.

The diameters and carrier densities of SWNTs are comparable to the sizes and surface-charge densities of bio-macromolecules. Therefore SWNTs serve as an ultrasensitive sensing layer in biosensors. m-SWNT based biosensors have chemiresistor structures [105–107]. Due to hydrophobic interactions, p–p stacking interactions, and possibly the amino-affinity of SWNTs, biomolecules, such as DNAs and proteins, non-specifically bind to the surfaces of SWNT film [108, 109]. Thus the SWNTs themselves can function as labels for efficient label-free detection [110]. Selective detection of antibodies can be done with this structure by taking advantage of antigen antibody interactions.

7.4.3 Supercapacitors and solar cells

Supercapacitors and solar cells are another important category of flexible devices that utilize SWNTs. The SWNT based supercapacitors and solar cells have worse performances than the established designs. Nonetheless, the flexibility of these devices makes them useful for applications where the rigid established designed cannot be sustained, such as placing them on curved surfaces. Moreover, they can be fabricated using printing processes which would enable low-cost manufacturing. The use of SWNT films in solar cells has so far only been for fabricating anode and cathode electrodes. Yang *et al* fabricated counter electrodes with SWNT thin films

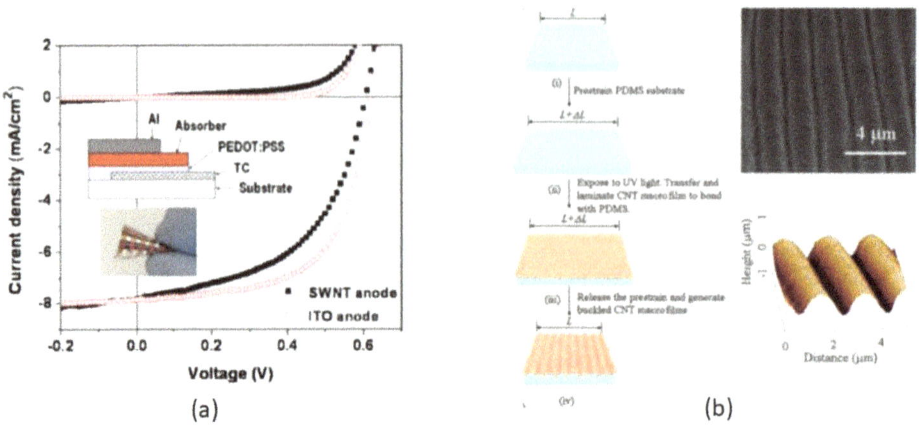

Figure 7.15. (a) Flexible organic solar cell with an SWNT based electrode. (Reproduced with permission from [112]. Copyright 2006 AIP Publishing.) (b) Stretchable supercapacitor with SWNT electrodes. (Reproduced with permission from [38]. Copyright 2009 Wiley.)

on flexible substrates in a dye synthesized solar cell [111]. They tested the mechanical integrity by bending the solar cell for many cycles. There was no observable structural damage or change in conductivity in the solar cell. In [112] the researchers fabricated anodes of a flexible organic solar cell using an SWNT film. This solar cell can be seen in figure 7.15(a). They vacuum filtered SWNTs from a solution and then transferred the SWNT films onto the flexible substrate using a patterned PDMS stamp. The roughness of the SWNT films can cause shorting in the device. To reduce the roughness of the SWNT film, they coated the SWNT film with PEDOT:PSS. The solar cells show more flexibility than conventional devices using an ITO anode and a power conversion efficiency of 2.5%. Flexible films of SWNT have also been utilized for supercapacitors [113, 114]. In [115] the researchers spray-coated thin films of SWNTs and used them as both electrodes and charge collectors in combination with a printable gel electrolyte. These devices demonstrated good performance with an energy density of 6 W hkg^{-1} and a power density of 23 KW kg^{-1}. The researchers in [38] fabricated supercapacitors using a buckled SWNT film. The supercapacitor, as shown in figure 7.15(b), is stretchable up to 30 times and has excellent power and energy densities.

7.5 Conclusion and future prospects

It is clear that the flexible and printed electronics market will continue to grow and SWNTs will become a dominant material for that market. According to a report by IDTechEx, the market share for flexible and printed electronics was $9.46 billion in 2012. It will reach $63.28 billion by 2022 [116]. Although great progress has been achieved in research for SWNT based flexible and printed devices, no SWNT based flexible and printed device is commercially available at this time [117]. Researchers and manufacturers need to overcome several challenges to bring SWNT based electronic devices and systems to consumer markets. From the material perspective, semiconductor enriched SWNTs are commercially available in large amounts. However, they are still largely inhomogeneous in terms of nanotube length and chirality. This leads to a lack of uniformity in device performance. Other challenges to obtaining uniform device performance are to find cost-effective and scalable ways to separate metallic and semiconducting SWNTs, and to process and deposit SWNTs in a controllable manner on flexible substrates. As presented in this chapter, research is ongoing to tackle these issues. Moreover, long SWNTs are required to reduce the number of tube-to-tube junctions. This results in improvement of device mobility. The main challenge to using long nanotubes is the difficulty with the dissolution and separation of long SWNTs. The effects of surfactants on the electrical characteristics of devices need to be investigated as well. Although surfactants help with the dispersion of SWNTs, they are difficult to remove after deposition. The surfactants can reduce electronic conduction and increase channel resistance. Exploring surfactant free methods to dissolve and deposit SWNTs without damaging or shortening them is an important area for future research [118, 119]. Other challenges with SWNTS are obtaining air-stable n-type SWNT films, and improving the yield and stability of SWNT devices. Although printing is a promising method for large-scale manufacturing at low cost, the electrical performance and uniformity of printed devices are still far inferior to their counterparts fabricated using traditional microfabrication techniques. This is due to the poor reproducibility, accuracy, and low resolution of currently available printing processes. Flexible devices are relatively new and have attracted interest in many application areas. Although both sc-SWNT and m-SWNT networks and thin films have good flexibility, obtaining flexible dielectrics and reliable flexible interfaces are real bottlenecks in this area. It is predicted that the first commercial use of SWNTs will be for transparent electrodes in touchscreens or in other simple devices. SWNT TFT based cheap devices such as sensors and RFIDs also have the potential to be successful very soon. More complicated SWNT TFT based systems such as flexible memory and integrated circuits are less likely to be commercialized in the near future due to the challenges associated with SWNTs and printing. Finally, graphene and other 2D semiconducting materials are also showing a lot of promise for flexible and printed electronics. Incorporating SWNTs with these materials may result in excellent performance of flexible and printed devices.

References

[1] Forrest S 2004 The path to ubiquitous and low-cost organic electronic appliances on plastic *Nature* **428** 911–18

[2] Kim S *et al* 2011 Low-power flexible organic light-emitting diode display device *Adv. Mater.* **23** 3511–16

[3] Yoon B, Ham D, Yarimaga O, An H, Lee C and Kim J 2011 Inkjet printing of conjugated polymer precursors on paper substrates for colorimetric sensing and flexible electrothermochromic display *Adv. Mater.* **23** 5492–97

[4] Gao W *et al* 2016 Fully integrated wearable sensor arrays for multiplexed *in situ* perspiration analysis *Nature* **529** 509–14

[5] Chen L, Tee B, Chortos A, Schwartz G, Tse V, Lipomi D J, Wong H, McConnell M and Bao Z 2014 Continuous wireless pressure monitoring and mapping with ultra-small passive sensors for health monitoring and critical care *Nat. Commun.* **5** 5028
[6] Imani S, Bandodkar A, Mohan A, Kumar R, Yu S, Wang J and Mercier P 2016 A wearable chemical–electrophysiological hybrid biosensing system for real-time health and fitness monitoring *Nat. Commun.* **7** 11650
[7] Zhang H *et al* 2015 A flexible and implantable piezoelectric generator harvesting energy from the pulsation of ascending aorta: *in vitro* and *in vivo* studies *Nano Energy* **12** 296–304
[8] Khan Y, Ostfeld A, Lochner C, Pierre A and Arias A 2016 Monitoring of vital signs with flexible and wearable medical devices *Adv. Mater.* **28** 4373–95
[9] Fang H *et al* 2017 Capacitively coupled arrays of multiplexed flexible silicon transistors for long-term cardiac electrophysiology *Nat. Biomed. Eng.* **1** 0055
[10] Arias A, MacKenzie J, McCulloch I, Rivnay J and Salleo A 2010 Materials and applications for large area electronics: solution-based approaches *Chem. Rev.* **110** 3–24
[11] Singh M, Haverinen H, Dhagat P and Jabbour G 2010 Inkjet printing-process and its applications *Adv. Mater.* **22** 673–85
[12] Wang C, Takei K, Takahashi T and Javey A 2013 Carbon nanotube electronics—moving forward *Chem. Soc. Rev.* **42** 2592–609
[13] Donaldson K, Aitken R, Tran L, Stone V, Duffin R, Forrest G and Alexander A 2006 Carbon nanotubes: a review of their properties in relation to pulmonary toxicology and workplace safety *Toxicol. Sci.* **92** 5–22
[14] Cui Z, Zhou C, Qiu S, Chen Z, Lin J, Zhao J, Ma C and Su W 2016 *Printed Electronics.* (Singapore: Wiley)
[15] Wang C M, Roy Chowdhury A N, Koh S J A and Zhang Y Y 2014 Molecular dynamics simulation and continuum shell model for buckling analysis of carbon nanotubes *Modeling of Carbon Nanotubes, Graphene and their Composites Springer Series in Materials Science* ed K Tserpes and N Silvestre vol 188 (Cham: Springer)
[16] Okimoto H, Takenobu T, Yanagi K, Miyata Y, Shimotani H, Kataura H and Iwasa Y 2010 Tunable carbon nanotube thin-film transistors produced exclusively via inkjet printing *Adv. Mater.* **22** 3981–86
[17] Ha M, Xia Y, Green A, Zhang W, Renn M, Kim C, Hersam M and Frisbie C 2010 Printed, sub-3V digital circuits on plastic from aqueous carbon nanotube inks *ACS Nano* **4** 4388–95
[18] Chen P, Fu Y, Aminirad R, Wang C, Zhang J, Wang K, Galatsis K and Zhou C 2011 Fully printed separated carbon nanotube thin film transistor circuits and its application in organic light emitting diode control *Nano Lett.* **11** 5301–8
[19] Beecher P *et al* 2007 Ink-jet printing of carbon nanotube thin film transistors *J. Appl. Phys.* **102** 043710
[20] Cao X, Chen H, Gu X, Liu B, Wang W, Cao Y, Wu F and Zhou C 2014 Screen printing as a scalable and low-cost approach for rigid and flexible thin-film transistors using separated carbon nanotubes *ACS Nano* **8** 12769–76
[21] Ha M, Seo J, Prabhumirashi P, Zhang W, Geier M, Renn M, Kim C, Hersam M and Frisbie C 2013 Aerosol jet printed, low voltage, electrolyte gated carbon nanotube ring oscillators with sub-5 μs stage delays *Nano Lett.* **13** 954–60
[22] Jung M *et al* 2010 All-printed and roll-to-roll-printable 13.56-MHz-operated 1-bit RF tag on plastic foils *IEEE Trans. Electron Devices* **57** 571–80

[23] Lau P, Takei K, Wang C, Ju Y, Kim J, Yu Z, Takahashi T, Cho G and Javey A 2013 Fully printed, high performance carbon nanotube thin-film transistors on flexible substrates *Nano Lett.* **13** 3864–69

[24] Wei B, Vajtai R and Ajayan P 2001 Reliability and current carrying capacity of carbon nanotubes *Appl. Phys. Lett.* **79** 1172–74

[25] Dürkop T, Getty S, Cobas E and Fuhrer M 2004 Extraordinary mobility in semiconducting carbon nanotubes *Nano Lett.* **4** 35–9

[26] Zhou X, Park J, Huang S, Liu J and McEuen P 2005 Band structure, phonon scattering, and the performance limit of single-walled carbon nanotube transistors *Phys. Rev. Lett.* **95**

[27] Liu B, Wang C, Liu J, Che Y and Zhou C 2013 Aligned carbon nanotubes: from controlled synthesis to electronic applications *Nanoscale* **5** 9483

[28] Cao Q and Rogers J 2009 Ultrathin films of single-walled carbon nanotubes for electronics and sensors: a review of fundamental and applied aspects *Adv. Mater.* **21** 29–53

[29] Hu L, Hecht D and Grüner G 2010 Carbon nanotube thin films: fabrication, properties, and applications *Chem. Rev.* **110** 5790–844

[30] Li Z, Xie D, Dai R, Xu J, Sun Y, Sun M, Zhang C and Li X 2016 High-performance heterogeneous complementary inverters based on n-channel MoS_2 and p-channel SWCNT transistors *Nano Res.* **10** 276–83

[31] Lipomi D, Vosgueritchian M, Tee B, Hellstrom S, Lee J, Fox C and Bao Z 2011 Skin-like pressure and strain sensors based on transparent elastic films of carbon nanotubes *Nat. Nanotechnol.* **6** 788–92

[32] Yamada T, Hayamizu Y, Yamamoto Y, Yomogida Y, Izadi-Najafabadi A, Futaba D and Hata K 2011 A stretchable carbon nanotube strain sensor for human-motion detection *Nat. Nanotechnol.* **6** 296–301

[33] Cohen D, Mitra D, Peterson K and Maharbiz M 2012 A highly elastic, capacitive strain gauge based on percolating nanotube networks *Nano Lett.* **12** 1821–25

[34] Liang J, Li L, Niu X, Yu Z and Pei Q 2013 Elastomeric polymer light-emitting devices and displays *Nat. Photonics* **7** 817–24

[35] Yu Z, Niu X, Liu Z and Pei Q 2011 Intrinsically stretchable polymer light-emitting devices using carbon nanotube–polymer composite electrodes *Adv. Mater.* **23** 3989–94

[36] Feng C, Liu K, Wu J, Liu L, Cheng J, Zhang Y, Sun Y, Li Q, Fan S and Jiang K 2010 Flexible, stretchable, transparent conducting films made from superaligned carbon nanotubes *Adv. Funct. Mater.* **20** 885–91

[37] Niu Z, Dong H, Zhu B, Li J, Hng H, Zhou W, Chen X and Xie S 2012 Highly stretchable, integrated supercapacitors based on single-walled carbon nanotube films with continuous reticulate architecture *Adv. Mater.* **25** 1058–64

[38] Yu C, Masarapu C, Rong J, Wei B and Jiang H 2009 Stretchable supercapacitors based on buckled single-walled carbon-nanotube macrofilms *Adv. Mater.* **21** 4793–97

[39] Park S, Vosguerichian M and Bao Z 2013 A review of fabrication and applications of carbon nanotube film-based flexible electronics *Nanoscale* **5** 1727

[40] Cai L and Wang C 2015 Carbon nanotube flexible and stretchable electronics *Nanoscale Res. Lett.* **10** 320

[41] Dan B, Irvin G and Pasquali M 2009 Continuous and scalable fabrication of transparent conducting carbon nanotube films *ACS Nano* **3** 835–43

[42] Wang C, Zhang J, Ryu K, Badmaev A, De Arco L and Zhou C 2009 Wafer-scale fabrication of separated carbon nanotube thin-film transistors for display applications *Nano Lett.* **9** 4285–91
[43] Öncel Ç and Yürüm Y 2006 Carbon nanotube synthesis via the catalytic CVD method: a review on the effect of reaction parameters *Fuller. Nanotub. Carbon Nanostructures* **14** 17–37
[44] Yu W, Chae S, Lee S, Duong D and Lee Y 2011 Ultra-transparent, flexible single-walled carbon nanotube non-volatile memory device with an oxygen-decorated graphene electrode *Adv. Mater.* **23** 1889–93
[45] Cao Q, Hur S, Zhu Z, Sun Y, Wang C, Meitl M, Shim M and Rogers J 2006 Highly bendable, transparent thin-film transistors that use carbon-nanotube-based conductors and semiconductors with elastomeric dielectrics *Adv. Mater.* **18** 304–9
[46] Kang S, Kocabas C, Kim H, Cao Q, Meitl M, Khang D and Rogers J 2007 Printed multilayer superstructures of aligned single-walled carbon nanotubes for electronic applications *Nano Lett.* **7** 3343–48
[47] Kocabas C, Meitl M, Gaur A, Shim M and Rogers J 2004 Aligned arrays of single-walled carbon nanotubes generated from random networks by orientationally selective laser ablation *Nano Lett.* **4** 2421–26
[48] Huang S, Woodson M, Smalley R and Liu J 2004 Growth mechanism of oriented long single walled carbon nanotubes using 'fast-heating' chemical vapor deposition process *Nano Lett.* **4** 1025–28
[49] Zhang Y, Chang A, Cao J, Wang Q, Kim W, Li Y, Morris N, Yenilmez E, Kong J and Dai H 2001 Electric-field-directed growth of aligned single-walled carbon nanotubes *Appl. Phys. Lett.* **79** 3155–57
[50] Ismach A, Kantorovich D and Joselevich E 2005 Carbon nanotube graphoepitaxy: highly oriented growth by faceted nanosteps *JACS* **127** 11554–55
[51] Ding L, Yuan D and Liu J 2008 Growth of high-density parallel arrays of long single-walled carbon nanotubes on quartz substrates *JACS* **130** 5428–29
[52] Krupke R 2003 Separation of metallic from semiconducting single-walled carbon nanotubes *Science* **301** 344–47
[53] Ishikawa F, Chang H, Ryu K, Chen P, Badmaev A, Gomez De Arco L, Shen G and Zhou C 2008 Transparent electronics based on transfer printed aligned carbon nanotubes on rigid and flexible substrates *ACS Nano* **3** 73–9
[54] Bergin S *et al* 2008 Towards solutions of single-walled carbon nanotubes in common solvents *Adv. Mater.* **20** 1876–81
[55] Islam M, Rojas E, Bergey D, Johnson A and Yodh A 2003 High weight fraction surfactant solubilization of single-wall carbon nanotubes in water *Nano Lett.* **3** 269–73
[56] McCarthy M, Liu B, Donoghue E, Kravchenko I, Kim D, So F and Rinzler A 2011 Low-voltage, low-power, organic light-emitting transistors for active matrix displays *Science* **332** 570–73
[57] Vaillancourt J *et al* 2008 All ink-jet-printed carbon nanotube thin-film transistor on a polyimide substrate with an ultrahigh operating frequency of over 5 GHz *Appl. Phys. Lett.* **93** 243301
[58] Li J *et al* 2011 Ink-jet printed thin-film transistors with carbon nanotube channels shaped in long strips *J. Appl. Phys.* **109** 084915

[59] Kordás K, Mustonen T, Tóth G, Jantunen H, Lajunen M, Soldano C, Talapatra S, Kar S, Vajtai R and Ajayan P 2006 Inkjet printing of electrically conductive patterns of carbon nanotubes *Small* **2** 1021–25

[60] Han X, Janzen D, Vaillancourt J and Lu X 2007 Printable high-speed thin-film transistor on flexible substrate using carbon nanotube solution *Micro Nano Letters* **2** 96

[61] Collins P 2001 Engineering carbon nanotubes and nanotube circuits using electrical breakdown *Science* **292** 706–9

[62] Noh J, Jung M, Jung K, Lee G, Lim S, Kim D, Kim S, Tour J and Cho G 2011 Integrable single walled carbon nanotube (SWNT) network based thin film transistors using roll-to-roll gravure and inkjet *Org. Electron.* **12** 2185–91

[63] Shekhar S, Stokes P and Khondaker S 2011 Ultrahigh density alignment of carbon nanotube arrays by dielectrophoresis *ACS Nano* **5** 1739–46

[64] Park J, Meitl M, Hur S, Usrey M, Strano M, Kenis P and Rogers J 2006 *In situ* deposition and patterning of single-walled carbon nanotubes by laminar flow and controlled flocculation in microfluidic channels *Angew. Chem. Int. Edn* **45** 581–85

[65] Roberts M, LeMieux M, Sokolov A and Bao Z 2009 Self-sorted nanotube networks on polymer dielectrics for low-voltage thin-film transistors *Nano Lett.* **9** 2526–31

[66] Shimoda H, Oh S, Geng H, Walker R, Zhang X, McNeil L and Zhou O 2002 Self-assembly of carbon nanotubes *Adv. Mater.* **14** 899

[67] Hur S, Kocabas C, Gaur A, Park O, Shim M and Rogers J 2005 Printed thin-film transistors and complementary logic gates that use polymer-coated single-walled carbon nanotube networks *J. Appl. Phys.* **98** 114302

[68] Cao Q, Xia M-G, Shim M and Rogers J 2006 Bilayer organic–inorganic gate dielectrics for high-performance, low-voltage, single-walled carbon nanotube thin-film transistors, complementary logic gates, and p–n diodes on plastic substrates *Adv. Funct. Mater.* **16** 2355–62

[69] Hur S, Yoon M, Gaur A, Shim M, Facchetti A, Marks T and Rogers J 2005 Organic nanodielectrics for low voltage carbon nanotube thin film transistors and complementary logic gates *JACS* **127** 13808–9

[70] Robert-Peillard A and Rotkin S 2005 Modeling hysteresis phenomena in nanotube field-effect transistors *IEEE Trans. Nanotechnol.* **4** 284–88

[71] Lee J, Ryu S, Yoo K, Choi I, Yun W and Kim J 2007 Origin of gate hysteresis in carbon nanotube field-effect transistors *J. Phys. Chem.* C **111** 12504–7

[72] Hur S, Park O and Rogers J 2005 Extreme bendability of single-walled carbon nanotube networks transferred from high-temperature growth substrates to plastic and their use in thin-film transistors *Appl. Phys. Lett.* **86** 243502

[73] Cao Q, Kim H, Pimparkar N, Kulkarni J, Wang C, Shim M, Roy K, Alam M and Rogers J 2008 Medium-scale carbon nanotube thin-film integrated circuits on flexible plastic substrates *Nature* **454** 495–500

[74] Sun D, Timmermans M, Tian Y, Nasibulin A, Kauppinen E, Kishimoto S, Mizutani T and Ohno Y 2011 Flexible high-performance carbon nanotube integrated circuits *Nat. Nanotechnol.* **6** 156–61

[75] Wang C, Hwang D, Yu Z, Takei K, Park J, Chen T, Ma B and Javey A 2013 User-interactive electronic skin for instantaneous pressure visualization *Nat. Mater.* **12** 899–904

[76] Zhang J, Fu Y, Wang C, Chen P, Liu Z, Wei W, Wu C, Thompson M and Zhou C 2011 Separated carbon nanotube macroelectronics for active matrix organic light-emitting diode displays *Nano Lett.* **11** 4852–58

[77] Zhang J, Wang C and Zhou C 2012 Rigid/flexible transparent electronics based on separated carbon nanotube thin-film transistors and their application in display electronics *ACS Nano* **6** 7412–19
[78] Takahashi T, Takei K, Gillies A, Fearing R and Javey A 2011 Carbon nanotube active-matrix backplanes for conformal electronics and sensors *Nano Lett.* **11** 5408–13
[79] Takahashi T, Yu Z, Chen K, Kiriya D, Wang C, Takei K, Shiraki H, Chen T, Ma B and Javey A 2013 Carbon nanotube active-matrix backplanes for mechanically flexible visible light and x-ray imagers *Nano Lett.* **13** 5425–30
[80] Sekitani T, Noguchi Y, Hata K, Fukushima T, Aida T and Someya T 2008 A rubberlike stretchable active matrix using elastic conductors *Science* **321** 1468–72
[81] Lee D, Hong H, Lee C, Park C and Min N 2011 Microfabrication and characterization of spray-coated single-wall carbon nanotube film strain gauges *Nanotechnology* **22** 455301
[82] Li Z, Dharap P, Nagarajaiah S, Barrera E and Kim J 2004 Carbon nanotube film sensors *Adv. Mater.* **16** 640–43
[83] Dharap P, Li Z, Nagarajaiah S and Barrera E 2004 Nanotube film based on single-wall carbon nanotubes for strain sensing *Nanotechnology* **15** 379–82
[84] Li X, Levy C and Elaadil L 2008 Multiwalled carbon nanotube film for strain sensing *Nanotechnology* **19** 045501
[85] Loh K, Lynch J and Kotov N 2005 Conformable single-walled carbon nanotube thin film strain sensors for structural monitoring *Proc. 5th Int. Workshop on Structural Health Monitoring (Stanford, CA, September 12–14, 2005)*
[86] Vemuru S, Wahi R, Nagarajaiah S and Ajayan P 2009 Strain sensing using a multiwalled carbon nanotube film *J. Strain Anal. Eng. Design* **44** 555–62
[87] Yang X, Zhou Z, Zheng F, Zhang M, Zhang J and Yao Y 2009 A high sensitivity single-walled carbon-nanotube-array-based strain sensor for weighing Proc. Solid-State Sensors Actuators Microsyst. Conf. pp 21–5
[88] Kim K, Shin K, Han J, Lee K, Kim W, Park K, Ju B and Pak J 2011 Deformable single wall carbon nanotube electrode for transparent tactile touch screen *Electron. Lett.* **47** 118
[89] Choi E, Kim J, Chun S, Kim A, Lee K, Jeong M, Lim C, Isoshima T, Hara M and Lee S 2011 Fabrication of a flexible and transparent touch sensor using single-walled carbon nanotube thin-films *J. Nanosci. Nanotechnol.* **11** 5845–49
[90] David S, David T, Liangbing H, Corinne L, Tom L, Youngbae P, Glen I and Paul D 2009 Introduction *Inf. Disp.* **17** 941–6
[91] Kong J, Franklin N, Zhou C, Chapline M, Peng S, Cho K and Dai H 2000 Nanotube molecular wires as chemical sensors *Science* **287** 622–25
[92] Collins P, Bradley K, Ishigami M and Zettl A 2000 Extreme oxygen sensitivity of electronic properties of carbon nanotubes *Science* **287** 1801–4
[93] Besteman K, Lee J, Wiertz F, Heering H and Dekker C 2003 Enzyme-coated carbon nanotubes as single-molecule biosensors *Nano Lett.* **3** 727–30
[94] Qi P, Vermesh O, Grecu M, Javey A, Wang Q, Dai H, Peng S and Cho K 2003 Toward large arrays of multiplex functionalized carbon nanotube sensors for highly sensitive and selective molecular detection *Nano Lett.* **3** 347–51
[95] Mahar B, Laslau C, Yip R and Sun Y 2007 Development of carbon nanotube-based sensors —a review *IEEE Sensors J.* **7** 266–84
[96] Snow E, Perkins F, Houser J, Badescu S and Reinecke T 2005 Chemical detection with a single-walled carbon nanotube capacitor *Science* **307** 1942–45

[97] Snow E, Perkins F and Robinson J 2006 Chemical vapor detection using single-walled carbon nanotubes *Chem. Soc. Rev.* **35** 790

[98] Novak J P, Snow E S, Houser E J, Park D, Stepnowski J L and McGill R A 2003 Nerve agent detection using networks of single-walled carbon nanotubes *Appl. Phys. Lett.* **83** 4026

[99] Li B, Hahm M, Kim Y, Jung H, Kar S and Jung Y 2011 Highly organized two- and three-dimensional single-walled carbon nanotube–polymer hybrid architectures *ACS Nano* **5** 4826–34

[100] Woo C, Lim C, Cho C, Park B, Ju H, Min D, Lee C and Lee S 2007 Fabrication of flexible and transparent single-wall carbon nanotube gas sensors by vacuum filtration and poly (dimethyl siloxane) mold transfer *Microelectron. Eng.* **84** 1610–13

[101] Wang X, Li G, Liu R, Ding H and Zhang T 2012 Reproducible layer-by-layer exfoliation for free-standing ultrathin films of single-walled carbon nanotubes *J. Mater. Chem.* **22** 21824

[102] Su P, Lee C, Chou C, Cheng K and Chuang Y 2009 Fabrication of flexible NO_2 sensors by layer-by-layer self-assembly of multi-walled carbon nanotubes and their gas sensing properties *Sens. Actuators* B **139** 488–93

[103] Sun Y, Wang H and Xia M 2008 Single-walled carbon nanotubes modified with Pd nanoparticles: unique building blocks for high-performance, flexible hydrogen sensors *J. Phys. Chem.* C **112** 1250–59

[104] Zhu L, Sun Y, Hess D and Wong C 2006 Well-aligned open-ended carbon nanotube architectures: an approach for device assembly *Nano Lett.* **6** 243–47

[105] Kim S, Rusling J and Papadimitrakopoulos F 2007 Carbon nanotubes for electronic and electrochemical detection of biomolecules *Adv. Mater.* **19** 3214–28

[106] Allen B, Kichambare P and Star A 2007 Carbon nanotube field-effect-transistor-based biosensors *Adv. Mater.* **19** 1439–51

[107] Gruner G 2005 Carbon nanotube transistors for biosensing applications *Anal. Bioanal. Chem.* **384** 322–35

[108] Zheng M, Jagota A, Semke E, Diner B, Mclean R, Lustig S, Richardson R and Tassi N 2003 DNA-assisted dispersion and separation of carbon nanotubes *Nat. Mater.* **2** 338–42

[109] Bradley K, Briman M, Star A and Grüner G 2004 Charge transfer from adsorbed proteins *Nano Lett.* **4** 253–56

[110] Gui E, Li L, Lee P, Lohani A, Mhaisalkar S, Cao Q, Kang S, Rogers J, Tansil N and Gao Z 2006 Electrical detection of hybridization and threading intercalation of deoxyribonucleic acid using carbon nanotube network field-effect transistors *Appl. Phys. Lett.* **89** 232104

[111] Yang Z, Chen T, He R, Guan G, Li H, Qiu L and Peng H 2011 Aligned carbon nanotube sheets for the electrodes of organic solar cells *Adv. Mater.* **23** 5436–39

[112] Rowell M, Topinka M, McGehee M, Prall H, Dennler G, Sariciftci N, Hu L and Gruner G 2006 Organic solar cells with carbon nanotube network electrodes *Appl. Phys. Lett.* **88** 233506

[113] Izadi-Najafabadi A, Yamada T, Futaba D, Yudasaka M, Takagi H, Hatori H, Iijima S and Hata K 2011 High-power supercapacitor electrodes from single-walled carbon nanohorn/nanotube composite *ACS Nano* **5** 811–19

[114] Hu S, Rajamani R and Yu X 2012 Flexible solid-state paper based carbon nanotube supercapacitor *Appl. Phys. Lett.* **100** 104103

[115] Kaempgen M, Chan C, Ma J, Cui Y and Gruner G 2009 Printable thin film supercapacitors using single-walled carbon nanotubes *Nano Lett.* **9** 1872–76

[116] Harrop Peter and Das Raghu 2012 Printed Electronics - Customer Sourcebook & Routes to Profit Case studies, opportunities in the value chain, and progress at over 1000 organizations IDTechEx, 2012.
[117] De Volder M, Tawfick S, Baughman R and Hart A 2013 Carbon nanotubes: present and future commercial applications *Science* **339** 535–39
[118] Davis V *et al* 2009 True solutions of single-walled carbon nanotubes for assembly into macroscopic materials *Nat. Nanotechnol.* **4** 830–34
[119] Fogden S, Howard C, Heenan R, Skipper N and Shaffer M 2011 Scalable method for the reductive dissolution, purification, and separation of single-walled carbon nanotubes *ACS Nano* **6** 54–62

Chapter 8

Flexible strain sensors using graphene and its composites

Rajanna Konandur, M M Nayak, Suresh Nuthalapati, Vijay Shirhatti and Vaishakh Kedambaimoole

8.1 Introduction

Over the past decade the domain of flexible sensors has seen phenomenal progress owing to the discovery of new materials and methods. While silicon technology has ushered in a promising range of sensors via micro-electro-mechanical systems (MEMS), these sensors are largely unsuitable for flexible and wearable applications. The rigid silicon substrates do not comply to the requirements of flexible devices that need to conform to the curvature and deformations of the human body or humanoid robots. The progress of flexible wearable sensors has been propelled mainly by the discovery of the first two-dimensional (2D) material, i.e. graphene, by Geim and Novoshelov in 2004. The distinct properties of graphene, such as high conductivity, mechanical, thermal and chemical stability, optical transparency, and high flexibility, have all combined to provide an ideal material suitable for the development of flexible sensors [1, 2].

Among the 2D materials, graphene has attracted a lot of research interest due to its unique and superior electrical, mechanical, and optical characteristics. Pristine monolayer graphene has a zero band gap and electron mobility around 250 000 $cm^2\ Vs^{-1}$ at room temperature. It exhibits high mechanical strength with a Young's modulus of 1 TPa, and a breaking strength of 42 Nm^{-1} [3, 4]. These exceptional mechanical and electrical properties have motivated the research community to develop graphene based flexible electronics [5]. Graphene monolayers were first produced via mechanical exfoliation of graphite. As graphene gained popularity, various synthesis procedures such as chemical vapor deposition (CVD) [6], epitaxial growth on SiC, etc, were adopted to produce graphene in large quantities. While the extraordinary properties of graphene are specific to pristine monolayers, the synthesis of such material is very expensive with a low production

doi:10.1088/978-0-7503-3439-6ch8

© IOP Publishing Ltd 2021

yield. Solution processed graphene via oxidation and exfoliation of graphite turned out to be the most viable method to mass-produce graphene at low cost. Although the synthesis process introduced oxygen functional groups and defects into the graphene lattice affecting its physical properties, the graphene oxide (GO) and reduced graphene oxide (rGO) gained popularity among sensor researchers. The functional groups in GO aided the formation of printable inks on various solvents providing new fabrication strategies for sensor development.

Although numerous studies on monolayer graphene based strain sensors have shown promising results, the synthesis and transfer of monolayer graphene still remains a costly and complicated procedure. Flexible strain sensors using chemically derived graphene such as GO and rGO have garnered high interest among researchers due to the low cost of synthesis and facile device fabrication techniques. GO and rGO obtained through various reduction procedures and methods such as chemical treatment, laser, thermal, spark, and microwave reduction have been used effectively for the fabrication of flexible strain sensors. Different sensing mechanisms have been reported in these types of devices, such as the tunneling effect between adjacent graphene flakes, sliding of graphene flakes changing the effective conducting surface area, changing percolative networks of graphene, etc. Graphene based inks have been printed on different types of substrates such as the polymers PDMS, PMMA, PET, polyimide, PU, etc, and other complex materials such as cotton, silk, polyester fabrics, cellulose filter paper, etc, that are suitable for flexible sensors. A study has shown the use of rGO or other derived graphene in the form of graphene foams, monoliths, or aerogels. These types of materials are synthesized by making a suspension of GO or rGO with a settling polymer, which is then cast into a foam or monolith structure using a suitable sacrificial template. As the suspension settles, or is freeze dried, it attains a defined sponge/foam-like structure exhibiting high structural flexibility and strain sensitivity [7–11]. The crumpled, wrinkled, and porous nature of the rGO provides extensive scope for physical deformations and strain sensing behavior [12–17]. The sensors were also designed as skin conformal, bio-compatible, and flexible electronic skin, useful for measuring physiological parameters. While some sensors encapsulated the rGO in a bio-compatible polymer, others involved the fabrication of rGO polymer composites, where the rGO flakes formed an intercalated network with the flexible polymer, establishing an electrical nexus within the material [18]. In another approach, rGO was realized on a suitable substrate by direct reduction of the GO layer to obtain reduced functionalized graphene. These methods include laser treatment to reduce GO [19–21], thermal reduction or spark induced reduction of GO [22, 23], thermal reduction of GO–paper composite to obtain graphene oxide paper [24], etc. These approaches provide the advantage of the custom selection of a variety of substrates with the use of derived graphene material to meet any flexible sensor applications.

The following sections provide a review of the studies reported for the realization and applications of flexible strain sensors using rGO and its composites.

8.2 Graphene–metal nanocomposites for flexible sensor applications

Flexible strain sensors based on graphene–metal nanocomposites have displayed enhanced sensitivity due to the influence of the metal nanoparticles embedded in inter-layer graphene sheets. One such study discussed here is based on a nanocomposite composed of rGO and palladium nanoparticles [25, 26]. Palladium has demonstrated the highest strain sensitivity among noble metals [27]. A strain sensor with a thickness of 100 μm was fabricated on a stainless steel (SS304) substrate with rGO–Pd nanocomposite using screen printing technology, as shown in figure 8.1(a). A micro-universal test machine (UTM) was used to characterize the mechanical response of the as-fabricated sensor. The change in electrical resistance (ΔR) of the sensor was recorded as 10.75 Ω for loading and unloading under the tensile strain of 166.67 microstrains for 100 cycles, as shown in figure 8.1(b). The sensor's response was observed to be stable and the same ΔR was maintained for all the cycles. Similarly, different loads of 83.33 and 250 microstrain responses were applied on the same sensor and the electrical response was compared with the previous load of 166.67 microstrains, as shown in figure 8.1(c). The ΔR of the sensor was noted as 6.92, 10.75, and 14.55 Ω for strains of 83.33, 166.67, and 250 microstrains, respectively. The sensor's response was proven to be linear and stable with different loads. The as-fabricated sensor was tested at higher loads up to 3000 microstrains. The response at different loads was observed to be linear and stable. The proposed sensor demonstrated high sensitivity for broad strain values. The average gauge factor was found to be higher compared to traditional metal-type strain sensors.

In general, the working principle of GO and rGO based strain sensors follow different mechanisms such as change in the overlapping area of adjacent sheets [28, 29], structural deformation [30, 31], and tunneling mechanism [32]. In the reported study, the rGO–Pd sensor works on the principle of a change in the overlapping area of graphene nanosheets with respect to the applied strain leading to a change in the sensor resistance. The response of the rGO–Pd nanocomposite was observed to be superior when compared to standalone rGO strain sensors. The detailed working

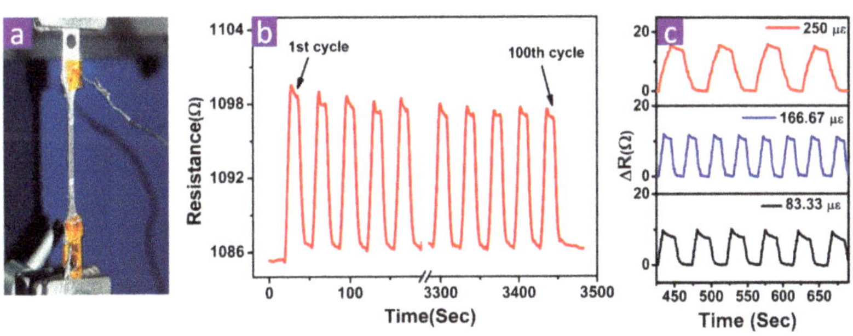

Figure 8.1. (a) The tested strain sensor. (b) The response of the sensor for a load of 166.67 microstrains. (c) The sensor's response comparison for different loads [25], copyright 2019 IOP Publishing. Reproduced with permission. All rights reserved.

principle and the Pd nanoparticles' contribution to enhance the sensitivity of the sensor has been explained in the reported work [25, 26].

8.3 Pulse measurement using PDMS encapsulated rGO–Pd sensors

The application of rGO–Pd nanocomposite was demonstrated as a wearable sensor for pulse measurements. A commercially viable flexible strain sensor with a biocompatible substrate of polydimethylsiloxane (PDMS) was fabricated. The schematic diagram of the sensor is shown in figure 8.2(a). A PDMS layer of uniform thickness was spin coated on a glass substrate. Further, the rGO–Pd nanocomposite was fabricated with screen printing technology. After making the electrical contacts, the rGO–Pd film was sandwiched using a PDMS layer to ensure tight packing and to avoid moisture absorption by the film. The sensor was transferred onto a healthy 35-year-old person to measure the pulse at a rest position, as shown in figure 8.2(b). The sensor was highly skin conformal and was found to be sensitive to muscle movements. A change in resistance (ΔR) of around 450 Ω was observed, as shown in figure 8.2(c), and it can be perceived from the graph that the sensor measured 11 pulses in 10 s, which is a resting heart rate of 66 pulse min^{-1}. The response of the proposed sensor to finger tapping, wrist bending, and finger bending incidents was observed and it was noted to be stable and linear with high sensitivity.

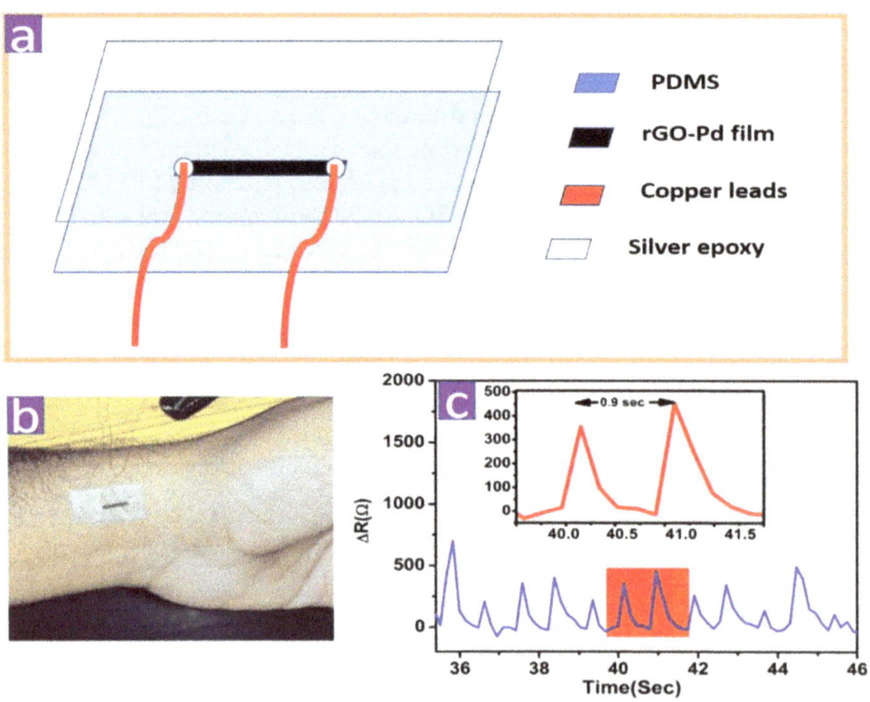

Figure 8.2. (a) Schematic diagram of an rGO–Pd composite with a flexible substrate. (b) Pulse measurement. (c) Change in resistance w.r.t. human pulse.

8.4 Graphene capacitive strain sensor

The following work explored the use of graphene in the form of a capacitor as a flexible strain sensor [33]. An innovative approach towards the development of graphene–electrolyte capacitive sensor (GECS) using a supercapacitor model has been demonstrated in this work. Capacitive sensors are one of the prominent types of sensors used in numerous applications, mainly in strain and force detection type devices. Capacitive sensors have proved superior to other types of sensors in aspects such as high sensitivity, low power consumption, and fast response, but fall short in areas such as noise immunity and circuit complexity [34–37]. Most recent developments in capacitive sensing are focused toward reducing noise by means of electronic design and shielding technology, but with limited improvement at higher expenditure. Most capacitive sensors have a base capacitance in the range of femtofarads to picofarads and a corresponding change in capacitance to applied input is in the range of femtofarads and even attofarads in some cases. This minute magnitude of signal requires complex electronics for measurement and is prone to noise signals, which are in the range of picofarads. This work aimed to improve the base capacitance of the sensor which is the primary reason for issues related to capacitive sensors. The reported work set out to address this issue through the use of a graphene supercapacitor as a sensor.

Graphene has been widely researched and adopted in supercapacitors due to its properties which are suited to an electrode material. Graphene has emerged as a wonder material owing to its outstanding electrical, thermal, optical, and mechanical properties. Supercapacitors realized using graphene have proved exceedingly superior to its competitors due to its high electric double layer capacitance (EDLC) property. EDLC is referred to as the effective capacitance developed at the electrode–electrolyte interface due to opposite charges collecting near the phase separation. The atomic scale distance between these charges renders high capacitance to the EDLC which is useful in energy storage technology. However, it is seen that this high EDLC of graphene has not been exploited for sensor applications. The high specific surface area of graphene (theoretically 2675 m^2 g^{-1}) and large specific capacitance value (theoretically 550 F g^{-1}) [38, 39] can provide a solution for the shortcomings seen in capacitive sensors, and hence has been chosen as the sensor material.

The GECS consisted of a thermally reduced microwave exfoliated graphene oxide (rMEGO) based interdigitated electrode (IDE) capacitor on a polymethyl methacrylate (PMMA) substrate (figures 8.3(a) and (b)). The rMEGO was characterized via standard techniques such as x-ray diffraction, Raman spectroscopy, and field emission scanning electron microscopy (FE-SEM). The IDE has a chromium–gold thin-film underlayer for easy device contacts. The IDE was coated with PVA–H_2SO_4 gel electrolyte GE which gave the necessary EDLC. Polydimethylsiloxane (PDMS) polymer was coated on the sensor as protection against environmental factors. Experimental studies were carried out in a three-point bending set-up to test the capacitance response of GECS on the application of known strain.

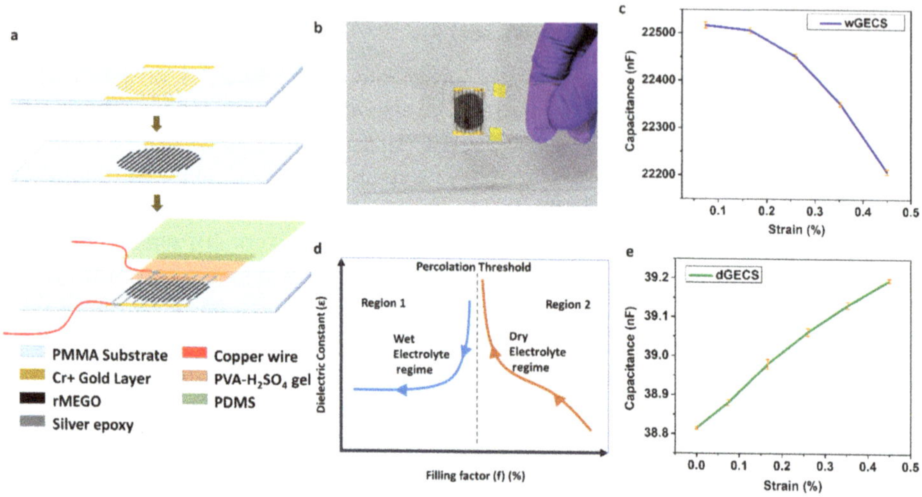

Figure 8.3. (a) Schematic representation of GECS device fabrication. (b) Image of the as-prepared GECS device. (c) wGECS capacitance response with respect to percentage strain induced in the sensor. (d) Dielectric constant versus biaxial strain (schematic view only). Arrows depict a decrease in filling factor due to strain. (e) dGECS capacitance response with respect to percentage strain induced in the sensor [33], copyright 2019 IOP Publishing. Reproduced with permission. All rights reserved.

Upfront, a million-fold increase in the base capacitance was achieved, due to the EDLC incorporated in the sensor. The device showed a very high base capacitance of ~22 uF in the no-load condition and unprecedented high operating output range for the corresponding strain input. The operating range of the sensor was observed to be ~300 nF for the given input 4500 $\mu\varepsilon$ (figure 8.3(c)). The GECS exhibited an impressive device sensitivity of 11.24 nF kPa^{-1} which helps in detection of the slightest strain magnitudes. The study also makes a comparison of a metal thin-film capacitive sensor of the same dimensions as the GECS and the latter shows an improvement of a million times in the base capacitance, from the picofarad range to the microfarad range. The high base capacitance means a requirement for a simple inexpensive electronic circuit to measure the capacitance and the high operating range simplifies the measurement of the change in capacitance, which is in the nanofarad range. Two regimes of operation, the wet-electrolyte and dry-electrolyte regime, are identified based on the procedure of device fabrication, both of which can have specific advantages and applications. The wet-electrolyte GECS (wGECS) and dry-electrolyte GECS (dGECS) devices were observed to show varying behavior which could be explained based on the percolation theory. The dGECS showed a reduced magnitude of base capacitance (nanofarad range) primarily due to loss of electrolyte moisture and hence lower EDLC. Further, the capacitance variation on the application of strain was seen to be contrary in dGECS compared to wGECS, which follows the graph showing a change of the dielectric constant with respect to varying filling factor, which is caused by the applied strain (figures 8.3(d) and (e)), as reported earlier by Gaiser *et al* [40]

The reliability of the devices was tested by subjecting them to loading–unloading cycles and repeatable results were seen even after hundreds of such cycles

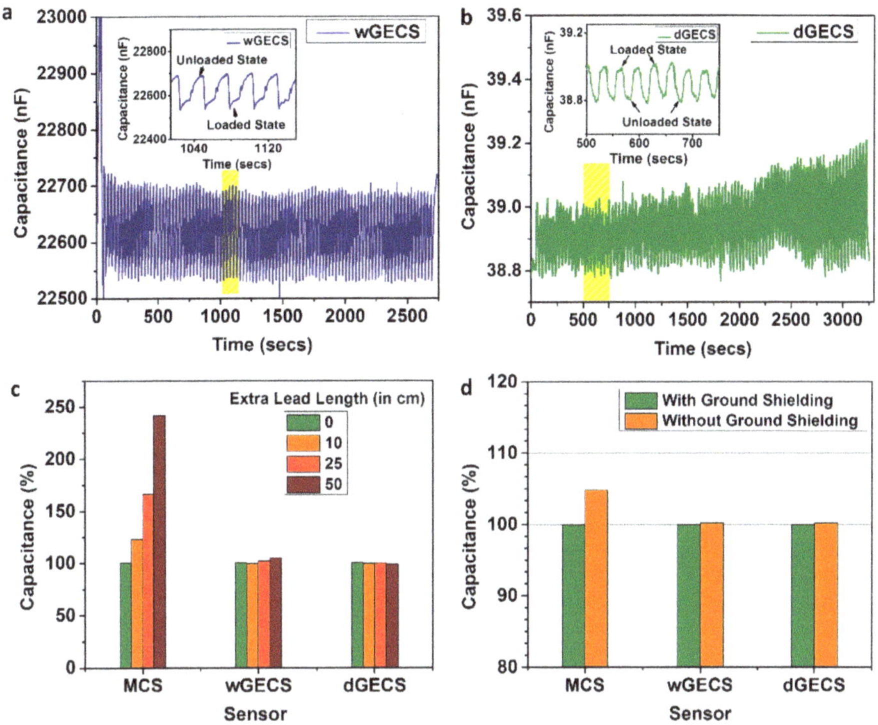

Figure 8.4. (a) A cyclic loading experiment (100 cycles) on a wGECS in the linear operation range. (b) A cyclic loading experiment (100 cycles) on a dGECS. (c) The lead capacitance effect: the effect of change in the lead wire length of the sensors on the capacitance measurement. (d) The stray capacitance effect: the effect of the presence and absence of a ground shield around sensors on the capacitance measurement [33], copyright 2019 IOP Publishing. Reproduced with permission. All rights reserved.

(figures 8.4(a) and (b)). The device was tested for noise susceptibility and excellent immunity towards noise was observed. A simple experiment to gage the effect of lead capacitance was carried out by varying the lead length and observing the changing capacitance of the sensor. Similarly, the effect of grounding and operation without a ground shield on the sensor capacitance was also studied. The major sources of noise, i.e. lead capacitance and stray capacitance, only affected the sensor performance by 1.6% (with a 500% increase in lead length) and 0.19%, respectively (figures 8.4(c) and (d)). Therefore, the prime deterrents in any capacitive sensor are rendered insignificant. The transient time response of the device was found to be as fast as 20 ms, much better than those available in literature. The fabrication techniques involved in realizing the device are inexpensive non-cleanroom-method based steps which can be scaled easily to an industrial environment. The substrate can be modified easily to suit any sensing application keeping the original design as a template. Traditionally capacitive sensors were known for low power and high sensitivity, but this work has added more desirable features such as low cost, low noise, and the requirement for a simple measurement circuit to give a complete sensor module.

The device can be utilized for numerous applications, more specifically as biomedical sensors. One such biomedical application of breath pattern monitoring has been tested

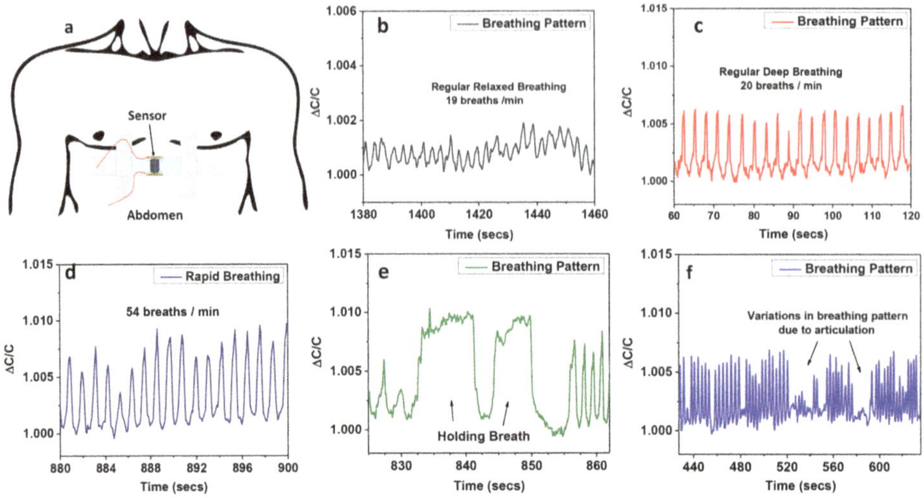

Figure 8.5. (a) Biomedical application of a GECS: the dGECS is firmly strapped on the abdomen to detect the breathing pattern, showing different types of breathing: (b) relaxed breathing, (c) deep breathing, (d) rapid breathing, and the effect of (e) holding breath and (f) talking on the breathing pattern [33], copyright 2019 IOP Publishing. Reproduced with permission. All rights reserved.

and demonstrated (figures 8.5(a)–(f)). The experiments show the detection of the breathing rate of the subject in the normal condition and in a rapid breathing condition, and different modes such as shallow breathing, deep breathing, holding of breath, and breath pattern during speech. This low cost, facile fabricated device for continuous breath pattern monitoring has been tested and can has potential applications in human respiratory system supervision. A study of the state-of-the-art in biomedical sensing devices shows that piezoresistive and capacitive principles are used widely in the majority of systems. Even though piezoresistive sensors have a simple design and easy readability, they suffer from low sensitivity, slow response, high power consumption, and drift to external factors. The reviewed work has provided capacitive sensors with all the desirable properties, such as simple design, high sensitivity, high range, low power consumption, fast response, and easy readability. The device has been demonstrated as a flexible, body-wearable, continuous breath monitoring system which can be used for respiratory system analysis while carrying out all routine daily activities.

8.5 Graphene based flex sensor on textile

The following work demonstrates a graphene based strain sensor on a textile substrate realized by fast spontaneous reduction of GO, allowing the opportunity for large-scale sensor fabrication [41]. Sensors can be woven into stretchy fabrics to create smart garments that are capable of tracking vital signs such as temperature, respiration, hydration level, and heart rate on the human body [42]. The fabrics are made out of materials ranging from cotton and polyester, to nylon, that are flexible and offer great potential to integrate electronic devices for a wide range of applications. Sportswear brands such as Adidas, Nike, and Under Armor are working on textile embedded electronics for athletic applications to track the body parameters and performance of athletes. Similarly, tech giants such as Samsung, Alphabet, etc, are developing smart textiles for energy harvesting that could power various gadgets. A growing number of companies are developing pressure, strain, and temperature monitoring devices for garment-based sensor applications.

Graphene and its derivatives (GO and rGO) form functionalized inks in various solvents and assist in printing on cloth. Graphene based inks demonstrate versatile characteristics with tunable conductivity, chemical stability, mechanical flexibility, transparency, and unique rheology to accommodate a wide range of printing strategies. Solution processed graphene can be dispersed in a suitable solvent to obtain printable inks of desired characteristics. Although these materials yield lower electrical conductivity compared to pristine graphene, the low manufacturing cost and good mechanical and chemical stability have made them a prominent alternative to conventional conductive inks.

Two methods can be followed to integrate solution processed graphene on cloth to form a conducting layer for sensing applications:

1. Coating rGO inks using various printing and coating methods. rGO, being hydrophobic in nature, requires special solvents and additives to form uniform dispersion inks (figure 8.6(a)) [43].
2. GO on the other hand is hydrophilic due to the presence of oxygen functional groups. GO readily disperses in various solvents including water and can be

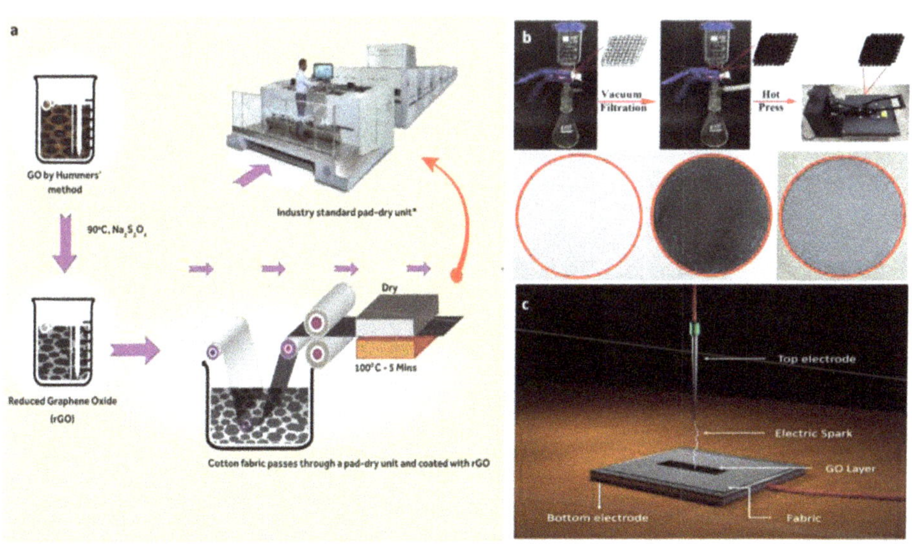

Figure 8.6. Integration of graphene on textile: (a) Preparation of rGO inks and transferring the material onto fabric using a pad dry unit for smart textile applications. (Reproduced with permission from [43]. Copyright 2017 American Chemical Society.) (b) GO solution deposited onto the fabric via a vacuum filtration method and subjected to thermal reduction. The reduction of GO was confirmed by a change in the color on GO deposition on the fabric. (Reproduced with permission from [22]. Copyright 2017 Elsevier.) (c) Schematic illustrating the spark reduction of GO. The set-up consists of GO coated fabric placed on a flat bottomed electrode and subjected to an electric spark from a pointed top electrode. (Reproduced with permission from [41]. Copyright 2020 American Chemical Society.)

easily transferred onto the fabric. The GO coated cloth is later subjected to reduction to remove the functional groups from the GO.

Transferring GO onto fabric and on-substrate reduction to form rGO is a widely used methodology due to the simple fabrication steps involved. Common reduction techniques used for GO reduction on fabric are thermal annealing (figure 8.6(b)) and chemical treatments [22, 44]. While these methods offer low-cost and large-scale reduction of GO, care should be taken to minimize the thermal or chemical damage caused to the fabric substrate. Recently, a new method was devised to reduce graphene oxide on fabric using an electric spark [41]. Here, GO coated fabric was subjected to a stream of electric sparks, reducing it instantly without causing any damage to the substrate. Figure 8.6(c) demonstrates the set-up for spark reduction consisting of a flat bottomed electrode and a spark dispensing top electrode interceded by the GO coated fabric. The method also provides active control over the extent of reduction, offering different techniques to form conducting patterns on the fabric substrate.

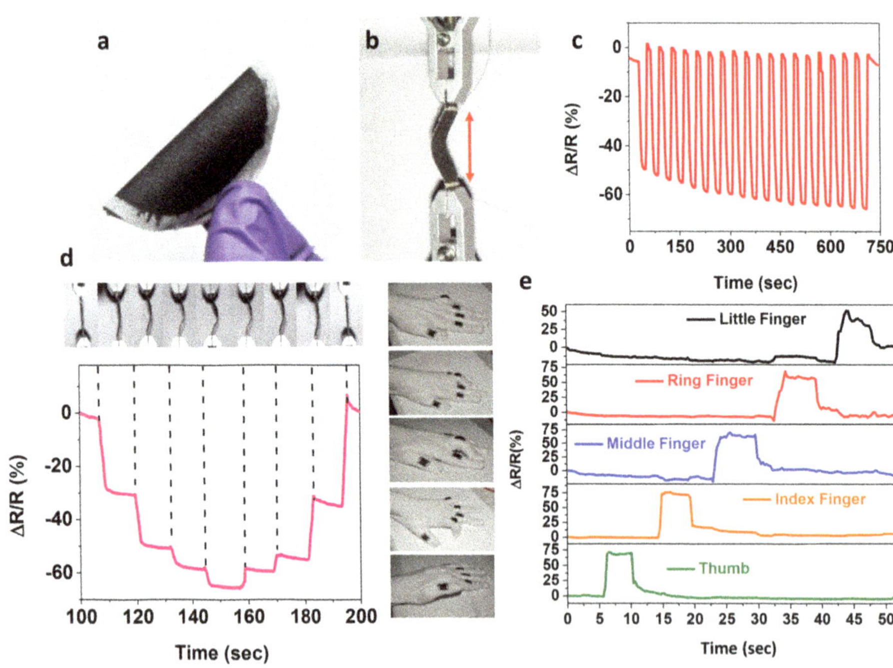

Figure 8.7. SrGO coated fabric for flex sensing: (a) The SrGO film on fabric displayed excellent flexibility. (b) To determine the sensor response for different bending conditions, the sample was tested using a μUTM. (c), (d) The sensor exhibited a large change in resistance for different bending angles with high sensitivity, stable response, and good repeatability. (e) SrGO films were attached to the gloves to sense finger movements. The response of the sensor for individual finger movement was plotted. (Reproduced with permission from [41]. Copyright 2020 American Chemical Society.)

The spark reduced graphene oxide (SrGO) film exhibited excellent conformity to the fabric substrate and high flexibility, as shown in figure 8.7(a). To study the flex sensing performance, a rectangular piece of SrGO fabric was subjected to different bending conditions using a μUTM analyzer, as shown in figure 8.7(b). Bending at different angles introduced specific strain into the sensing layer thereby changing the device's resistance. The response of the sensor was found to be highly stable and repeatable, as shown in figures 8.7(c) and (d). The mechanically characterized SrGO fabric was demonstrated as a wearable flex sensor by attaching it onto commercially available gloves to sense finger movements. Figure 8.7(e) illustrates the change in sensor resistance with respect to individual finger bending.

The economic benefits of spark reduction can be channeled toward the industrially scalable production of smart wearables for various applications in the domains of healthcare and robotics.

8.6 Summary and conclusions

This chapter demonstrates different techniques to fabricate graphene based flexible strain sensors. Chemically derived graphene provides a low-cost and facile method to fabricate high performance, bio-compatible, and flexible sensors. The material properties can be tuned through the formation of nanocomposites to improve the sensing performance for target applications. Graphene's ability to form uniform dispersions in various solvents produces functionalized inks that aid in various printing techniques. The versatile fabrication process and methodologies offer numerous application possibilities towards wearable sensing. Strain sensors are effectively employed to detect several body parameters such as muscle movements, pulse monitoring, and tactile sensing for medical and robotic applications. The emergence of this area of research has led to several works which are on the verge of materializing into market-oriented, highly efficient sensors for a variety of applications. The above studies show the versatility of this material and the scope of future progress in graphene based flexible sensors.

References

[1] Novoselov K S, Fal'ko V I, Colombo L, Gellert P R, Schwab M G and Kim K 2012 A roadmap for graphene *Nature* **490** 192–200

[2] Geim A K and Novoselov K S 2007 The rise of graphene *Nat. Mater.* **6** 183–91

[3] Novoselov K S, Geim A K, Morozov S V, Jiang D, Zhang Y, Dubonos S V, Grigorieva I V and Firsov A A 2004 Electric field effect in atomically thin carbon films *Science* **306** 666–9

[4] Lee C, Wei X, Kysar J W and Hone J 2008 Measurement of the elastic properties and intrinsic strength of monolayer graphene *Science* **321** 385–8

[5] Liu H *et al* 2018 Electrically conductive polymer composites for smart flexible strain sensors: a critical review *J. Mater. Chem.* C **6** 12121–41

[6] Reina A, Jia X, Ho J, Nezich D, Son H, Bulovic V, Dresselhaus M S and Kong J 2009 Large area, few-layer graphene films on arbitrary substrates by chemical vapor deposition *Nano Lett.* **9** 30–5

[7] Liu S *et al* 2017 Ultrafast dynamic pressure sensors based on graphene hybrid structure *ACS Appl. Mater. Interfaces* **9** 24148–54

[8] Zhang B-X, Hou Z-L, Yan W, Zhao Q-L and Zhan K-T 2017 Multi-dimensional flexible reduced graphene oxide/polymer sponges for multiple forms of strain sensors *Carbon* **125** 199–206

[9] Samad Y A, Li Y, Alhassan S M and Liao K 2015 Novel graphene foam composite with adjustable sensitivity for sensor applications *ACS Appl. Mater. Interfaces* **7** 9195–202

[10] Davoodi E *et al* 2020 3D-printed ultra-robust surface-doped porous silicone sensors for wearable biomonitoring *ACS Nano* **14** 1520–32

[11] Wan S, Bi H, Zhou Y, Xie X, Su S, Yin K and Sun L 2017 Graphene oxide as high-performance dielectric materials for capacitive pressure sensors *Carbon* **114** 209–16

[12] Jia J, Huang G, Deng J and Pan K 2019 Skin-inspired flexible and high-sensitivity pressure sensors based on RGO films with continuous-gradient wrinkles *Nanoscale* **11** 4258–66

[13] Weng M, Sun L, Qu S and Chen L 2020 Fingerprint-inspired graphene pressure sensor with wrinkled structure *Extreme Mech. Lett.* **37** 100714

[14] Xu M, Qi J, Li F and Zhang Y 2018 Highly stretchable strain sensors with reduced graphene oxide sensing liquids for wearable electronics *Nanoscale* **10** 5264–71

[15] Oren S, Ceylan H, Schnable P S and Dong L 2017 High-resolution patterning and transferring of graphene-based nanomaterials onto tape toward roll-to-roll production of tape-based wearable sensors *Adv. Mater. Technol.* **2** 1700223

[16] Kedambaimoole V, Shirhatti V, Neella N, Ray P, Rajanna K and Nayak M M 2017 PDMS encapsulated graphene–nickel composite film as flexible tactile sensor *IEEE Sensors* (Piscataway, NJ: IEEE) pp 1–3

[17] Shirhatti V, Nuthalapati S, Neella N, Kedambaimoolc V, Nayak M M and Rajanna K 2018 Supercapacitor based approach towards development of graphene sensors *IEEE 13th Annual Int. Conf. on Nano/Micro Engineered and Molecular Systems (NEMS)* (Piscataway, NJ: IEEE) pp 267–70

[18] Qiu J, Guo X, Chu R, Wang S, Zeng W, Qu L, Zhao Y, Yan F and Xing G 2019 Rapid-response, low detection limit, and high-sensitivity capacitive flexible tactile sensor based on three-dimensional porous dielectric layer for wearable electronic skin *ACS Appl. Mater. Interfaces* **11** 40716–25

[19] Das P S, Chhetry A, Maharjan P, Rasel M S and Park J Y 2019 A laser ablated graphene-based flexible self-powered pressure sensor for human gestures and finger pulse monitoring *Nano Res.* **12** 1789–95

[20] Sun B, McCay R N, Goswami S, Xu Y, Zhang C, Ling Y, Lin J and Yan Z 2018 Gas-permeable, multifunctional on-skin electronics based on laser-induced porous graphene and sugar-templated elastomer sponges *Adv. Mater.* **30** 1804327

[21] Zhu Y, Cai H, Ding H, Pan N and Wang X 2019 Fabrication of low-cost and highly sensitive graphene-based pressure sensors by direct laser scribing polydimethylsiloxane *ACS Appl. Mater. Interfaces* **11** 6195–200

[22] Ren J, Wang C, Zhang X, Carey T, Chen K, Yin Y and Torrisi F 2017 Environmentally-friendly conductive cotton fabric as flexible strain sensor based on hot press reduced graphene oxide *Carbon* **111** 622–30
[23] Kedambaimoole V, Kumar N, Shirhatti V, Nuthalapati S, Nayak M M and Konandur R 2020 Electric spark induced instantaneous and selective reduction of graphene oxide on textile for wearable electronics *ACS Appl. Mater. Interfaces.* **12** 15527–37
[24] Tao L-Q, Zhang K-N, Tian H, Liu Y, Wang D-Y, Chen Y-Q, Yang Y and Ren T-L 2017 Graphene–paper pressure sensor for detecting human motions *ACS Nano* **11** 8790–5
[25] Nuthalapati S, Shirhatti V, Kedambaimoole V, Neella N, Nayak M M, Rajanna K and Takao H 2019 Highly sensitive, scalable reduced graphene oxide with palladium nano-composite as strain sensor *Nanotechnology* **31** 035501
[26] Nuthalapati S, Shirhatti V, Neella N, Kedambaimoole V, Gaddam V, Rajanna K and Nayak M M 2018 RGO–PD nano-composite based films for strain measurement using screen printing technique *IEEE 13th Annual Int. Conf. on Nano/Micro Engineered and Molecular Systems (NEMS)* (Piscataway, NJ: IEEE) pp 381–4
[27] Guo J 1994 Noble metal alloys as strain gauge materials *Platinum Metals Rev.* **38** 98
[28] Hempel M, Nezich D, Kong J and Hofmann M 2012 A novel class of strain gauges based on layered percolative films of 2D materials *Nano Lett.* **12** 5714–8
[29] Li X *et al* 2012 Stretchable and highly sensitive graphene-on-polymer strain sensors *Sci. Rep.* **2** 870
[30] Fu X-W *et al* 2011 Strain dependent resistance in chemical vapor deposition grown graphene *Appl. Phys. Lett.* **99** 213107
[31] Wang Y, Yang R, Shi Z, Zhang L, Shi D, Wang E and Zhang G 2011 Super-elastic graphene ripples for flexible strain sensors *ACS Nano* **5** 3645–50
[32] Zhao J, He C, Yang R, Shi Z, Cheng M, Yang W, Xie G, Wang D, Shi D and Zhang G 2012 Ultra-sensitive strain sensors based on piezoresistive nanographene films *Appl. Phys. Lett.* **101** 063112
[33] Shirhatti V, Kedambaimoole V, Nuthalapati S, Neella N, Nayak M M and Rajanna K 2019 High-range noise immune supersensitive graphene–electrolyte capacitive strain sensor for biomedical applications *Nanotechnology* **30** 475502
[34] Tiwana M I, Redmond S J and Lovell N H 2012 A review of tactile sensing technologies with applications in biomedical engineering *Sens. Actuators Phys.* **179** 17–31
[35] Wan Y, Wang Y and Guo C F 2017 Recent progresses on flexible tactile sensors *Mater. Today Phys.* **1** 61–73
[36] Puers R 1993 Capacitive sensors: when and how to use them *Sens. Actuators Phys.* **37–38** 93–105
[37] Saccomandi P, Schena E, Oddo C M, Zollo L, Silvestri S and Guglielmelli E 2014 Microfabricated tactile sensors for biomedical applications: a review *Biosensors* **4** 422–48
[38] Karnan M, Subramani K, Sudhan N, Ilayaraja N and Sathish M 2016 Aloe vera derived activated high-surface-area carbon for flexible and high-energy supercapacitors *ACS Appl. Mater. Interfaces* **8** 35191–202
[39] Xu Y, Lin Z, Huang X, Liu Y, Huang Y and Duan X 2013 Flexible solid-state super-capacitors based on three-dimensional graphene hydrogel films *ACS Nano* **7** 4042–9
[40] Gaiser P, Binz J, Gompf B, Berrier A and Dressel M 2015 Tuning the dielectric properties of metallic-nanoparticle/elastomer composites by strain *Nanoscale* **7** 4566–71

[41] Kedambaimoole V, Kumar N, Shirhatti V, Nuthalapati S, Nayak M M and Konandur R 2020 Electric spark induced instantaneous and selective reduction of graphene oxide on textile for wearable electronics *ACS Appl. Mater. Interfaces* **12** 15527–37
[42] Wicaksono I, Tucker C I, Sun T, Guerrero C A, Liu C, Woo W M, Pence E J and Dagdeviren C 2020 A tailored, electronic textile conformable suit for large-scale spatiotemporal physiological sensing *in vivo NPJ Flex. Electron.* **4** 1–13
[43] Karim N, Afroj S, Tan S, He P, Fernando A, Carr C and Novoselov K S 2017 Scalable production of graphene-based wearable e-textiles *ACS Nano* **11** 12266–75
[44] Karim M R and Hayami S 2017 Chemical, thermal, and light-driven reduction of graphene oxide: approach to obtain graphene and its functional hybrids *Graphene Materials—Advanced Applications* (Rijeka: Intech)

IOP Publishing

Printed and Flexible Sensor Technology
Fabrication and applications
Subhas Chandra Mukhopadhyay and Anindya Nag

Chapter 9

Screen printed electrochemical and impedance biosensors

Chirasree Roychaudhuri and Naresh Mandal

In this chapter, different aspects of the screen printing method are elaborated along with the progress in research on various electrochemical and impedance biosensors, involving the detection of environmental contaminants, food toxins, protein molecules, bacteria, and viruses. Different cases are presented in a set of comprehensive tables, showing an overall review of published approaches. Finally, significant achievements and challenges are outlined along with future prospects.

9.1 Introduction

The reliable on-site analysis of various biological species, ranging from environmental contaminants and food toxins to DNA, bacteria, and viruses, is possible using electrochemical sensing due to its simplicity, small size, and automation. Additionally, the possibility of coupling screen printing technology has augmented their applicability for point-of-care diagnostics. Screen printed electrodes exhibit ductility in various shapes, can be composed of different materials, and can be modified suitably using various modifiers [1–5]. Modifications can be incorporated using different nanomaterials to enhance the surface-to-volume ratio and electron transfer rates. These improvements result in increased sensitivity of detection. Further, screen oriented electrodes can be applied for impedance biosensing. In this chapter we focus on a fundamental understanding of screen printing technology and some applications in electrochemical and impedance biosensing. Finally, challenges are also highlighted.

9.2 A fundamental understanding of screen printing technology

Screen printing or thick film technology is a commonly used method to produce low cost sensors and biosensors. Screen printed electrodes can be fabricated with various conductive inks, shapes, and measurements, depending on the analytical need. Other

doi:10.1088/978-0-7503-3439-6ch9

© IOP Publishing Ltd 2021

than their manufacturing, these platforms are suitable to be customized with a variety of (nano)materials and bio-elements for application in the ecological, food, and clinical fields. Despite the fact that the mechanical procedures associated with screen printing methods are largely uniform for all applications, the composition of the inks used within the printing processes varies extraordinarily. Screen printing is a stencil procedure where the screen includes a closed non-image area and open image area. The ink is forced through the screen by a resilient squeegee. The screen comprises a frame covered with screen fabric, and a stencil containing the printed data. The stencil/plate is generally produced using a light-sensitive emulsion which is photographically exposed, so that the printing zones are washed away, while the non-image areas are made permanent. The stencil is positioned on a fine fabric, which holds the parts of the design in place. The screen is a fine fabric made of natural silk, plastic, or metal fibers. Ink is transferred through the open mesh that is not covered by the stencil. The screen printing plate is, therefore, a combination of the screen and stencil [6]. The stencil on the fabric defines the actual printed picture. The stencil is on the side of the screen opposite to the side on which the squeegee (blade) works, to avoid damage and wear to the stencil [6]. Additionally, an important parameter in platemaking is the frame. Today, screen printing frames are usually made of aluminum, although some companies produce wooden and steel frames. The selection of the screen frame depends on the frame size, durability, stability, cost, and stretching method [7]. A squeegee is an apparatus used to push the ink through the stencil delivered on the screen. They are usually made from rubber or polymer and have a wooden or plastic handle in the case of manual printing. The squeegee should be an inch or two shorter than the width of the screen [8].

Screen printing procedures. Printing methods can be classified as contact and noncontact techniques, depending on whether they include a direct connection, often with high pressure, between the substrate and the surface from which the material is transferred, or whether the transfer of materials occurs by ejection [9, 10]. Advances in printing technologies have incorporated modification of traditional contact methods utilized for newspaper production and graphic art reproduction (e.g. screen and gravure printing) to develop more recent noncontact processes oriented towards highly detailed and exact purposes. The deposition of biological samples on biosensor platforms by inkjet printing or the manufacture of paper-based analytical devices (PADs) assisted by wax-ink printing are some examples of the most recent accomplishments of noncontact printing techniques.

1. *Printing media*: The use of a printing medium (i.e. ink or paste) in the screen printing process is fundamental for the transfer of the printed design. Normally these inks or pastes are manufactured utilizing conductive particles inside a solvent/binder mixture to permit the transfer of the particulate onto the substrate. Such inks can be made readily available and are very helpful to the user as they have the capacity to be controlled to yield different working electrode compositions. Wang *et al* show that the choice of carbon ink should depend on the current application, because of the differences within the electrochemical signals seen in a range of inks [11]. Some examples of

printable inks include the fabrication of platinum [12] and gold [13] screen printed sensors which have been applied in the electroanalytical detection of chromium species (VI and II), hydrazine, and hydrogen peroxide [12, 13]. Crucially, it has been determined that the requirement for electrode potential cycling prior to usage has been alleviated in the case of screen printed sensors owing to the noble metal utilized within the screen printing process being in the form of an oxide [14, 15]. For electrochemical sensing applications, these electrodes are modified by an array of raw carbon materials, such as carbon nanotubes [16], graphene [17], mediated carbon structures [18], and nano-particles [19], to give some examples. Through being able to modify the electrodes, the electron transfer capabilities are enhanced at the electrode surface, thereby improving the sensor response. Figure 9.1 shows the correlation of the voltammetric behavior of unmodified screen printed electrodes with that of polymer modified screen printed electrodes.

Choudhry *et al* have shown that this can be used for the electrolytic modification of various electrocatalytic metals, where the bespoke electrode surface acts as a micron-sized template [20]. It is to be noted that the modification uses less metal and hence has a reasonable cost, particularly where precious materials are utilized. Further, the method also has the benefit of creating a metal ensemble. The term ensemble is used to indicate that the spacing between each metal deposit, namely the metal domain, is not uniform, as in the case of arrays [20]. When an ensemble is used over an electrolytically modified macro-electrode a dramatic change in the mass transport occurs, facilitating low detection limits and improvements in the analytical performance, even though only a fraction of the electrode surface is covered. In addition to pure metallic modifications as reported above, the use of metallic oxides can be utilized. Bismuth oxide is a typical oxide structure that can be used in a printable ink for the recognition of metal ions, with such a change being much the same as using an exceptionally toxic mercury electrode systems.

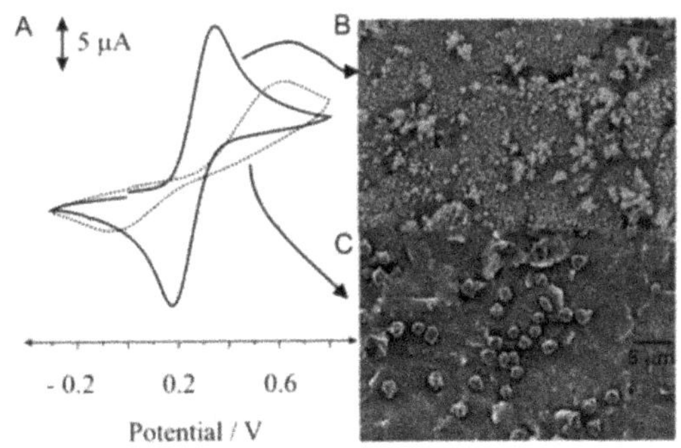

Figure 9.1. (A) Cyclic voltammetric profiles obtained in 1 mM potassium ferrocyanide/1 M KCl using a copper plated standard SPE (solid line) with that of a bespoke electrochemical platform (dashed line). Scans recorded at 100 mV s^{-1} versus SCE. SEM images of the copper plated (B) standard and (C) bespoke screen printed electrochemical platforms. (Reproduced with permission from [20]. Copyright 2010 Elsevier.)

2. *Substrate*: Screen printed electrodes are generally printed on ceramic [21, 22] or plastic substrates [16, 23, 27] and the requirement for ultra-flexible sensors has emerged for various wearable applications. Yang *et al* [28] developed the direct screen printing of amperometric carbon sensors onto clothing and the favorable electrochemical behavior of such textile-based sensors has been used to detect chemicals such as hydrogen peroxide and NADH [28], a case in which ultra-flexible sensors are fundamental in achieving an ideal electrode for such applications. Wang *et al* [24] developed wearable electrochemical sensors on underwater garments using synthetic rubber neoprene [24]. The neoprene-based sensor was assessed towards the voltammetric detection of trace heavy metal contaminants and nitro-aromatic explosives in seawater samples, with further applications including the primary case of enzyme (tyrosinase) immobilization on a wearable substrate towards the amperometric biosensing of phenolic contaminants in seawater. Moreover, depicted in figure 9.2 is the screen printed fabrication of electrochemical tattoos. Wang and co-workers have demonstrated that temporary transfer tattoo (T3) electrochemical sensors can be fabricated for physiological and security monitoring of chemical constituents, leading to the demonstration of 'electronic skin' [25]. Additionally, Wang and co-workers have studied the analysis of sodium within perspiration upon the epidermis, using a comparable arrangement to that shown in figure 9.2.

Bandodkar and co-workers have demonstrated the fabrication, characterization and application of an epidermal temporary-transfer tattoo-based potentiometric sensor, coupled with a miniaturized wearable wireless transceiver, for real-time monitoring of sodium in human perspiration [26]. Other work has investigated the impact of mechanical contortion and stress on polymeric sensors comprising Mylar®, polyethylene naphthalate, and Kapton [27]. In general, a good substrate is simply classed as something that has a decent adhesion to the screen printed ink.

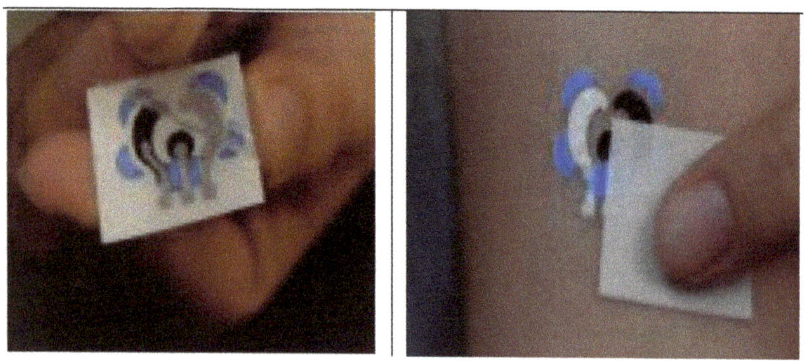

Figure 9.2. Optical images of an electrochemical tattoo upon the substrate and skin. (Reproduced from [25]. Copyright 2012 The Royal Society of Chemistry.)

3. *Other components*:
    (a) *Frame*: The frames may be made of wood, metal, or environmentally friendly recycled plastic. These are lightweight, easy to clean, and easy to lock into hinge clamps. The frames should not soak up water, decay, or twist.
    (b) *Screen work*: Different mesh sizes are utilized for various applications in the screen printing procedure. Multifilament and monofilament polyester are materials often utilized for screen mesh. The mesh count is a number which describes the fineness of the mesh. The mesh count for monofilament is 63, which means there are 63 openings for each centimeter. A wide scope of materials can be extended over the frame to make the screen. Materials such as cotton, silk, nylon, mono- and multifilament polyester, and metal are commonly used today [8]. The thread count is one of the most significant parameters that influences printing quality. It corresponds to the number of threads per centimeter square. Fabrics can be obtained with levels of fineness from 10 to 200 threads cm^{-1}, despite of fact that the most commonly utilized fabrics are in the range of 40 to 120 threads cm^{-1}. Fabrics with higher thread counts provide a better print quality [29]. Moreover, the fabric quality is also determined by the thread thickness used. It is indicated by four thicknesses that range from 'light' to 'heavy'. The fabric count determines the mesh width, and consequently the open screen area in threads cm^{-1}, and the thread thickness in micrometers. This estimation and the stencil's thickness affect the thickness of the ink film transferred [6].
    (c) *Screen prep tape*: This plastic tape is applied around the outer edges of the screen, making clean up a lot simpler.
    (d) *XR blockout*: This compound is perfect for touching up pinholes or register marks.
    (e) *Artist's knife*: This is a basic instrument for masking the film.
    (f) *Squeegee*: An essential piece of equipment (a rubber-like blade in a handle) utilized for flooding the screen with the ink and carrying out the printing. These blades have varying degrees of hardness (durometer) for different printing jobs. It could be said that squeegee is one of the most significant screen printing components. The squeegee is a rubber or plastic blade fixed to a handle for manual printing or to a clamp in mechanized screen printing. For larger runs, typically over 200 prints, the squeegee blades are composed of synthetic materials, such as polyvinyl and polyurethane, rather than rubber, as rubber tends to lose its edge more quickly. The squeegee blade is used to force ink into the mesh and through the printing areas of the stencil, thus applying it to the substrate. The squeegee controls the spread of ink across the screen; it pushes the screen onto the substrate, adapting the screen to the surface of the substrate. It also removes the excess ink from the mesh. The thickness of the printed ink film can be controlled by the amount of pressure applied to the squeegee [8]. Certain

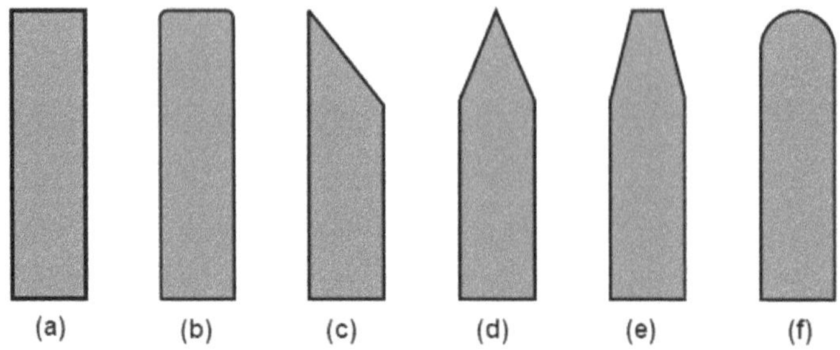

Figure 9.3. Squeegee shapes: (a) square edge, (b) square with rounded edge, (c) single-sided bevel edge, (d) double-sided bevel edge (knife edge), (e) blunt, and (f) round. Reprinted from [29] Copyright (2016), with permission from Elsevier.

squeegee shapes are recommended depending on the planned use. Squeegees can be categorized by their edge shape, as shown in figure 9.3.

During a screen-print cycle, an exceptionally significant factor is the constant movement of the printing medium toward the substrate, at an adequate speed and pressure. This is accomplished by the use of the squeegee. The squeegee applies pressure to force the ink through the stencil design, creating an even print upon the substrate in a controlled way. For the squeegee to complete its necessary tasks, it must be manufactured with materials that are flexible and resilient. During the fabrication of the squeegees many grades of softness are created, it is commonly understood that a softer material makes better contact with the screen. The angle at which the squeegee applies its pressure is likewise a crucial parameter for the successful transfer of the print medium through the stencil design. If the angle is too high the transfer of the printing medium will not be adequate to fill the print area, thereby creating a thinner application of ink/paste to the substrate below. If the squeegee is at too shallow an angle the hydrodynamic pressure increases, therefore transferring too much of the ink/paste onto the substrate and potentially blocking and hindering the controlled snap-off. Figure 9.4 shows a selection of angles and their advantages/disadvantages in the screen printing process.

(g) *Hinge clamps*: These feature a wide-wing thumbscrew for firm and easy fastening to the screen frame. Easy to install, the clamp offers positive locking for perfect registry and precise control.
(h) *Spray adhesive*: This is applied to the clean printing table to keep the paper in place and wrinkle-free during printing.
(i) *Screen strip*: This mixes with water to form a solution that removes the stencil after printing.

Figure 9.4. Angles of the squeegee and their effects on the screen-print.

Upon integration of the above mentioned prerequisites, a print cycle can occur; figure 9.5 presents the process in three simple steps. The initial step consists of the placement of the printing medium (i.e. ink, emulsion, or paste) on the mesh screen, where it can be seen that the screen is not in contact with the substrate. Such contact could potentially damage or spoil the print as the substrate would uncontrollably 'snap-off' the screen, creating an unclean print. The contact only occurs when the squeegee applies pressure over the screen, forcing the printing medium through the stencil design, creating the desired pattern or design in a controlled and efficient manner.

9.2.1 Electrochemical biosensors based on screen printed electrodes

In electrochemical biosensors, an electrode is used as the transduction component. As indicated by the 1999 IUPAC guidelines, an electrochemical biosensor is a self-contained integrated device that is capable of providing specific quantitative or semi-quantitative analytical information using a biological recognition element (biochemical receptor) that is held in direct spatial contact with an electrochemical transduction component. Electrochemical biosensors measure the current produced from oxidation and reduction reactions. Electrochemical biosensors are suited to accomplishing a direct reaction of a biological event to an electrical signal, which makes them quite desirable and attractive for analyzing the content/concentration of a biological sample or analyte of interest. These rely on the detection of an electrical property (resistance, current, potential, capacitance, impedance), which are measured utilizing various strategies such as potentiometry, conductometry, amperometry, or voltammetry. Different screen printed electrodes are deployed in such electrochemical biosensors, which are discussed in the following sections.

1. *Modified screen printed electrodes for biological species and pharmaceutical compound detection*:
    (a) *Electrochemical biosensors for antioxidant detection*: It is essential to detect and control oxidative stress within biological systems. The enzymatic antioxidant activities can be observed from the molecular behavior, where molecules that adopt the role of scavengers are oxidized sacrificially via reactive oxygen or nitrogen species (ROS/RNS). There are two sorts of sacrificial molecules; exogenous e.g.

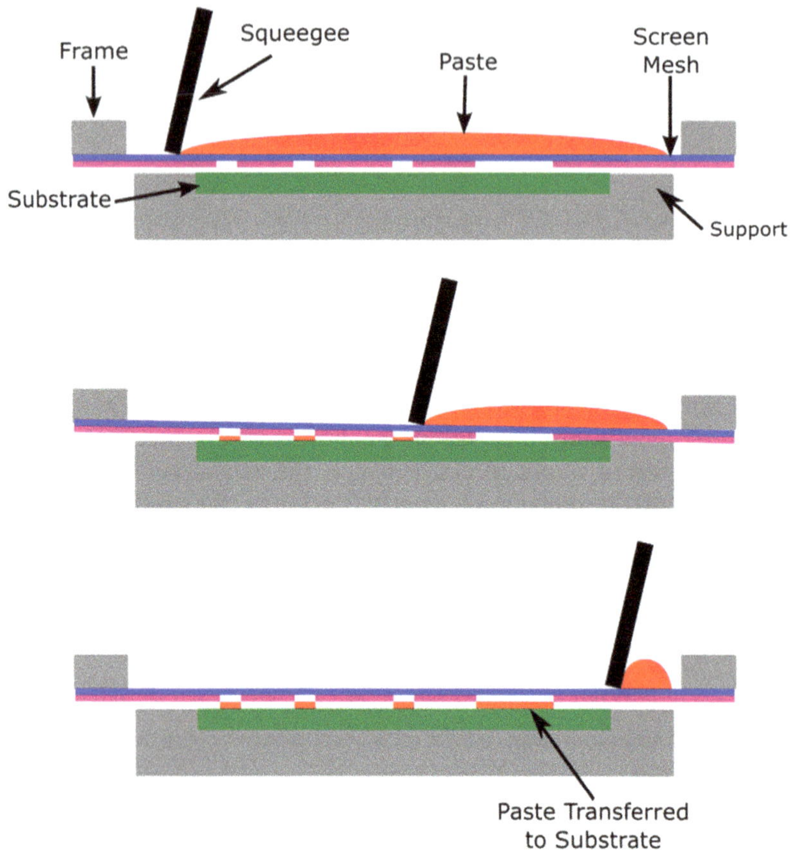

Figure 9.5. Schematic of the screen printing process. (A) Placement of the printing medium (i.e. ink, emulsion, or paste) on the mesh screen, where it can be seen that the screen is not in contact with the substrate. (B) The printing medium is passed through the openings in the screen mesh in a controlled manner. (C) The print is finished and the stencil design has been transferred onto the substrate.

ascorbic acid, or endogenous, e.g. glutathione. Molecules are oxidized via hydrogen atom receptors, e.g. ROO°, or via electron acceptors, e.g. Fe^{3+}, from a mechanistic perspective [30–36]. Carbon nanotube screen printed electrodes of a multi-walled nature have been utilized to demonstrate the idea of determining homocysteine and glutathione. In order to detect glutathione, calibration curves have been determined, such that in the 0.1–60.0 μM linear range the sensitivity is 0.0019 $\mu A\ \mu M^{-1}$ and the detection limit is ~0.11 μM. This study inspires the utilization of re-useable and/or disposable implementation of sensors for medical and biological purposes along with their potential benefits [37]. Glutathione has been identified using a Prussian blue modified screen printed electrode with a 2×10^{-6} M detection limit at an implemented potential of 200 mV versus Ag/

AgCl. The glutathione recoveries ranged from 92% to 103% in blood [38]. Cobalt phthalocyanine electrocatalyst incorporated screen printed carbon electrons (SPCEs) have been produced with the potential to fabricate microband electrodes to determine reduced glutathione. These studies are the first aimed at finding the cyclic voltammetric activity of such electrodes with potassium ferrocyanide. There is a linear chronoamperometric reaction between the 50 μM and 8.0 mM range with a 7.3 μM detection limit [39]. An electroanalytical strategy to determine reduced and overall glutathione has been recorded by Lee *et al*, by implementing square wave voltammetry on a screen printed electrode modified using carbon nanotubes. This study demonstrated that by means of cyclic voltammetry a sensitivity of 0.64 ± 0.008 μM has been determined with a detection limit of 3.0 ± 0.07 μM. The use of this method on human saliva indicated that the glutathione values and errors are consistent with commercially accessible enzyme assay experiments [40]. An electrochemical sensor has been created by Wang *et al* to recognize glutathione using a thymine–Hg^{2+}–thymine (T–Hg^{2+}–T) combination paired with mesoporous silica nanoparticles (MSNs). The interaction of Hg^{2+} ions with thymine bases occurs to create stable and resilient T–Hg^{2+}–T complexes. The methylene blue (MB) electroactive molecules are secured within the MSN inner pores using a specific sequential thymine containing DNA in the presence of Hg^{2+}. After adding the target, the created duplex DNA T–Hg^{2+}–T complex can turn into a single-strand DNA via a competitive deposition response, causing the MB to be released from the MSN. Then the quantity of target dependent released MB can be determined easily using a dodecanethiol modified screen printed electrode, as depicted in figure 9.6.

The proposed system presented an wide detection range of 1.0 nM to 1.0 μM with a 0.6 nM detection limit when applying glutathione as a model target. This is a straightforward and universal approach for different analytes with various competitive disposition ligands and these sensors possess significant potential to test biotools for commercial uses [41].

Ganjali *et al* manufactured a voltammetric sensor on the basis of a graphite screen printed electrode modified using a ZnO/Al_2O_3 nanocomposite to detect ascorbic acid. In the optimal state, the modified electrode displayed linear reactions towards ascorbic acid within the 1.0–100 μM range and a 0.6 μM detection limit with a signal-to-noise ratio of 3. This sensor has been produced with the potential to recognize ascorbic acid within real specimens [42]. Raveendran *et al* presented a disposable non-enzymatic sensor to be used for ascorbic acid which was achieved by adjusting an SPE using $Cu(OH)_2$ nanorods. This sensor exhibited a linear response in the ascorbic acid

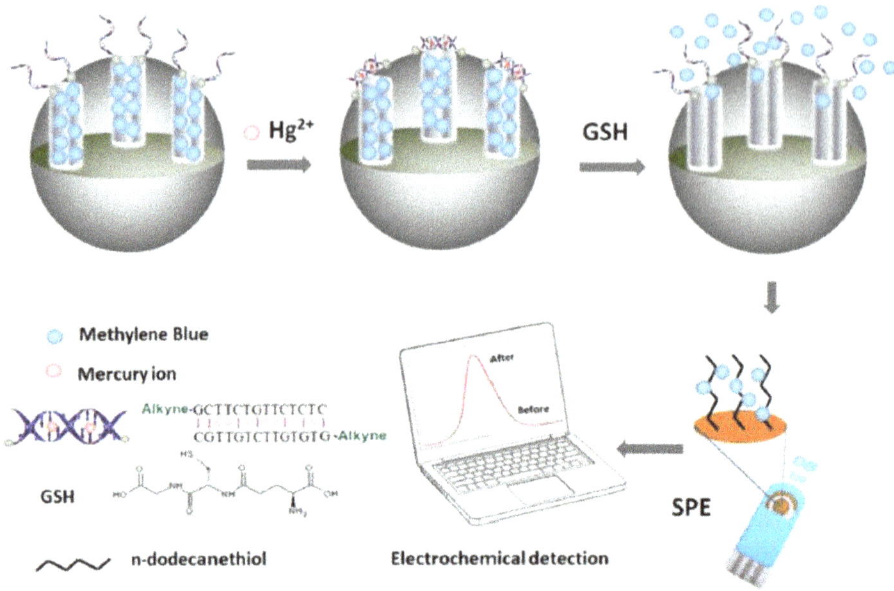

Figure 9.6. Schematic illustration of the principle of electrochemical assay for glutathione detection based on a target-responsive DNA molecule gated system. (Reproduced with permission from [41]. Copyright 2017 Elsevier.)

concentration range of 0.0125 to 10 mM. It has been utilized to determine ascorbic acid within urine and vitamin C tablets [43]. Asuncion Alonso-Lomillo *et al* have shown the ability of a modified screen printed electrode using gold nanoparticles to electrochemically detect ascorbic acid within serum specimens. Their results demonstrated linear responses exhibited by the electrode towards ascorbic acid within the 1.9–16.6 μM range. However, ascorbic acid was not distinguished from impeding compounds, namely uric acid and dopamine, when implementing this electrode [44]. Zhao *et al* manufactured an electrochemical biomimetic sensor to detect ascorbic acid within orange juice specimens by electropolymerizing phenylenediamine at the exterior of an SPE. The imprinted sensor's response current presented a linear correlation to ascorbic acid concentration within the 0.45–409.10 μM range. The relevant detection limit is 0.11 μM with a signal-to-noise ratio of 3. This sensor has been used to distinguish ascorbic acid within commercially available orange juice [45].

Guiberteau-Cabanillas *et al* proposed electroanalytical methods to estimate the overall phenolic compound content of low oxidation potential within cork boiled water, shown in the form of gallic acid equivalent g^{-1} dry cork, via an SPE which was modified utilizing graphene. The obtained detection limit was 0.05 ppm within the 0.05–0.1 ppm range of linear responses towards gallic acid [46]. Cheng and

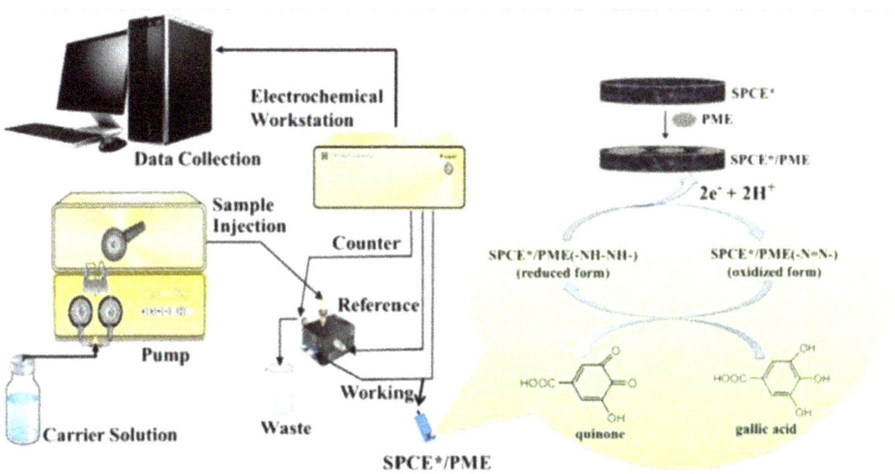

Figure 9.7. Schematic illustration of the determination of gallic acid in green tea samples based on a screen printed electrode with poly(melamine) film. (Reproduced with permission from [47]. Copyright 2015 Elsevier.)

Su looked into electrochemical sensors which were combined with an efficient flow-injection amperometry system to determine the gallic acid content (figure 9.7). This sensor had a fine electroactive poly (melamine) film that was immobilized on an SPCE. The proposed assay displayed an extensive linear reaction towards gallic acid in pH 3 and pH 7 phosphate buffer solutions at optimal flow-injection amperometry (FIA). The related detection limit at a signal-to-noise ratio of 3 was 0.21 mM for the pH 7 solution. When conducting 57 successive measurements for 50 mM of gallic acid, the relative standard deviation (RSD) was 3.9% in the pH 7 solution [47].

Piermarini *et al* produced a uricase biosensor on an SPE modified utilizing Prussian blue paired with mobile instrumentation which exhibited a 0.03–0.3 mM working range at a 0.01 mM detection limit. This biosensor is beneficial for analyzing uric acid within human blood serum outside conventional diagnostic laboratories [48]. Kanyong *et al* produced a screen printed carbon electrode incorporated with electrocatalyst cobalt phthalocyanine, CoPC, to determine uric acid. This electrode was utilized to investigate the urine of a healthy subject. The working range for the biosensor was from 0.015 to 0.25 mM with a 2.10 μA mM^{-1} sensitivity [49].

(b) *Electrochemical biosensors for amino acid detection*: Amino acid identification utilizing direct electrochemical strategies on modified and unmodified screen printed electrodes is prevalent among the studies conducted by various scientists. Su and Cheng studies cysteine electrocatalytic oxidation on an SPE modified using electrogenerated poly(3,4-ethylenedioxythiophene) film (PEDOT). The adjusted electrode reduced the over-potential and improved cysteine oxidation activity compared to typical SPEs according to cyclic voltammetric assessments. Figure 9.8 depicts the reaction mechanism. Outstanding diagnostic qualities (for example a high sensitivity of 0.0527 μA μM^{-1}), a suitable dynamic range (0.05–200 μM), and a low detection limit (0.03 μM) were acquired utilizing flow-injection amperometry in an optimal state [50].

Vasjari *et al* analyzed the response of an SPE to tyrosine and cysteine. It showed a linear response within a $5 \times 10^{-5} \times 10^{-4}$ M range with a sensitivity of 2.95 ± 0.03 $mAmM^{-1}$ for cysteine. The correlation coefficient was over 0.9980. This sensor was utilized to analyze economically accessible pharmaceutical specimens with acceptable results [51]. Hernandez-Ibanez *et al* manufactured an electrochemical sensor to determine cysteine precisely using electroanalytical methods on the basis of a Co (II)-phthalocyanine nanoparticle (CoPc) screen printed graphite electrode. The cysteine SWV (square wave voltammetry) reaction exhibited a 2.6–200 μM linear range and a 4 μM detection limit [52]. Lee *et al* used SPEs modified with carbon nanotubes to determine glutathione and homocysteine. To determine glutathione and homocysteine, calibration curves were used with sensitivities of 0.019 μA μM^{-1} and 0.0019 μA μM^{-1}, detection limits of ~1.2 μM and 0.11 μM within the linear regions of 0–40 and 0–60 μM, respectively [37]. Boonchingma *et al* proposed a selective electrochemical sensor to determine L-tryptophan. The SPE surface was modified utilizing p-phenylenediamine (PPD)

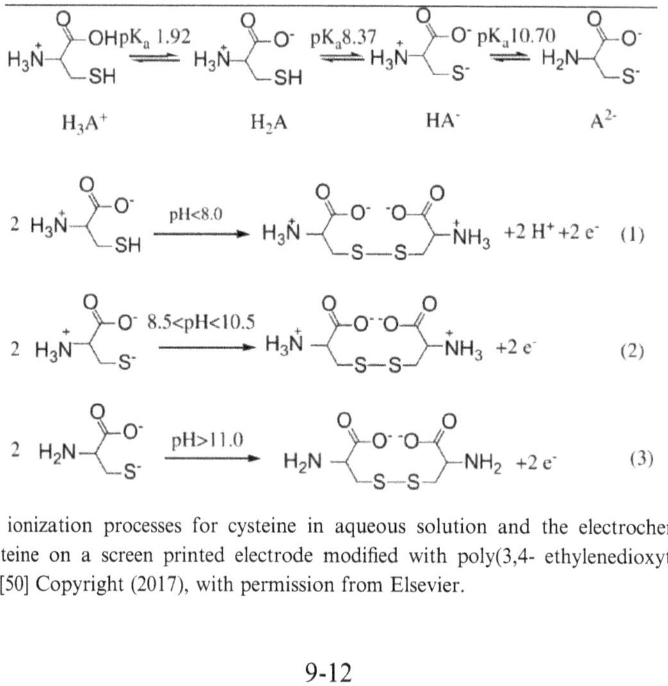

Figure 9.8. The ionization processes for cysteine in aqueous solution and the electrochemical oxidation processes of cysteine on a screen printed electrode modified with poly(3,4- ethylenedioxythiophene) film. Reprinted from [50] Copyright (2017), with permission from Elsevier.

electropolymerization. Its polymer film was connected covalently with cysteamine caped cadmium sulfide quantum dots (Cys–CdS QDs) where glutaraldehyde (GA) was used as a cross-linker producing an organic–inorganic hybrid composite film (figure 9.9).

This electrode played the role of a working electrode to recognize tryptophan. The modified electrode can be applied for tryptophan detection at a 14.74 μM detection limit showing good accuracy and less than 3.7% relative standard deviation. This electrode was applied for tryptophan detection in beverages with favorable recovery within the 91.9% to 104.9% range [53]. Gomez-Mingot *et al* proposed direct electrochemical methionine oxidation on screen printed graphite electrodes within aqueous solutions. For reasons of economy and its use as potential point-of-care sensor, methionine quantification was examined on SPEs, permitting a linear range of 0.05 to 5.0 mM and a 95×10^{-6} M detection limit within model solutions. This sensor was applied to determine methionine within a pharmaceutical product consisting of a diverse combination of chelated minerals, amino acids, and vitamins [54].

(c) *Electrochemical biosensors for catecholamine detection*:

Recently, catecholamines have been a focus of research due to their crucial role within the central nervous system. Catecholamines affect almost every tissue within human organs and influence numerous neuronal, endocrine, metabolic, and cardiovascular processes, while also affecting the immune system and the intestinal barrier [55–60]. Zhang *et al* constructed a dopamine electrochemical sensor on the basis of magnetism supported adjustment of an SPE using magnetic multi-walled carbon nanotubes (mMWCNTs). Figure 9.10(a) shows the synthesis of these mMWCNTs. Figure 9.10(b) depicts the modification procedure for the SPE to determine dopamine. The exhibited linearity had a range within 5–180 μM with a 0.43 μM detection limit. This electrode showed remarkable selectivity toward the detection of dopamine with approximately no intercession by ascorbic acid, which also exists with dopamine in several biological samples and results in interference. This electrode was implemented for dopamine

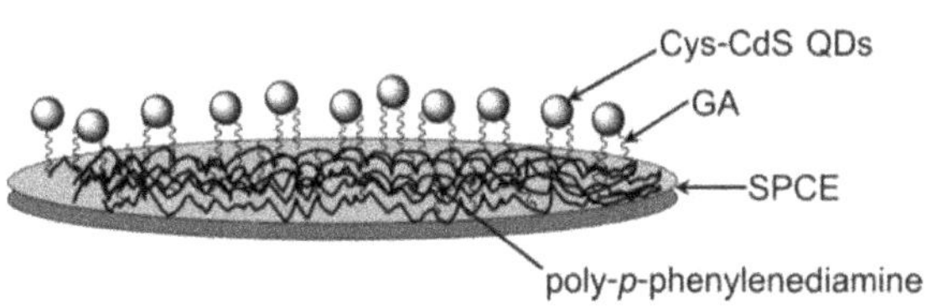

Figure 9.9. Schematic representation of a modified screen printed electrode with cadmium sulfide quantum dots/glutaraldehyde/p-phenylenediamine. (Reproduced with permission from [53]. Copyright 2014 Wiley.)

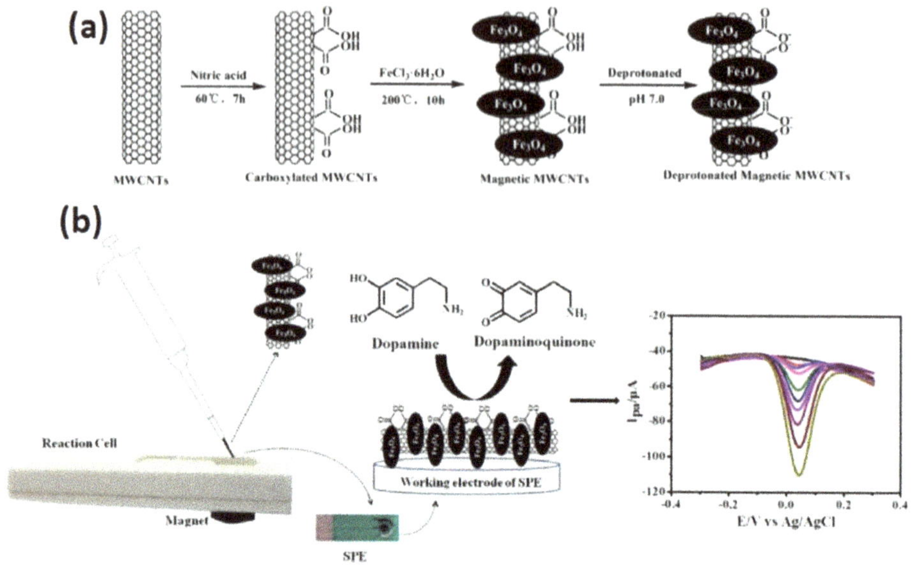

Figure 9.10. (a) Synthetic route of magnetic multi-walled carbon nanotube production. (b) Magnetism assisted modification of screen printed electrodes with magnetic multi-walled carbon nanotubes and the electrochemical response of dopamine on the modified electrode. (Reproduced with permission from [61]. Copyright 2017 Elsevier.)

concentration detection within spiked human blood serum with favorable recovery within a 97.43%–102.94% range and an RSD less than 2.27% [61].

Beitollahi *et al* assessed the creation of a GO/Fe_3O_4@SiO_2 core–shell nanocomposite modified graphite SPE to determine dopamine and uric acid. When implementing this modified electrode, dopamine was determined within a 0.1 to 600.0 μM concentration range with a 8.9×10^{-8} M detection limit. Effective real sample analysis was carried out using this adjusted electrode [62]. Beitollahi *et al* carried out another study where a modified SPE with graphene quantum dots (GQD) sensed dopamine in the presence of tyrosine. The implementation of this electrode caused a 435 mV separation at tyrosine and dopamine oxidation peaks. The related calibration range was between 0.1–1000.0 μM with a 0.05 μM detection limit for dopamine, at a signal-to-noise ratio of 3 [63]. Singh *et al* developed an modified SPE utilizing a multi-walled nanotube based on a nanostructured composition of N′-phenyl-p-phenylenediamine (PPDA) and nafion to determine dopamine. The dopamine oxidation current showed linearity within the 1–110 μM range with a 0.01 μM detection limit [64]. Ping *et al* fabricated a unique screen printing ink containing cellulose acetate, graphite, and an ionic liquid, n-octylpyridinum hexafluorophosphate (OPPF). This graphite cellulose acetate system was applied as an essential ink framework that was printed directly on the polyvinyl-chloride substrate. Dopamine was determined in a linear range from 1.0 μM to 2.5 mM with a 0.5 μM detection limit [65]. Ku *et al* developed a dopamine electrochemical sensor on a previously treated graphite/nafion composite modified SPCE. This developed electrode displayed favorable electrocatalytic oxidation for dopamine with a linear reaction range from 0.5 to 70 μM and a 0.023 μM detection limit. This modified electrode showed favorable recovery with regard to commercially available pharmaceutical dopamine specimens [66]. Valentini *et al* utilized oxidized single walled carbon nanohorns (o-SWCNHs) to produce chemically modified SPEs to detect epinephrine electrochemically in the presence of uric acid, acetaminophen, ascorbic acid, norepinephrine, dopamine, and serotonine. Differential pulse voltammetry was utilized to detect epinephrine within an extensive 2–2500 μM concentration range with a 0.1 μM low detection limit [67]. Soofiabadi *et al* reviewed a nanosensor to determine epinephrine via an SPE modified using a Cu (II) nanocomplex which showed unusual electrocatalytic behavior to oxidize epinephrine. For epinephrine trace determination, the differential pulse voltammetry approach was adopted. There was a linear dependence between the current versus concentration within the 10.0 to 600.0 μM range while the epinephrine detection limit was 2.5 ± 0.05 μM [68]. Baniasadi *et al* studied epinephrine determination using the voltammetric approach using a ZnO–graphene oxide nanocomposite for a modified graphite SPE. This electrode exhibited a 0.07 μM detection limit within a 0.5–500.0 μM dynamic linear range for epinephrine. This adjusted electrode was applied to determine epinephrine within urine specimens and epinephrine injection [69]. Apetrei and Apetrei clarified the sensing characteristics of a carbon based SPE modified using three carbonaceous substances: graphene, carbon microfibers, and carbon nanotubes. The modified SPE utilizing carbon nanofibers exhibited optimal performance with regard to stability and kinetics. The minimal detection point for dopamine and epinephrine was studied. Within the 1–60 μM concentration range, the sensors showed linearity towards epinephrine. These sensors were utilized for the determination of catecholamines within plasma [70]. Huang *et al* created an SPE modified using polyacrylic covered multi-walled carbon nanotubes (PAA–MWCNTs) to simultaneously determine ascorbic acid, uric acid, and norepinephrine. A peak current linear dependence on concentration was proved using the differential pulse voltammetry approach within the 0–10 μM range with a 0.131 μM detection limit for norepinephrine [71]. Wang *et al* constructed an electrochemically modified SPE using MWNT–ZnO/chitosan compounds to simultaneously determine serotonin and norepinephrine. Significant peak potentials were observed at approximately 280 mV and 90 mV, respectively. There was a linear dependence of norepinephrine peak currents within the 0.5–30 μM range with a 0.2 μM detection limit. This electrode can be preserved in a stable condition for a minimum of three months at a temperature of 4 °C inside a refrigerator [72]. Beitollahi and Nejad created an electrochemical sensor with a graphene oxide/ZnO nanorod (GR/ZnO) nanocomposite and graphite SPEs to simultaneously detect levodopa and tyrosine. In an optimal state, square wave voltammetry was utilized which resulted in a linear dynamic range from 1.0×10^{-6} to 1.0×10^{-3} M with a detection limit of 4.5×10^{-7} M for levodopa [73]. Rabinca *et al* proposed a voltammetric approach based on a disposable electrochemically modified screen printed carbon electrode utilizing carbon nanoparticles to determine levodopa. In an optimal state of differential pulse voltammetry, a linear dynamic range was exhibited from 1.0×10^{-6}–1.0×10^{-4} M with a detection limit of 3.6×10^{-7} M for levodopa. This modified electrode was applied to determine levodopa in commercially available pharmaceutical compounds [74]. A review of the uses of SPEs for pharmaceutical formulations and biological species is presented in table 9.1, which collects the trends and progress discussed in previous sections.

2. *Modified screen printed electrodes for environment pollution detection*: SPEs have been implemented to support the production of mobile electrochemical sensors which have the potential for use in environmental analysis, namely water quality experiments for heavy metals, organic pollutants, and other species.
   (a) *Electrochemical biosensors for pesticide detection*: Pesticides are a huge and heterogeneous group of agricultural chemicals, which have been applied broadly to eliminate or ward off pests in various fields. The widespread use of pesticides has prompted inevitable repercussions concerning public health in countries where they are implemented [75, 76]. The damaging effects are related to systems such as the immune, cardiovascular, renal, respiratory, and endocrine systems via the irrevocable effects of acetylcholine esterase, which causes the progression of detrimental human diseases, such as renal failure, diabetes, multiple sclerosis, Alzheimer's, and Parkinson's disease. Thus, the production of dependable, speedy, and cost-effective analytical tools for detecting small quantities of pesticides within the environment is vital [77–81]. Pelle *et al* fabricated an electrochemical screening assay to detect phenyl carbamates, i.e. fenobucarb, isoprocarb, carbofuran, and carbaryl. To construct an efficient delocalizable SPE, nano carbon black (CB) was applied strategically. The analytes exhibited linearity in the 1.0×10^{-7} to 1.0×10^{-4} M range with an LOD $\leqslant 8.0 \times 10^{-8}$ M [82]. Hassani *et al* conducted a study to fabricate a label-free electrochemical aptasensor to detect diazinon, which is a common

Table 9.1. Selected applications of screen printed electrodes in biological species and pharmaceutical compound analysis.

Reference	Analyte	Modifier	Electrochemical method	Linear range	Detection limit	Concomitant compounds
[37]	Glutathione	MWCNTs	CV (cyclic voltammetry)	0.1–60.0 μM	0.11 μM	Homocysteine
[38]	Glutathione	Prussian blue	Amperometry	2.0×10^{-3}–0.5 mM	2×10^{-6} M	Cysteamine
[39]	Glutathione	Cobalt phthalocyanine	CHA (chronoamperometry)	50 μM–8.0 mM	7.3 μM	
[40]	Glutathione	Carbon nanotubes	CV	10.0–100.0 μM	3.0 μM	
[41]	Glutathione	T DNA/MSN	DPV (differential pulse voltammetry)	1.0 nM–1.0 μM	0.6 nM	
[42]	Ascorbic acid	ZnO/Al_2O_3	DPV	1.0–100 μM	0.6 μM	
[43]	Ascorbic acid	$Cu(OH)_2$ nanorods	DPV	0.0125–10 mM		
[44]	Ascorbic acid	AuNPs	DPV	1.9–16.6 μM		Uric acid, dopamine
[45]	Ascorbic acid	Phenylenediamine	SWV	0.45–409.10 μM	0.11 μM	
[46]	Gallic acid	Graphene	DPV	0.05–0.1 ppm	0.05 ppm	Ellagic acid
[47]	Gallic acid	Poly(melamine)	FIA	1–1000	0.21	
[48]	Uric acid	Prussian blue	CHA	0.03–0.3 mM	0.01 mM	
[49]	Uric acid	CoPC	CV	0.015–0.25 mM		
[50]	Cysteine	PEDOT	FIA	0.05–200 μM	0.03 μM	
[51]	Cysteine		LSV (linear sweep voltammetry)	5×10^{-5}–5×10^{-4} M		Tyrosine
[52]	Cysteine	CoPc	SWV	2.6–200 μM	4 μM	
[37]	Homocysteine Glutathione	CNTs	CV	0–40 μM 0–60 μM	1.2 μM 0.11 μM	
[53]	Tryptopha	QDs/GA/PPD	SWV	0.1–0.5 μM	14.74 μM	Tyrosine
[54]	Methionine		DPV	0.05–5.0 mM	95×10^{-6} M	Vitamin C, vitamin B6

(*Continued*)

Table 9.1. (*Continued*)

Reference	Analyte	Modifier	Electrochemical method	Linear range	Detection limit	Concomitant compounds
[61]	Dopamine	mMWCNTs	DPV	5–180 μM	0.43 μM	Ascorbic acid
[62]	Dopamine	GO/Fe_3O_4@SiO_2	DPV	0.1–600.0 μM	8.9×10^{-8} M	Uric acid
[63]	Dopamine	GQDs	DPV	0.1–1000.0 μM	0.05 μM	Tyrosine
[64]	Dopamine	MWCNTs/Nafion/PPDA	CV	1–110 μM	0.01 μM	
[65]	Dopamine	Graphite/OPPF (n-octylpyridinumhexa-fluorophosphate)	DPV	1.0 μM–2.5 mM	0.5 μM	Ascorbic acid, uric acid
[66]	Dopamine	Graphite/Nafion	LSV	0.5–70 μM	0.023 μM	
[67]	Epinephrine	o-SWCNHs	DPV	2–2500 μM	0.1 μM	Serotonine, dopamine, norepineprhine, ascorbic acid, acetaminophen, uric acid
[68]	Epinephrine	Cu(II) nanocomplex	DPV	10.0–600.0 μM	2.5 ± 0.05 μM	
[69]	Epinephrine	ZnO–GO	DPV	0.5–500.0 μM	0.07 μM	
[70]	Epinephrine	CNTs CNF GR	DPV	DPV 1.0–60.0 μM	1.98 μM 1.06 μM 3.52 μM	Dopamine
[71]	Norepinephrine	PAA–MWCNTs	DPV	0–10 μM	0.131 μM	Uric acid, ascorbic acid
[72]	Norepinephrine	MWNTs–ZnO/chitosan	SWV	0.5–30 μM	0.2 μM	Serotonin
[73]	Levodopa	GR/ZnO	SWV	1.0×10^{-6}–1.0×10^{-3} M	4.5×10^{-7} M	Tyrosine
[74]	Levodopa	Carbon nanoparticles	DPV	1.0×10^{-6}–1.0×10^{-4} M	3.6×10^{-7} M	

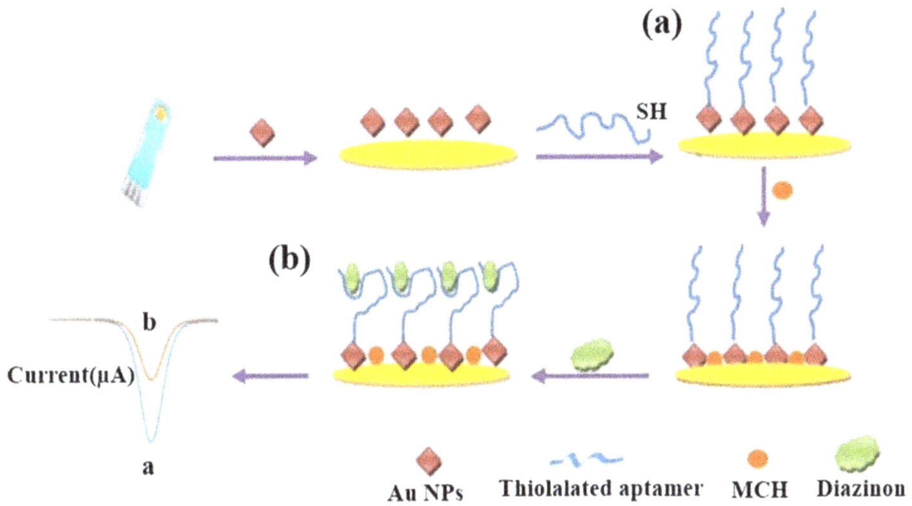

Figure 9.11. Schematic presentation of aptasensor fabrication for the detection of diazinon. (a) AuNP modified screen printed electrode. (b) Bare screen printed electrode. (Reproduced with permission from [83]. Copyright 2018 Elsevier.)

organophosphorus compound. This aptasensor was created using an SPE modified with thiolated aptamers that were immobilized on gold nanoparticles (AuNPs). Figures 9.11(a) and (b) depict the AuNP modified SPE and bare SPE. The current fluctuations were analyzed within a 0.1 to 1000 nM concentration range for diazinon. This modified aptasensor exhibited a detection limit of 0.0169 nM [83]. Govindasamy *et al* reported a screen printed carbon electrode adjusted using graphene oxide nanoribbons (GONRs) to sensitively determine methyl parathion.

This sensor showed linearity from the 100 nM to 100 μM range with a 0.5 nM detection limit and a signal-to-noise ratio of 3. This electrode was applied effectively to determine methyl parathion in ugli, tomato, broccoli, and beetroot, showing exceptional practical use [84]. Selvoline *et al* recommended an electrochemical DNA aptasensor to detect profenofos, an organophosphorus pesticide, on the basis of a competitive format of disposable graphite SPEs. The immobilization of a thiol tethered DNA capture probe on AuNP/polyaniline composite film modified electrodes was carried out, which supplemented the selected aptamer sequence. Within the 0.10 μM to 10 μM range a dose reaction curve was achieved with a 0.27 μM detection limit [85]. Khairy *et al* studied nickel oxide nanoplatelet modified SPE application to detect an organophosphate pesticide (Parathion). When differential pulse voltammetry was conducted with the modified electrode in the presence of parathion, there was a characteristic peak current at −1.0 V (versus Ag/AgCl). Parathion determination was achieved within the 0.1–30 μM concentration range with a 0.024 μM detection limit [86]. Jirasirichote *et al* manufactured an electrochemical sensor using an SPCE modified using graphene oxide and AuNPs to determine carbofuran. This methodology showed a broad linear range within 1–250 μM for detecting carbofuran via the differential pulse voltammetry method. The related detection limit and quantitation were obtained as 0.22 and 0.72 μM, respectively [87]. Zhang *et al* created magnetic nanoparticles of Fe_3O_4@TiO_2 to electrochemically quantify organophosphorylated butyrylcholinesterase (BChE) within plasma. The synthesis of Fe_3O_4@TiO_2 nanoparticles via tetrabutyltitanat hydrolysis on the exterior of Fe_3O_4 magnetic nanospheres is depicted in figure 9.12. In order to conduct selective enrichment of the phosphorylated moiety as an alternative to the phosphoserine antibody, as done in conventional sandwich immunoassays, the functionalization of Fe_3O_4@TiO_2 nanoparticles was carried out. Secondary determination has been conducted via quantum dot (QD)-tagged anti-BchE antibody (QD–anti-BChE). By utilizing a magnet, the final sandwich complex, Fe_3O_4@TiO_2/BChE/QDs–anti-BchE, was separated effortlessly from the specimen solutions and a dispensable SPE was used to detect the released cadmium ions. Within a 0.02 to 10 nM BchE concentration range, the immunosensor exhibited a linear reaction with a 0.01 nM detection limit [88].

Kumar and Souza proposed the immobilization of *Escherichia coli* on SPCEs through glutaraldehyde to determine methyl parathion. The sensor had a detection range from 2 to 80 μM of methyl parathion

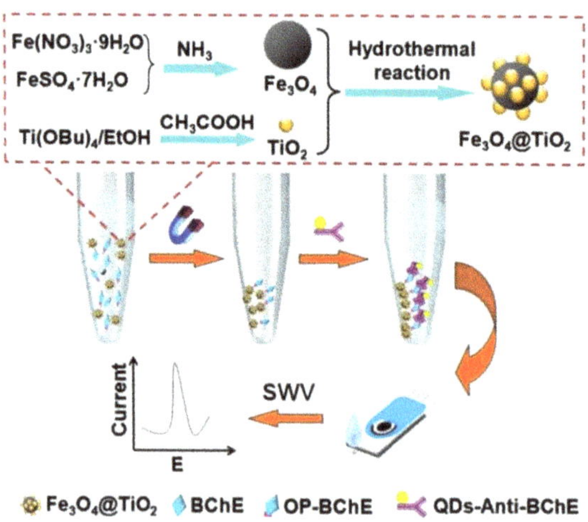

Figure 9.12. Schematic illustration of sandwich-like immunoassays for electrochemical quantification of organophosphorylated butyrylcholinesterase. (Reproduced with permission from [88]. Copyright 2013 Elsevier.)

concentration [89]. Gan *et al* carried out a study where Fe_3O_4/Au and ZrO_2 NP composites were used for the immobilization of acetylcholinesterase enzyme on the SPCE surface to determine dimethoate. There was a linear response within the 1.0×10^{-3}–10.0 ng ml^{-1} range with a 0.6×10^{-4} ng mL^{-1} detection limit. In Chinese cabbage, the recovery rates were observed to be within the 88%–105% range [90]. Chen *et al* created a balanced SPE on the basis of dicyclohexyl phthalate (DCHP) and MWCNTs to detect chlorpyrifos. This electrode displayed an extensive range within 0.05 to 1.0×105 $\mu g\ L^{-1}$ with a 0.05 $\mu g\ L^{-1}$ detection limit [91]. Fu *et al* fabricated an electrochemical acetylcholinesterase (AChE) biosensor that was modified using cerium oxide–chitosan (CeO_2–CS) and ordered mesoporous carbon–chitosan (OMC–CS) through an SPCE to detect methamidophos and chlorpyrifos. The characteristics of the adjusted electrode were determined using differential pulse voltammetry and cyclic voltammetry approaches. In an optimal state, linearity was exhibited within the 0 to 105 $\mu g\ L^{-1}$ range for chlorpyrifos with a 0.01 $\mu g\ L^{-1}$ detection limit [92].

(b) *Electrochemical biosensors for heavy metal detection*: Urbanization along with quick industrial advancements have increased environmental contaminants. Of these contaminants, heavy metals are the most inescapable. Upon their release into the environment some of these heavy metals, which are vital for industrial utilizations, may pass into the human body through the food chain and cause detrimental effects to human health [93–99]. Bobrowski *et al* assessed the voltammetric functionality of an *in situ* coated antimony film SPCE (SbF–SPE) within a tartrate buffer to detect copper (II) using differential pulse anodic stripping voltammetry (DP-ASV). The linear range was within 2.5–150 $\mu g\ L^{-1}$ with a detection limit of 0.14 $\mu g\ L^{-1}$ for copper (II). This strategy was investigated practically on soil conditioner specimens [100]. Yao *et al* modified an SPE using SWCNHs. This sensor showed clear and separate stripping peaks with regard to lead and cadmium. In an optimal state, the linearity of the SWCNH film modified electrode varied within the 1.0 to 60.0 $\mu g\ L^{-1}$ range for both of the heavy ions. The related detection limits for lead (II) and cadmium (II) ions have been observed to be 0.4 $\mu g\ L^{-1}$ and 0.2 $\mu g\ L^{-1}$, respectively. This electrode was used to detect lead (II) and cadmium (II) ions within milk and honey samples [101]. Adarakatti *et al* created caliaxarene bulk modified SPEs to simultaneously measure mercury (II), copper (II), and toxic metal lead (II) ions. By implementing varied concentrations within the 100–2400 $\mu g\ L^{-1}$ range for lead, copper, and mercury, calibration plots were produced. The detection limits were observed to be 38, 40, and 48 $\mu g\ L^{-1}$, respectively [102]. Perez Rafols *et al* fabricated two different kinds of silver nanoparticles of various shapes and sizes, namely nanoprisms and nanoseeds, via a carbon nanofiber adjusted SPE which was used to simultaneously determine copper (II) and lead (II). The most favorable analytical performance was evident when combining Ag nanoseeds with the carbon nanofiber modified SPE. The created sensor enabled copper (II) and lead (II) to be determined simultaneously at trace levels. Its usage on natural specimens was tested utilizing a certified ground water reference substance which resulted in significant accuracy and reproducibility [103]. Jeromiyas *et al* synthesized a 3D hierarchical network consisting of bismuth nanoparticle coated graphene carbon nanotube nanocomposites (Bi NPs@Gr–CNTs), which were applied to detect mercury (II) using an electrocatalytic approach. In an optimal state, the adjusted electrode showed exceptional mercury (II) sensing features within a 1.0 nM–217.4 μM range at a 0.2 nM detection limit. The applicability of this technique was tested with urine specimens, human serum, fish oil tablets, and tap water with satisfactory results [104]. Wang *et al* investigated an SPE to detect cadmium (II) on the basis of a differential pulse voltammetry method, coated with ionic liquid (IL) and magnetite nanoparticle (Fe_3O_4). Bismuth film (Bi) was also deposited. In an optimal state, cadmium (II) concentrations exhibited linearity with current reactions within the 0.5 to 40 $\mu g\ L^{-1}$ range with a 0.05 $\mu g\ L^{-1}$ limit of detection and a signal-to-noise ratio of 3. This electrode was used to detect cadmium (II) within soil specimens at recovery rates within the 91.77%–107.83% range [105]. Shoub *et al* carried out a study where an SPE was modified using AuNPs and ionophores to manufacture an electrochemical sensor to determine lead (II) ions via linear sweep anoding sweep voltammetry (LSASV), which is depicted in figure 9.13. In an optimal state, i.e. a deposition time of 240 s, a deposition potential of −1.2 V in the presence of Tris-HCl (0.1 mol L^{-1}, pH 2), and a 0.1 Vs^{-1} scan rate, the linear sweep voltammogram was related to lead (II) concentration within the 0.4 to 20 $\mu g\ mL^{-1}$ range and a 0.082 $\mu g\ mL^{-1}$ detection limit [106]. Tyszczeuk-Rotko and Domanska implemented bismuth films on adjusted SPCEs to determine lead (II) and cadmium (II). For a deposition period of 190 s, the associated detection limits were obtained as 1.41×10^{-10} M and 1.35×10^{-9} M for lead (II) and cadmium (II), respectively [107]. The applications of SPEs in the analysis of environmental contaminants are provided in table 9.2. The collected information in table 9.2 outlines the individual trends and advancements examined in the preceding text.

3. *Modified screen printed electrodes for bacteria detection*: Recently, *Salmonella* species infection has posed a serious threat to public health, particularly in developing countries, where people use poultry and other food-producing animals without control measures [108–110]. In Vietnam it is reported that *Salmonella*, a bacterial pathogen, is present in most animal-origin foods including poultry, ovines, porcines, fish, and seafood, and their

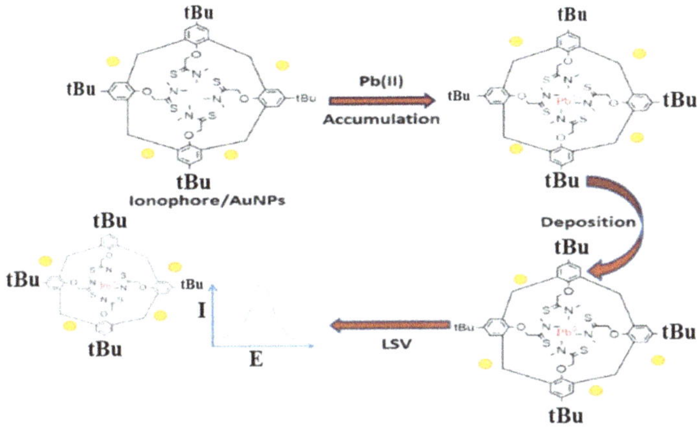

Figure 9.13. Schematic diagram of a screen printed electrode modified with AuNPs and an ionophore to provide an electrochemical sensor for determination of lead (II). (Reproduced with permission from [106]. Copyright 2017 MYU K K.)

food products [111]. Furthermore, some *Salmonella* contaminated fruits and fresh vegetables are also reported to be associated with salmonellosis [112–114]. However, detecting *Salmonella* is time-consuming and labor-intensive, because the conventional methods require isolation, pre-enrichment, or genome amplification [115]. The development of a rapid and reliable method, in particular biochips for electrochemical detection of the presence of these pathogens at a low concentration, is also a very important aim for point-of-care applications. Nguyen *et al* developed a new type of electrochemical biochip for quick and accurate detection of bacterial pathogens [116]. They designed and fabricated compact ZnO NR-modified platinum (Pt) electrodes by integration of counter, working, and pseudoreference electrodes on a chip. In the design, Pt was used as both the counter and pseudoreference electrodes, while vertically grown ZnO NRs acted as the working electrode. They also investigated the stability of the electrochemical signal of the ZnO NR-modified Pt SPEs before and after antibody immobilization and the detection of *Salmonella* bacteria [116]. The success of this study will help in fabricating multimodified sensors on a biochip for further direct detection of bacterial pathogens. The design of the platinum SPEs includes the Cr/Pt counter, working, and pseudoreference electrodes. A double layer of Cr/Pt (20/150 nm thickness) was deposited on a Si wafer covered with 300 nm of SiO_2 using a sputtering system. The thin Cr layer promotes adhesion of the Pt layer. Figure 9.14 describes the entire preparation process of ZnO NR-modified Pt SPEs for electrochemical detection of bacterial pathogens. After the growth of ZnO NRs on the Pt working electrode, they were functionalized by crosslinkers to provide active sites for antibody immobilization. Then, nonspecific binding sites were blocked by bovine serum albumin (BSA).

Table 9.2. Selected applications of screen printed electrodes in environmental pollution analysis.

Ref.	Analyte	Modifier	Electrochemical method	Linear range	Detection limit	Concomitant compounds
[82]	Carbaryl Carbofuran Isoprocarb Fenobucarb	CB	DPV	1.0×10^{-7}–1.0×10^{-4} M 1.0×10^{-7}–1.0×10^{-4} M 1.0×10^{-7}–1.0×10^{-4} M 1.0×10^{-7}–1.0×10^{-4} M	4.8×10^{-8} M 4.9×10^{-8} M 7.9×10^{-8} M 8.0×10^{-8} M	
[83]	Diazinon	AuNPs/aptamers	DPV	0.1–1000 nM	0.0169 nM	
[84]	Methyl parathion	GONRs	Amperometry	100 nM–100 μM	0.5 nM	
[85]	Profenofos	AuNPs/polyaniline	DPV	0.10–10 μM	0.27 μM	
[86]	Parathion	NiO NPs	DPV	0.1–30 μM	0.024 μM	
[87]	Carbofuran	AuNPs/GO	DPV	1–250 μM	0.22 μM	
[88]	Butyrylcholinesterase	Fe_3O_4/TiO_2/QD–anti-BChE	SWV	0.02–10 nM	0.01 nM	
[89]	Methyl parathion	*Escherichia coli*	CV	2–80 μM	0.5 μM	
[90]	Dimethoate	$Fe_3O_4/Au/ZrO_2$NPs	DPV	1.0×10^{-3}–10.0 ng mL^{-1}	5.6×10^{-4} ng mL^{-1}	
[91]	Chlorpyrifos	DCHP/MWCNTs	CV	0.05–1.0 × 105 μg L^{-1}	0.05 μg L^{-1}	
[92]	Chlorpyrifos	AChE/OMC–CS/CeO_2–CS	DPV	0–10^5 μg L^{-1}	0.01 μg L^{-1}	Methamidophos
[100]	Copper(II)	SBF	DP-ASV	2.5–150 μg L^{-1}	0.14 μg L^{-1}	
[101]	Cadmium (II) Lead (II)	SWCNH	SWASV (square wave anodic stripping voltammetry)	1.0–60.0 μg L^{-1}	0.2 μg L^{-1} 0.4 μg L^{-1}	

[102]	Lead (II) Copper (II) Mercury (II)	Calixarene	DP-ASV	100–2400 $\mu g\ L^{-1}$	38 $\mu g\ L^{-1}$ 40 $\mu g\ L^{-1}$ 48 $\mu g\ L^{-1}$
[103]	Lead (II) Copper (II)	Ag nanoseeds Ag nanoprisms Ag nanoseeds Ag nanoprisms	Stripping voltammetry	6.6–53.5 $\mu g\ L^{-1}$ 7.8–53.5 $\mu g\ L^{-1}$ 10.0–77.0 $\mu g\ L^{-1}$ 8.3–100.7 $\mu g\ L^{-1}$	1.98 $\mu g\ L^{-1}$ 2.35 $\mu g\ L^{-1}$ 2.99 $\mu g\ L^{-1}$ 2.49 $\mu g\ L^{-1}$
[104]	Mercury (II)	Bi NPs@Gr–CNTs	DPV	1.0 nM–217.4 μM	0.2 nM
[105]	Cadmium (II)	Bi/Fe_3O_4/IL	DPV	0.5–40 $\mu g\ L^{-1}$	0.05 $\mu g\ L^{-1}$
[106]	Lead (II)	AuNPs/ionophore	LSASV	0.4–20 $\mu g\ mL^{-1}$	0.082 $\mu g\ mL^{-1}$
[107]	Cadmium (II) Lead (II)	Bismuth films	SWASV	5×10^{-9}– 1×10^{-6} M 5×10^{-10}– 5×10^{-7} M	1.35×10^{-9} M 1.41×10^{-10} M

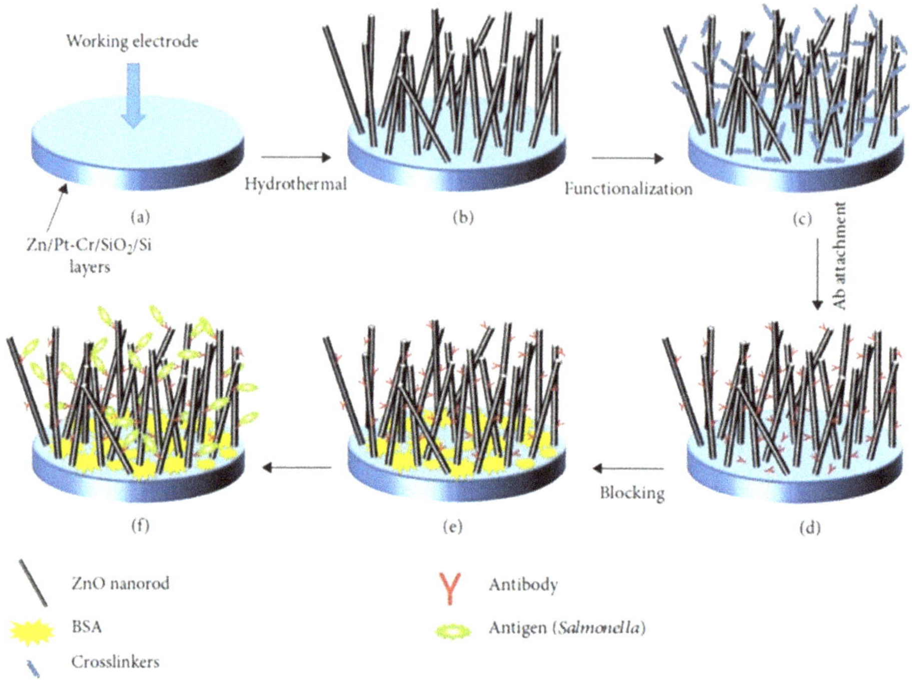

Figure 9.14. A schematic outline of the fabrication of on-chip electrochemical electrodes based on ZnO NRs: (a) deposition of Zn/Pt electrode on the silicon substrate, (b) growth of ZnO NRs, (c) functionalization of ZnO NRs with silane and GMBS, (d) antibody immobilization, (e) blocking the unspecific ZnO NR sites using BSA, and (f) binding of bacterial antigens on the surface of the working electrode via specific antibodies immobilized in advance. (Reproduced with permission from [116]. Copyright 2019 Hindawi.)

Finally, bacterial pathogens were added on the working electrode for testing of the electrochemical measurements. Setterington *et al* presented a novel electrochemical detection technique and applied it to *Bacillus cereus* and *E. coli* O157:H7, with lower detection limits of 40 CFU mL^{-1} and 6 CFU mL^{-1}, respectively, from pure culture, when pathogen concentration is in the range of 10^0 to 10^2 CFU mL^{-1}. (Samples containing pathogen concentrations greater than 10^3 CFU mL^{-1} generated false negative signals, and needed to undergo dilution prior to detection for accurate results) [117]. The immunomagnetic separation procedure is depicted in figure 9.15.

4. *Modified screen printed electrodes for protein detection*:Razzino *et al* carried out a study into a sensitive strategy for the determination of tau protein, a hallmark of Alzheimer's disease (AD), involving a sandwich immunoassay and amperometric detection with disposable SPCEs modified with a gold nanoparticle–poly(amidoamine) (PAMAM) dendrimer nanocomposite (3D-Au–PAMAM) covalently immobilized onto electrografted p-aminobenzoic acid (p-ABA). The capture antibody (CAb) was immobilized by crosslinking with glutaraldehyde (GA) on the amino groups of the 3D-Au–PAMAM–p-ABA–SPCE, where tau protein was sandwiched with a secondary antibody

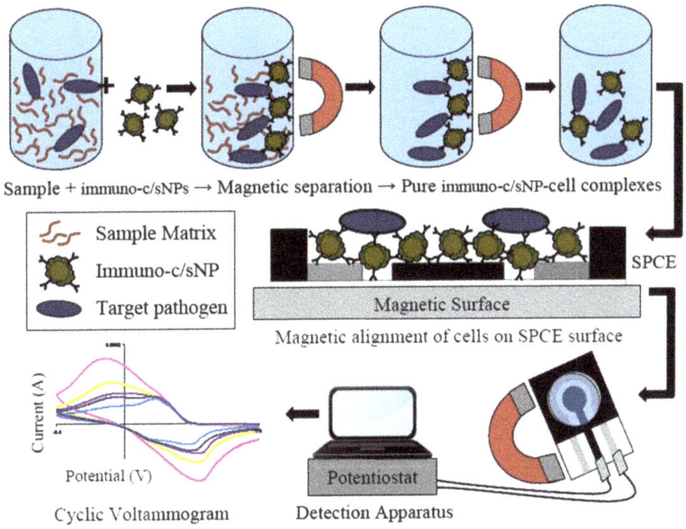

Figure 9.15. Immunomagnetic separation from the sample matrix, magnetic alignment on the SPCE surface, and electrochemical detection of *B. cereus* and *E. coli* O157:H7 cells. (Reproduced with permission from [117]. Copyright 2012 MDPI.)

labeled with horseradish peroxidase (HRP-DAb). Amperometry at −200 mV (versus the Ag pseudoreference electrode) upon the addition of hydroquinone (HQ) as electron transfer mediator and H_2O_2 as the enzyme substrate has been utilized to detect the immunocomplex development. The incredible analytical performance of the immunosensor in terms of selectivity and a low limit of detection (LOD) (1.7 pg mL^{-1}) permitted the direct determination of the target protein in raw plasma samples and in brain tissue extracts from healthy individuals and post-mortem diagnosed AD patients, utilizing a straightforward and fast protocol [118]. The various steps associated with the preparation of the HRP-DAb–tau-CAb–3D-Au–PAMAM–p-ABA–SPCE immunosensor and the reactions involved in the amperometric transduction are shown in figure 9.16.

Glycoprotein hormones have likewise been determined using biosensors based on screen printed carbon transducers with suitable antibodies. Bagel *et al* fabricated a two-site heterogeneous enzyme immunoassay for human chorionic gonadotrophin (hCG), based on an ion-exchange (Nafion) film-coated SPCE adapted to the bottom of a polystyrene microwell. Following competition between soluble biotinylated monoclonal anti-hCG and antibody-coated microwells for binding of hCG in the sample, streptavidin-labeled ALP conjugate was brought into the wells. The analytical signal was obtained following presentation of the enzyme substrate, a phosphorylated form of [(4-hydroxyphenyl) aminocarbonyl]-cobaltocenium, and its hydrolysis to an anionic product which was collected in the Nafionw membrane.

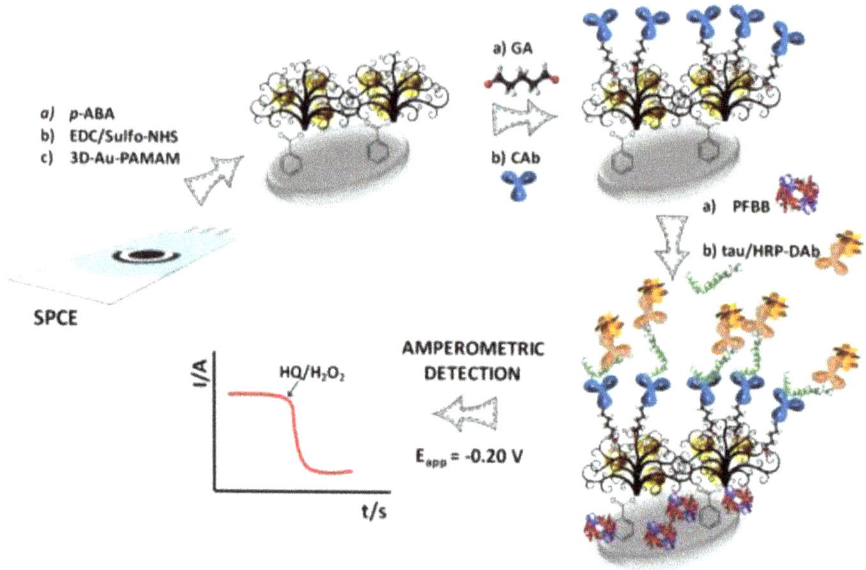

Figure 9.16. Fabrication and amperometric transduction involved in the development of a HRP-DAb–tau-CAb–3D-Au-PAMAM–p-ABA–SPCE immunosensor for tau protein. (Reproduced with permission from [118]. Copyright 2020 Elsevier.)

Analytical signals were obtained utilizing cyclic voltammetry by measuring peak current responses at +0.5 V versus Ag/AgCl. The hCG calibration plot was linear over the range 200–10 000 mIU mL^{-1}, with a detection limit of 100 mIU (1 mg = 9825 IU) [119]. Wang *et al* have presented a sensor for insulin based on a potentiometric stripping study and including adsorptive aggregation of the hormone, followed by oxidation of the tyrosine residues. This methodology permitted detection down to 20 nM level in physiological buffer solution [120].

9.2.2 Advantages of electrochemical biosensors based on screen printed electrodes

In the past few years, screen printed components implemented in sensors and biosensors have experienced significant progress with a considerable number of scientific papers being published on this topic. SPEs present various advantages compared to typical electrodes since they can be utilized efficiently for low volumes, leading to the development of reproducible, precise, and sensitive sensors [121–125]. Appropriate modifying compounds can be used within the carbon ink or placed on the electrode surface to achieve increased selectivity and/or sensitivity for numerous applications [126–129]. They permit the development of mass produced portable, precise, and reproducible sensors [130]. They are additionally cost effective, disposable, and of miniaturized form, which make them amenable for large-scale production. Further, the utilization of arrays of SPEs offers the advantages of short times, and the possibility to perform the calibration and analysis of several unknown samples simultaneously [2].

9.2.3 Impedance biosensors based on screen printed electrodes

Screen printed impedance biosensors are a class of electrical biosensors that show promise for point-of-care and other applications because of their low cost, simplicity of scaling down, and label-free operation. Unlabeled DNA and protein targets can be recognized by observing changes in surface impedance when a target molecule binds to an immobilized probe. Different possible mechanisms for impedance change upon target binding have been reported.

Screen printed impedance biosensors include the application of a small amplitude AC voltage to the sensor electrode and measurement of the in-/out-of-phase current response as a function of frequency.

1. *Impedance biosensors for food toxin detection on screen printed electrodes*:

   Samanta *et al* applied screen printed nanocrystalline silicon oxide based impedance immunosensors for food toxin detection in real food samples procured from two distinct geographical locations of India. It was observed that processing the sensor parameters with an incremental fuzzy algorithm enables estimation of AFB1 (aflatoxin B1), AFB2, and OTA (ochratoxin A) down to 0.1 fg mL^{-1} with 93% precision in complex matrices and with a negligible effect of food matrix interference. This system is thus expected to be suitable in the food processing industry since the performance is fundamentally improved compared to existing technologies and also does not require skilled technicians for operation. Further, the low detection limit will enable screening of food samples at the production site with toxin contamination, which is more than three orders of magnitude lower than the safe range. This guarantees, regardless of whether the toxins multiply during transportation, that the food samples that arrive at the receiver end will still be safe and consumable [131]. An interdigitated electrode pattern with 0.5 mm spacing, as shown in figure 9.17(a), was used since it provides the highest area usage factor for a low concentration of biomolecules. At one time, four sensors were interfaced with a PC through a common sensor holder set-up, indicated schematically in figure 9.17.

   Ghosh and Roychaudhuri *et al* reported an electrochemically fabricated silicon oxide nanoporous structure for ultrasensitive recognition of AFB1 in food through a shift in peak frequency corresponding to maximum sensitivity. It has been observed that the impedance sensitivity changes from 19% to 40% (which is only double), whereas the peak frequency shifts from 500 Hz to 50 kHz for a change in concentration from 1 fg mL^{-1} to 1 pg mL^{-1}. This has been ascribed to the combined impact of significant pore narrowing with increasing AFB1 concentration and the opposing nature of impedance change within the nanopores and the conducting substrate immediately

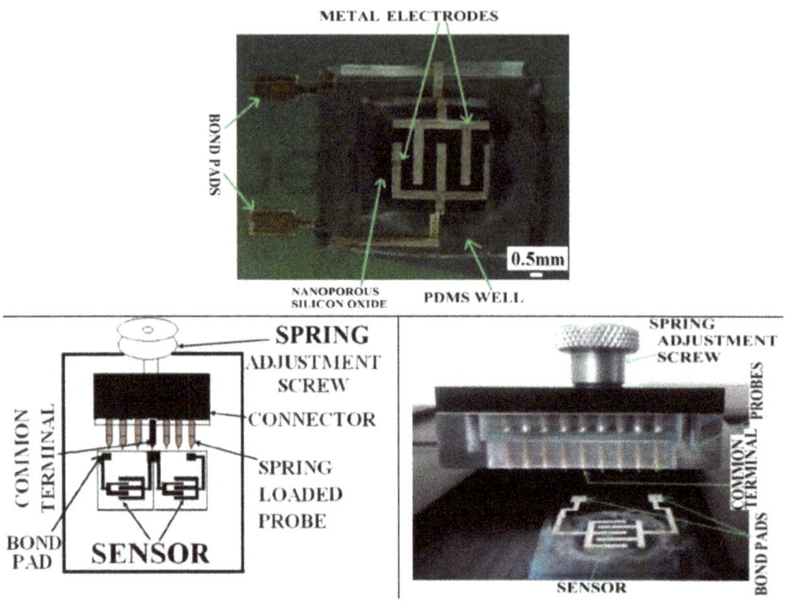

Figure 9.17. (a). Picture of the sensor. (b) Schematic of the sensor connection. (c) Picture of the sensor holder. (Reproduced with permission from [131]. Copyright 2016 IEEE.)

below the nanoporous layer [132]. For metallization, lateral metal contacts with aluminum paste were screen printed on nanostructured silicon oxide columns, details of which are reported in [133, 134].

2. *Impedance biosensors for bacteria and virus detection using screen printed electrodes*:

Chakraborty *et al* reported the quantitative differentiation of Hep-B and Hep-C surface antigens in blood, utilizing a nanoporous silicon oxide immunosensor array with high sensitivity over a wide range. The peak frequency output (f_p) from the steady state sensitivity characteristics and the first cut-off frequency (f_c) from the transient characteristics of the two sensor chips, functionalized with anti-Hep-B and anti-Hep-C antibodies, were considered as inputs to the two identical multilayer artificial neural networks (ANNs). The execution of a few classifier blocks in the ANN architecture, coupled with a suitable choice of learning rate and loss function to avoid over-fitting but give a reasonable amount of rigidity to the network for accommodating the dispersion associated with the input frequencies, has enabled the quantification of viruses with an accuracy of around 95% in the range of 0.04 fM to 1 pM and with an accuracy of around 90% beyond 1 pM and within 25 nM in different blood samples [135]. The sensitive area of the sensor is the nanocrystalline silicon oxide columns surrounding the electrodes within the white border of the circles shown in figure 9.18. An image of the sensor array chip with the PDMS well is shown in figure 9.18 and a close up

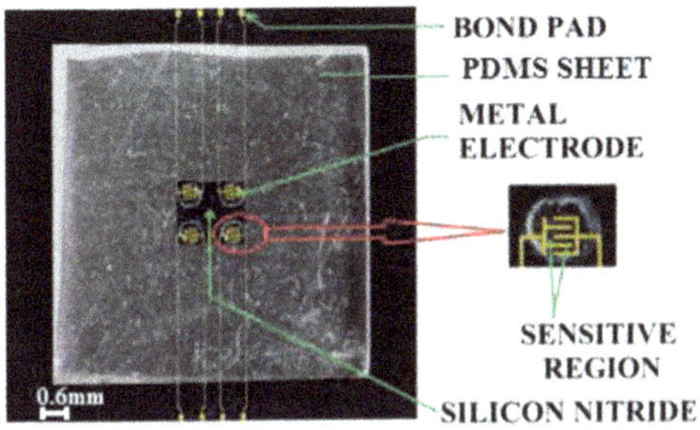

Figure 9.18. Image of the sensor chip. (Reproduced with permission from [135]. Copyright 2017 Elsevier.)

view of a single sensor showing the screen printed electrodes and the active region is shown in the inset of figure 9.18.

Das *et al* reported the optimization of screen printed electrode geometry on a macroporous silicon biosensor for improved bacteria identification. Macroporous silicon of 8 μm thickness and 55% porosity was fabricated on a 10–20 Ω-cm wafer utilizing hydrofluoric acid and dimethyl sulfoxide. The electrode configurations chosen for optimization were coplanar rectangular types with simple two-electrode, interdigitated comb finger-like electrodes, and a combination of simple electrodes and comb-shaped electrode designs. With these set-ups, 14 different patterns were produced with varying lengths and spacing within the technological constraints of a low cost screen printing method of electrode fabrication with a wide range of effective exposed area from 0.04 to 0.6 mm^2 by techniques from design of experiments. The sensitivity of bacteria detection for all the patterns was assessed by calculation of an area utilization factor (AUF), which was characterized as the ratio of the effective area occupied by the captured bacteria to the effective exposed area available for capture. From the AUF values, eight patterns were chosen, which are above a lower cut-off threshold for experimentation with *E. coli* O157 in the range of 10^2 CFU mL^{-1} to 10^6 CFU mL^{-1}. It was seen that the pattern with the highest AUF is capable of detecting 10^2 CFU mL^{-1} *E. coli* O157 with a sensitivity of 11% without any pre-concentration [136, 137]. Images of the sensor with four different typical electrode patterns are shown in figure 9.19.

For all the electrode configurations, the dimensions are selected such that the impedance in the presence of the control solution is not too high to cross the limit of the measuring equipment and furthermore not too low to reduce the range of bacterial concentration measurement. In the electrode configuration shown in figure 9.19(a), L is the length, W is the spacing, and t is the thickness. Because of the technological limitations of low cost screen printing techniques, the minimum values of L and t are 1 mm, considering the thin gold coating around the screen printed electrodes, and that of W is 0.5 mm.

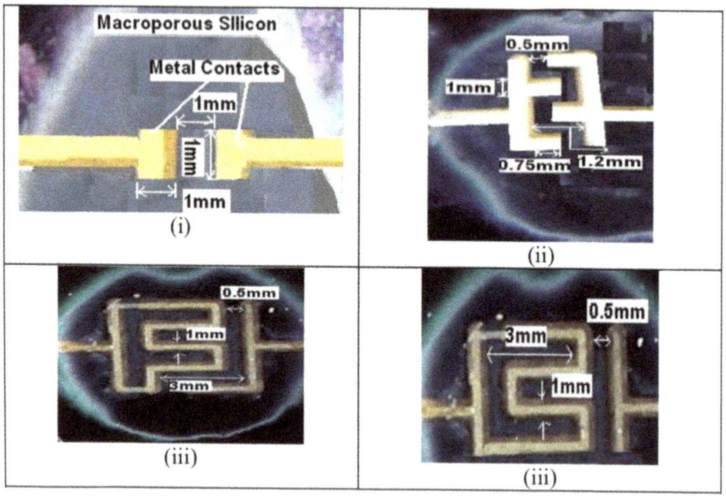

Figure 9.19. Images of the sensors with four different electrode patterns. (Reproduced with permission from [136]. Copyright 2012 IEEE.)

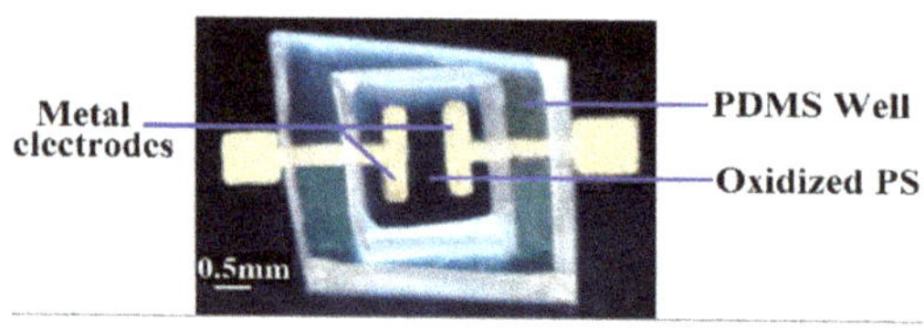

Figure 9.20. Final appearance of the sensor. (Reproduced with permission from [138]. Copyright 2016 IEEE.)

Further, researchers have looked into the reliability aspects of screen printed nanoporous silicon oxide impedance sensors for the detection of virus in blood. The improved selectivity of these sensors is dependent on the pore morphology and surface roughness of the nanoporous silicon oxide. Again, surface roughness can change the contact angle which can affect the degree of contamination upon long term exposure to buffer solution [138]. A picture of the sensor is shown in figure 9.20, which shows that the inter-electrode spacing is of the order of 500 μm. A schematic representation of the biosensor, demonstrating the various components, is shown figure 9.21.

Das *et al* reported, for the first time, IDE (interdigitated electrode)-less impedance detection of *E. coli* O157 in phosphate buffer solution with three different pore geometries of screen printed macroporous silicon, utilizing optimized antibody immobilization. To examine the effect of the pore geometry on the sensitivity of the impedance sensing, a transmission line-based distributed electrical model of macroporous silicon with an IDE-less electrode structure was accounted for. The electrical results obtained analytically were matched with the CoventorWare results. The experimental results

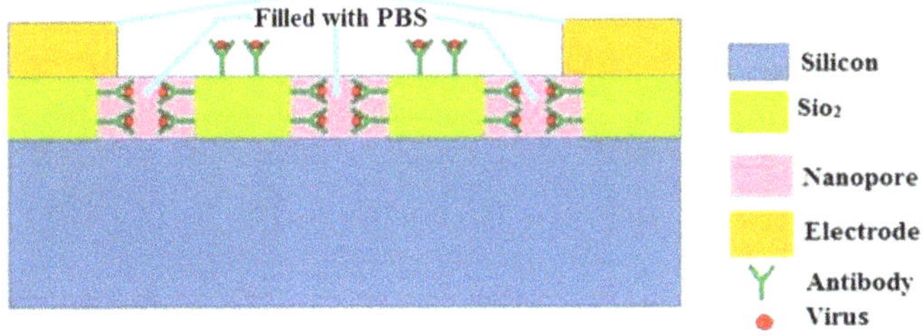

Figure 9.21. Impedance biosensor schematic. (Reproduced with permission from [138]. Copyright 2016 IEEE.)

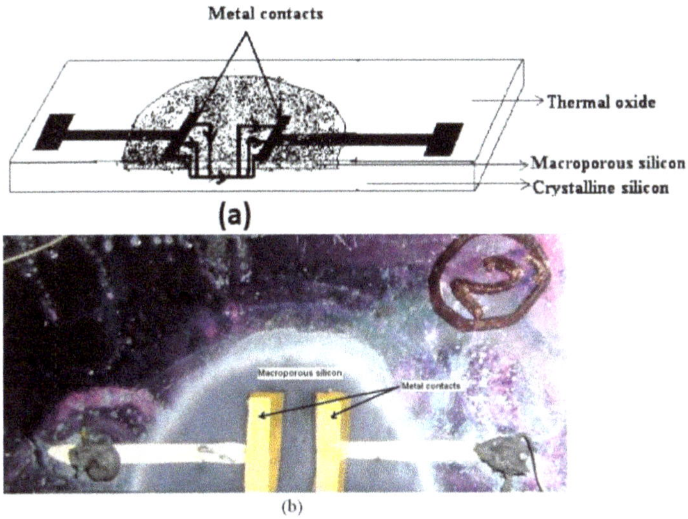

Figure 9.22. (a) Schematic diagram of the sensor. (b) Picture of the sensor. (Reproduced with permission from [139]. Copyright 2011 IEEE.)

were coordinated subjectively with the theoretical prediction. The optimized sensor was discovered to be reproducible, explicit, and suited for detection down to 10^3 CFU mL^{-1} with a straightforward structure. The deposition of metal paste by screen printing techniques results in an almost continuous metal film of around 10 μm thickness, which enables reproducible utilization of macroporous silicon for bacteria detection [139]. An schematic diagram and image of the sensor are presented in figures 9.22(a) and (b), respectively.

3. *Screen printed impedance cytosensors*:

   Mondal *et al* investigated the real-time sensing of the interactions of HaCaT cells using label-free impedance spectroscopy on oxidized porous silicon of two pore diameters (50 and 500 nm) with screen printed electrodes. One of the significant prerequisites for effective impedance spectroscopy measurement is to restrict the channeling of electric field lines through the pores. It was observed from the impedance measurements that the

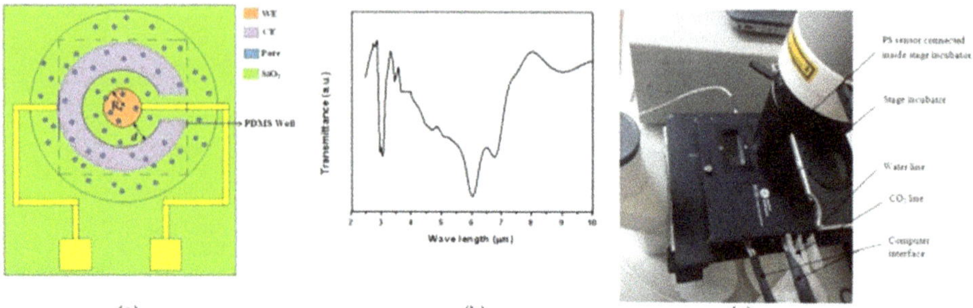

(a) (b) (c)

Figure 9.23. (a) Schematic of the sensor. (b) FTIR of PLL. (c) Picture of the impedance spectroscopy measurement set-up. (Reproduced with permission from [140]. Copyright 2015 American Institute of Physics.)

proliferation rate increases for 50 nm pores but decreases for 500 nm pores compared to that for planar substrates. The rate of decrease in cell–substrate separation in the initial stage is more than the rate of increase in cell–cell junction resistance corresponding to the initial adhesion phase of cells. Consequently, the impedance spectroscopy combined with the proposed theoretical framework for porous silicon substrates with screen printed electrodes can sense and quantify cellular interactions [140]. The working electrode (WE) was around 200 μm in diameter and its spacing with the counterelectrode (CE) was around 300 μm, as shown in figure 9.23(a). The functionalization with PLL (poly-l-lysine) was confirmed by Fourier transform infrared (FTIR) studies, as shown in figure 9.23(b). The peaks for amine groups are clearly visible in the PLL spectrum at 3.05 μm (3274 cm^{-1}) and 6.04 μm (1656 cm^{-1}) [141]. An image of the whole set-up is shown in figure 9.23(c).

Further, continuous monitoring of wound healing phenomena using screen printed porous silicon substrates with two pore morphologies using impedance measurement has been reported. It was observed from impedance measurements and models that the wound closure is faster on 50 nm pores followed by 500 nm pores and planar substrates. This was confirmed subjectively by the extracted time dependent cell–cell and cell–substrate interaction parameters using a well-known mechanical model of the wound healing process [142]. A picture of the fabricated sensor is shown in figure 9.24(a). A schematic diagram (cross sectional view) of the sensor in the presence of cells is shown in figure 9.24(b).

9.2.4 The advantages of impedance biosensors based on screen printed electrodes

The most encouraging applications of screen printed impedance biosensors are situations where low cost, small instrument size, and speed of analysis are crucial,

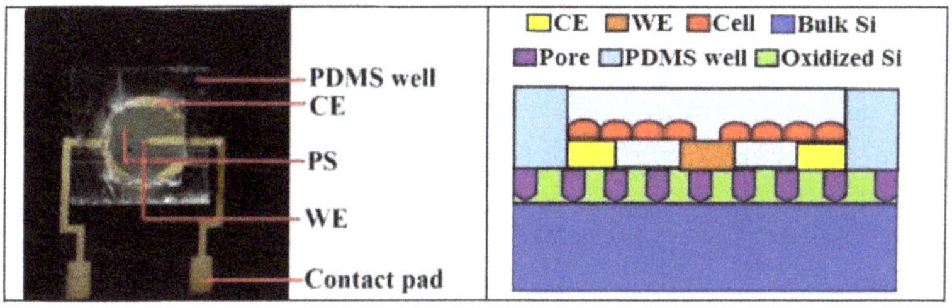

Figure 9.24. (a) Picture of the fabricated sensor. (b) Schematic diagram (cross sectional view) of the PS sensor in the presence of cells. (Reproduced with permission from [142]. Copyright 2016 IEEE.)

but cutting-edge accuracy and detection limits are not. Point-of-care diagnostics—testing and diagnosis at the bedside, in an ambulance, or during a clinic visit—is a promising application. If the cost and time per data point is reduced, screening for different cancer and disease markers utilizing an electrical biosensor could become standard for routine medical check-ups. Other applications include biowarfare agent detection, consumer test kits, bioprocess monitoring, and water quality testing. One key inquiry is whether impedance biosensors can have adequate selectivity for use in real-world applications, because actual samples typically contain an uncontrolled but significant amount of nontarget molecules. This can be partially addressed by incorporating intelligent data processing algorithms.

9.2.5 Challenges associated with real sample analysis using screen printed electrode based biosensors

The examples of electrochemical biosensors applied so far for the detection of allergens in wine, egg white, chocolate, biscuits, snacks, cakes, etc (table 9.1), prove the potential of these bioanalytical devices for the analysis of real samples. Electrochemical interference has been eliminated by a combination of interfaces well-protected against nonspecific adsorption, and adequate analysis protocols, including washing steps and sample preparation. Nonspecific binding of compounds present in the sample matrix to the transducer surface results in a change of the electrical properties at the interface, and potential interference with the biosensor response has been prevented by blocking using small thiols such as mercaptohexanol (Au electrodes and electrodes with AuNPs), bovine serum albumin, whole milk, and casein. The analysis protocols typically involved transferring the biosensor after incubation with the sample into a solution containing the electrochemical reporter probe, thus avoiding interference from electrochemically active compounds that might be present in the sample solution. Several studies reporting on new biosensors for allergens estimated the accuracy and suitability of the method for real samples by analyzing intentionally spiked 'allergen-free' food. Very good recovery factors in the range of 95%–110% were found for samples spiked at three concentration levels. It is worth mentioning that the spiking should be performed at the very beginning, before applying any sample treatment, cleaning, or dilution procedure, otherwise the effectiveness of sample preparation and the effect of sample preparation on allergen–bioreceptor affinity will remain unknown. Other biosensor studies included comparisons using current standard ELISA [143, 144] and a few even described detailed methods of validation [145]. Apart from the previously mentioned issues experienced in the improvement of biosensors, allergen examination faces some extra difficulties. These are related to sample preparation, changes suffered by allergen proteins during food processing, the availability of standards and reference materials, and a lack of established maximum allowed levels for each allergen. In some cases, the sensitivity that will be required for a newly developed method can be derived from current labeling rules. Sample preparation for allergen analysis with biosensors generally follows the well-established protocols for ELISA, consisting of extracting the sample with an appropriate buffer while heating, followed by centrifugation and dilution. In samples where the allergen protein is bound to phenolic compounds (chocolate, wine) this interaction must be broken first. However, there is a significant variability between results provided by laboratories performing ELISA with different extraction cocktails and protocols and sample preparation remains an area where improvements will undoubtedly follow in the future. For biosensors, the sample preparation procedure, while ensuring quantitative extraction of allergens from the food matrix, should not cause changes in protein structure that will affect the binding to its bioreceptor. Aptamer affinity for the target analyte is influenced dramatically by ionic strength and the nature of ions in the sample solution [146]. One approach solving both the problem of electrochemical interference and ensuring the preservation of aptamer affinity, as seen with standard allergen solutions, is to use a high enough dilution factor. This procedure, using sensitive biosensors, could, however, introduce errors and should be used with caution. Food processing involving heating, fermentation, and hydrolysis can lead to major changes in protein structure and extensive fragmentation. Antibodies developed for the native gliadin, for example, show little affinity to deamidated gliadin and to the small peptide fragments present in hydrolyzed products, and thus appropriate bioreceptors should be considered for analysis of processed food. Additionally, small peptide fragments are not suitable for sandwich formats as these require accommodating two epitopes—another aspect to consider in biosensor development.

The detection of contaminates in water is very significant since high pollution (heavy metals, pesticide and antibiotic residues, etc) or the presence of pathogens (infectious micro-organisms such as viruses, bacteria, and fungi) can seriously endanger human health. A few specific difficulties hinder the development and fabrication of biosensors: limitations in detecting small molecules, reusability, and sufficient stability for repeated testing [147, 148]. However, the most crucial issue is whether impedance biosensors have adequate selectivity for their application in real samples, which typically contain an unknown amount of nontarget molecules. There are two primary kinds of impedimetric biosensors—with or without a specific biorecognition element [149]. The most widely used biorecognition elements are specific antibodies [150, 151]. The way to achieve the fruitful application of electrochemical impedance spectroscopy (EIS)-based biosensors is through reducing nonspecific bindings and increasing the selectivity. Selectivity is particularly significant in real samples where the analyte concentration can be much smaller than the concentration of nontarget molecules. Nonspecific binding is typically ascribed to proteins contained in a sample matrix attaching to the sensor interface through an undesirable process not including bimolecular recognition [152]. One option to reduce nonspecific binding is to use blocking agents such as BSA, cysteine, or ethanolamine [153–155]. One of the primary difficulties is the sensitivity, which depends on the thickness of the sensing layer [156–160]. If the sensing layer is excessively thin, the electrode surface might be uncovered, which would reduce the signal-to-noise ratio and decrease the sensitivity. If the sensing layer is too thick, the detected AC impedance current decreases, meaning that the electron transfer between layers is hindered and the sensitivity is reduced. For example, Groß *et al* studied the effect of the thickness on the base resistance in the range of 30 to 150 μm and found that the sensitivity decreased along with the sensitive layer becoming thicker [156]. They also found that there is a trade-off between a wide linear range and high sensitivity. Likewise, the sensor signal became slower as the thickness of the sensitive layer increased [156]. Functionalization of the electrodes with high-affinity biomolecules improves both the selectivity and sensitivity of the system. Accordingly, EIS is very often combined with various nanostructured interfaces in order to increase the amount of biorecognition material on the surface and therefore to improve the sensor sensitivity and extend its linear working [161–166]. This improvement is related to the dimensions of the nanomaterials, which embues them with a large surface–volume ratio and high specific area, enabling them to immobilize a larger amount of biomolecules on the biosensor surface [167]. The reusability of antibody-based biosensors can be problematic because of the strength and irreversibility of antibody–antigen binding, and the regeneration of these surfaces without harming the antibody layer can be complicated because of harsh conditions [168]. For impedance biosensors, the extreme pH values of strong acids or bases may not be compatible with the chemistries employed for the protein immobilization, meaning that the reusability of a biosensor can be problematic. Finally, it can be challenging to detect small molecules such as heavy metals, pesticides, or antibiotic residues with EIS due to the exponential increment of the charge–transfer resistance through the polymer–protein layer to the underlying electrode surface [152, 169]. Small molecules (less than kDa) alone usually induce very small detectable responses, which can be very difficult to quantify, particularly in real samples where the concentration of the target molecule can be very low [152]. One possibility to improve the detection of small molecules is to conjugate these via a functional group to a larger carrier molecule (i.e. a protein) or with electrochemically bright metal and semiconductor nanomaterials, as changes because of the binding of large molecules can be identified more easily.

9.2.6 Future outlook and concluding remarks

Screen printed sensors are simple, user-friendly, and disposable, and require reduced analyte volume (microliters). More advanced structures attempt to implement smaller and more well-defined components, namely screen printed microelectrode arrays and 'lab-on-a-chip' schemes. The screen printed electrodes are considered a beneficial analytical tool to meet the increasing demand for rapid, selective, sensitive, and economical analysis. These electrodes may be applied directly or in addition to other methods to fulfill specific uses. In this chapter we discussed the various forms of screen printed electrodes. Modifications with nanomaterials are performed to enhance the sensitivity and selectivity of the SPEs. Using these modified SPEs, the performance and stability of sensors can be improved considerably. We believe that this field will be an area of growing research in coming years for analyzing analytes of environmental and biological relevance.

References

[1] Metters J P *et al* 2011 New directions in screen printed electroanalytical sensors: an overview of recent developments *Analyst* **136** 1067

[2] Li M *et al* 2012 Recent developments and applications of screen-printed electrodes in environmental assays—a review *Anal. Chim. Acta* **734** 31–44

[3] Tan S N *et al* 2010 Paper disk on screen printed electrode for one-step sensing with an internal standard *Anal. Chem.* **82** 8844–7

[4] Määttänen A *et al* 2012 Inkjet-printed gold electrodes on paper: characterization and functionalization *ACS Appl. Mater. Interfaces* **4** 955–64

[5] Injang U *et al* 2010 Determination of trace heavy metals in herbs by sequential injection analysis–anodic stripping voltammetry using screen-printed carbon nanotubes electrodes *Anal. Chim. Acta* **668** 54–60

[6] Kipphan H (ed) 2001 *Handbook of Print Media, Technologies and Production Methods* 1st edn (Berlin: Springer)

[7] Ingram S 1999 *Screen Printing Primer* 2nd edn (Pittsburgh, PA: GATF)

[8] Novakovic D and Kasikovic N 2013 *Propusna Stampa* 1st edn (Novi Sad: Faculty of Technical Sciences, University of Novi Sad)

[9] Barbulovic-Nad I *et al* 2006 Bio-microarray fabrication techniques—a review *Crit. Rev. Biotechnol.* **26** 237–59

[10] Dias A *et al* 2014 Recent advances in bioprinting and applications for biosensing *Biosensors* **4** 111–36

[11] Wang J *et al* 1998 Performance of screen-printed carbon electrodes fabricated from different carbon inks *Electrochim. Acta* **43** 3459–65

[12] Metters J P *et al* 2012 Platinum screen printed electrodes for the electroanalytical sensing of hydrazine and hydrogen peroxide *Anal. Methods* **4** 1272

[13] Metters J P *et al* 2012 Electroanalytical sensing of chromium (III) and (VI) utilising gold screen printed macro electrodes *Analyst* **137** 896

[14] Wang J *et al* 1997 Microfabricated electrochemical sensor for the detection of radiation-induced DNA damage *Anal. Chem.* **69** 1457–60

[15] Ren R *et al* 2010 A chronocoulometric DNA sensor based on screen-printed electrode doped with ionic liquid and polyaniline nanotubes *Biosens. Bioelectron.* **25** 2089–94

[16] Randviir E P *et al* 2013 Electrochemical impedance spectroscopy versus cyclic voltammetry for the electroanalytical sensing of capsaicin utilising screen printed carbon nanotube electrodes *Analyst* **138** 2970

[17] Randviir E P *et al* 2014 The fabrication, characterisation and electrochemical investigation of screen-printed graphene electrodes *Phys. Chem. Chem. Phys.* **16** 4598–611

[18] Foster C *et al* 2014 Cobalt phthalocyanine modified electrodes utilised in electroanalysis: nano-structured modified electrodes versus bulk modified screen-printed electrodes *Sensors* **14** 21905–22
[19] Tiwari I *et al* 2015 Design of screen-printed bulk modified electrodes using anthraquinone–cysteamine functionalized gold nanoparticles and their application to the detection of dissolved oxygen *Anal. Methods* **7** 2020–7
[20] Choudhry N A *et al* 2010 Screen printed electrodes provide micro-domain sites for fabricating disposable electro-catalytic ensembles *Electrochem. Commun.* **12** 406–9
[21] Salazar P *et al* 2012 Improvement and characterization of surfactant-modified Prussian blue screen-printed carbon electrodes for selective H_2O_2 detection at low applied potentials *J. Electroanal. Chem.* **674** 48–56
[22] Matemadombo F *et al* 2012 Comparison of carbon screen-printed and disk electrodes in the detection of antioxidants using CoPc derivatives *Sens. Actuators* B **166-67** 457–66
[23] Ramdani O *et al* 2013 Forensic electrochemistry: sensing the molecule of murder atropine *Analyst* **138** 1053
[24] Malzahn K *et al* 2011 Wearable electrochemical sensors for *in situ* analysis in marine environments *Analyst* **136** 2912
[25] Windmiller J R *et al* 2012 Electrochemical sensing based on printable temporary transfer tattoos *Chem. Commun.* **48** 6794
[26] Bandodkar A J *et al* 2014 Epidermal tattoo potentiometric sodium sensors with wireless signal transduction for continuous non-invasive sweat monitoring *Biosens. Bioelectron.* **54** 603–9
[27] Cai J *et al* 2009 Flexible thick-film electrochemical sensors: impact of mechanical bending and stress on the electrochemical behavior *Sens. Actuators* B **137** 379–85
[28] Yang Y-L *et al* 2010 Thick-film textile-based amperometric sensors and biosensors *Analyst* **135** 1230
[29] Novakovic D *et al* 2011 *Tehnike Stamped—Praktikum za Vezbe* 1st edn (Novi Sad: Faculty of Technical Sciences, University of Novi Sad)
[30] Falco B D *et al* 2018 Metabolomic analysis by UAE-GC MS and antioxidant activity of *Salvia hispanica*(L.) seeds grown under different irrigation regimes *Ind. Crops Prod.* **112** 584–92
[31] Narasimhaiah M *et al* 2018 Organic zinc and copper supplementation on antioxidant protective mechanism and their correlation with sperm functional characteristics in goats *Reprod. Domest. Anim.* **53** 644–54
[32] Avalos-Llano K R *et al* 2018 Effect of pulsed light treatments on quality and antioxidant properties of fresh-cut strawberries *Food Chem.* **264** 393–400
[33] Yuan S *et al* 2017 Optimization of decolorization process in agar production from *Gracilaria lemaneiformis* and evaluation of antioxidant activities of the extract rich in natural pigments *3 Biotech* **8** 8
[34] Vione D *et al* 2018 Effects of the antioxidant moieties of dissolved organic matter on triplet-sensitized phototransformation processes: implications for the photochemical modeling of sulfadiazine *Water Res.* **128** 38–48
[35] Marín D *et al* 2018 Freeze-dried phosphatidylcholine liposomes encapsulating various antioxidant extracts from natural waste as functional ingredients in surimi gels *Food Chem.* **245** 525–35

[36] Beta T and Hwang T 2018 Influence of heat and moisture treatment on carotenoids, phenolic content, and antioxidant capacity of orange maize flour *Food Chem.* **246** 58–64

[37] Lee P *et al* 2014 The use of screen-printed electrodes in a proof of concept electrochemical estimation of homocysteine and glutathione in the presence of cysteine using catechol *Sensors* **14** 10395–411

[38] Ricci F *et al* 2006 Glutathione amperometric detection based on a thiol–disulfide exchange reaction *Anal. Chim. Acta* **558** 164–70

[39] Honeychurch K C and Hart J P 2012 The chronoamperometric and voltammetric behaviour of glutathione at screen-printed carbon micro-band electrodes modified with cobalt phthalocyanine *Adv. Anal. Chem.* **2** 46–52

[40] Lee P T *et al* 2015 Electrochemical determination of free and total glutathione in human saliva samples *Sens. Actuators* B **221** 962–8

[41] Wang Y *et al* 2017 Electrochemical detection of glutathione by using thymine-rich DNA-gated switch functionalized mesoporous silica nanoparticles *Biosens. Bioelectron.* **87** 459–65

[42] Ganjali M R *et al* 2017 Highly sensitive voltammetric sensor for determination of ascorbic acid using graphite screen printed electrode modified with ZnO/Al_2O_3 nanocomposite *Int. J. Electrochem. Sci.* **12** 3231–40

[43] Raveendran J *et al* 2017 Voltammetric determination of ascorbic acid by using a disposable screen printed electrode modified with $Cu(OH)_2$ nanorods *Microchim. Acta* **184** 3573–9

[44] Alonso-Lomillo M A *et al* 2017 Determination of ascorbic acid in serum samples by screen-printed carbon electrodes modified with gold nanoparticles *Talanta* **174** 733–7

[45] Zhao X *et al* 2014 Disposable electrochemical ascorbic acid sensor based on molecularly imprinted poly(o-phenylenediamine)-modified dual channel screen-printed electrode for orange juice analysis *Food Anal. Methods* **7** 1557–63

[46] Guiberteau-Cabanillas A *et al* 2014 Electroanalytical behavior of gallic and ellagic acid using graphene modified screen-printed electrodes. Method for the determination of total low oxidation potential phenolic compounds content in cork boiling waters *Electroanalysis* **27** 177–84

[47] Su Y-L and Cheng S-H 2015 Sensitive and selective determination of gallic acid in green tea samples based on an electrochemical platform of poly(melamine) film *Anal. Chim. Acta* **901** 41–50

[48] Piermarini S *et al* 2013 Uricase biosensor based on a screen-printed electrode modified with Prussian blue for detection of uric acid in human blood serum *Sens. Actuators* B **179** 170–4

[49] Kanyong P *et al* 2012 Development of a sandwich format, amperometric screen-printed uric acid biosensor for urine analysis *Anal. Biochem.* **428** 39–43

[50] Su W-Y and Cheng S-H 2008 Electrocatalysis and sensitive determination of cysteine at poly(3,4-ethylenedioxythiophene)-modified screen-printed electrodes *Electrochem. Commun.* **10** 899–902

[51] Vasjari M *et al* 2005 Amino acid determination using screen-printed electrochemical sensors *Microchim. Acta* **150** 233–8

[52] Hernández-Ibáñez N *et al* 2016 l-cysteine determination in embryo cell culture media using Co (II)-phthalocyanine modified disposable screen-printed electrodes *J. Electroanal. Chem.* **780** 303–10

[53] Boonchiangma S *et al* 2014 A highly selective electrochemical sensor forl-tryptophan based on a screen-printed carbon electrode modified with poly-p-phenylenediamine and CdS quantum dots *J. Appl. Polym. Sci.* **131** 403562014

[54] Gómez-Mingot M *et al* 2011 Direct oxidation of methionine at screen printed graphite macroelectrodes: towards rapid sensing platforms *Sens. Actuators* B **155** 831–6
[55] Sominsky L *et al* 2010 Neonatal overfeeding increases capacity for catecholamine biosynthesis from the adrenal gland acutely and long-term in the male rat *Mol. Cell. Endocrinol.* **470** 295–303
[56] Askri D *et al* 2018 Sub-acute intravenous exposure to Fe_2O_3 nanoparticles does not alter cognitive performances and catecholamine levels, but slightly disrupts plasma iron level and brain iron content in rats *J. Trace Elem. Med. Biol.* **50** 73–9
[57] Gegg M and Piston D 2018 The role of DJ-1 complexes and catecholamine metabolism: relevance for familial and idiopathic Parkinson's disease *Neural Regen. Res.* **13** 815
[58] Ritter S 2018 Why I can't say 'no' to hindbrain catecholamine neurons *Appetite* **126** 210
[59] Shin M *et al* 2018 Hemostatic swabs containing polydopamine-like catecholamine chitosan-catechol for normal and coagulopathic animal models *ACS Biomater. Sci. Eng.* **4** 2314–8
[60] Dai J B *et al* 2018 EGR-1 expression in catecholamine-synthesizing neurons reflects auditory learning and correlates with responses in auditory processing areas *Neuroscience* **379** 415–27
[61] Zhang Y-M *et al* 2017 Magnetism-assisted modification of screen printed electrode with magnetic multi-walled carbon nanotubes for electrochemical determination of dopamine *Mater. Sci. Eng.* C **74** 62–9
[62] Beitollahi H *et al* 2017 GO/Fe_3O_4@SiO_2 core–shell nanocomposite-modified graphite screen-printed electrode for sensitive and selective electrochemical sensing of dopamine and uric acid *Anal. Methods* **9** 5541–9
[63] Beitollahi H *et al* 2018 Voltammetric determination of dopamine in the presence of tyrosine using graphite screen-printed electrode modified with graphene quantum dots *Ionics* **24** 4023–31
[64] Singh M *et al* 2018 Highly sensitive and selective determination of dopamine using screen-printed electrodes modified with nanocomposite of N′-phenyl-p-phenylenediamine/multi-walled carbon nanotubes/nafion *Mater. Res. Bull.* **101** 253–63
[65] Ping J *et al* 2010 Development of an ionic liquid modified screen-printed graphite electrode and its sensing in determination of dopamine *Electrochem. Commun.* **12** 1738–41
[66] Ku S *et al* 2013 Highly selective dopamine electrochemical sensor based on electrochemically pretreated graphite and nafion composite modified screen printed carbon electrode *J. Colloid Interface Sci.* **411** 182–6
[67] Valentini F *et al* 2014 Highly selective detection of epinephrine at oxidized single-wall carbon nanohorns modified screen printed electrodes (SPEs) *Biosens. Bioelectron.* **59** 94–8
[68] Soofiabadi F *et al* 2017 Application of Cu(II) nanocomplex modified graphite screen printed electrode to improve the sensitivity and selectivity for epinephrine detection *Anal. Bioanal. Electrochem.* **9** 340–50
[69] Baniasadi M *et al* 2017 Voltammetric determination of epinephrine based on ZnO nanoparticles assisted graphene oxide nanosheets *Adv. Mater.* **9** 718–28
[70] Apetrei I M *et al* 2015 Study of different carbonaceous materials as modifiers of screen-printed electrodes for detection of catecholamines *IEEE Sensors J.* **15** 3094–101
[71] Huang S-H *et al* 2010 Simultaneous determination of norepinephrine, uric acid, and ascorbic acid at a screen printed carbon electrode modified with polyacrylic acid-coated multi-wall carbon nanotubes *Biosens. Bioelectron.* **25** 2351–5

[72] Wang Y *et al* 2015 A disposable electrochemical sensor for simultaneous determination of norepinephrine and serotonin in rat cerebrospinal fluid based on MWNTs–ZnO/chitosan composites modified screen-printed electrode *Biosens. Bioelectron.* **65** 31–8

[73] Beitollahi H and Garkani Nejad F 2016 Graphene oxide/ZnO nano composite for sensitive and selective electrochemical sensing of levodopa and tyrosine using modified graphite screen printed electrode *Electroanalysis* **28** 2237–44

[74] Rabinca A A *et al* 2015 Electrochemical behaviour and rapid determination of l-dopa at electrochemically pretreated screen printed carbon electrode *Electroanalysis* **27** 2275–9

[75] Arduini F *et al* 2006 Detection of carbamic and organophosphorous pesticides in water samples using a cholinesterase biosensor based on Prussian blue-modified screen-printed electrode *Anal. Chim. Acta* **580** 155–62

[76] Andreescu S 2002 Screen-printed electrode based on AChE for the detection of pesticides in presence of organic solvents *Talanta* **57** 169–76

[77] Albuquerque Y D T D and Ferreira L F 2007 Amperometric biosensing of carbamate and organophosphate pesticides utilizing screen-printed tyrosinase-modified electrodes *Anal. Chim. Acta* **596** 210–21

[78] Li H *et al* 2011 A novel photoelectrochemical sensor for the organophosphorus pesticide dichlofenthion based on nanometer-sized titania coupled with a screen-printed electrode *Anal. Chem.* **83** 5290–5

[79] Lezi N and Economou A 2015 Voltammetric determination of neonicotinoid pesticides at disposable screen-printed sensors featuring a sputtered bismuth electrode *Electroanalysis* **27** 2313–21

[80] Chai Y *et al* 2013 Carbamate insecticide sensing based on acetylcholinesterase/Prussian blue-multi-walled carbon nanotubes/screen-printed electrodes *Anal. Lett.* **46** 803–17

[81] Upadhyay S *et al* 2011 Application of bimetallic nanoparticles modified screen printed electrode for the detection of organophosphate compounds using an enzyme inhibition approach *Anal. Methods* **3** 2246–53

[82] Pelle F D *et al* 2018 Nano carbon black-based screen printed sensor for carbofuran, isoprocarb, carbaryl and fenobucarb detection: application to grain samples *Talanta* **186** 389–96

[83] Hassani S *et al* 2018 Novel label-free electrochemical aptasensor for determination of Diazinon using gold nanoparticles-modified screen-printed gold electrode *Biosens. Bioelectron.* **120** 122–8

[84] Govindasamy M *et al* 2017 Graphene oxide nanoribbons film modified screen-printed carbon electrode for real-time detection of methyl parathion in food samples *J. Electrochem. Soc.* **164** B403–8

[85] Selvolini G *et al* 2018 DNA-based sensor for the detection of an organophosphorus pesticide: profenofos *Sensors* **18** 2035

[86] Khairy M *et al* 2018 Non-enzymatic electrochemical platform for parathion pesticide sensing based on nanometer-sized nickel oxide modified screen-printed electrodes *Food Chem.* **255** 104–11

[87] Jirasirichote A *et al* 2017 Voltammetric detection of carbofuran determination using screen-printed carbon electrodes modified with gold nanoparticles and graphene oxide *Talanta* **175** 331–7

[88] Zhang X *et al* 2013 Preparation, characterization of Fe_3O_4 at TiO_2 magnetic nanoparticles and their application for immunoassay of biomarker of exposure to organophosphorus pesticides *Biosens. Bioelectron.* **41** 669–74

[89] Kumar J and D'souza S 2011 Microbial biosensor for detection of methyl parathion using screen printed carbon electrode and cyclic voltammetry *Biosens. Bioelectron.* **26** 4289–93
[90] Gogol E *et al* 2000 Amperometric biosensors based on nafion coated screen-printed electrodes for the determination of cholinesterase inhibitors *Talanta* **53** 379–89
[91] Chen D *et al* 2017 Electrochemical acetylcholinesterase biosensor based on multi-walled carbon nanotubes/dicyclohexyl phthalate modified screen-printed electrode for detection of chlorpyrifos *J. Electroanal. Chem.* **801** 185–91
[92] Fu J *et al* 2012 Sensitive acetylcholinesterase biosensor based on screen-printed carbon electrode modified with cerium oxide–chitosan/mesoporous carbon–chitosan for organophosphorus pesticide residue detection *Int. J. Electrochem. Sci.* **13** 9231–41
[93] Nuapia Y *et al* 2018 Assessment of heavy metals in raw food samples from open markets in two African cities *Chemosphere* **196** 339–46
[94] Fang L *et al* 2018 A novel method for the sequential removal and separation of multiple heavy metals from wastewater *J. Hazard. Mater.* **342** 617–24
[95] Tang J *et al* 2018 Biosurfactants enhanced heavy metals removal from sludge in the electrokinetic treatment *Chem. Eng. J.* **334** 2579–92
[96] Karri V *et al* 2017 Comparative *in vitro* toxicity evaluation of heavy metals (lead, cadmium, arsenic, and methylmercury) on HT-22 hippocampal cell line *Biol. Trace Elem. Res.* **184** 226–39
[97] Wang T *et al* 2020 Adsorption of heavy metals from aqueous solution by UV-mutant *Bacillus subtilis* loaded on biochars derived from different stock materials *Ecotox. Environ. Saf.* **148** 285–92
[98] Nguyen T C *et al* 2017 Adsorptive removal of five heavy metals from water using blast furnace slag and fly ash *Environ. Sci. Pollut. Res.* **25** 20430–8
[99] Wen J *et al* 2018 Investigating organic matter properties affecting the binding behavior of heavy metals in the rhizosphere of wetlands *Ecotoxicol. Environ. Safety* **162** 184–91
[100] Bobrowski A *et al* 2017 Determination of copper(II) through anodic stripping voltammetry in tartrate buffer using an antimony film screen-printed carbon electrode *Anal. Lett.* **50** 2920–36
[101] Yao Y *et al* 2019 Simultaneous determination of Cd(II) and Pb(II) ions in honey and milk samples using a single-walled carbon nanohorns modified screen-printed electrochemical sensor *Food Chem.* **274** 8–15
[102] Adarakatti P S *et al* 2017 Calixarene bulk modified screen-printed electrodes (SPCCEs) as a one-shot disposable sensor for the simultaneous detection of lead(II), copper(II) and mercury(II) ions: application to environmental samples *Sens. Actuators* A **267** 517–25
[103] Pérez-Ràfols C *et al* 2017 Ag nanoparticles drop-casting modification of screen-printed electrodes for the simultaneous voltammetric determination of Cu(II) and Pb(II) *Sensors* **17** 1458
[104] Jeromiyas N *et al* 2019 Bismuth nanoparticles decorated graphenated carbon nanotubes modified screen-printed electrode for mercury detection *J. Taiwan Inst. Chem. Eng.* **95** 466–74
[105] Wang H *et al* 2017 Screen-printed electrode modified by bismuth/Fe_3O_4 nanoparticle/ionic liquid composite using internal standard normalization for accurate determination of Cd(II) in soil *Sensors* **18** 6
[106] Shoub S A B *et al* 2017 Gold nanoparticles/ionophore-modified screen printed electrode for detection of Pb(II) in river water using linear sweep anodic stripping voltammetry *Sensors Mater.* **29** 555–56

[107] Tyszczuk-Rotko K and Domańska K 2017 Simple and sensitive voltammetric procedure for determination of Cd(II) and Pb(II) using bismuth-coated screen-printed carbon electrode prepared with mediator *J. Electrochem. Soc.* **164** H537–44
[108] Phu Huong Lan N *et al* 2016 Invasive non-typhoidal *Salmonella* infections in Asia: clinical observations, disease outcome and dominant serovars from an infectious disease hospital in Vietnam *PLoS Negl.Trop. Dis.* **10** e0004857
[109] Toan L Q *et al* 2013 Risk assessment of *Salmonella* in pork in Hanoi, Vietnam *Vietnam. J. Prev. Med.* **23** 10–7
[110] Huong L Q *et al* 2006 Prevalence of *Salmonella* in retail chicken meat in Hanoi, Vietnam *Ann. NY Acad. Sci.* **1081** 257–61
[111] Nguyen D T A *et al* 2016 Prevalence, antibiotic resistance, and extended-spectrum and AmpC β-lactamase productivity of *Salmonella* isolates from raw meat and seafood samples in Ho Chi Minh City, Vietnam *Int. J. Food Microbiol.* **236** 115–22
[112] Park S *et al* 2012 Risk factors for microbial contamination in fruits and vegetables at the preharvest level: a systematic review *J. Food Prot.* **75** 2055–81
[113] Guard-Petter J 2001 The chicken, the egg and *Salmonella enteritidis Environ. Microbiol.* **3** 421–30
[114] Nandakumar V *et al* 2008 A methodology for rapid detection of *Salmonella typhimurium* using labelfree electrochemical impedance spectroscopy *Biosens. Bioelectron.* **24** 1039–42
[115] Cinti S *et al* 2017 Electrochemical biosensors for rapid detection of foodborne *Salmonella*: a critical overview *Sensors* **17** 1910
[116] Nguyen T H P *et al* 2019 Stable electrochemical measurements of platinum screen-printed electrodes modified with vertical ZnO nanorods for bacterial detection *J. Nanomater.* **2019** 2341268
[117] Setterington E B and Alocilja E C 2012 Electrochemical biosensor for rapid and sensitive detection of magnetically extracted bacterial pathogens *Biosensors* **2** 15–31
[118] Razzino C A *et al* 2020 An electrochemical immunosensor using gold nanoparticles–PAMAM-nanostructured screen-printed carbon electrodes for tau protein determination in plasma and brain tissues from Alzheimer patients *Biosens. Bioelectron.* **163** 112238
[119] Bagel O *et al* 2000 Enzyme affinity assays involving a single-use electrochemical sensor. Applications to the enzyme immunoassay of human chorionic gonadotropin hormone and nucleic acid hybridization of human cytomegalovirus DNA *Electroanalysis* **12** 1447–52
[120] Wang J *et al* 2002 Single-use thick-film electrochemical sensor for insulin *Electroanalysis* **14** 1365–8
[121] Chiu M H *et al* 2009 Determination of lincomycin in urine and some foodstuffs by flow injection analysis coupled with liquid chromatography and electrochemical detection with a preanodized screen-printed carbon electrode *J. Chromatogr.* B **877** 991–4
[122] Asadollahi-Baboli M and Mani-Varnosfaderani A 2014 Rapid and simultaneous determination of tetracycline and cefixime antibiotics by mean of gold nanoparticles–screen printed gold electrode and chemometrics tools *Measurement* **47** 145–9
[123] Centi S *et al* 2010 Development of an electrochemical immunoassay based on the use of an eight-electrodes screen-printed array coupled with magnetic beads for the detection of antimicrobial sulfonamides in honey *Electroanalysis* **22** 1881–8
[124] Conzuelo F *et al* 2014 Rapid screening of multiple antibiotic residues in milk using disposable amperometric magnetosensors *Anal. Chim. Acta* **820** 32–8

[125] Gilmartin M A T and Hart J P 1994 Rapid detection of paracetamol using a disposable, surface-modified screen-printed carbon electrode *Analyst* **119** 2431

[126] Stulík K *et al* 2000 Microelectrodes. Definitions, characterization, and applications (technical report) *Pure Appl. Chem.* **72** 1483–92

[127] Konash A *et al* 2008 Theoretical and experimental evaluation of screen-printed tubular carbon ink disposable sensor well electrodes by dc and Fourier transformed ac voltammetry *J. Solid State Electrochem.* **13** 551–63

[128] Mann T S and Mikkelsen S R 2008 Antibiotic susceptibility testing at a screen-printed carbon electrode array *Anal. Chem.* **80** 843–8

[129] Corgier B P *et al* 2007 Direct electrochemical addressing of immunoglobulins: immuno-chip on screen-printed microarray *Biosens. Bioelectron.* **22** 1522–6

[130] Biscay J *et al* 2012 Amperometric fructose sensor based on ferrocyanide modified screen-printed carbon electrode *Talanta* **88** 432–8

[131] Samanta N and Roychaudhuri C 2016 Nanocrystalline silicon oxide impedance immunosensors for sub-femtomolar mycotoxin estimation in corn samples by incremental fuzzy approach *IEEE Sensors J.* **16** 1069–78

[132] Ghosh H and Roychaudhuri C 2013 Ultrasensitive food toxin biosensor using frequency based signals of silicon oxide nanoporous structure *Appl. Phys. Lett.* **102** 243701

[133] Das R D *et al* 2009 Macroporous silicon based simple and efficient trapping platform for electrical detection of *Salmonella typhimurium* pathogens *Biosens. Bioelectron.* **24** 3215–22

[134] Ghosh H and Roychaudhuri C 2015 Noise spectroscopy as an efficient tool for impedance based sub-femtomolar toxin detection in complex mixture using nanoporous silicon oxide *Biosens. Bioelectron.* **67** 757–62

[135] Chakraborty W *et al* 2017 Quantitative differentiation of multiple virus in blood using nanoporous silicon oxide immunosensor and artificial neural network *Biosens. Bioelectron.* **98** 180–8

[136] Das R D *et al* 2012 Optimized electrode geometry for an improved impedance based macroporous silicon bacteria detector *IEEE Sensors J.* **12** 1868–77

[137] Deng L *et al* 2002 Effects of geometrical factors of coplanar electrodes on biomedical measurements as exemplified by the electrical impedance measurement of saline and blood *Proc. Second Joint 24th Annual Conf. and the Annual Fall Mtg of the Biomedical Engineering Society, Engineering in Medicine and Biology* (Piscataway, NJ: IEEE) pp 1667–8

[138] Das N and Roychaudhuri C 2015 Reliability study of nanoporous silicon oxide impedance biosensor for virus detection: influence of surface roughness *IEEE Trans. Device Mater. Reliab.* **15** 402–9

[139] Das R D *et al* 2011 Interdigitated electrode-less high-performance macroporous silicon structure as impedance biosensor for bacteria detection *IEEE Sensors J.* **11** 1242–52

[140] Mondal D *et al* 2015 Real-time sensing of epithelial cell–cell and cell–substrate interactions by impedance spectroscopy on porous substrates *J. Appl. Phys.* **118** 044701

[141] Amorim S *et al* 2014 Hyaluronic acid/poly-l-lysine bilayered silica nanoparticles enhance the osteogenic differentiation of human mesenchymal stem cells *J. Mater. Chem.* B **2** 6939–46

[142] Mondal D *et al* 2016 Real-time monitoring of wound healing on nano-patterned substrates: non-invasive impedance spectroscopy technique *IEEE Trans. Nanotechnol.* **15** 791–800

[143] Montiel R V *et al* 2016 Simultaneous determination of the main peanut allergens in foods using disposable amperometric magnetic beads-based immunosensing platforms *Chemosensors* **4** 11

[144] Montiel V R V *et al* 2015 Sensitive and selective magnetoimmunosensing platform for determination of the food allergen Ara h 1 *Anal. Chim. Acta* **880** 52–9

[145] Alves R C *et al* 2015 Detection of the peanut allergen Ara h 6 in foodstuffs using a voltammetric biosensing approach *Anal. Bioanal. Chem.* **407** 7157–63

[146] Potty A S R *et al* 2011 Biophysical characterization of DNA and RNA aptamer interactions with hen egg lysozyme *Int. J. Biol. Macromol.* **48** 392–7

[147] Heo J and Hua S Z 2009 An overview of recent strategies in pathogen sensing *Sensors* **9** 4483–502

[148] Bogomolova A *et al* 2009 Challenges of electrochemical impedance spectroscopy in protein biosensing *Anal. Chem.* **81** 3944–49

[149] Wang Y *et al* 2012 New trends in impedimetric biosensors for the detection of foodborne pathogenic bacteria. *Sensors.* **12** 3449–71

[150] Baryeh K *et al* 2017 Introduction to medical biosensors for point of care applications *Medical Biosensors for Point of Care (POC) Applications* ed R J Narayan (Sawston: Woodhead) pp 3–25

[151] Morales M A and Halpern J M 2018 Guide to selecting a biorecognition element for biosensors *Bioconjugate Chem.* **29** 3231–39

[152] Radhakrishnan R *et al* 2014 Impedance biosensors: applications to sustainability and remaining technical challenges *ACS Sustain. Chem. Eng.* **2** 1649–55

[153] Frederix F *et al* 2004 Reduced nonspecific adsorption on covalently immobilized protein surfaces using poly(ethylene oxide) containing blocking agents *J. Biochem. Biophys. Methods* **58** 67–74

[154] Riquelme M V *et al* 2016 Optimizing blocking of nonspecific bacterial attachment to impedimetric biosensors *Sens. Bio-Sensing Res.* **8** 47–54

[155] Contreras-Naranjo J E and Aguilar O 2019 Suppressing non-specific binding of proteins onto electrode surfaces in the development of electrochemical immunosensors *Biosensors* **9** 15

[156] Groß A *et al* 2012 The effect of the thickness of the sensitive layer on the performance of the accumulating NO_x sensor *Sensors* **12** 12329–46

[157] Comeau K D and Meli M V 2012 Effect of alkanethiol chain length on gold nanoparticle monolayers at the air–water interface *Langmuir* **28** 377–81

[158] Campuzano S *et al* 2006 Characterization of alkanethiol-self-assembled monolayers-modified gold electrodes by electrochemical impedance spectroscopy *J. Electroanal. Chem.* **586** 112–21

[159] Bhadra P *et al* 2015 Studies on varying n-alkanethiol chain lengths on a gold coated surface and their effect on antibody–antigen binding efficiency *RSC Adv.* **5** 80480–87

[160] Braiek M *et al* 2012 An electrochemical immunosensor for detection of *Staphylococcus aureus* bacteria based on immobilization of antibodies on self-assembled monolayers-functionalized gold electrode *Biosensors* **2** 417–26

[161] Zhang W *et al* 2017 A 3D graphene-based biosensor as an early microcystin-LR screening tool in sources of drinking water supply *Electrochim. Acta* **236** 319–27

[162] Mehta J *et al* 2017 Graphene quantum dot modified screen printed immunosensor for the determination of parathion *Anal. Biochem.* **523** 1–9

[163] Bonanni A *et al* 2012 Graphene for impedimetric biosensing *Trends Anal. Chem.* **37** 12–21

[164] Aceta Y *et al* 2017 Graphene electrode platform for impedimetric aptasensing *Electrochim. Acta* **229** 458–66
[165] Shao Y *et al* 2010 Graphene based electrochemical sensors and biosensors: a review *Electroanalysis* **22** 1027–36
[166] Jiang D *et al* 2015 Silver nanoparticles anchored on nitrogen-doped graphene as a novel electrochemical biosensing platform with enhanced sensitivity for aptamer-based pesticide assay *Analyst* **140** 6404–11
[167] Tîlmaciu C M and Morris M C 2015 Carbon nanotube biosensors *Front. Chem.* **27** 1–2
[168] Goode J A *et al* 2015 Biosensor regeneration: a review of common techniques and outcomes *Langmuir* **31** 6267–76
[169] Bradbury C R *et al* 2008 Distance-independent charge-transfer resistance at gold electrodes modified by thiol monolayers and metal nanoparticles *J. Phys. Chem.* C **112** 10153–60

Chapter 10

Cellulose paper for flexible electronics: design and technology

Supriya Yadav, N N Sharma and Jamil Akhtar

10.1 Introduction

Paper has been used for more than 2000 years in different civilizations. Paper was first introduced in China by Tazi Lun. The word 'paper' is derived from the Latin word papyrus, for the plant *Cyperus papyrus*. Papyrus is a thick material produced from the pith of the plant. Paper production is one of the largest manufacturing sectors in the world. Cellulose paper is porous and hygroscopic in nature due to the hydroxyl groups of cellulose fibres and has been used in sensing applications for a long time [1]. Cellulose is the most abundant biopolymer of glucose on Earth. The cellulosic fibres are attached to each other by lignin and are separated through manufacturing processes which may involve mechanical pulping or chemical pulping [2].

Cellulose paper was first introduced as a microfluidic analytical device by the Whitesides group in 2007 [3, 4]. Such a simple, emerging development encouraged researchers to integrate cellulose paper as a substrate material where cost effectiveness, flexibility, and bendability are the critical parameters for the fabrication of electronic devices [5]. Electronic components or sensors [6, 7] are integrated into the paper to make paper-based electronics which are very flexible, cost effective, recyclable, and renewable in nature [8–10]. The fabrication of such devices is based on patterning using a hydrophobic material such as wax [11] or PDMS [12] on hydrophilic cellulose paper.

Generally, silicon [13–15] and glasses [16] have been used for the fabrication of electronic devices, but these substrates have drawbacks such as high cost, weak affinity between the functional materials, and the need for a well-established clean room for fabrication. In contrast, cellulose paper substrates are eco-sustainable, renewable, lightweight, and offer promising alternatives to conventional substrates. Paper has been traditionally used for manual writing using inks, and later with the

doi:10.1088/978-0-7503-3439-6ch10 © IOP Publishing Ltd 2021

more advanced technology of printing, for keeping records [17]. With the development of advanced electronics, cellulose paper has been used as a substrate, with the unique 3D hierarchical structure of cellulose fibres providing a versatile and cost effective technological platform for next-generation smart systems. Thus, the use of electronic devices on flexible cellulose paper or flexible electronics is predicted to gain popularity in the near future to solve the problem of environmental pollution caused by electronic waste. In this chapter we will discuss the basic structure of a capillary design on cellulose paper and also discuss the application of a design on cellulose paper using processing technology and using cellulose paper as a substrate for the fabrication of flexible electronics.

10.2 Cellulose paper structure and fabrication

Cellulose is the most common and abundant polysaccharide, and is synthesized by trees and other plants. Cellulose is present in paper and is used in our daily lives for various purposes such as writing, printing, wiping, wrapping, and packaging. Paper is produced in a mat- or sheet-like structure and various raw materials are used in the fabrication process, such as wood (printing paper), cotton (filter paper and chromatography paper), jute, grass, bamboo, bagasse, straw, hemp, etc. Paper is manufactured using the various properties of the raw materials with different techniques on a paper machine. As shown in figure 10.1, paper is made up of cellulose fibres. In cellulose the molecule, a hydroxyl group is present on inner and outer side of the structure, which is why cellulose is hydrophilic in nature and the surface of paper is rough (porous and fibrous). Paper production has amplified globally and will continue to increase in the near future and is already one of the largest manufacturing sectors in the world [18]. These cellulose fibres are separated from wood by debarking the wood for making pulp. Pulp is a base material for the manufacture of any paper. The debarked wood is further processed through

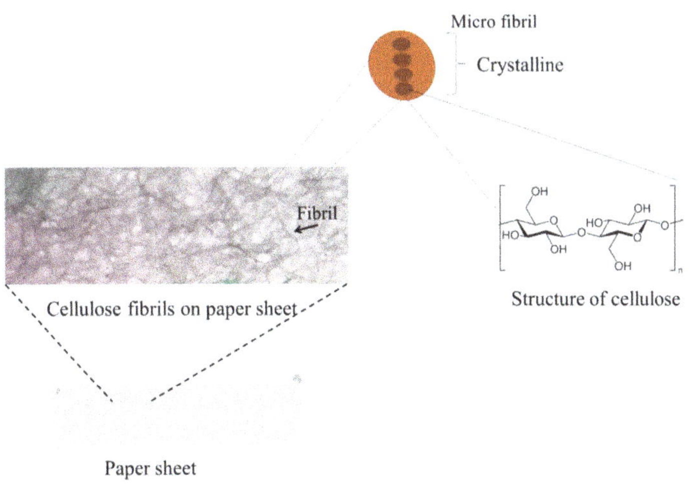

Figure 10.1. Schematic representation of the structure of cellulose.

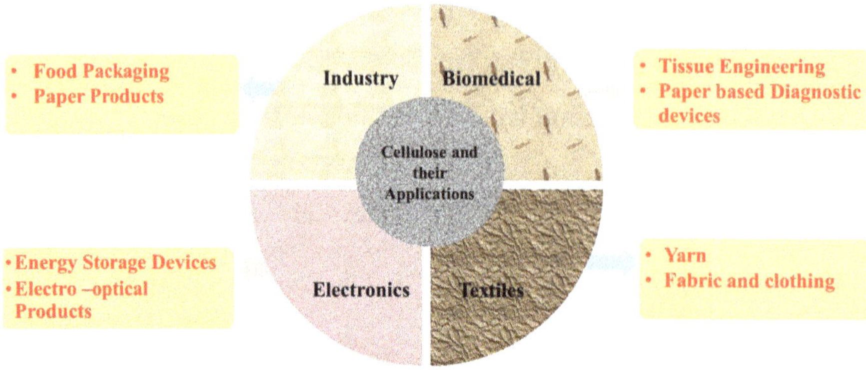

Figure 10.2. Applications of cellulose paper.

mechanical (pulped to fibre) or chemical processes (pulped to chips), depending on which type or grade of paper is desired.

Paper made using a mechanical process is very weak and yellowish in colour, because the lignin is not separated from the cellulose fibre, thus the yield and opacity of the paper are very high but the mechanical strength and brightness are low, like newsprint paper. Printing paper requires high opacity, so a mechanical process is applied for the production of printing paper [19]. Through the chemical pulping process, the lignin is separated from the cellulose fibre to form wood-free paper and the yielding is very low and the paper prepared by this method is very hygroscopic in nature (like filter paper) [20]. After the initial processing the pulp is bleached to remove all the impurities present in the pulp and make the paper white (increase the brightness), which is also necessary for printing paper. Then the pulp is poured into the paper making machine and forms a sheet of paper. Then the drying, pressing, and finishing/sizing and coating of paper is carried out. For the production of paper according to our use, there are some additives like phenolic resins (filter paper) or fillers (printing paper) mixed with pulp for refining. Some additives such as phenolic resins (filter paper) and fillers (printing paper) are added in the production of paper according to its intended use. In the paper machine the suspension reaches a wire pit via the head box. In the wire pit the paper sheet formation is started and leads two types of sheet formation depending upon the motion of the fibre in the suspension: the first type is for filtration and the second is for thickening. If we are making filter paper then the paper will proceed further for drying and pressing and then sizing, packaging, and storage. If we are making printing paper, after thickening the paper will proceed further for coating and calendaring, and then sizing, packaging, and storage. Cellulose paper can be used in various applications, as shown in figure 10.2, such as in electronics, biomedical devices, textiles, and industry [21–23].

10.3 A basic capillary structure design on cellulose paper

Cellulose paper is a flexible fibrous and porous material. In a cellulose fibre matrix, the fibre diameter is 1–100 μm with and average pore size of 1–10 μm. A fibre can be defined as an elongated material with a high aspect ratio of its length (l) to its

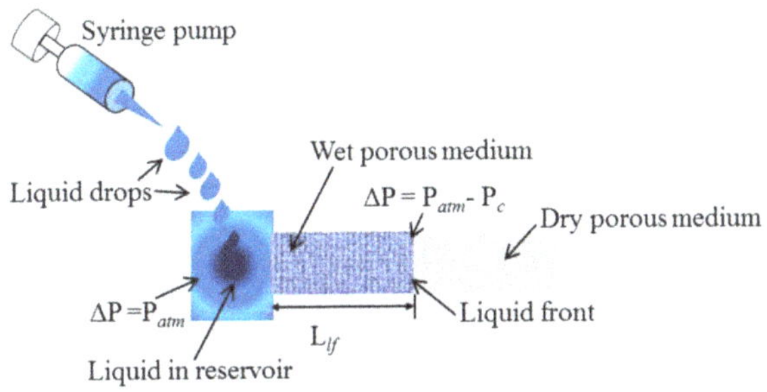

Figure 10.3. Capillary structure on cellulose paper.

diameter or thickness (d). For a given mass (m) or volume of fibre, its length (l) varies inversely as the square of the fibre diameter (d):

$$\frac{l}{m} = \frac{4}{\rho \pi d^2}. \tag{10.1}$$

Hence, from equation (10.1) we can say that if we halve the fibre diameter and density of the fibre material (ρ), we will increase its length fourfold. A fibre is flexible when it can be bent to an uninformed (random) radius [24]. Formerly fibre bending depended on the elastic modulus or stiffness of the fibre and the fibre diameter or fibre area or cross section [25]. The flexibility of a fibre can be express by the equation

$$\frac{M}{I} = \frac{E}{R} = \frac{E\pi d^4}{64}. \tag{10.2}$$

In this equation M is the applied bending moment, R is the radius of the curved fibre, E is the elastic modulus, and I is the second cross sectional area of the fibre. From equation (10.2) the inverse product of the bending moment (M) and the radius of curvature (R) are termed the measures of flexibility.

The cellulose molecule is hydrophilic and porous in nature. The hydroxyl group present on the outer and inner sides of the molecule is responsible for the intra- and inter-fibre bonding in a cellulose network. These hydroxyl groups are bound through hydrogen bonds. When a fluid drop enters into a porous cellulose paper, capillarity (capillary pressure) is the basic transport mechanism to move the fluid further. This capillarity mechanism on cellulose paper is shown in figure 10.3. Figure 10.3 shows that the pressure drop in the liquid front is accountable for the flow of fluid further from the wet porous medium to the dry porous medium. The capillary pressure, which is equal to the pressure difference at the liquid–solid (paper) interface in the wetted region, is governed by the Laplace pressure [26].

10.4 The application of designs and processing technologies on cellulose paper for flexible electronics

Paper is an emergent material in flexible electronics. The use of paper is global and varies depending on our needs. All papers are made of cellulose, but these papers

Figure 10.4. An SEM image of cellulose fibres on different types of papers: (a) 1250 qualitative filter paper, (b) Whatman filter paper grade 1, and (c) A4 sheet paper.

differ from each other in their properties such as pore size, porosity, stiffness, wet strength, chemical reactivity, and surface area. During the manufacture of paper, these properties can be varied by the manufacturer and produce different papers for different applications. In figure 10.4 scanning electron microscopy (SEM) images of different paper substrates are shown. The properties of the paper depend on the morphology of the cellulose. All these properties make the cellulose paper surface rough and porous. The structure and composition of the cellulose, i.e. the morphology of the cellulose, are very important to consider when exploiting the surface of cellulose paper in the design of flexible electronics.

Cellulose paper is low cost, biodegradable, flexible, and biocompatible in nature. Due to these properties, cellulose paper is a green biomaterial and suitable for developing disposable and degradable electronic devices. Currently, sensors and actuators are fabricated simply on cellulose paper substrates for making flexible electronic devices.

Cellulose paper is hydrophilic in nature due to the intramolecular and intermolecular hydrogen bonds on the cellulose molecule. They provide a good mechanical strength to paper allowing it to become a versatile substrate candidate for the fabrication of sensors, solar cells, tribolectric nanogenerators (TENGs), and field-effect transistor (FET) devices. The abundant hydroxyl groups on the cellulose molecule provide a variety of hybridization possibilities for paper to work as a smart active material to develop cellulose-based electrodes for electrochemical energy storage (EES) devices. The wettability and thermal stability of cellulose paper make it attractive for dielectric materials or gel electrolytes for flexible organic light-emitting diodes (OLEDs), foldable batteries, and printed photovoltaic cells [27].

The unique surface properties of cellulose paper make it suitable for producing pattern designs as well as for the large-scale manufacture of flexible electronics. The fabrication of patterns through modern processing techniques, such as spray-coating, flexographic printing, microfluidic spinning, electrostatic spinning, and 3D printing, is possible only through systematically modifying and controlling the surface/interface chemistry on cellulose paper [28]. Recently, paper became an emerging material for producing high-performance, next-generation, and low-cost flexible electronics.

Paper is used widely in our daily lives, such as in books, newspapers, and household products. However, the use of computers and the Internet has opened a new and faster way of transferring information, which has led to a decrease in the usage of paper. Paper is an environmentally friendly, flexible, and renewable cellulose material. Due to its flexible and recyclable nature, paper is now used for making flexible electronic devices. In flexible electronic devices the conductors typically require conductive electrodes to supply and collect charge. If the coatings on conductors are thick, they can crack during bending, and a paper substrate is very significant in this case because of its flexibility.

10.5 Conclusion

Paper is an emergent material in flexible electronics, while it remains ubiquitous in day-to-day life. Paper is a low-cost substrate which portable and degradable in nature. The surface of paper is porous and also fibrous. Different papers are used for different applications such as printing and packaging. However, applying electronics on a paper substrate is challenging because of its fibrous surface. Most electronic devices manufactured directly onto paper substrates are compatible with the paper surface. In conclusion, paper is the best material choice for the fabrication of low-cost flexible electronics devices and promising for organic electronics in the near future. However, existing paper electronics required continued attention to make them industry/consumer-ready electronics.

References

[1] Alava M and Niskanen K 2006 The physics of paper *Rep. Prog. Phys.* **69** 669

[2] Tsuen-Hsuin T 1985 *Chemistry and Chemical Technology, Part I: Paper and PrintingScience and Civilisation in China vol 5* ed J Needham (Cambridge: Cambridge University Press)

[3] Martinez A W, Phillips S T, Butte M J and Whitesides G M 2017 Patterned paper as a platform for inexpensive, low-volume, portable bioassays *Angew. Chem. Int. Ed.* **46** 1318–20

[4] Martinez A W, Phillips S T, Whitesides G M and Carrilho E 2009 Diagnostics for the developing world: microfluidic paper-based analytical devices *Anal. Chem.* **82** 3–10

[5] Grau G 2017 Low-cost fabrication of paper-based systems: microfluidics, sensors, electronics and deployment *2017 IEEE 60th Int. Midwest Symp. on Circuits and Systems (MWSCAS)* (Piscataway, NJ: IEEE) pp 84–7

[6] Gelinck G, Heremans P, Nomoto K and Anthopoulos T D 2010 Organic transistors in optical displays and microelectronic applications *Adv. Mater.* **22** 3778–98

[7] Guo Y, Yu G and Liu Y 2010 Functional organic field-effect transistors *Adv. Mater.* **22** 4427–47

[8] Aswal D K, Aswal D K and Yakhmi J V 2010 *Molecular and Organic Electronics Devices* (Hauppauge, NY: Nova Science)

[9] Tobjörk D and Österbacka R 2011 Paper electronics *Adv. Mater.* **23** 1935–61
[10] Hyun W J, Secor E B, Kim C H, Hersam M C, Francis L F and Frisbie C D 2017 Scalable, self-aligned printing of flexible graphene micro-supercapacitors *Adv. Energy Mater.* **7** 1700285
[11] Lu Y, Shi W, Jiang L, Qin J and Lin B 2017 Rapid prototyping of paper-based microfluidics with wax for low-cost, portable bioassay *Electrophoresis* **30** 1497–500
[12] Sia S K and Whitesides G M 2018 Microfluidic devices fabricated in pol(dimethylsiloxane) for biological studies *Electrophoresis* **24** 3563–76
[13] Joyce R, Yadav S, Sharma A K, Panwar D K, Bhatia R R, Varghese S and Akhtar J 2016 Pattern transfer of microstructures between deeply etched surface for MEMS applications *Mater. Sci. Semicond. Process.* **56** 373–80
[14] Yadav S, Joyce R, Sharma A K, Sharma H, Sharma N N, Varghese S and Akhtar J 2017 Process sequence optimization for digital microfluidic integration using EWOD technique *AIP Conf. Proc.* **1724** 020085
[15] Kumar M, Yadav S, Kumar A, Sharma N N, Akhtar J and Singh K 2019 MEMS impedance flow cytometry designs for effective manipulation of micro entities in health care applications *Biosens. Bioelectron.* **142** 111526
[16] Mazurczyk R, Mansfield C D and Lygan M 2015 Glass microstructure capping and bonding techniques *Microfluidic Diagnostics* (Totowa, NJ: Humana) pp 141–51
[17] Chen J, Ku L, Xie J, Wang Y, Pan L and Zu Q 2007 The effect of laser inkless eco-printing on the carbonized microstructure of paper *Cellul. Chem. Technol.* **50** 101–8
[18] Martin J and Haggith M 2018 The state of the global paper industry Environmental Paper Network 2 https://environmentalpaper.org/wp-content/uploads/2018/04/StateOfTheGlobalPaperIndustry2018_FullReport-Final-1.pdf
[19] Printerland 2010 The definitive guide to printing paper https://printerland.co.uk/blog/printer-buying-guide/in-depth-guide-to-printer-paper/
[20] Whatman Ltd 1983 *US Patent* US4524103A https://patents.google.com/patent/US4524103A/en
[21] Bharimalla A K, Deshmukh S P, Vigneshwaran N, Patil P G and Prasad V 2017 Nanocellulose–polymer composites for applications in food packaging: current status, future prospects and challenges *Polym.-Plast. Technol. Eng.* **56** 805–23
[22] Park G and Thayer J F 2014 From the heart to the mind: cardiac vagal tone modulates top-down and bottom-up visual perception and attention to emotional stimuli *Front. Psychol.* **5** 278
[23] Prathapan R, McLiesh H, Garnier G and Tabor R F 2017 Surface engineering of transparent cellulose nanocrystal coatings for biomedical applications *ACS Appl. Bio. Mater.* **1** 728–37
[24] Dresher W H 1969 The age of fibers *JOM* **21** 17–26
[25] Chawla K K 2012 *Composite Materials: Science and Engineering* 3rd edn (New York: Springer)
[26] Washburn E W 1921 The dynamics of capillary flow *Phys. Rev.* **17** 273
[27] Zhao D, Zhu Y, Cheng W, Chen W, Wu Y and Yu H 2020 Cellulose-based flexible functional materials for emerging intelligent electronics *Adv. Mater.* **33** 2000619
[28] Tobjörk D and Österbacka R 2011 Paper electronics *Adv. Mater.* **23** 1935–61

IOP Publishing

Printed and Flexible Sensor Technology
Fabrication and applications
Subhas Chandra Mukhopadhyay and Anindya Nag

Chapter 11

Graphene-based implantable electrodes for neural recording/stimulation

Sara Khademi, Md Eshrat E Alahi and Kamran Foroutani

In neural engineering implantable electrodes are an exciting recent topic of research for neural recording and stimulation to understand the neural behavior of the central nervous system (CNS). The implantable electrodes should be highly flexible, biocompatible, highly conductive, have low impedance, and be able to transfer the charge effectively during stimulation. Graphene is the right candidate to fulfill most of these characteristics and has been used for neural recording and stimulation. Various synthesizing methods for enhancing capabilities can improve these characteristics. In this chapter the synthesizing and characterizing methods for printing/fabricating graphene-based implantable electrodes are discussed for various neural recording and stimulation processes.

11.1 Introduction

Neural recording and stimulation have played a significant role in understanding the basic neural processes and neurological behaviors. Current clinical practices depend on the penetration of an electrode inside the brain, and sharp silicon-based electrodes are used in most cases. However, there are significant side effects and concerns arise, such as inflammation and implantation damage, which can be avoided using minimally invasive approaches by implanting flexible microelectrode arrays. Graphene has played a significant role in recent developments. Robust and flexible microelectrode arrays with efficient recording and stimulation capabilities can significantly affect our understanding of cortical microcircuits. An excellent cycling endurance can enhance the spatial resolution significantly for clinical therapies and brain–computer interface (BCI) applications.

An ideal implantable electrode for neural recording and stimulation should have specific properties, such as having minimally invasive, flexible, and robust interfacing, with low impedance and high charge transfer capacity. Conventional electrodes

doi:10.1088/978-0-7503-3439-6ch11

© IOP Publishing Ltd 2021

are made from metal alloys, and platinum is used in most clinical research. However, the charge injection capacity (CIC) is limited, and titanium nitride, iridium oxide, PEDOT, and carbon nanotubes are used as modified materials to improve the CIC for further improving the simulation capabilities. Several studies have reported delamination issues during neural stimulation, and poor adhesion is the primary cause of this delamination. In recent years, graphene has emerged as one of the essential materials for neuro-interfacing applications, given its outstanding physical and chemical properties. It is formed of a two-dimensional (2D) hexagonal carbon lattice connected by strong covalent in-plane σ–σ bonds. Additional π–π bonds in electronic orbitals oriented vertically to the graphene plane lead to the delocalization of electrons [9, 10], which is vital for electrochemical sensing applications [11]. Graphene has a zero-gap with a small overlap between its valence and conduction bands. It shows a strong ambipolar electric field effect with high charge carrier mobilities of $\approx$ 10 000 $cm^2\ V^{-1}\ s^{-1}$ at room temperature [9, 12]. The white light absorbance of a suspended graphene monolayer is 2.3% (or a transmittance of 97.7%) with a negligible reflectance of <0.1% [9]. Moreover, graphene exhibits excellent thermal stability (3000 $\approx$ 5000 $W\ m^{-1}\ K^{-1}$) [9], with high mechanical strength (Young's modulus $\approx$ 1 TPa) [9, 12], and a large theoretical specific surface area (SSA) of 2630 $m^2\ g^{-1}$ [9]. It has excellent electrical conductivity [13–15], thermal conductivity [13], mechanical, thermal, and optical characteristics [13, 16], transferability [13], broad-spectrum transparency [13, 15], biocompatibility [11–14], flexibility [11, 13, 15, 17], mass-producibility [13], and protection against corrosion [11]. Thus all of its characteristics make it an ideal candidate for applications in various areas, such as acting as the cellular interface of stem cell differentiation [18], sensing biomolecules [19], promoting cell growth [20], and in electrical stimulation [15].

The direct contact and exposure of neurons to graphene-based electrodes is crucial to promote tight adhesion between cell membranes and interfacing electrodes, which is an essential requirement to detect small (tens of μV) signals during extracellular recordings and tissue stimulation, thus improving charge transfer [14]. The high electrical conductivity and low noise of graphene can provide an encouraging route for scaling the electrode size to single-cell dimensions by improving the signal-to-noise ratio (SNR), reducing the electrical impedance, and improving the charge injection capacity [11]. The graphene sheet exhibits excellent chemical and mechanical stability and is highly compatible with flexible substrates, which is essential for soft biological tissue interfaces and is a vital factor for the longevity of prostheses. This chapter explains the synthesizing and characterization of graphene-based electrodes for neural recording/stimulation.

11.2 Synthesis of the graphene sheet

There are several synthesize methods to prepare graphene and they can be categorized into top-down and bottom-up methods. The top-down methods are based on weakening the van der Waals forces between the graphene layers in graphite. Mechanical exfoliation using adhesive tape, and chemical and electrochemical exfoliation are top-down methods. Conversely, in the bottom-up methods, graphene is synthesized by depositing carbon atoms into single or few-layer graphene on the substrate. In this method the graphene can be provide by chemical vapor deposition (CVD) of an organic carbon precursor on the catalyst surface, such as Cu, Ni, and Pt. The characteristics of graphene depend strongly on the preparation methods. Here we discuss mechanical exfoliation and CVD graphene growth due to the wide use of such graphene in implantable electrodes in biomedical applications.

11.2.1 Mechanical exfoliation of graphite

In 2004 Geim and Novoselvo found that single-layer graphene can be produced by mechanical cleavage of highly oriented pyrolytic graphite (HOPG) [21]. This method is known as the 'Scotch-tape method,' where the single-layer graphene is obtained by peeling off an adhesive tape from HOPG, and it can be transferred onto an arbitrary substrate through wet or dry transfer techniques. Pristine graphene is a high-quality graphene that can be achieved by following this method. In addition to the low reproducibility of graphene, it is difficult to achieve large-scale graphene by this method. Hence, it can only be utilized for basic research.

11.2.2 Chemical vapor deposition (CVD)

The CVD method is a bottom-up method that synthesizes graphene sheets from small organic molecules. The CVD method provides the cost-effective continuous growth of graphene films with excellent electronic properties [9, 22]. The rigid substrate should withstand high temperatures up to approximately 900 °C. CVD growth of graphene has been performed on some transition metals. Appropriate substrates for CVD process are Pd [23], Ru [24], Pt [25, 26], Ir [27], Ni [10, 28], and Cu [29]. Carbon precursors are reacted at high temperatures, about 900 °C–1100 °C, in the presence of thin metal film or foil (as a catalyst). The key parameters that influence single-layer graphene growth by a CVD method, such as the substrate and growth conditions such as temperature and pressure, will be clarified in the following sections.

11.2.2.1 The effect of the substrate on graphene CVD growth

Graphene growth has been performed on metal and insulator substrates, and transition metals are preferred for a variety of reasons:

(i) Single-layer large-scale graphene with high uniformity and few defects is feasible.
(ii) A wide variety of transition metals can be used as a catalyst, and the role of various process parameters can be investigated.
(iii) Integration with high volume microelectronic technologies is possible.
(iv) Understanding of the graphene synthesis mechanism is possible using metal catalysts [30].

Table 11.1. The affinity and solubility of carbon on different transition metals. The affinity decreases from Fe to Cu and from top to bottom (adapted from [29] with permission of The Royal Society of Chemistry).

Carbon affinity of different transition metals			
Fe ([Ar]$3d^6 4s^2$)	Co ([Ar]$3d^7 4s^2$)	Ni ([Ar]$3d^8 4s^2$)	Cu ([Ar]$3d^{10} 4s^1$)
Ru ([Kr]$4d^7 5s^1$)	Rh ([Kr]$4d^8 5s^1$)	Pt ([Xe]$4f^{14}$ $5d^9 6s^1$)	Ag ([Kr]$4d^{10} 5s^1$)
Os ([Xe]$4f^{14}$ $5d^9 6s^2$)	Ir ([Xe]$4f^{14}$ $5d^{10} 6s^2$)	Pd ([Kr]$4d^{10}$)	Au ([Xe]$4f^{14}$ $5d^{10} 6s^1$)

Carbon solubility in different transition metals		
Cu: 0.001– 0.008 wt % at 1084 °C	Ni: 0.6 wt % at 1326 °C	Co: 0.9% wt at 1320 °C

The nature of the substrate affects the final synthesized characteristics of graphene directly, such as quality, uniformity, and continuity. Several transition metals, such as Ru, Ir, Co, Re, Ni, Pt, Pd, and Cu, are utilized as the catalyst for graphene growth because of the fact that these atoms have high coordination numbers and are highly reactive. The most appropriate catalyst should have a low affinity towards carbon and stabilize carbon on its surface by forming weak bonds. Table 11.1 lists the carbon affinity of different transition metals. Cu has the lowest affinity to carbon because it does not form any carbide phases and has the lowest solubility in comparison to Ni and Co (table 11.1). The low reactivity of Cu may arise from the fact that the d orbital of Cu is full of electrons; this implies that the electron distribution is symmetrical, which minimizes mutual repulsions. Consequently, Cu can only form weak bonds with carbon via charge transfer from the p electrons in the sp^2 hybridized carbon to the empty 4s states of copper [29]. Therefore, the very low carbon affinity of Cu, along with the formation of intermediate soft bonds, makes copper an appropriate catalyst for graphene growth. Cu and Ni, due to the high quality of the grown graphene, their low cost, and high commercial availability, are used widely to synthesize graphene.

Graphene synthesized on a Ni substrate is non-uniform multilayer graphene, in particular on polycrystalline Ni. These phenomena occur because, at high temperatures, carbon atoms are more soluble in Ni than Cu. Hence, during the cooling process carbon atoms moved from the bulk to the Ni substrate and crystallized at the surface, producing polycrystalline and non-uniform graphene.

11.2.2.2 The effect of carbon precursors on CVD growth of graphene

Several carbon precursors, including gas, liquid, and solid sources, can be used to synthesize CVD growth of graphene. Compared to liquids and solids, hydrocarbon gas precursors are highly pure; hence, they are the most popular carbon source. Methane gas (CH_4) is typically used as a precursor due to the ease of controlling the gas flow rate in a wide range. Furthermore, the structure, morphology, and size of the graphene can be controlled precisely. Graphene can be doped using NH_3 and PH_3 via N and P atoms [31, 32]. The high activation energy of CH_4 leads to a synthesized uniform single-layer graphene that occurs at a high temperature of 1000 °C.

Thus several researchers have attempted to reduced the synthesis temperature by using various hydrocarbons with low activation energy such as alcohols and polycyclic aromatic hydrocarbons (PAHs).

The growth temperature of graphene on Cu using PAHs as a solid precursor is reduced drastically due to the low activation energy of these materials for dehydrogenation and nucleation on the catalyst surface, but Raman spectroscopy results showed that graphene has numerous vacancy-type defects, which affect the performance of graphene in electronic devices (figures 11.1(a) and (c)) [33, 34]. To solve this problem, Lee *et al* synthesized graphene on Cu using some different PAHs with aliphatic hydrocarbon sources at 400 °C–600 °C. It is clear from Raman spectroscopy, as shown in figures 11.1(b) and (d), that defects decrease dramatically after using aliphatic hydrocarbon [34].

In another experiment, by Li *et al*, polymeric precursors such as PMMA and polystyrene, and benzene were utilized as solid and liquid hydrocarbons in the same CVD device to overcome the high-temperature problem in graphene growth. High-quality uniform and single-layer graphene was achieved, and the process temperature was reduced significantly to 300 °C and 400 °C using benzene and solid hydrocarbon, respectively [35]. Uniform single-layer graphene was achieved using benzene at atmospheric pressure without the presence of oxygen. Graphene

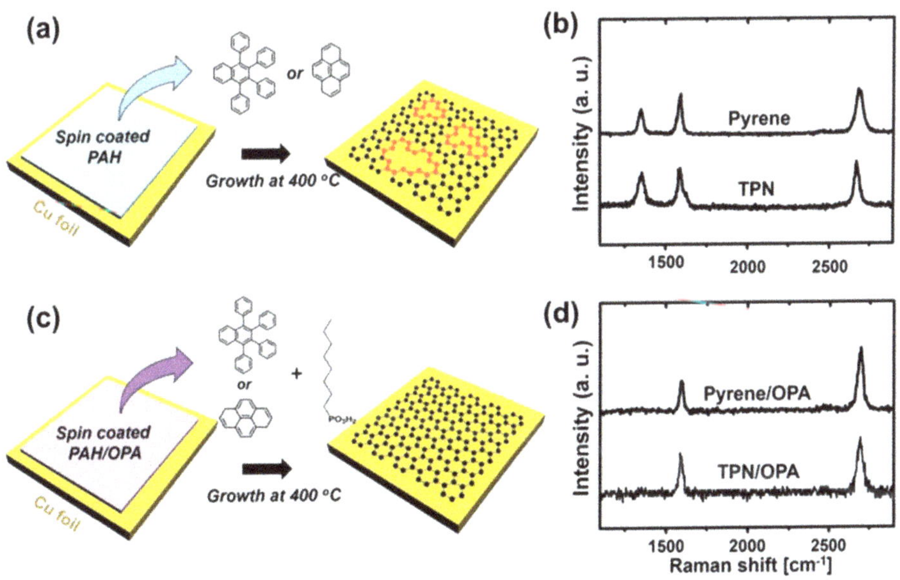

Figure 11.1. (a) Schematic representations of defective graphene growth on PAH-coated Cu at a minimal temperature, (b) Raman spectrum of transferred pyrene- and TPN-synthesis graphene on SiO_2 substrate, (c) schematic illustrations of defect free graphene growth on Cu surface using PAH/OPA heterogeneous solid carbon sources with presence of 1-octylphosphonic acid (OPA) at low temperature, and d) Raman spectrum of transferred graphene were synthesized by pyrene/OPA and TPN/OPA on SiO_2 substrate. (Adapted with permission from [34], copyright (2016) John Wiley & Sons.)

synthesized in the absence of residual oxidizing impurities by introducing pumping and purging cycles before growth led to continuous monolayer graphene films with full surface coverage and excellent quality, comparable to those of high-temperature CVD synthesized graphene. The transmittance, surface coverage, and carrier mobilities of graphene grown under ambient pressure at 300 °C were 97.6%, 100%, and 1900–2500 $cm^2 V^{-1} s^{-1}$, respectively [36]. Briefly, the activation energy of the hydrocarbon sources for dehydrogenation is a critical point that can lead to a temperature reduction of graphene growth. Ring-shaped hydrocarbon sources (such as benzene, Pyridine, PAHs, etc) and solid hydrocarbon sources (such as PMMA and polystyrene) have a lower activation energy than CH_4, which may form graphene layers at temperatures lower than 1000 °C. It is essential to note that aromatic hydrocarbon such as benzene, which led to the atmospheric pressure synthesis of graphene at low temperatures, are toxic and require high vacuum conditions. In figure 11.2 schematic diagrams of graphene sheet synthesized by using benzene at different reaction conditions is shown.

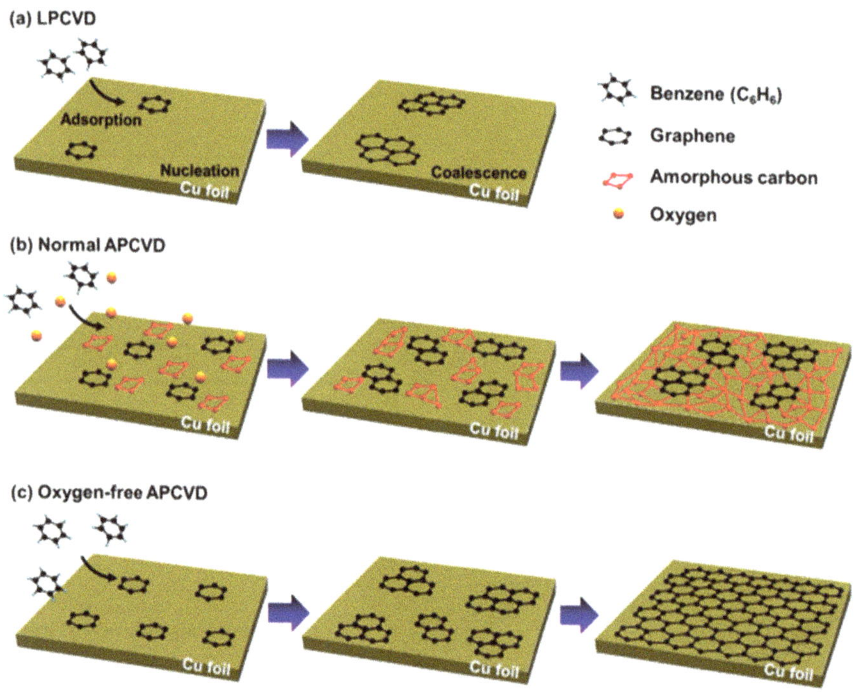

Figure 11.2. Schematic diagrams showing the growth mechanisms of graphene on a Cu surface from benzene by: (a) LPCVD, where the probability of the adsorption of benzene molecules onto the Cu surface is reduced at low pressures, which leads to the reduction of both dehydrogenation and C–C bond formation, thus limiting the domain growth of graphene; (b) normal APCVD, where nucleation occurs sufficiently at ambient pressure, but defective carbon films with unwanted amorphous and oxidized carbon regions are synthesized because of the existence of oxygen species during growth; and (c) the growth of continuous graphene films by the oxygen-free APCVD method. (Reproduced with permission from [36]. Copyright 2015 Nature.)

11.2.2.3 The effect of pressure on CVD growth of graphene

One of the key thermodynamic factors in the graphene CVD growth process is chamber pressure. A wide range of CVD chamber pressures from a few atmospheres to several millitorrs or even lower can be utilized, and this influences the gas flow behavior greatly. At low pressure the gas velocity and gas volume flow are much enhanced for the same molar flow, while the precursor concentration decreases, based on the ideal gas equation $PV = nRT$. Therefore, the reaction is more controllable at low concentration and high velocity of the precursor mass feed. Researchers around the world have investigated the CVD growth of graphene at low pressure (LPCVD) and atmospheric pressure (APCVD). The gas species diffusion rate is the main difference between LPCVD and APCVD. The mechanism of graphene growth on the Cu substrate in LPCVD is limited due to the rapid mass-transport process. In LPCVD the carbon source diffuses rapidly to the Cu surface and leads to surface reaction limited growth. Therefore, at a wide range of methane concentrations up to 95% monolayer coverage can be achieved.

11.2.2.4 Epitaxial growth of graphene

Epitaxial growth provides high-quality multilayer graphene to interact strongly with the substrate. The self-assembly of soluble graphene sheets demonstrates the possibility of low-cost synthesis and the fabrication of large-scale transparent films. However, these assembled graphene films show relatively low electrical conductivity owing to the poor contact resistance of the interlayer junction and the formation of structural defects during the vigorous exfoliation and reduction processes [37].

11.2.3 Transfer methods of graphene onto the target surface

Although the CVD is useful for large-scale growth of graphene, for use in a wide variety of applications, such as optoelectronic devices, electronics, sensors, energy storage devices, separation membranes, and biological and barrier coatings, graphene needs to be transferred from the as-grown catalytic surface to the relevant substrate [22]. In these applications an intact and clean graphene sheet is essential to improve device performance. Therefore graphene transfer is a critical process to fabricate a high-quality device with desirable performance. The transferring methods affect the quality of graphene because of the possibility of tearing and ripping the graphene sheets during the transfer. Graphene transfer methods are classified into wet and dry transfer methods.

11.2.3.1 The wet transfer method

Wet transfer methods cover the graphene sheet with a protective polymeric layer such as PMMA or PDMS. Subsequently, the polymeric support is floated for etching of the catalytic layer, such as Cu, Ni, and other similar metals. Then the floating film can be transferred onto an arbitrary substrate. Ultimately, the polymeric support should be removed by detaching (PDMS) or dissolving (PMMA) in acetone or acetic acid. Then the support should be rinsed in a mixture of methanol and water. Iron chloride ($FeCl_3$), hydrochloric acid (HCl), nitric acid (HNO_3), iron (III), nitrate ($Fe(NO_3)_3$, ammonium persulfate ($(NH_4)_2S_2O_8$), and copper chloride ($CuCl_2$) can be used as the etchant. The wet transferring method can be performed without any polymeric support, but the ultra-thin graphene is apt to tearing and ripping during the etching and transferring process. Thus, the polymeric support can protect graphene during the transferring process.

PDMS is a polymeric support which was used as a protective layer in light of the fact that it is durable, unreactive, moldable, has low surface free energy ($\approx$ 22–25 mN m^{-1}) [38], is resistant against many solvents [39], and is useful for soft lithography [40]. It can be used for conformal contact even with surface irregularities due to its suitable flexibility [41, 42]. It has low surface energy and weak van der Waals forces. Graphene adhesion to PDMS is low, hence the graphene sheet is quickly and preferably released from the PDMS and transfer to the target substrate with a relatively higher surface free energy [37]. Figure 11.3 shows a schematic diagram of the PDMS transfer process.

PMMA is another widely used polymer support for transferring graphene from CVD growth. PMMA can be easily spin-coated on graphene grown on metals. Unlike PDMS, the PMMA coating forms covalent bonds with graphene. By using PMMA as polymeric support after etching the metal and lifting the PMMA/graphene film to an arbitrary substrate, the PMMA can be dissolved in acetone or acetic acid, and be rinsed with DI water and methanol. Li *et al* reported that after

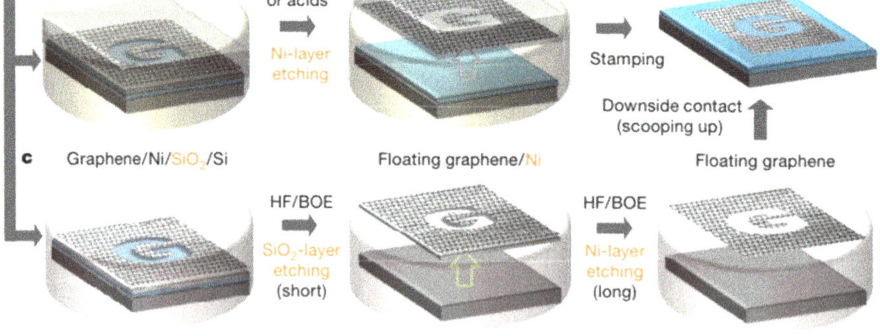

Figure 11.3. Synthesis, etching, and transfer processes of patterned graphene films. (a) Synthesis of graphene thin film by CVD on fine nickel layers. (b) Dry transfer process of CVD graphene thin film using PDMS. (c) Wet etching transfer process of CVD graphene thin film using BOE or hydrogen fluoride (HF) solution. (Adapted from [37] by permission from Springer Nature, copyright (2009).)

drying the PMMA film, it ultimately holds the graphene rigid. Therefore, the ripples and folds on graphene that may have formed during the growth and cooling process remain intact [43]. Thus, these ripples impede sufficient contact between the graphene and substrate. To address this issue, applying a second layer of PMMA coating before washing off the first PMMA layer is proposed. Using this additional PMMA coating improved the transferred graphene, with improved quality, lower sheet resistance, and high optical transmittance [44]. If the Ni is etched by a strong acid such as HNO_3, the graphene sheet can be damaged by the hydrogen bubbles produced. Kim *et al* reported the wet etching of a Ni substrate via an aqueous $FeCl_3$ solution (1 M), and no bubbles or precipitates were created. After a few minutes, the Ni substrate was dissolved, the graphene sheet floated on the surface solution, and could then be transferred onto the required substrate. The etching reaction is

$$2\,Fe^{3+}\,(aq) + Ni(s) \rightarrow 2Fe^{2+} + Ni^{2+}. \tag{11.1}$$

However, the graphene sheet is fragile and apt to tear and rip during etching and transferring. To solve this problem, they also reported a dry transfer technique using PDMS as polymer support. The polymer support ensures a safe transfer and large area graphene transfer [37].

Graphene surfaces are usually contaminated with ionic impurities from the etchant and metallic residues from deficient etching during the metal etching process. These contaminants are trapped at the interface between the graphene and target substrate and lead to electrical property degradation and a reduction in device performance [7, 45]. Additional cleaning steps can address this issue after metal etching with harsh chemicals, such as a mixture of hydrochloric acid and hydrogen peroxide [45]. However, further surface cleaning makes the graphene transfer process more complicated and increases the risk of mechanical and chemical damage during the cleaning process. Therefore direct graphene delamination from the as-grown metal eliminates metal etching and an additional cleaning process.

11.2.3.2 Dry transfer methods

Several researchers have attempted to transfer graphene through dry transfer. Dry graphene transfer is based on a direct delamination technique from a metal substrate. Accordingly, it eliminates the metal etching process. Therefore the risk of physical damage, such as tearing, ripping, and folding during the etching process can be addressed. Yang *et al* reported a dry transfer method with a high-quality graphene sheet, free from ionic and metal residues, to eliminate the metal etching step in the transfer process. They transferred the graphene sheet by pre-treating the graphene with a poly(vinyl acetate) (PVAc) solution. First, the graphene growth substrate was immersed in an aqueous PVAc solution to form a 20 μm PVA film on it. The PVAc acted as a carrier and protected the graphene during the transfer. Subsequently, to release the graphene from the Cu substrate, a PDMS stamp was brought in contact with the PVAc/graphene/Cu and retracted with high speed. Then, the PDMS/PVAc/graphene was put on a target substrate followed by retracting the

PDMS stamp from the target substrate slowly upon heating. Ultimately, the PVAc layer was removed in DI water [2]. In another study, a polyisobutylene (PIB) layer was used as the graphene-support polymer, and the graphene was delaminate from Cu using a PDMS stamp followed by transferring it to the target substrate. Then, the PIB layer was dissolved in an aliphatic solvent, namely hexane. This process is depicted in figure 11.4(A) [46].

Another group reported H_2 bubbling induced using electrochemical delamination [7]. Gas bubbles are very effective in separating solid layers uniformly. PMMA was spun-coated on graphene from CVD growth. Then, using epoxy glue, PET film was attached to the PMMA/graphene and was used as a cathode, because Cu may be electrochemically oxidized and etched if used as the anode. Then, by applying a current, the graphene was delaminated from the Cu substrate. This film can be transfer easily to an arbitrary substrate. This process is illustrated in figure 11.4(B). Raman spectroscopy showed high-quality graphene transfer. They found that

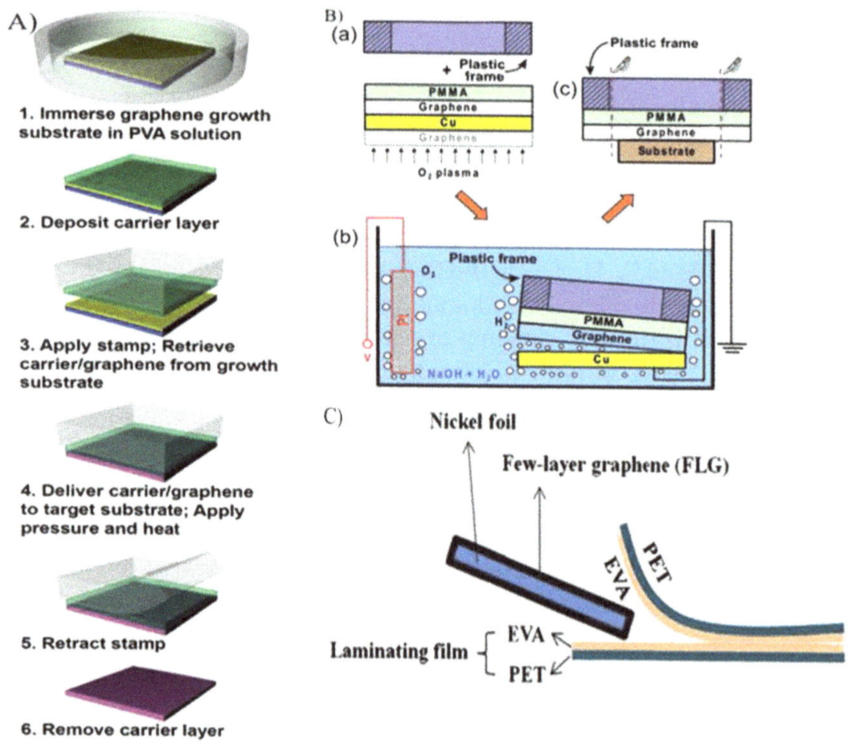

Figure 11.4. (A) Direct graphene delamination schematic and transfer onto a SiO_2 substrate using PVAc/PMMA stamp, steps 1–6. (Adapted from [2], copyright (2015) John Wiley & Sons.) (B) Graphene delamination with electrochemically induced H_2 bubbles ((a) and (b)) and transferring onto an arbitrary substrate (c). (Adapted from [7] with permission of AIP Publishing.) (C) Schematic illustration of graphene transfer using PET/EVAc as a supporting layer and an office laminator. Adapted from [8] Copyright (2014), with permission from Elsevier.

two-sided graphene transferring is possible using these methods [7]. The reaction equations leading to H_2 bubbling delamination are

$$4H_2O_{(aq)} + 4e^- \rightarrow 2H_{2(g)} + 4OH^-_{(aq)} \tag{11.2}$$

$$2H_2O_{(aq)} - 4e^- \rightarrow O_{2(g)} + 4H^+_{(aq)}. \tag{11.3}$$

Another group studied the electrochemical delamination of graphene. They found the D band in Raman spectroscopy, which indicates damage to the graphene via bubbles during electrochemical etching of Cu foil [47]. Abhay *et al* proposed a simple dry graphene transfer method using an office laminator, a polyvinyl alcohol (PVA) sheet, and water. The PVA/graphene can easily be stored and used whenever needed. It can be transferred onto the target substrate by placing the two material in contact and performing the lamination at 110 °C again. Using this method, a 25×30 cm^2 graphene sheet was successfully detached from copper. The properties of the graphene transferred by this method presented a higher carrier density and mobility than the wet etching and bubbling methods [48].

Furthermore, Chen and Hsieh, as shown in figure 11.4(C) transferred graphene from Ni foil by using a photolaminator and commercial laminating film (PET/EVAc). The PET film was placed on the graphene from the PVAc side and was heated using the photolaminator [8]. These easy and economical graphene transfer techniques can greatly open up the prospects of graphene-based applications in an environmentally friendly way. Another group also reported transfer of a graphene sheet directly from metal to the target substrate, but chemical etching of metal was still required [17, 49].

11.3 Graphene characterization methods

11.3.1 Raman spectroscopy

Raman spectroscopy is one of the most fundamental and non-destructive characterization tools, and is used widely to characterize graphene film. Carbon allotropes have unique Raman characteristic peaks at around 1350, 1580, and 2700 cm^{-1}, which are known as the D, G, and 2G peaks, respectively [50, 51]. Thus researchers can determine the layer number and the quality of the graphene layers synthesized and transferred. The D band arises from the disorder in the sp^2-hybridized carbon atoms and is characteristic of lattice distortions and defects in graphene sheets. The G band displays the tangential stretching (E_{2g}) mode of highly oriented pyrolytic graphite (HOPG). The 2D band arises from the second-order Raman scattering process [52–54]. The standard Raman spectra of graphene and graphite are shown in figure 11.5. By calculating the intensity ratio of 2D/G (I2D/IG) from the Raman spectrum, the quality of graphene can be evaluated. A comparably small amplitude of the D peak indicates that good quality graphene has been synthesized and transferred [51, 55–57]. The number of graphene layers can be estimated by comparing the G, G*, and 2D band intensities. Milenov *et al* showed that the D peak intensity changed for mono-, bi-, and trilayer graphene (figures 11.5B(a)–(c)) [58]. The band intensity increases with an increasing number of layers, but only up to seven layers and then decreases thicker layers. The G* band has a relatively smaller

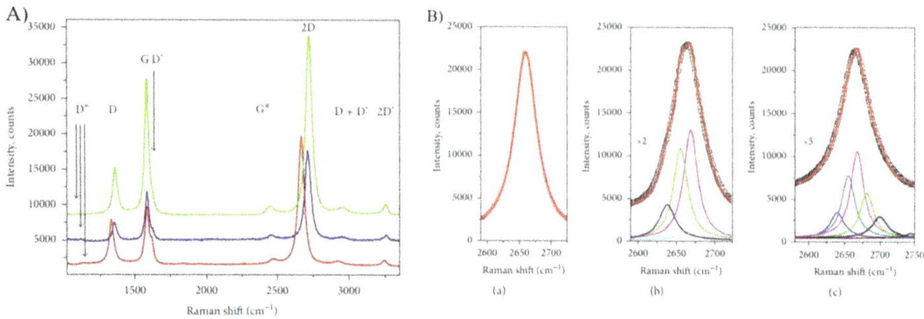

Figure 11.5. (A) The Raman spectra of graphene exhibited at three different laser wavelengths (633 nm—red, 514 nm—blue, and 488 nm—green) and (B) the effect of the number of layers on the 2D band in Raman intensity (single-layered, bilayer, and three-layered graphene are represented by (a), (b), and (c), respectively). (Adapted from [58] CC BY 4.0.)

intensity than G and 2D bands, and by increasing the number of graphene layers, G* is shifted slightly to lower Raman shift. In addition, the G* band of monolayer graphene is sharper than for few-layer graphene [51]. Monolayer graphene can be distinguished from few-layer graphene by a 2D band in Raman spectra. Monolayer graphene has a sharper and greater 2D peak intensity than the others [51, 59]. Full of half maximum (FWHM) is another route to calculate the number of graphene layers. FWHM is half of the 2D maximum value. Monolayer graphene has an FWHM of about 30 $cm^{-1,}$ and for multilayer graphene, it is more than 60 cm^{-1} [30, 60]. Milenov *et al* found that the laser wavelength effects the Raman spectra and figure 11.5 shows the effect of different laser wavelengths such as 633 nm, 514 nm, and 488 nm on D, G, and 2D [58].

11.3.2 FESEM and SEM

FESEM and SEM are widely used to characterize graphene morphology. Some defects, such as wrinkles, tearing, and folding, can be created during graphene synthesis and transfer and can be observed by FESEM and SEM images. FESEM can obtain high-resolution, clear, and less electrostatically distorted images, and it is a more efficient tool to study graphene surface morphology. Takahashi and co-workers employed SEM to monitor *in situ* graphene growth in a carbon segregation process [61].

11.3.3 TEM and HRTEM

An atomic-level image can be obtained by TEM [62], and point defects, Stone–Wales rotation, vacancy, and dislocations can be observed using HRTEM [63]. A TEM image is useful for identifying monolayer graphene, whereas the selected area electron diffraction (SAED) pattern reveals the typical hexagonal crystalline nature of graphene, as shown in figure 11.6 and figure 11.7.

Two to seven graphene layers can be clearly observed in a TEM image. TEM images for mono- and bilayer graphene are shown in figure 11.6.

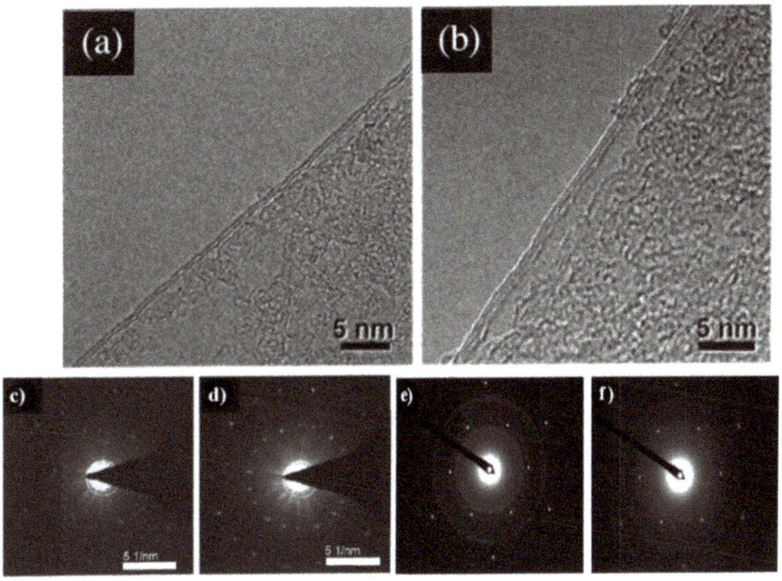

Figure 11.6. High-resolution TEM images of representative edges of (a) bilayer and (b) trilayer graphene. SAED images of (c) bilayer and (d) trilayer polycrystalline graphene. (Adapted from [3], copyright (2011) with permission from Elsevier.) (e) and (f) SAED images from graphene single crystal. (Adapted from [6], copyright (2011) with permission from Elsevier.)

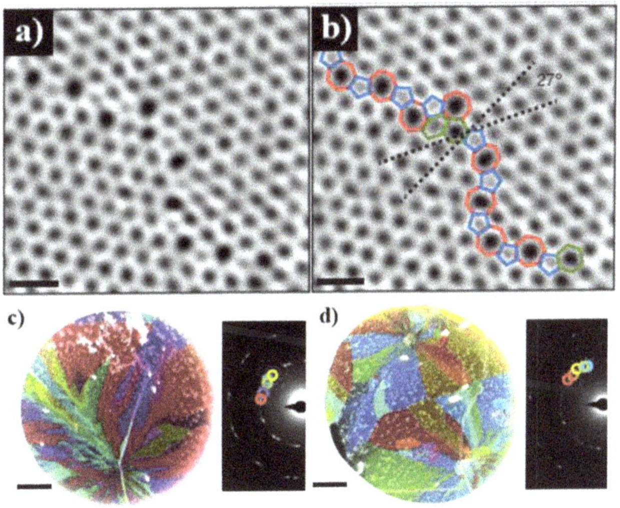

Figure 11.7. (a) and (b) Atomic-resolution ADF-STEM images of graphene crystals. Two different graphene crystals with a 27° relative rotation. The pentagons (blue), heptagons (red), and distorted hexagons (green) show the C–C structure observed at the grain boundary; scale bars 50 nm. (c) and (d) Large-scale grain imaging using DFTEM of polycrystalline graphene using several different aperture locations and color-coding them; scale bars 500 nm. (Adapted from [5], reprinted by permission from Springer Nature, copyright (2011).)

To characterize the graphene's crystallinity, the SAED pattern is useful. The SAED analysis in the inset images shows that the graphene sheets produced were polycrystalline [3]. As observed in the SAED of single-crystalline graphene in figure 11.6(e) and (f), only one set of symmetric hexagonal diffraction spots without rotation can be observed.

In polycrystalline materials, the size of their grains and the atomic structure of grain boundaries affect their properties directly. A line of defect can divide and disrupt a crystal in two-dimensional materials such as graphene. These issues are fundamental in graphene, which is a hexagonal, two-dimensional crystal. The misorientation of crystals and the atomic structure at grain boundaries can be detected by annular dark-field scanning transmission electron microscopy (ADF-STEM), as observed in figure 11.7(c) and (d). Relative misorientation of 27° in two crystals of the grain boundary stitched jointly by a series of pentagons, heptagons, and distorted hexagons is shown in figure 11.7(a) and (b).

11.3.4 UV–vis spectroscopy

UV–vis spectroscopy is based on the absorption and reflectance of radiation in the UV spectral range, in which the atoms of the substance absorb light in the visible, near UV, and infrared regions due to the electronic transitions. The optical transparency of graphene film can be determined by measuring the transmittance or absorbance characteristics. Several absorptions can determine the absorption or transmittance spectrum of an atom or transmittance bands [64]. The UV–vis spectrum is useful to identify the number of graphene layers—the optical transparency decreases by increasing the number of graphene layers observed in figure 11.8 [1]. The optical transmittance of graphene decreases by almost 2.2%–2.3% per increase in layer. Therefore, light can be passed through the graphene sheet while simultaneously providing better electrical conductivity. Accordingly, these unique properties are particularly fundamental in photovoltaic applications.

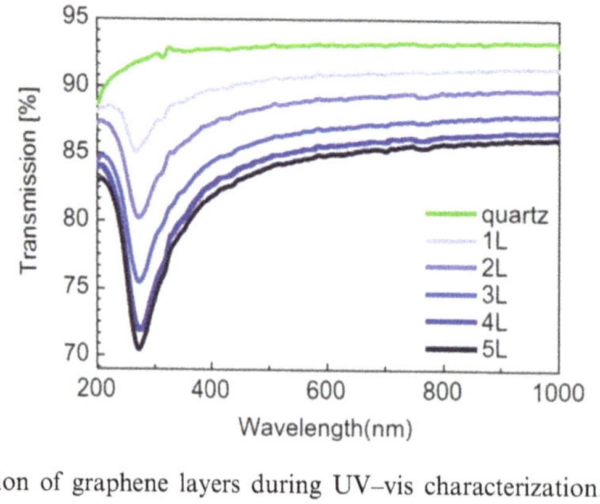

Figure 11.8. The absorption of graphene layers during UV–vis characterization. (Adapted from [1] under CC BY 3.0.)

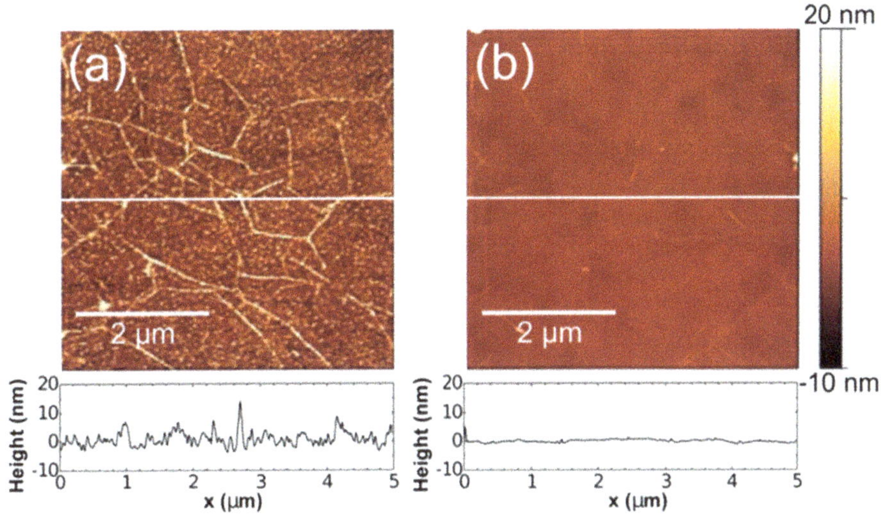

Figure 11.9. AFM images of (a) graphene transferred onto SiO_2 without treatment, the average surface roughness on the white line was about 4.6 nm, and (b) after graphene transfer annealed at 300 °C for 3 hours, with surface roughness reduced to 0.6 nm. (Adapted from [4] with the permission of AIP Publishing.)

11.3.5 AFM

AFM is another useful method for the characterization of graphene surface morphology and graphene transfer quality. Graphene surface roughness and graphene linear defects such as tearing, wrinkles, and folding, which take place during the synthesis and transfer processes, can be observed in AFM images. Pirkle *et al* investigated the effect of the annealing process after graphene transfer on graphene quality. As can be observed in figure 11.9, AFM images showed that graphene surface roughness significantly decreases after annealing treatment; therefore, graphene layer wrinkles are eliminated after the annealing process [4].

11.4 The chemically modified graphene electrode

To date, to broaden the properties of graphene and its applications, the chemical functionalization of graphene has been used widely by researchers. Generally, the chemical functionalization of graphene is classified as covalent [65, 66] and non-covalent [67] functionalization. Covalent bonds can achieve stronger and more robust linkage to the graphene sheets. Because graphene properties vary by chemical functionalization, such as the electrical and optical properties, new biological and chemical sensing, electrical, and optical devices, and hybrid and composite materials can be developed. Graphene electron transfer plays a critical role in reaction mechanisms and rates. Aryl diazonium functionalization, graphene selective oxidation, halogenation, cycloaddition, the Diels–Alder reaction, sialylation reaction, etc, have been investigated by researchers worldwide. The D peak in the Raman spectroscopy of graphene appears due to carbon atom hybrid changes from sp^2 to sp^3 during covalent functionalization. Aryl diazonium salts are usually used for electron transfer of graphene covalent functionalization. To functionalize graphene using aryl diazonium salt, two methods can be utilized. The first is by immersing graphene in diazonium solution under ambient conditions. Applying electrochemical bias is another method that can be used for the aryl functionalization of graphene. The reaction velocity is enhanced in this method in light of the fact that the diazonium ion concentration increases locally on the graphene surface, or by shifting the Fermi level of graphene to enhance the degree of orbital overlap for electron transfer.

The substrate can play a fundamental role in graphene reactivity during electrochemical reactions. A highly doped substrate, in comparison to a low doped substrate, can induce a greater electron–hole puddle in the graphene sheet, and graphene reactivity can be increased. The van der Waals interactions between graphene and the substrate can affect the graphene's reactivity. Graphene edges, due to the dangling bonds and high density of states (DOS), are more reactive than the basal plane. Functionalized graphene properties and surface morphology can be varied by using different substituent groups in the diazonium salt. The functionalization location of diazonium salts with an extended aliphatic group can be controlled as well-defined rows on graphene obtained by the covalent attachment of 4-docosyloxy-benzene diazonium (DBT) molecules on graphene. The physical properties, such as electronic and magnetic properties, can be changed by covalent functionalization. Graphene has a zero bandgap, therefore, by covalent functionalization, sp^2 hybridization changes to sp^3 hybridization in the carbon atom, leading to the bandgap opening. Unlike charge carrier mobility, which usually decreases through covalent functionalization, some researchers found that even at high functionalization concentrations, electrical conductivity still is high due to an increase in carrier concentration while mobility decreases. Substituent groups in diazonium salts can affect the electronic and transport properties.

11.5 Graphene-based microelectrode arrays

Microelectrode technologies have progressed through several modifications and innovations to improve their functionality and usability. Microelectrode arrays (MEAs) are used for various purposes, including different stimulation implants (cochlear, retinal, cortical, and spinal), signal recording for better understanding neuronal network functions, and pharmacological studies into dissociated neuronal networks (REF). The highly efficient recording and stimulation of robust and flexible microelectrode arrays can considerably affect the understanding of neural circuits and greatly enhance the spatial resolution for clinical therapies. An ideal device for neural recording and stimulation needs to be minimally invasive, robust, flexible, and biocompatible, while providing low impedance and high charge transfer capacity. The recorded signals are typically of some tens of microvolts for a single pulse activity to a maximum amplitude of around 80 mV for the intracellular potentials measured. On the other hand, neural stimulation sometimes requires relatively high electrode voltages and current densities, sometimes as high as several volts. Therefore, it may lead to high energy transfer, triggering chemical reactions that involve corrosion and changes in electrode properties [9].

11.5.1 Material requirements for neural implants

For a facile surgical procedure, and efficient and consistent activity, neural implants should be as minimally invasive as possible. Implantation duration can vary from several hours for experiments that are known as acute implantation, such as cortical recording for brain mapping of epileptic lesions, to several years for chronic applications, such as cochlear and retinal implants or deep brain stimulation electrodes. The essential requirements for neural implants are:

(i) Biocompatibility—neural implants must be biocompatible with the lowest possible inflammatory responses from peripheral tissue.
(ii) Sufficient charge injection capacity to effectively stimulate neural cells.
(iii) A signal-to-noise ratio (SNR) as high as possible can detect neural signals.
(iv) Mechanical compatibility for reducing inflammatory responses.

Electrical neural interfaces should provide outstanding biocompatibility and mechanical compliance with surrounding neural tissue to minimize the foreign-body reaction. It is essential to reduce inflammatory responses and mechanical mismatch between the nervous tissues and the implantable devices should be decreased for chronic neural implantation. In table 11.2 the Young's moduli for different human tissues are listed.

Due to its superior properties, such as high electrical conductivity, outstanding mechanical strength and flexibility, high chemical stability in liquids and electrolytes, transparency, and biocompatibility in a single material, graphene has become a desirable material for neural applications. More specifically, graphene simultaneously offers mechanical compatibility, cell adhesion, low toxicity properties, a high SNR, and corrosion protection [69]. Because it is feasible to tune the electrical, physical, and chemical properties of graphene through covalent and noncovalent functionalizations, graphene is a popular (universal) and better candidate for bioelectronics.

11.5.2 Graphene-based microelectrodes for stimulation

Effective stimulation is required for stimulus electrodes, such as in the deep brain, cortical and intra-cortical, or spinal cord stimulation, cochlear and retinal implants, as well as in peripheral nervous system applications. To achieve this goal, a minimum level of charge injection capacitance is needed to depolarize the membrane

Table 11.2. Young's moduli of different human tissues [68].

Tissue	Young's modulus (MPa)
Stratum corneum	12000
Viable epidermis	16
Dermis	12
The average value for brain tissue	8.863

of excitable cells in the vicinity of the stimulating electrode. Generally, they rely on the neural tissue to be stimulated, and the electrode material should provide adequate charge, which varies from tens of $\mu C.cm^{-2}$ to several $mC.cm^{-2}$, within a pulse of 100 μs to 1 ms. The charge injection capacitance depends on the electrolytic double-layer capacitance of the material and its potential window in water. The charge injection capacity for CVD grown graphene is about 5 to 20 $\mu F.cm^{-2}$ for a 1 V potential window in water, which is slightly lower than for standard noble metal electrodes such as Pt and Au, and well below that of the alternative novel materials proposed for neural stimulation, such as PEDOT–CNT or IrOx [70]. The electrochemical characteristics for CVD grown graphene in different studies are summarized in table 11.3.

Heo *et al* fabricated a graphene/PET electrode for controlling neural cell-to-cell interactions *in vitro* via non-contact electrical field stimulation with a 32 min stimulation period. The graphene/PET electrode was assembled in a 35 mm culture dish, in which the graphene electrode distance was 2 mm and was connected to the stimulator by Au wire, as illustrated in figures 11.10(a) and (b). They investigated the cell-to-cell interactions of SHSY5Y human neuroblastoma cells using different electric field stimulations and optical microscopy. They reported that the cell-to-cell interaction was influenced by electric field stimulation and the cell-to-cell interactions were classified into four major groups: (i) cell waving (CW), (ii) cell-to-cell decoupling (CD), in which, after stimulation, cells separated from each other, (iii) cell-to-cell coupling (CC), which was divided into newly formed cell-to-cell coupling (NCC) in which new connections between cells formed after stimulation and strengthened cell-to-cell coupling (SCC), and (iv) cell shrinking (CS) (see figure 11.10(c)) [15].

Due to the high transparency of graphene in a broad spectrum of light from IR to UV, it is feasible to combine electrophysiology with other methods, such as calcium imaging and optogenetics, in which electrode transparency is very important. In terms of an *in vivo* experiment, Park *et al* electrically stimulated GCaMP6f mouse brains using CVD grown graphene, depicted in figure 11.11. Simultaneously, imaging of the spatiotemporal neural responses to electrical stimulation were performed, which are impossible in opaque electrodes such as Pt and gold. They found that the graphene neural electrodes could safely deliver stimulation currents of up to 116.07–174.10 $\mu C\ cm^{-2}$ to neural tissue. Although this charge density is lower than for other neural electrodes such as Pt and PEDOT:PSS, it is still within the documented limits for safely activating tissue (10–800 $\mu C\ cm^{-2}$ per phase) [76, 77].

11.5.3 Graphene-based microelectrodes for neural recording

The impedance of graphene electrodes is limited by quantum capacitance, which arises from the graphene's low density of states around the Dirac point [78]. Cohen-Karni *et al* reported a graphene-based transistor for cellular interfaces and electrical recordings for the first time in 2010 [79]. A flexible graphene-based microelectrode was fabricated by Chen *et al* in 2013 for neural and cardiac recording. The graphene layer had been treated by mild steam plasma (SP) to increase its hydrophilicity by

Table 11.3. Overview of CVD grown graphene in the literature.

Application	Electrode size (μm^2)	Substrate	Impedance @ 1 kHz in KΩ	SNR	CSC $\mu C\ cm^{-2}$	CIC $\mu C\ cm^{-2}$	Tissue type	Number of graphene layers	References
Recording	13500	SU-8	0.7	20			Heart tissue	1	[71]
Stimulation	7 854 (D = 100) 17 671 (D = 150) 31 416 (D = 200)	Parylene C	286.4 ± 62.6 284.7 ± 125.0 215.7 ± 120.4		87.8	57.13	*In vivo* (brain)	4	[72]
Stimulation	310 000	Borosilicate glass	10		1248.7 ± 41.5			1	[73]
Recording	314	Quartz glass	170	10				1	[74]
Neural imaging and optogenetics	31 400	Parylene C	243				*In vivo* (brain)	4	[13]
Neuroimaging and recording	2 500	Polyimide	541	31			*In vivo* (brain)	1	[11]
Electrochemical characterization	7854 (D = 100)	Borosilicate glass	2.650.26 MΩ	35.8	910 ± 0.13	150 ± 0.05	Cortical rat neuron culturing	1	[75]

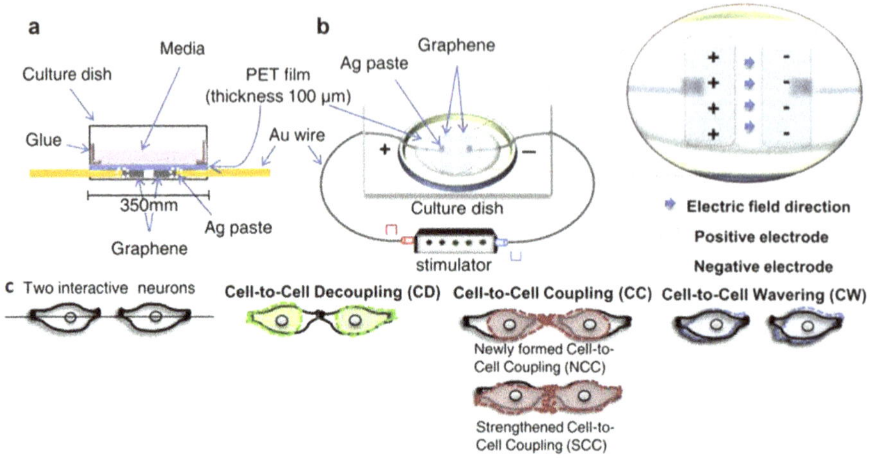

Figure 11.10. Schematic illustrating (a) a side view and (b) a top view of graphene/PET electrode in a culture dish. Neural cells were placed between two graphene electrodes on the PET surface and the electrical field formed among two cathode and anode graphene electrode. (c) The cell-to-cell interaction was classified into cell-to-cell decoupling (CD), cell-to-cell coupling (CC), and cell-to-cell wavering (CW), with the highest percentage of them occurring for no electrode stimulation. Adapted from [15] Copyright (2011), with permission from Elsevier.

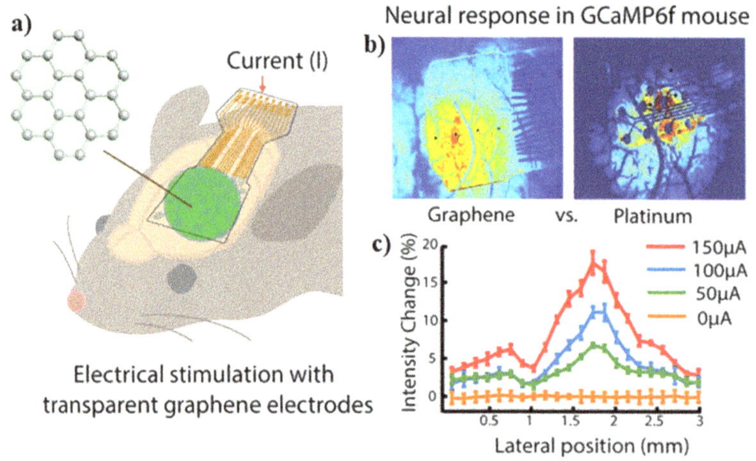

Figure 11.11. (a) Electrical stimulation of the GCaMP6f mouse sensorimotor cortex by a transparent graphene MEA. (b) Visualization of the intensity of the neural response to 100 μA and 500 μA for graphene- and Pt-based MEAs, respectively. (c) Spatial response analysis by plotting the fluorescence intensity versus different lateral positions at different stimulation currents. The maximum fluorescence peak is enhanced by increasing the stimulation current. (Adapted from [72], copyright (2018) American Chemical Society.)

introducing C–OH, C=O, and OH–C=O bonds on the graphene surface. Accordingly, the graphene impedance decreased (from 7216 to 5424 Ω mm^{-2}) and the SNR increased (from 20.30± 3.31 dB to 27.81 ± 4.03 dB) during electrochemical detection and maintained a stable state bias voltage without drift. Hence the performance of the treated graphene was increased, and the identification of their distinct shapes was feasible due to the neural signals being precisely separated. The amplitude of the P wave, QSR complex, and T wave in ECG of the zebrafish heart were enhanced remarkably because of the increasing SNR of the treated graphene. After SP treatment, the interfacial bonding energy of the graphene surface was more robust due to the van der Waals forces. Therefore, the graphene adhesion on the electrode was improved, which is essential for long-term implantation (figure 11.12) [71].

The transparent, flexible graphene electrode array with simultaneous electrophysiology and optical neuroimaging was fabricated by Kuzum *et al* in 2014 [80] for the first time. CVD graphene grown on Cu was transferred onto a polyimide substrate which had previously been patterned with gold. Then the graphene was patterned using plasma etching. The whole of the electrode, except for the graphene sites, were insulated using SU-8. Then, to reduce the sheet resistance of the graphene, the electrodes were doped with nitric acid, leading to adsorption of NO_3^- groups on the graphene surface resulting in p-type doping. The phase angle in EIS spectroscopy for the doped-G electrode was constant (−50°) for a wide range of frequencies, which represents more complex charge transfer rather than for the Au electrode. The nature of the doped-G electrode was capacitive. In neural recording electrodes, the electrode noise emanating from resistive charge transfer can be reduced with a large interface capacitance. Therefore, the doped-G electrode could prevent electronic noise due to its small charge transfer resistance paired with a large capacitance. The doped-G electrode had a high charge storage capacity that could increase the amount of charge transfer, which is essential for neural stimulation electrodes.

In another work in 2014, Park *et al* fabricated transparent graphene electrodes for neural imaging and optogenetic applications. Four-layer CVD grown graphene was

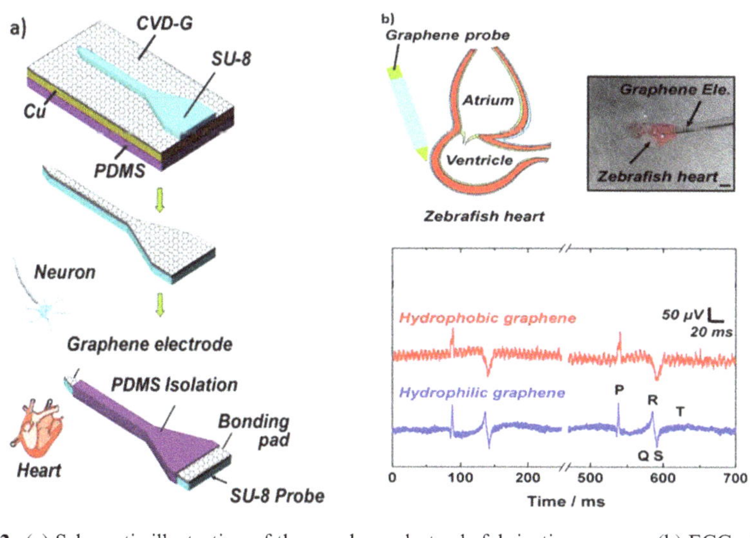

Figure 11.12. (a) Schematic illustration of the graphene electrode fabrication process. (b) ECG of a zebrafish heart. Adapted from [71] Copyright (2018), with permission from Elsevier.

transferred onto a parlyne-C substrate and then patterned. In order to achieve a good mechanical connection to the zero insertion force printed circuit board (PCB), the connection pads and initial parts of the tracks were patterned in Au. But the stimulation electrodes sites which contacted with brain tissue were graphene. The doped four-layered graphene electrode had 90% transmittance with the minimum sheet resistance. The transmittance of doped graphene is superior to ITO and ultra-thin metals. Therefore, optogenetic stimulation and both optical coherence tomography (OCT) and fluorescence imaging were performed successfully directly beneath the graphene electrode sites (figure 11.13), which were artifact-free when high-intensity light was applied directly onto an electrode site [13]. Another research group in China manufactured biocompatible graphene MEAs and demonstrated that the MEA had excellent long-term stability in aqueous solutions [74].

Impedance can be reduced by increasing the geometric or surface areas—electrode transparency decreases by increasing the surface roughness but promotes the recording of action potentials. Kshirsagar *et al* improved the electrochemical impedance of graphene electrodes successfully by electrodepositing PEDOT:PSS, but sacrificed the electrode transparency. The impedance at 1 kHz was reduced by about 70% from 2.9 ± 0.4 MΩ for bare graphene to 42 ± 2 kΩ for the PEDOT:PSS electrodeposited electrode. On the other hand, the transparency decreased as electrodeposition time increased. The optimum impedance (166 ± 13 kΩ, 18-fold reduction) with high transparency (84.4 ± 4%) was obtained by 1 sec PEDOT:PSS deposition. Recordings of cardiac field potentials for a graphene electrode coated with PEDOT:PSS showed amplitudes up to 3.8 mV peak-to-peak and noise under

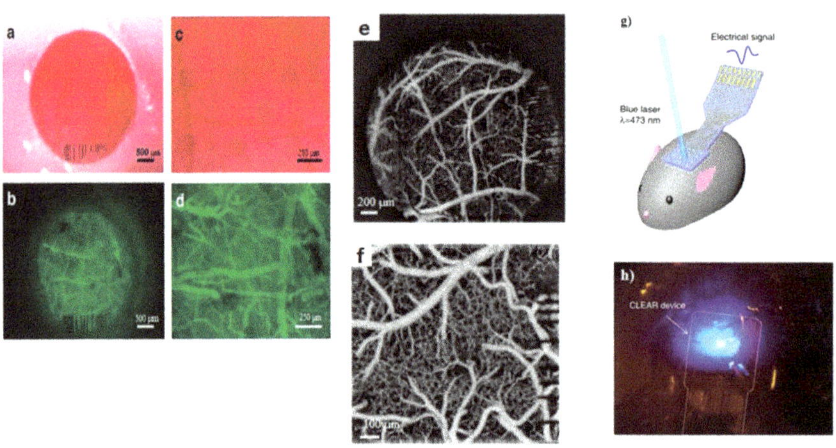

Figure 11.13. *In vivo* images of the cortical vasculature under the CLEAR device. (a) and (c) Bright-field images of the graphene electrode implanted on the cerebral cortex beneath a cranial window. (b) and (d) Fluorescence image of the same device as shown in (a) and (c). (e) and (f) OCT angiogram showing the cortical vasculature visible through the graphene electrode. (g) Schematic of optical stimulation with blue light with a 473 nm wavelength on the cerebral cortex, which can be passed through the transparent graphene MEA. (h) Recording neural signal evoked by blue light via the transparent graphene MEA. (Adapted from [13] by permission from Springer Nature, copyright (2014).)

20 μV peak-to-peak. Due to the high transparency of the coated electrode, neuro-imaging of the transgenic mouse retina with EYP-labeled ganglion cells was performed using fluorescence microscopy [81].

Other groups found that the impedance of the graphene electrode decreased 100-fold by electrodepositing of Pt nanoparticles (PtNPs), which can overcome the quantum capacitance limit. The electrode transparency was not decreased by deposited PtNPs. *In vivo* assays on transgenic mouse models showed that this MEA was useful for combining electrophysiology with optical imaging, which cannot be achieved using opaque electrodes [82].

11.6 Conclusion

This chapter discussed the advantages of using graphene for fabricating or printing flexible implantable electrodes for neural recording and stimulation. The synthesizing methods of graphene were also explained. The CVD growth method is better than the other available methods. The graphene characterization methods are also discussed to provide an overview for the readers. Finally, the recording and stimulation using various flexible implantable electrodes are also discussed along with the related research findings.

Funding information

This work is supported by the National Natural Science Foundation of China (61950410613) and the CAS President International Fellowship Initiative (2019PT0008) from the Chinese Academy of Sciences.

References

[1] Wu B *et al* 2014 Microwave absorption and radiation from large-area multilayer CVD graphene *Carbon* **77** 814–22

[2] Yang S Y *et al* 2015 Metal-etching-free direct delamination and transfer of single-layer graphene with a high degree of freedom *Small* **11** 175–81

[3] Tu Z *et al* 2014 Controllable growth of 1–7 layers of graphene by chemical vapour deposition *Carbon* **73** 252–8

[4] Pirkle A *et al* 2011 The effect of chemical residues on the physical and electrical properties of chemical vapor deposited graphene transferred to SiO_2 *Appl. Phys. Lett.* **99** 122108

[5] Huang P Y *et al* 2011 Grains and grain boundaries in single-layer graphene atomic patchwork quilts *Nature* **469** 389

[6] Chen X *et al* 2015 Chemical vapor deposition growth of 5 mm hexagonal single-crystal graphene from ethanol *Carbon* **94** 810–5

[7] De La Rosa C J L *et al* 2013 Frame assisted H_2O electrolysis induced H_2 bubbling transfer of large area graphene grown by chemical vapor deposition on Cu *Appl. Phys. Lett.* **102** 022101

[8] Chen C-S and Hsieh C-K 2014 An easy, low-cost method to transfer large-scale graphene onto polyethylene terephthalate as a transparent conductive flexible substrate *Thin Solid Films* **570** 595–8

[9] Huang X, Zeng Z, Fan Z, Liu J and Zhang H 2012 Graphene-based electrodes *Adv. Mater.* **24** 5979–6004

[10] Obraztsov A, Obraztsova E, Tyurnina A and Zolotukhin A 2007 Chemical vapor deposition of thin graphite films of nanometer thickness *Carbon* **45** 2017–21

[11] Kuzum D *et al* 2014 Transparent and flexible low noise graphene electrodes for simultaneous electrophysiology and neuroimaging *Nat. Commun.* **5** 5259

[12] Chen C *et al* 2011 A graphene-based microelectrode for recording neural signals *16th Int. Solid-State Sensors, Actuators and Microsystems Conf. (TRANSDUCERS)* (Piscataway, NJ: IEEE) pp 1883–6

[13] Park D-W *et al* 2014 Graphene-based carbon-layered electrode array technology for neural imaging and optogenetic applications *Nat. Commun.* **5** ncomms6258

[14] Fabbro A *et al* 2016 Graphene-based interfaces do not alter target nerve cells *ACS Nano* **10** 615–23

[15] Heo C *et al* 2011 The control of neural cell-to-cell interactions through non-contact electrical field stimulation using graphene electrodes *Biomaterials* **32** 19–27

[16] Tian H-C *et al* 2014 Graphene oxide doped conducting polymer nanocomposite film for electrode–tissue interface *Biomaterials* **35** 2120–9

[17] Bae S *et al* 2010 Roll-to-roll production of 30-inch graphene films for transparent electrodes *Nat. Nanotechnol.* **5** 574

[18] Park S Y *et al* 2011 Enhanced differentiation of human neural stem cells into neurons on graphene *Adv. Mater.* **23** H263–7

[19] Liu Y, Yu D, Zeng C, Miao Z and Dai L 2010 Biocompatible graphene oxide-based glucose biosensors *Langmuir* **26** 6158–60

[20] Ruiz O N *et al* 2011 Graphene oxide: a nonspecific enhancer of cellular growth *ACS Nano* **5** 8100–7

[21] Novoselov K S *et al* 2004 Electric field effect in atomically thin carbon films *Science* **306** 666–9

[22] Lee Y *et al* 2010 Wafer-scale synthesis and transfer of graphene films *Nano Lett.* **10** 490–3

[23] Hamilton J and Blakely J 1980 Carbon segregation to single crystal surfaces of Pt, Pd and Co *Surf. Sci.* **91** 199–217

[24] Sutter P W, Flege J-I and Sutter E A 2008 Epitaxial graphene on ruthenium *Nat. Mater.* **7** 406–11

[25] Vitchev R *et al* 2010 Initial stages of few-layer graphene growth by microwave plasma-enhanced chemical vapour deposition *Nanotechnology* **21** 095602

[26] Land T, Michely T, Behm R, Hemminger J and Comsa G 1992 STM investigation of single layer graphite structures produced on Pt (111) by hydrocarbon decomposition *Surf. Sci.* **264** 261–70

[27] Coraux J, Diaye A T N, Busse C and Michely T 2008 Structural coherency of graphene on Ir (111) *Nano Lett.* **8** 565–70

[28] Eizenberg M and Blakely J 1979 Carbon monolayer phase condensation on Ni (111) *Surf. Sci.* **82** 228–36

[29] Mattevi C, Kim H and Chhowalla M 2011 A review of chemical vapour deposition of graphene on copper *J. Mater. Chem.* **21** 3324–34

[30] Bhaviripudi S, Jia X, Dresselhaus M S and Kong J 2010 Role of kinetic factors in chemical vapor deposition synthesis of uniform large area graphene using copper catalyst *Nano Lett.* **10** 4128–33

[31] Wei D, Liu Y, Wang Y, Zhang H, Huang L and Yu G 2009 Synthesis of N-doped graphene by chemical vapor deposition and its electrical properties *Nano Lett.* **9** 1752–8

[32] Lv R *et al* 2012 Nitrogen-doped graphene: beyond single substitution and enhanced molecular sensing *Sci. Rep.* **2** 586
[33] Wan X, Chen K, Liu D, Chen J, Miao Q and Xu J 2012 High-quality large-area graphene from dehydrogenated polycyclic aromatic hydrocarbons *Chem. Mater.* **24** 3906–15
[34] Lee E *et al* 2016 Heterogeneous solid carbon source-assisted growth of high-quality graphene via CVD at low temperatures *Adv. Funct. Mater.* **26** 562–8
[35] Li Z *et al* 2011 Low-temperature growth of graphene by chemical vapor deposition using solid and liquid carbon sources *ACS Nano* **5** 3385–90
[36] Jang J *et al* 2015 Low-temperature-grown continuous graphene films from benzene by chemical vapor deposition at ambient pressure *Sci. Rep.* **5** 17955
[37] Kim K S *et al* 2009 Large-scale pattern growth of graphene films for stretchable transparent electrodes *Nature* **457** 706
[38] Clarson S and Semlyen J 1993 *Siloxane Polymers* (Englewood Cliffs, NJ: Prentice Hall)
[39] Lee J N, Park C and Whitesides G M 2003 Solvent compatibility of poly (dimethylsiloxane)-based microfluidic devices *Anal. Chem.* **75** 6544–54
[40] Gates B D, Xu Q, Stewart M, Ryan D, Willson C G and Whitesides G M 2005 New approaches to nanofabrication: molding, printing, and other techniques *Chem. Rev.* **105** 1171–96
[41] Rolland J P, Hagberg E C, Denison G M, Carter K R and De Simone J M 2004 High-resolution soft lithography: enabling materials for nanotechnologies *Angew. Chem. Int. Ed.* **43** 5796–9
[42] Kang J, Shin D, Bae S and Hong B H 2012 Graphene transfer: key for applications *Nanoscale* **4** 5527–37
[43] Li X *et al* 2009 Large-area synthesis of high-quality and uniform graphene films on copper foils *Science* **324** 1312–4
[44] Li X *et al* 2009 Transfer of large-area graphene films for high-performance transparent conductive electrodes *Nano Lett.* **9** 4359–63
[45] Liang X *et al* 2011 Toward clean and crackless transfer of graphene *ACS Nano* **5** 9144–53
[46] Bouša M, Kalbáč M, Jirka I, Kavan L and Frank O 2015 Experimental study of PIB-based CVD graphene transfer efficiency *NANOCON 2015*
[47] Yang X, Peng H, Xie Q, Zhou Y and Liu Z 2013 Clean and efficient transfer of CVD-grown graphene by electrochemical etching of metal substrate *J. Electroanal. Chem.* **688** 243–8
[48] Shivayogimath A *et al* 2019 Do-it-yourself transfer of large-area graphene using an office laminator and water *Chem. Mater.* **31** 2328–36
[49] Martins L G, Song Y, Zeng T, Dresselhaus M S, Kong J and Araujo P T 2013 Direct transfer of graphene onto flexible substrates *Proc. Natl Acad. Sci.* **110** 17762–7
[50] Gupta A, Chen G, Joshi P, Tadigadapa S and Eklund P 2006 Raman scattering from high-frequency phonons in supported *n*-graphene layer films *Nano Lett.* **6** 2667–73
[51] Yoon D, Moon H, Cheong H, Choi J S, Choi J A and Park B H 2009 Variations in the Raman spectrum as a function of the number of graphene layers *J. Korean Phys. Soc.* **55** 1299–303
[52] Li W *et al* 1997 Raman characterization of aligned carbon nanotubes produced by thermal decomposition of hydrocarbon vapor *Appl. Phys. Lett.* **70** 2684–6
[53] Eklund P, Holden J and Jishi R 1995 Vibrational modes of carbon nanotubes; spectroscopy and theory *Carbon* **33** 959–72
[54] Wang Y Y *et al* 2008 Raman studies of monolayer graphene: the substrate effect *J. Phys. Chem.* C **112** 10637–40

[55] Kumar C S 2012 *Raman Spectroscopy for Nanomaterials Characterization* (Berlin: Springer)

[56] Rodriguez C L, Kessler F, Dubey N, Rosa V and Fechine G J 2017 CVD graphene transfer procedure to the surface of stainless steel for stem cell proliferation *Surf. Coat. Technol.* **311** 10–8

[57] Barin G B, Song Y, de Fátima Gimenez I, Souza Filho A G, Barreto L S and Kong J 2015 Optimized graphene transfer: influence of polymethylmethacrylate (PMMA) layer concentration and baking time on graphene final performance *Carbon* **84** 82–90

[58] Milenov E V T I and Popov V N 2017 Raman spectroscopic study of as-deposited and exfoliated defected graphene grown on (001) Si substrates by CVD *J. Spectrosc.* **2017** 3495432

[59] Ferrari A C *et al* 2006 Raman spectrum of graphene and graphene layers *Phys. Rev. Lett.* **97** 187401

[60] Kalbacova M, Broz A, Kong J and Kalbac M 2010 Graphene substrates promote adherence of human osteoblasts and mesenchymal stromal cells *Carbon* **48** 4323–9

[61] Takahashi K, Yamada K, Kato H, Hibino H and Homma Y 2012 *In situ* scanning electron microscopy of graphene growth on polycrystalline Ni substrate *Surf. Sci.* **606** 728–32

[62] Robertson A W and Warner J H 2013 Atomic resolution imaging of graphene by transmission electron microscopy *Nanoscale* **5** 4079–93

[63] Banhart F, Kotakoski J and Krasheninnikov A V 2011 Structural defects in graphene *ACS Nano* **5** 26–41

[64] Lee H C *et al* 2017 Review of the synthesis, transfer, characterization and growth mechanisms of single and multilayer graphene *RSC Adv.* **7** 15644–93

[65] Criado A, Melchionna M, Marchesan S and Prato M 2015 The covalent functionalization of graphene on substrates *Angew. Chem. Int. Ed.* **54** 10734–50

[66] Hirsch A, Englert J M and Hauke F 2013 Wet chemical functionalization of graphene *Acc. Chem. Res.* **46** 87–96

[67] Georgakilas V *et al* 2016 Non-covalent functionalization of graphene and graphene oxide for energy materials, biosensing, catalytic, and biomedical applications *Chem. Rev.* **116** 5464–519

[68] Gabran S R 2006 Design and optimization methodology of sub-dermal electroencephalography dry spike-array electrode *Thesis* University of Waterloo

[69] Reina G, González-Domínguez J M, Criado A, Vázquez E, Bianco A and Prato M 2017 Promises, facts and challenges for graphene in biomedical applications *Chem. Soc. Rev.* **46** 4400–16

[70] Venkatraman S *et al* 2011 *In vitro* and *in vivo* evaluation of PEDOT microelectrodes for neural stimulation and recording *IEEE Trans. Neural Syst. Rehabil. Eng.* **19** 307–16

[71] Chen C-H *et al* 2013 A flexible hydrophilic-modified graphene microprobe for neural and cardiac recording *Nanomed. Nanotechnol. Biol. Med.* **9** 600–4

[72] Park D-W *et al* 2018 Electrical neural stimulation and simultaneous *in vivo* monitoring with transparent graphene electrode arrays implanted in GCaMP6f mice *ACS Nano* **12** 148–57

[73] Koerbitzer B, Krauss P, Nick C, Yadav S, Schneider J J and Thielemann C 2016 Graphene electrodes for stimulation of neuronal cells *2D Mater.* **3** 024004

[74] Du X *et al* 2015 Graphene microelectrode arrays for neural activity detection *J. Biol. Phys.* **41** 339–47

[75] Körbitzer B, Krauß P, Belle S, Schneider J J and Thielemann C 2019 Electrochemical characterization of graphene microelectrodes for biological applications *ChemNanoMat* **5** 427–35

[76] McCreery D B, Agnew W F, Yuen T G and Bullara L 1990 Charge density and charge per phase as cofactors in neural injury induced by electrical stimulation *IEEE Trans. Biomed. Eng.* **37** 996–1001
[77] Shannon R V 1992 A model of safe levels for electrical stimulation *IEEE Trans. Biomed. Eng.* **39** 424–6
[78] Fang T, Konar A, Xing H and Jena D 2007 Carrier statistics and quantum capacitance of graphene sheets and ribbons *Appl. Phys. Lett.* **91** 092109
[79] Cohen-Karni T, Qing Q, Li Q, Fang Y and Lieber C M 2010 Graphene and nanowire transistors for cellular interfaces and electrical recording *Nano Lett.* **10** 1098–102
[80] Kuzum D *et al* 2014 Transparent and flexible low noise graphene electrodes for simultaneous electrophysiology and neuroimaging *Nat. Commun.* **5** 1–10
[81] Kshirsagar P *et al* 2019 Transparent graphene/PEDOT:PSS microelectrodes for electro-and optophysiology *Adv. Mater. Technol.* **4** 1800318
[82] Lu Y *et al* 2018 Ultralow impedance graphene microelectrodes with high optical transparency for simultaneous deep two-photon imaging in transgenic mice *Adv. Funct. Mater.* **28** 1800002

IOP Publishing

Printed and Flexible Sensor Technology

Fabrication and applications

Subhas Chandra Mukhopadhyay and Anindya Nag

Chapter 12

Screen printed electrode based sensor for biological and chemical species detection

Kalyan Kumar Mistry

Screen printed electrodes (SPEs) fabricated using thick-film technology represent a promising electrochemical sensor strip for the detection and quantification of different biological and chemical species. The basis of the measurement system is a screen printed electrode (SPE) combined with an electronic interface device capable of interrogating the SPE to provide a reading of the sample contamination level. The SPE is an electrochemical device and is coated with a material that responds electrically when a sample of fluid containing the contaminant of interest is placed upon it. The SPE based sensors are disposable in nature, once a sensor is used to record the signal it is discarded and a new fresh untarnished sensor is mounted for another reading. SPE based sensors are commonly used in healthcare and have revolutionized human health diagnosis tools. SPE based sensors are also used for the detection of trace-level chemical species such as pathogens, heavy metals in water, food, and other liquid samples. A common SPE comprises a chemically inert and electrically insulated substrate on which three electrodes are printed to make a electrochemical cell system. The three electrodes are called the working electrode (WE), reference electrode (RE), and counter electrode (CE), and are screen printed on the same substrate, forming a circular well on which the liquid sample solution dipenses. Among these three, WE is the main sensing platform or sensing electrode, on which the electrochemical reaction takes place after binding or recognition of biological or chemical species. SPE based disposable electrochemical methods are potentially the most promising reliable techniques to provide very accurate measurements of trace-level biochemical species in a sample solution with rapid and cheap analysis.

12.1 Introduction

Screen printed electrode based sensors are beneficial for researchers who are involved in the development of biosensors and chemosensors for the monitoring

 © IOP Publishing Ltd 2021

of trace-level (micro- to nanogram per milliliter) biological or chemical species, sometimes at the ppm or ppb level. SPE based sensors have had a great impact by providing a field-ready tool which can be deployed to collect data by interfacing electronics circuitry. The sensor measurement system will allow a greater amount of reliable data to be gathered and analyzed rapidly, rather than optical or microscopic measurement tools. This additional data will provide a greater understanding of food or water contamination transfer in an inherently dynamic system. SPE based sensors can be mass-produced readily by screen printing techniques using low-cost materials to produce a physical sensor. Through the treatment of the electrode surface (the sensing electrode WE), it is possible to sensitize the electrode to respond to specific chemical or biological species. The liquid sample to be measured is dispensed or dropped in the electrode well, and the response of the sensor is examined by, for example, using cyclic voltammetry (CV), linear sweep voltammetry (LSV), strip voltammetry measurements, etc. The voltage to current characteristics of the CV indicate the presence and concentration of a specific species by comparing the calibrated plot of the peak current response.

An SPE is an electrically conductive substrate produced by a screen printing technique on a synthetically dormant and electrically isolated solid platform made of materials such as ceramic [1–5], plastic [6–8], glass [9], polyester film [8–12], etc. In an electrochemical sensing framework, when a realized potential is applied in the middle of the RE and WE as control inputs, a current is produced and flows between the AE and WE, which are used as measuring outputs. At a known control input potential, the analyte of the cell system starts to interact at the electrode surface and produces an electrical signal. The electrochemical signal of the system depends on various parameters, among which the amount of analyte concentration in the electrochemical system is very important. Using an electrochemical system the precise evaluation of an analyte concentration of even a few ng mL^{-1} to a few fg mL^{-1} is possible for a specific analyte and electrode under some defined hypothetical environmental conditions.

The screen printing technique produces an electrode by depositing the shape of the electrode in conducting paste, pigment, or ink. The stencil is attached to a screen mesh, through which the ink moves due to mechanical rubbing pressure and the designed pattern is transferred to a solid surface.

12.2 Screen printed electrode (SPE) fabrication

Screen printing is a very popular printing method for the production of chemosensors and biosensors for the detection of trace-level analytes in a liquid sample or medium. Screen printing is the oldest technique used for the production of printed electronics circuits. It has been used since the 1940s for the mass-production of electronic circuits [13]. Currently, it is frequently used in the production of printed circuit boards and the metallization of silicon solar photovoltaic cells. In the screen printing technique, the thickness of the printed layers varies from 10 μm to up to 500 μm, as screen printing produces thick-film coating. The screen mesh is positioned at a specific close distance on top of the substrate, and ink is squeegeed by pressing onto the substrate against the screen mesh, causing the print to be produced by the ink from the screen mesh [14]. The moving squeegee pushes the ink through the open regions of the stencil. A schematic illustration of the screen printing technique is shown in figure 12.1. The schematic representation of a screen printed electrode is shown in figure 12.2, with its electrode terminals and the well for sample dispensing. Usually, high-viscous ink or pastes with shear-thinning properties are used for screen printing. This avoids any immediate flow through the mesh and allows for the movement of the ink when a force is applied. The mesh imprint on the accumulated ink is highly dependent on the properties of the ink and the drying procedure. The screen printed layer thickness depends on various process variables such as squeegee power, print speed, snap-off size, snap-off angle, and the rheology of the ink. The dry layer thickness d can be determined as

$$d = V_{sc}k_p\frac{c}{\rho},$$

where V_{sc} is the screen volume in $cm^3\ m^{-2}$, k_p is the final pick out ratio, c is the centralization of the solids in the ink in gm cm^{-3}, and ρ is the thickness of the material in the last film in gm cm^{-3}. When the printing resolution is less than 100 μm then a special screen mesh is required. The characteristics needed for the ink, i.e. ink

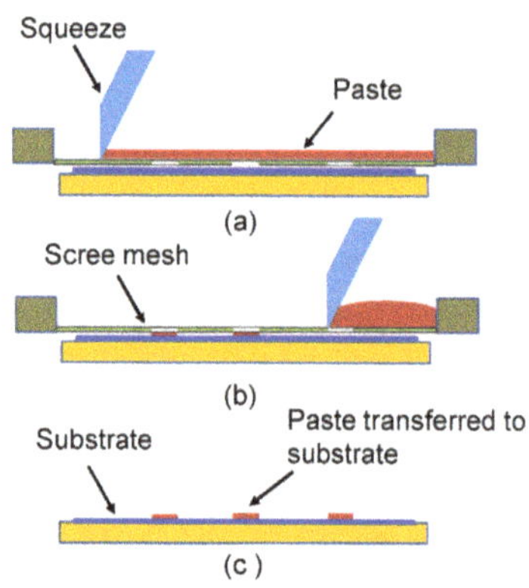

Figure 12.1. Process flow diagram of the screen print technique.

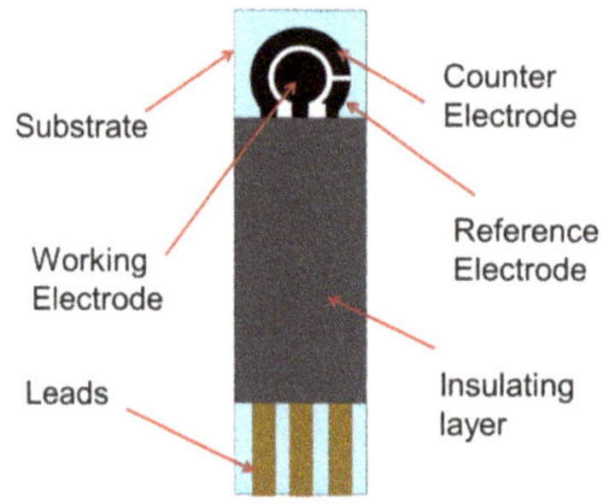

Figure 12.2. Schematic representation of an SPE.

or paste viscosity and instability, are both limiting factors for the widespread use of electrode manufacture [15–19]. To fabricate a screen printed electrode, selective conductive ink or paste materials are required for layer by layer printing on the substrate according to a design. The final production cycle includes three to four phases. A common method of printing electrodes is as follows. In the first step a silver track line is printed on the substrate surface using a silver conducting paste. The silver conducting layers need to air dry to removing the ink solvent, then they are cured in an air oven at a pre-defined temperature for a specific time. In the second step the printing of the working and counter electrodes is carried out using carbon paste (carbon material is most commonly used for the WE and AE). Similarly to the first step, the printed carbon layers need to be air dried and cured in an air oven at a defined temperature for a specific time depending upon the ink or paste properties.

The third step involves printing the reference electrode, using silver/silver chloride (Ag/AgCl) paste, followed by drying and curing, similarly to steps one and two. In the last step it is necessary to cover the unneeded conducting area using an insulating paste material to avoid any unwanted electrical short circuits or any chemical reactions during the application.

12.3 Theory and operation of an electrochemical sensor

The common concept of electrochemical sensing is a reaction conversion obtained from a chemical or biological substance with an electrically noticeable indication, the amplitude of which corresponds to the concentration of an analyte. The sample solution is moved onto the SPE surface where it forms a three-electrode electrochemical cell structure. A redox phenomenon occurs at the electrode surface at a given control sweeping potential as a result of electron transfer between the electrolyte and the electrode interface, and a current flows in between the active and auxiliary electrodes. The standard current–voltage plot indicates the electrochemical reaction strength that occurs at the working electrode (WE) site. The expected change of electrical signal during the reaction at a stable state of equilibrium is proportional to the increase in the concentration of a given analyte. There are three probable common electrochemical transduction modes: (i) potentiometric, (ii) conductometric, and (iii) amperometric. Potentiometric and conductometric measurements are required to reach an equilibrium state of reaction at the electrode interface, which takes a longer time to achieve a stable electrochemical response [20]. The amperometric type of transduction is a voltammetric calculation subset. The ideal voltammetric calculation is the calculation of the current applied to the device at a specified constant potential. It tests the peak current magnitude corresponding to the uncertain concentration of the analyte, either on the anodic or cathodic side.

In the voltammetric method, the current response is a measure of the analyte present in the cell and the applied potential is denoted by Nernst equation:

$$E = E^{\circ} + \frac{RT}{nF} \ln \frac{C_O}{C_R}, \tag{12.1}$$

where E is the potential of the cells (electromotive force) at a temperature of interest, E° is the standard cell potential, R is the universal gas constant ($R = 8.314\,472$ J K^{-1} mol^{-1}), and T is the absolute temperature.

C_O and C_R are the concentrations of oxidized and reduced species, respectively. The pace of electrode response is controlled by the rate at which the electroactive species are taken to the outside of the terminal utilizing mass exchange. The reaction obeys the thermodynamic relationship at the electrode and is a reversible reaction.

According to the first law of Fick, the substance diffuses from an area of higher fixation to an area of lower fixation, and the resultant flux is proportional to moles of fluid diffusion per unit area per unit time:

$$\text{Flux} = \frac{\text{moles}}{\text{area/time}} = \frac{1}{A}\frac{dn}{dt} = -D\frac{dc}{dx},$$

where A is the electrode active surface area and D is the diffusion coefficient of the species. The flux is proportional to the diffusion coefficient of the species. Once again, the electrochemical flux is quickly converted by Faraday's law into the current, which specifies that the number of moles of oxidized or reduced species is proportional to the oxidation or reduction rate:

$$Q = nFN$$

$$i = \frac{dq}{dt} = -nFAD\frac{dc}{dt}, \quad \text{at } x = 0.$$

This is the basic relation used in amperometric research to measure the current. Again, $\frac{\partial C}{\partial x}$ at $x = 0$ is not easy to calculate. It depends on the applied potential and how the potential varies with time, determined by $\frac{C_O}{C_R}$, the ratio at the electrode surface.

Heterogeneous reaction rates depend on the exchange of masses with the electrode surface and the electrode surface effect. The heterogeneous reaction rate is described as

$$\text{rate} = \frac{i}{nFA} = \frac{j}{nF}(\text{mol s}^{-1}\text{ cm}^{-2})$$

$k_a^r = \frac{i}{nFA}$, where k_a^r is the reaction rate.

Mass exchange assumes a major job in electrochemical dynamics. Mass transfer is the movement of species onto the electrode surface in three modes: (i) migration, (ii) diffusion, and (iii) convection. Migration is the movement of species influenced by an electric current. Dispersion is cell removal affected by substance expected angle, and convection occurs because of blending or hydrodynamic exchange. The mass exchange of species on the surface of the electrode is determined by the Nernst–Plank condition:

$$J_O(x) = -D_O\frac{\partial C_O(x)}{\partial x} - \frac{n_iF}{RT}D_OC_O\frac{\partial \phi(x)}{\partial x} + C_Ov(x), \tag{12.2}$$

and $J_O(x)$ is the flux of the oxidized classes at a space x from the electrode exterior, D_O is its diffusion coefficient, $\frac{\partial C_O(x)}{\partial x}$ and $\frac{\partial \phi(x)}{\partial x}$ are the concentration gradient and potential gradient at a distance x from the electrode surface, n is the number of charges, and $v(x)$ is the velocity.

Assuming that there is no influence of the electric field and no stirring of the electrolyte solution, and considering mass transfer due to the diffusion effect, the flux of a substance and its concentrations C_O and C_R in a solution are expressed as functions of distance x from the electrode surface and time t. By Fick's first laws of diffusion, the flux is proportional to the gradient of concentration $J_O(x)\alpha\frac{\partial C_O(x)}{\partial x}$

If the flux of a substance O at a given location x and at a time t is written as $J_O(x, t)$, then the net mass-transfer rate of O, assuming that no migration or convection of mass transfer occurs, and equation (12.2) is simplified as

$$-J_O(x, t) = D_O\frac{\partial C_O(x, t)}{\partial x}. \tag{12.3}$$

The initial conditions of concentration at the electrode surface as an element of separation from the electrode x are

$$f(x) = C_O(x, 0).$$

If initially no reduced species R is present in the bulk concentration then

$$C_R(x, 0) = 0. \tag{12.4}$$

Consider the reduction of species O at a cathode oxidized from O [$Fe_3(CN)_6^{3-}$] to R [$Fe_3(CN)_6^{4-}$], species O begins its concentration at the electrode surface C_O $(x = 0)$.

Assume there is no stirring, i.e. no convection of mass-transfer velocity [$v(x) = 0$] and no migration of mass transport due to the excess amount of supporting electrolytes present in the sample. The simply treated assumption that the electrode surface contains a stagnant layer (d_O) is called the Nernst diffusion layer.

The concentration gradient is proportional to the mass-transfer rate

$$\begin{aligned} &k_m^r\alpha\frac{\partial C_O}{\partial x}\bigg|_{x=0} \\ &k_m^r\alpha\frac{\partial C_O}{\partial x}\bigg|_{x=0} \end{aligned} \tag{12.5}$$

$$\because k_m^r = D_O\frac{\partial C_O}{\partial x}\bigg|_{x=0}. \tag{12.6}$$

Further, assuming δ is a linear concentration gradient within the diffusion layer, it can be written as

$$k_m^r = D_O[C_O^* - C_O(x = 0)]/\delta \tag{12.7}$$

$$k_m^r = k_m^0[C_O^* - C_O(x = 0)], \tag{12.8}$$

where C_O^* is the concentration of species O in the bulk solution, D_O is the diffusion coefficient, and $k_m^0 = D_O/\delta$ is called the mass-transfer coefficient (cm s^{-1}), i.e.

$$\frac{i}{nFA} = k_m^0[C_O^* - C_O(x = 0)]. \tag{12.9}$$

Under this condition a net cathodic reaction R is produced at the electrode surface $C_R\ (x = 0) > C_R^*$:

$$\frac{i}{nFA} = k_m^R[C_R(x = 0) - C_R^*], \tag{12.10}$$

where C_R and C_o, is R and O represent reduction and oxidation, whereas, $(x = 0)$ are a function of the electrode potential E and the value of the current under this condition is called the limiting current, denoted by

$$i_l = nFk_m^oC_O^R. \tag{12.11}$$

When a limiting current flows, the electrode process occurs at the maximum rate possible for a given mass-transfer conditions, meaning that species O is being reduced to R at fast rate. The condition of O and R at the electrode surface can be assumed to be at equilibrium with the electrode potential as governed by the Nernst equation for the half-reaction:

$$E = E_o' + \frac{RT}{nF}\ln\frac{C_O(x = 0)}{C_R(x = 0)}. \tag{12.12}$$

E_o' is the formal potential

$$E = E_o' + \frac{RT}{nF}\ln\frac{C_O}{C_R}. \tag{12.13}$$

If a system follows the Nernst equation then it is said to be a thermodynamically or electrochemically reversible reaction.

The boundary and initial conditions of the actual concentration at a distance x from the electrode surface and at an instant t is denoted by $C_O(x, t)$, i.e. at the time $(t = 0)$ $C_O(x, 0) = C_O^*$ and $C_R(x, 0) = 0$

$$C_O(0, t) = 0 \quad \text{for } t > 0$$
$$E = E_o' + \frac{RT}{nF}\ln\frac{C_O(0, t)}{C_R(0, t)}.$$

The flux balance equation is

$$\frac{\partial C_O(x, t)}{\partial t} = D_O\frac{\partial^2 C_O(x, t)}{\partial x^2}$$
$$\frac{\partial C_R(x, t)}{\partial t} = D_R\frac{\partial^2 C_R(x, t)}{\partial x^2}$$

and $D_O \frac{\partial^2 C_O(x,t)}{\partial x^2}\Big|_{x=0} + D_R \frac{\partial^2 C_R(x,t)}{\partial x^2}\Big|_{x=0} = 0.$

$J_O(0, t) = -J_R(0, t)$ and the flux at the electrode surface is directly proportional to the current. Rearranging from equations (12.3) and (12.9):

$$J_O(0, t) = -\frac{i(t)}{nFA} \tag{12.14}$$

$$O + ne \underset{k_b}{\overset{k_f}{\rightleftarrows}} R, \tag{12.15}$$

where k_f and k_b are the forward and backward reaction rates $\frac{k_f}{k_b} = k = \frac{C_B}{C_A}$.

The forward process rate is $R_{f,\text{ rate}} = k_f C_A = k_f C_O(0, t) = \frac{i_c}{nFA}$.

The backward process rate is $R_{b,\text{ rate}} = k_b C_B = k_b C_R(0, t) = \frac{i_a}{nFA}$.

i_c and i_a are the cathode and anode currents and t is time. The net current is represented as

$$\begin{aligned} R_{\text{net, rate}} &= R_{f,\text{ rate}} - R_{b,\text{ rate}} \\ &= k_f C_O(0, t) - k_b C_R(0, t) = \frac{i}{nFA} \end{aligned}$$

and the overall current is $i = i_c - i_a = nFA[k_f C_O(0, t) - k_b C_R(0, t)]$.

The cathodic and anodic currents in an electrochemical cell are denoted by the equation

$$i_c = nFAk_f C_O e^{\left[\frac{-\alpha n(E-E_o')}{RT}\right]} \tag{12.16}$$

and

$$i_a = -nFAk_b C_R e^{\left[\frac{(1-\alpha)n(E-E_o')}{RT}\right]}, \tag{12.17}$$

where E and E_o are the applied and equilibrium potentials, respectively, k_f and k_b are the forward and backward rate constants, respectively, and α is the transfer coefficient (asymmetry factor). The potential sweep, which is also called the control voltage or cyclic voltammetry (CV), is a reversible technique. Consider the reaction $O + ne \rightleftarrows R$, a linear diffusion and solution containing only species O. The potential sweep varies linearly as (V s^{-1}), so that the potential at every instant is $E(t) = E_i - vt$, where E_i is the initial potential and v is the scan rate:

$$f(t) = \frac{C_O(0, t)}{C_R(0, t)} = \exp\left[\frac{nF}{RT}\left(E_i - vt - E_o'\right]\right. . \tag{12.18}$$

Mathematically the sweep potential or CV is complicated. The problem was first taken into consideration by Randles and Sevcik and later by Nicholson and Shain at the boundary condition:

$$\frac{C_O(0,\ t)}{C_R(0,\ t)} = \theta e^{-\sigma t}, \tag{12.19}$$

where $\theta = \exp[(nF/RT)(E_i - E^o)]$ and $\sigma = vnF/RT$.

The concentration at the surface, related to the diffusion coefficients by Laplace transformation, is represented as

$$C_O(x,\ s) = \frac{C_O^*}{s} + A(s)\exp\left[-\left(\frac{s}{D_O}\right)^{1/2} x\right]. \tag{12.20}$$

The current in the whole process is related to the concentration at the surface and the diffusion coefficient by the Laplace transform is represented as

$$i(s) = nFAD_o\left[\frac{\partial C_O(x,\ s)}{\partial x}\right]\Bigg|_{x=0}. \tag{12.21}$$

Combining equation (12.20) and equation (12.21), and inverting the equation by the convolution theorem, gives

$$C_O(0,\ t) = C_O^* - [nFA(\pi D_o)^{1/2}]^{-1}\int_0^t i(\tau)(t-\tau)^{-1/2}d\tau, \tag{12.22}$$

and by simplifying equation (12.18) $f(\tau) = \frac{i(\tau)}{nFA}$, replacing $f(\tau)$ in equation (12.22):

$$C_O(0,\ t) = C_O^* - (\pi D_o)^{-1/2}\int_0^t f(\tau)(t-\tau)^{-1/2}d\tau. \tag{12.23}$$

Considering that R is initially absent, then the expression for $C_R(0,\ t)$ is

$$C_R(0,\ t) = (\pi D_R)^{-1/2}\int_0^t f(\tau)(t-\tau)^{-1/2}d\tau. \tag{12.24}$$

Combining equations (12.19), (12.23), and (12.24) we obtain

$$\int_0^t f(\tau)(t-\tau)^{-1/2}d\tau = \frac{C_O^*}{\theta s(t)(\pi D_R)^{-1/2} + (\pi D_O)^{-1/2}} \tag{12.25}$$

and

$$\int_0^t i(\tau)(t-\tau)^{-1/2}d\tau = \frac{nFA\pi^{1/2}D_R^{1/2}C_O^*}{\theta s(t)\xi + 1}, \tag{12.26}$$

where $\xi = (\frac{D_O}{D_R})^{1/2}$.

$i(t)$ is replaced by $i(E)$, accomplished by substituting

$$\sigma t = \frac{nFVt}{RT} = \frac{nF(E_i - E_o')}{RT}.$$

Letting $f(\tau) = g(\sigma\tau)$,

$z = \sigma\tau$ at $t = \tau$ and $z = 0$ at $t = 0$ we obtain the integral functions

$$\int_0^t f(\tau)(t - \tau)^{-1/2} d\tau = \int_0^t g(z)\left(t - \frac{z}{\sigma}\right)^{-1/2} \frac{dz}{\sigma} \tag{12.27}$$

$$\int_0^t g(z)(\sigma t - z)^{-1/2} \sigma^{-1/2} dz = \frac{C_O^*(\pi D_O)^{1/2}}{1 + \xi\theta s(\sigma t)}, \tag{12.28}$$

where

$$X(\sigma t) = X(z) = \frac{g(z)}{C_O^*(\pi D_O \sigma)^{1/2}} = \frac{i(\sigma t)}{nFAC_O^*(\pi D_O \sigma)^{1/2}}. \tag{12.29}$$

Equation (12.28) at any value of $s(\sigma t)$ is a function of E and the current can be obtained by rearrangement of equation (12.29):

$$i = nAFC_O^*(\pi D_O \sigma)^{1/2} X(\sigma t). \tag{12.30}$$

At some instant the function $X(\sigma t)$ is a pure number, so equation (12.30) gives the functional relation between any instant of the LSV curve and the variable, specifically current (i), is directly proportional to the bulk concentration C_O^*, the square root of the scan rate v. The solution of (12.29) is carried out mathematically by a series of solutions (Sevick, Reinmuth) and it is evaluated and observed analytically that when the value of a function $\pi^{1/2} X(\sigma t)$ reaches about 0.4463 then the current reaches a maximum peak value for the reversible reaction.

The Randles–Sevcik equation describes the current peak amplitude in a redox event depending on various parameters and is denoted by the equation

$$i_p = 0.4463\, nFAC_o^* \sqrt{\frac{nFD_o v}{RT}}, \tag{12.31}$$

where

i_p = current peak amplitude in amps
n = number of electrons transferred in the redox event
A = electrode surface area in cm^2
F = Faraday constant in C mol^{-1}
D = diffusion coefficient in cm^2 s^{-1}
C = concentration in mol cm^{-3}
ν = applied potential scan rate in V s^{-1}
R = gas constant in VC K^{-1} mol^{-1}
T = environmental temperature in K.

Replacing T by 298 K and F by 96 485.3365 C mol^{-1} in equation (12.31) we obtain

$$i_p = 2.687 \times 10^5 AC_o^* n^{3/2} \sqrt{D_o v}. \tag{12.32}$$

The peak current amplitude (i_p) is the measure of concentration (C) if all other parameters remain constant.

12.4 The SPE based biosensor

The WE is the main electrode of concern since it produces a binding site for the analyte molecules. For that, the WE surface's active side is cultivated with a specific species by placing a regulated quantity of antibody–antigen in the solution or by plunging the SPE into the solution at a predetermined temperature and environmental condition with a possible rate of scanning. Binding the antibodies to create a film on the electrode surface requires some amount of incubation time. In that case, the antigen–antibody interaction along with some reducing substance and enzymes produces an electrochemical signal which transduces corresponding electrical signal to the analyte [21]. A schematic representation of the process flow for biomolecular recognition in the preparation of an SPE based biosensor is shown in figure 12.3.

12.4.1 Immunosensor

An immunosensor is defined as a device for the detection of chemical compounds using proper biochemical reactions excited by some enzyme conjugates, tissues, organelles in blood [22], water [23], drink, food, or urine [24], usually in the form of an optical or electrical signal. The immunosensor is a biosensor category that uses an antigen or an antibody species combined with a signal transducer to identify supplementary species in blood serum or body fluids to trace diseases [25].

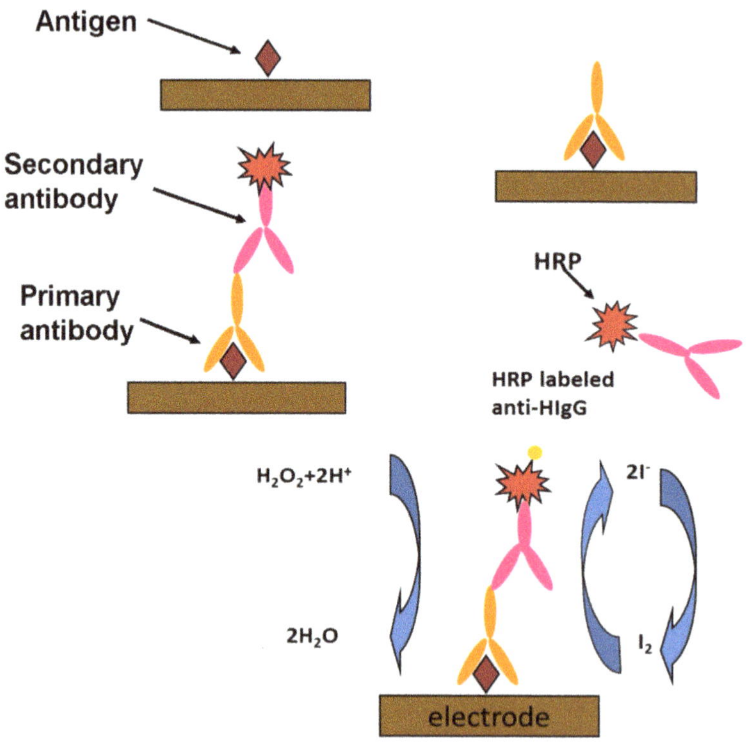

Figure 12.3. Schematic flow diagram of biosensor preparation.

The immunosensor performs two functions: (i) recognition and (ii) transduction. In the first, the biological molecules of interest interact selectively with the recognition or sensing element. The sensing element may comprise some chemically and biologically distinct recognition receptors. Due to the interaction of the biomolecular element of interest with the sensing receptors, the properties of the sensing element vary with a function of the biomolecular analyte concentration. This process is called signal transduction and is shown in figure 12.4. When this biological compound is produced due to immunological responses such as antigens, antibodies, enzymes, or hormones then the above-described system is called an immunosensor.

12.4.2 Immunoassay

An immunoassay is a biological identification and analysis technique using a particular antibody or immunoglobulin for biomolecules in a test solution. Immunoassays depend on an antibody's ability to identify a particular target, which in effect generates a detectable reaction that is used for identification with the assistance of a marker. The enzyme-linked immune sorbent assay (ELISA) is a popular technique for immunoassay. It is commonly used as a therapeutic screening instrument. The key idea involved is to immobilize an undefined quantity of antigen on a surface and then add a particular antibody to the surface to bind with that antigen.

12.4.3 The construction of an amperometric type immunosensor

The design of an SPE based immunosensor requires three stages [26]. The first stage is electrode manufacture, the second stage involves electrode surface functionalization, and the final stage is biomolecular immobilization using the ELISA format, which is effective for electrochemical transduction [2–24]. This process explains all the steps for the proper summary of the actual methods that are being introduced for the desired implementation of amperometric immunosensing devices based on SPEs [27–29].

12.4.4 Chemosensor

A chemical species that collaborates through an analyte to express a noticeable change of the electrical signal is called a chemosensor [30]. The chemosensor working principle is similar to that of biosensors, by the recognition and quantification of a liquid or

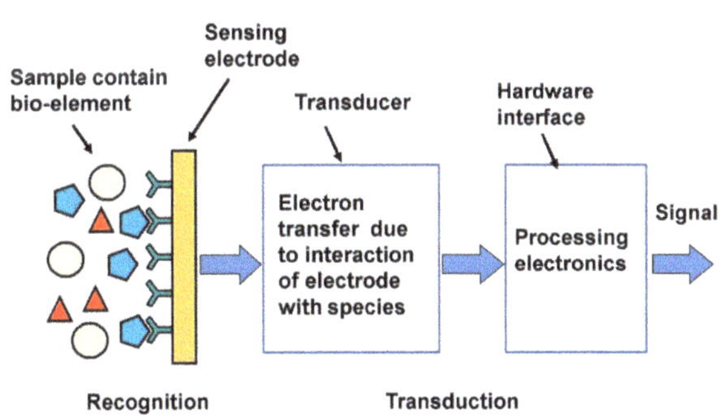

Figure 12.4. A schematic model of an immunosensor.

gaseous chemical analyte due to interaction with the electrode surface. Chemical sensors have many applications and are being used for food and drinking water testing, beverage quality monitoring, environmental pollutant detection, etc. Popular applications of disposable SPE based chemosensors for the detection of heavy metals such as lead, arsenic, mercury, cadmium, etc, in food and water is at present an area of high demand. SPE based sensors are comparatively easy to fabricate, have simple measurement capability, and are a portable and low-cost rapid test application compared to existing microscopic measurements. Many researchers have already established SPE based chemosensors for trace-level heavy metal detection, in particular in drinking water. A popular example is the Bi–Sb SPE electrode which is made by depositing bismuth (III) and antimony (III) with analytes on the SPE surface [31] for lead and arsenic detection. Another example of an SPE based chemosensor was prepared with bismuth precursor compound and graphite ink loaded on the working electrode surface for the detection of trace-level lead and cadmium [32].

12.4.5 Substrate and electrode materials

The SPE is printed on various types of substrate materials such as ceramics [1, 2], paper, plastic [6–8], glass [9], polyester [8–11], etc. The choice of substrate material is an important parameter. First, it must be an electrical insulator and a very good adhesive surface to print on it. Flexible polyester film is the best-suited material for bulk printing, as it is easy to cut, lightweight, and low cost.

The sensing electrode of an SPE is usually made of carbon material and the sensitivity of the electrode depends the carbon ink composition, particle size, and fabrication process parameters. In addition to the material and fabrication, some other important factors are the design parameters, electrode surface orientation, surface morphology, geometry of electrodes, edge-plane effects, etc. The surface area of the electrode active site can be increased by the correct carbon ink composition to allow greater edge-plane areas, holes, and rough surfaces to promote quicker transmission of an electron, which in effect enhances sensitivity. The quantity of organic binders used to create the carbon ink often prevents cracks on the printed surface and implicitly improves the transmission rate of electrons. The property of the wettability of the carbon electrode surface enhances hydrophobicity towards the analyte substance when the species of the sample solution are easily absorbed onto the electrode surface, which initiates electron transfer at the solid–liquid interface. A higher curing temperature can result in unintentional cracks and defects, which also increases the edge-plane effects and finally increases sensitivity.

The WE is usually made up of carbon ink, platinum paste, gold nanoparticles, etc. The structure of the ink, the shape of the ink, and the particle size affect the output of such SPEs [25, 26]. The SPE's sensing mechanism is based on the ink's material properties and the printing process. The carbon surface composition, shape, and surface orientation at the boundary of the electrode surface and electrolyte solution are essential in the transfer of electrons. Some different available SPEs are shown in figure 12.5, in which most of the electrode material is carbon base, but different substrates, dimensions, and ink formulations are used.

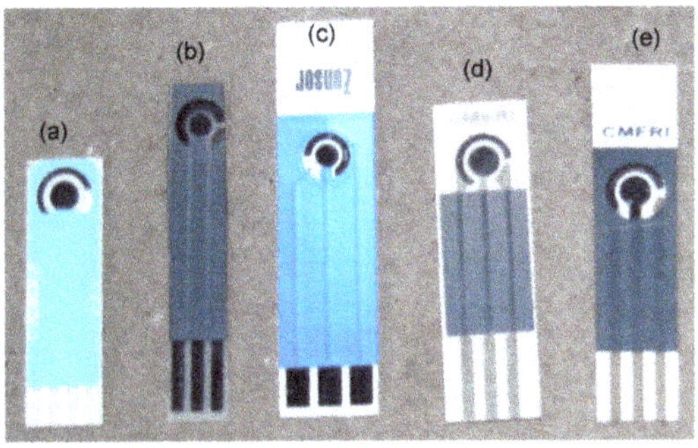

Figure 12.5. SPE sprinted on different substrates.

12.4.6 Electrode surface modification

An SPE works through a process where a bio-analyte or receptor molecule is detected and bound with the electrode surface followed by a catalyzed chemical process. An effective contract or binding mechanism between the analyte and the electrode surface should be built to enhance the amperometric reaction, which is a crucial factor that increases the sensitivity, stability, and specificity of the sensor. Numerous technologies have been implemented to improve the binding, surface orientation, and alignment of biomolecular or chemical receptors to optimize their performance for various applications. The polymerization of the electrode surface is one of the suitable techniques for the entrapment of enzymes. On the electrode surface, metallic nanoparticles develop good physical and chemical properties, such as a high surface-to-volume ratio, high electrical conductivity, solid adsorption power, and good surface characteristics. Gold nanoparticles (AuNPs) are usually used to enable the transfer of electrons between redox proteins and mass electrode materials due to their excellent electron transfer capability. Carbon nanotubes are carbon allotropes with a cylindrical nanostructure, providing special chemical, physical, and mechanical characteristics, such as maximum power, durability, and thermal stability, combined with satisfactory electrical characteristics, such as good conductivity and potential for electron transfer. The use of self-assembled monolayers (SAMs) is increasing quickly in the field of biosensors.

12.5 Screen printed electrode fabrication

The masks for a common SPE fabrication technique are shown in figure 12.6 for a clear understanding of each step. The ceramic/polyester substrate is placed beneath mask (a) of the specified designed screen mesh pattern. The conducting Ag paste is squeezed over the screen and mask (a) by a rubber squeegee. Due to pressure, the ink passes through the mask, and the pattern is transferred to the substrate. The conducting Ag tracking layer is first dried at room temperature to evaporate the solvent from the ink in an air dryer and it then needs to cure in a heating chamber at a set temperature and time, depending on the ink's properties, to make a zero resistance conducting track.

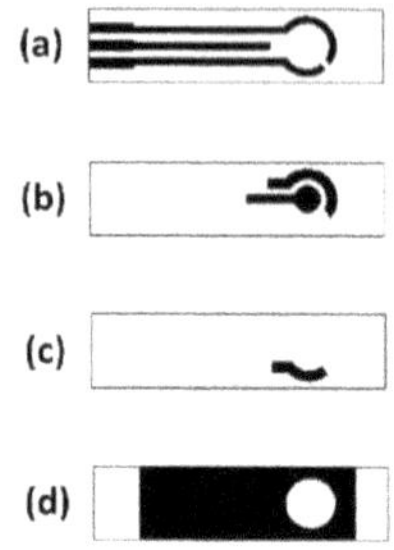

Figure 12.6. Typical screen masks for the (a) conducting track, (b) working and counter electrodes, (c) reference electrode, and (d) insulating layer.

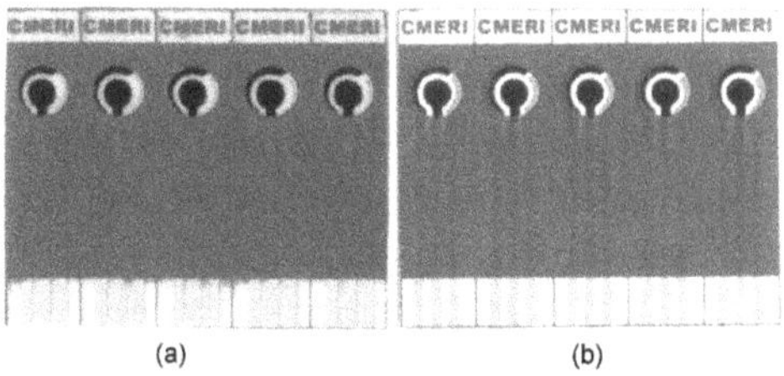

Figure 12.7. Flexible screen printed electrodes on: (a) polyester film and (b) PET coated polyester film.

The second step involves the printing of the working and counter electrode using mask (b) as per the designers choice. After completing the first step the substrate is placed beneath mask (b) in the screen print set-up and is aligned properly. The mask alignment needs to be perfect and its resolution depends upon the design dimension of the printed electrode gaps. The carbon paste is squeezed over on the screen mesh and transferred to the substrate over the silver conducting line to make a multi-layer connection according to the design pattern. Similarly to the first step, the carbon layer also follows the air drying and curing process depending on the carbon ink properties. This second step involves printing both the working and counter electrode because the same conducting material carbon is used, but if different ink materials are used another intermediate step needs to be performed for the printing of the counter electrode.

After the completion of the second step it is now necessary to print the reference electrode. Generally, Ag/AgCl is used to fabricate the reference electrode in the electrochemical cell system, thus Ag/AgCl ink is used to print the reference electrode. Similarly the second step, in the third step the Ag/AgCl ink is dried to remove the solvent and is heated in a chamber to cure a solid stable metallic layer. After printing and curing all the various conducting and sensing layers, most of the open areas need to be covered with an insulating layer, excluding the sensing well and lead connection. Typical design patterns using different masks and fabricated electrodes using two different flexible substrates, polyester film and PET coated polyester film, are shown in figures 12.7(a) and 12.7(b). Figure 12.8 shows an electrode fabricated using a hard ceramic (alumina) substrate, which need to be

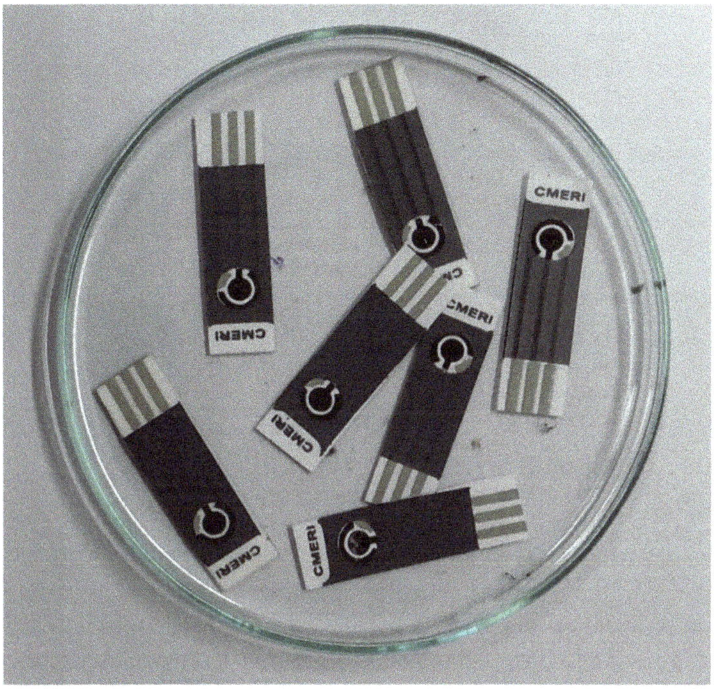

Figure 12.8. SPEs fabricated on a ceramic substrate.

printed one by one, but in the case of a flexible sheet of polyester film, a large number of electrodes can be produced for bulk production.

12.6 Electrochemical signal measurement

Typically the electrochemical signals of SPEs are recorded using a CV apparatus. The CV instrument is used to apply a predetermined ramp potential to the electrode to excite the electrochemical redox or oxidation reaction. According to the Randles–Sevcik equation, the current peak is increased with the ramp rate of the applied voltage, hence it is clear that the sensor response is better at a higher ramp rate. However, practically higher ramp rate current–voltage characteristics are not correlated with the theoretical results, so the scan rate should be as low as possible without compromising the sensitivity of an amperometric measurement.

12.6.1 A basic potentiostat circuit of electrochemical signal transduction

A basic potentiostat circuit and computer interface for electrochemical signal transformation for amperometric measurement is required for measuring the signal from SPEs. The potentiostat circuit is a type of potentiodynamic electrochemical measurement system [33], where a control ramped potential (varying linearly with time) is applied to the cell. In an amperometric type electrochemical measurement, the current–voltage characteristics are plotted for a redox system and the measured current peak amplitude determines the presence of the analyte or quantifies the

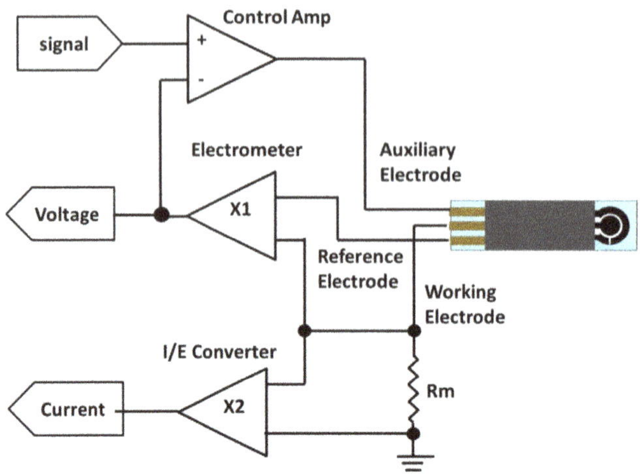

Figure 12.9. Basic potentiostat circuit.

analyte concentration level. The potentiostat circuit consists of a control signal generator to excite the electrochemical reaction on the electrode surface. A precise current amplifier circuit is required to produce a corresponding voltage level to generate an amperometric signal, which is ultimately fed to an interface card to store the signal data and plot.

The basic configuration of the potentiostat circuit consists of four blocks: (i) control signal, (ii) control amplifier, (iii) electrometer, and (iv) *I*/*E* converter (figure 12.9)

12.6.1.1 Control amplifier

The control amplifier is a servo amplifier which supplies the power to maintain a regulated voltage difference between both electrodes. It compares the voltage level between the measured cell voltage and the desired voltage and produces the current in the cell. To understand how it works, the electrical equivalent of the electrochemical cell and the current measuring resistor R_m can be regarded as two impedances Z_1 and Z_2 in figure 12.10. Z_1 is an equivalent impedance that involves R_m in series with counter-electrode interface impedance and solution resistance between the counter and the reference electrode, Z_2 reflects the interface impedance of the working electrode in line with the solution resistance between the electrodes. The control amplifier's function is to compare the input signal with the reference electrode voltage which is the negative feedback in inverting input. In an ideal situation, the cell current (I_c) can be written as

$$I_c = \frac{E_{\text{out}}}{Z_1 + Z_2} = \frac{E_r}{Z_2} \tag{12.33}$$

$$E_r = \left(\frac{Z_2}{Z_1 + Z_2}\right) E_{\text{out}} = \beta E_{\text{out}}, \tag{12.34}$$

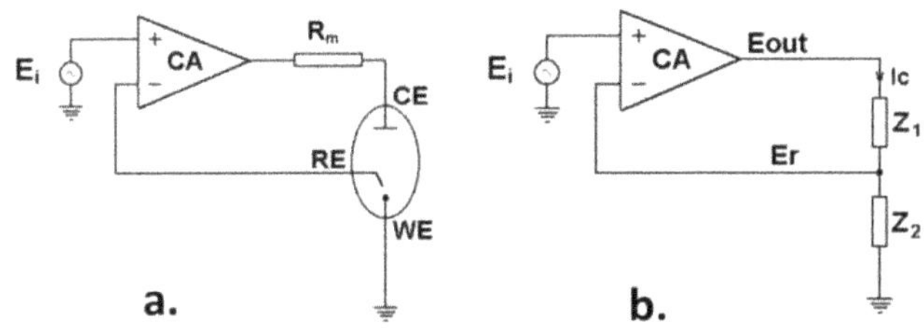

Figure 12.10. (a) The control amplifier and (b) electrical equivalent representation of an electrochemical cell.

where β is the feedback factor of the control amplifier returned to its negative input

$$\beta = \frac{Z_2}{Z_1 + Z_2}. \tag{12.35}$$

The relation of amplifier inverting and non-inverting inputs is

$$\frac{E_r}{E_i} = \frac{\beta A}{1 + \beta A}, \tag{12.36}$$

where A is the amplification gain of the control amplifier when the quantity βA becomes very large. Equation (12.36) is reduced to equation (12.37) as

$$E_r = E_i. \tag{12.37}$$

12.6.1.2 Electrometer

The electrometer is a comparator circuit that measures the potential variation between the two electrodes and the output is the feedback to the control amplifier input (a schematic diagram is shown in figure 12.11). An ideal electrometer has zero input current and infinite input impedance. The potential will change when the current passes through the reference electrode. Modern electrometers have input currents close to zero. Two key features of the electrometer are its bandwidth and input capacitance. The bandwidth of the electrometer is characteristic of AC frequencies. The bandwidth of the electrometer will be greater than that of the other electronic devices within the potentiostat.

12.6.1.3 I/E *converter*

The role of the *I/E* converter is to convert the cell current to the equivalent voltage. The *I/E* converter allows the current to travel through the measuring resistance (R_m) and the voltage developed across R_m is applied to the amplifier. In practical situations, where the current can often vary from very small to high magnitude levels, a single current calculating resistance (R_m) cannot be used to calculate the current across such a broad range. In such situation, an array of different R_m resistors can be switched ON/OFF under computer control to set up the desired current range. This allows the measurement of widely varying currents using a proper resistor. Typically, an *I/E* auto range optimization algorithm is used to identify the proper resistors.

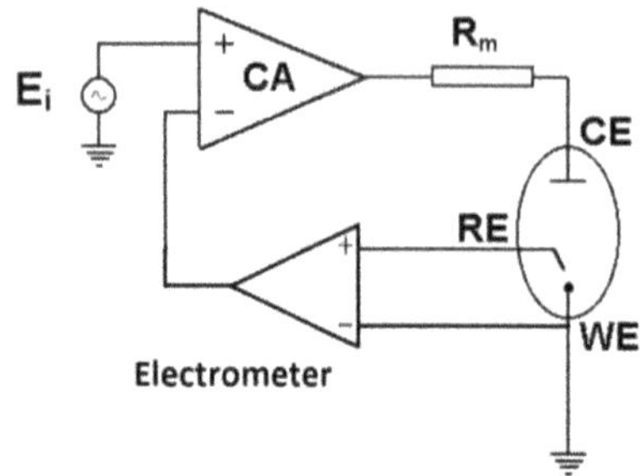

Figure 12.11. Schematic of the electrometer.

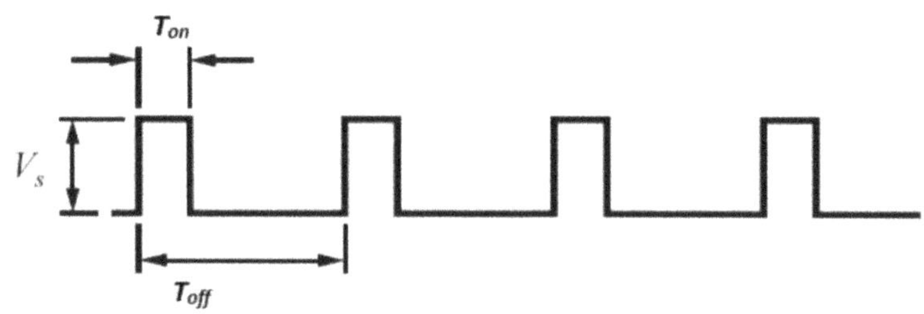

Figure 12.12. Schematic diagram of the PWM signal.

12.6.1.4 Signal generator

The potentiostat circuit is typically used in CV measurement and it requires control of the potential, which should be linearly varying with time. In the electrochemical cell system for redox and oxidation, a linearly varying analog ramp signal can be generated using a pulse width modulator (PWM) (figure 12.12). The PWM is a technique that offers electrical power to produce a slowly varied DC output voltage by varying its pulse width. The PWM is one of the methods that is used to produce a ramp signal in the potentiostat circuit.

If the output of a PWM channel is V_s, then the effective output voltage V_{eff} is

$$V_{\text{eff}} = V_s \frac{T_{\text{on}}}{T_{\text{off}}}.$$

If this signal is delivered as an input to a system with a reaction time that is far greater than T_{on}, the system can encounter the signal with an average of DC input with the voltage V_{eff}. By controlling the pulse width with time the effective DC voltage is varied from the preset initial value to the final value. In a CV experiment, the WE potential is sloped linearly versus time from a set of initial values to the final value. When cyclic voltammetry reaches the final voltage, the ramp signal is inverted to reach the initial value.

12.6.2 Cyclic voltammetry

The most common voltammetry method involves sweeping the electrode potential from an initial value (E_a) to a final value (E_c) at a function of time $\frac{\partial e}{\partial t}$. When working in the context of the electroanalytical method the rotating approach is called cyclic sweep voltammetry. CV is an electrochemical reaction process to determine the cell current generated in an electrochemical cell system when a control voltage is applied to the cell. Applying voltage excites the electrochemical redox and oxidation reaction. Normally in cyclic voltammetry, the electrode potential is a linearly time-varying voltage ramp signal. During each one of these phases, the rate of voltage change with time is considered as the scan rate (V s^{-1}). During the electrochemical reaction, the potential between the electrodes is calculated and converts the corresponding current signal. Figure 12.13 shows a time span from time t_1 to t_2, where the voltage ramp is a gradually increasing potential known as a forward scan. In the forward scan the cathodic current will increase over this time span t_1 to t_2 and a sudden sharp current peak indicates where the redox reaction takes place, depending upon the redox agent. Similarly, during the time span from t_2 to t_3, where the voltage ramp is a gradually decreasing potential known as the backward scan voltage, the anodic current is reducing with time and a sharp negative current peak indicates the potential at which the oxidation reaction takes place and vice versa for each rotating cycle. The common redox reagent is potassium ferricyanide(Fe-III)[$K_3Fe(CN)_6$] and is typically used to examine the electrode and potentiostat circuit by first observing the nature of the CV and the positions of the redox and oxidation potentials.

12.6.3 Linear sweep voltammetry

Working in the context of the electroanalytical method is a non-rotating manner is called linear sweep voltammetry (LSV). LSV is a one-directional potentiostat sweep method, which uses only one segment or half a cycle of cyclic voltammetry. In LSV the applied electrode potential is swept linearly from the initial value (E_i) to the final value (E_f) and the corresponding current is measured as a function of time.The common output of an LSV experiment is also current versus potential. LSV is often used to study the electron transfer reaction kinetics for organic and inorganic synthesis, sensor and biological system evaluation, etc. In figure 12.14 a basic linear sweep potential start from the value E_i at time t_1 and increases with a function of time $\frac{\partial e}{\partial t}$ at final voltage value E_f at time t_2. This means that linear sweeps are

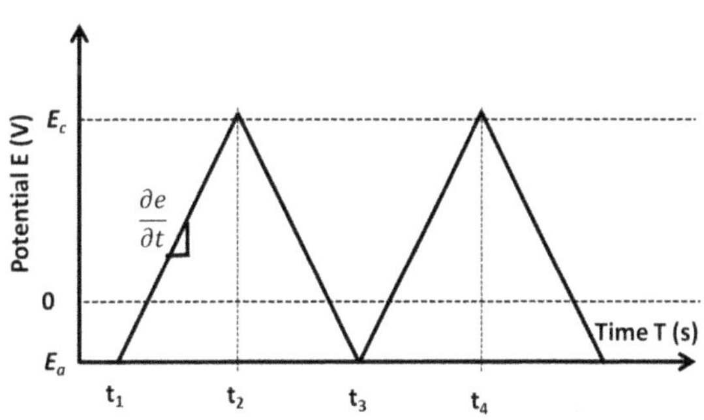

Figure 12.13. Cyclic voltammetry waveform.

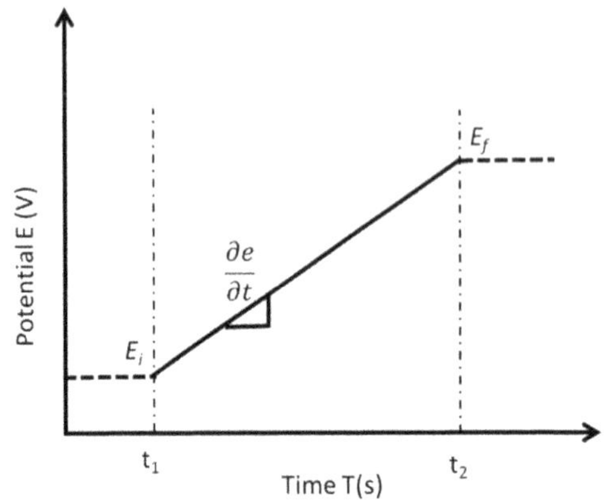

Figure 12.14. Linear sweep voltammetry.

approximated by a series of small stair steps, whose step size is defined by an analog to digital converter (ADC).

12.6.4 Pulse voltammetry

Pulse voltammetry or square wave voltammetry is a derivative of staircase or linear sweep voltammetry, where a series of linear sweeps or stair step voltage pulses are superimposed on the applied potential (shown in figure 12.15). The current is measured before and after each applied pulses. Pulse voltammetry incorporates a pulse height that is implemented at intermittent intervals, rising periodically. The current measurement is usually the same as CV or LSV. The potential between the working and reference electrode is changed corresponding to the pulse from an initial potential to an interlevel potential. Usually, the interlevel potential time is about 5 to 100 ms. After the interlevel potential, it changes to the final potential. The pulse is repeated, changing the final potential, and a constant difference is maintained between the initial and the interlevel potential. The values of the current between the working electrode and auxiliary electrode before and after the pulse are sampled and their differences are plotted. In normal pulse polarography (NPP) the potential is not a continuously increasing staircase potential ramp, but increases by square wave potential pulses with increasing height (ΔE_p), overlaid on a constant initial potential (figure 12.16). The pulse voltammetry technique is ideally suited to digital rather than analog generation and computer control, so much of this work has been coupled with the development of digital potential generators and computerized instrumentation. However, despite the different waveform, the shape of pulse voltammetry is similar to that of linear sweep curves, and this feature still detracts from the use of this technique.

12.6.5 Stripping voltammetry

Stripping voltammetry is a two-stage method where the first step involves the electrolytic deposition of a chemical substance at a fixed concentration over an inert

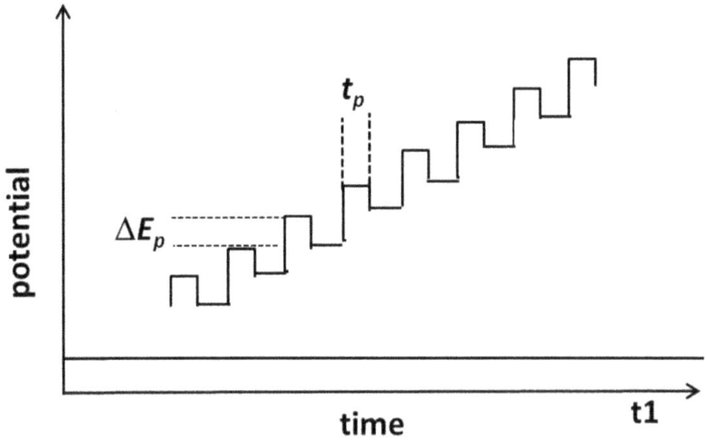

Figure 12.15. Pulse voltammetry or square wave voltammetry.

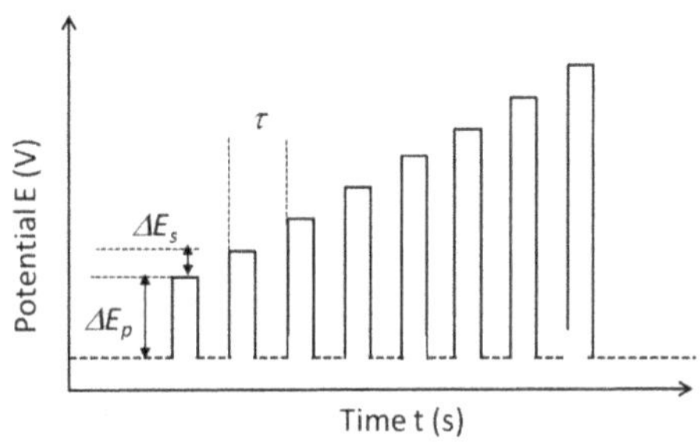

Figure 12.16. Normal pulse voltammetry.

electrode surface. Stripping voltammetry is mainly a method for analyzing traces. It may be used to achieve standard analytical decisions at the sub-ppm level, but this involves a high degree of care in the laboratory procedure and the processing of samples. In stripping voltammetry, the voltammetry electrode should be in a steady position and the optimal functioning electrode must have a reproducible surface area and small residual current.

12.7 Basic characteristics of some electrochemical signals

12.7.1 Electrochemical cell and signal

An example of a CV experiment is the application of a sweeping potential of −0.5 V to +0.5 V at a scanning rate of 50 mV s^{-1}. A common redox solution is potassium ferricyanide, prepared at various concentrations. The current–voltage data for each step is recorded and current versus applied voltage is plotted. The simple current–voltage plot of a standard CV using a potassium ferricyanide solution is shown in figure 12.17. The CV characteristics, irrespective of amplitude, indicate that the SPEs are responding to the electrochemical signal. There are two peaks in the forward and reverse directions, meaning that both redox and oxidation occur. In this

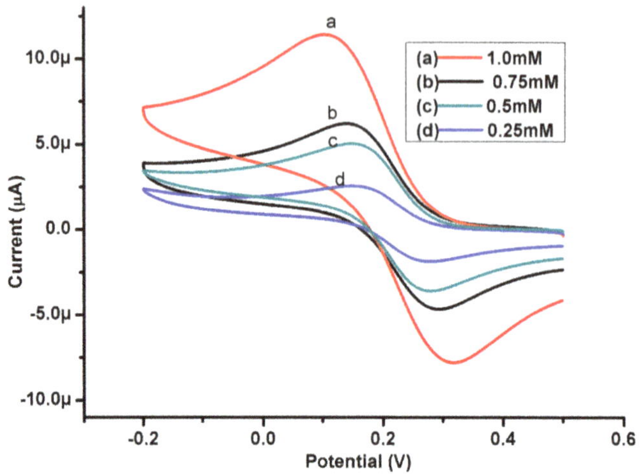

Figure 12.17. Current–voltage characteristics. (a), (b), (c), and (d) indicate the increases in the concentration of analyte.

example if the concentration of the potassium ferricyanide is increased while maintaining other parameters, then only the amplitude of the redox and oxidation current will be increased, which is proportional to the analyte concentration and must follow the Randles–Sevcik equation. The nature of the CV and redox gap between the redox and oxidation peak or voltage difference and position of the redox and oxidation reaction indicates the quality and sensitivity of the electrodes or sensors [34].

12.7.2 The electrochemical response of different SPEs

The electrochemical reactions and current responses of some commercial SPEs in various redox analytes are compared in tables 12.1–12.3. For a better understanding, three different analyte (potassium ferricyanide (Fe-III), ascorbic acid (AA), and nicotinamide adenine dinucleotide (NAD)) redox samples [35] are presented for three different SPEs from different manufacturers, with different shapes, sizes, and materials, namely the Dropsense SPE (C110), Kanichi electrode (KS540), and Zensor electrode (TE100). Various measured parameters for the three electrodes in the three different standard redoxs are tabulated for a better understanding of electrode performance in the application of a sensor. In table 12.1 the current density with respect to the geometric surface area of the WE is higher for C110 in a stable concentration of Fe-III, whereas the ratio of the anodic and cathodic peak current is better for KS540 and there is a narrow peak-to-peak potential separation for C110. In table 12.2 the comparison of electrode performance used 1 mM AA, and only anodic peak is present for all three electrodes. The current density is almost double for C110 compare to KS540 and TE100, and the redox reaction occurs at a comparatively lower potential (260 mV for C110). In table 12.3 the electrochemical performances of the three electrodes in 1 mM NAD

Table 12.1. Current response for a 1 mM Fe-III sample.

Redox sample	Electrode	Area of WE (mm^2)	I_{pa} (μA)	I_{pc} (μA)	I_{pa}/I_{pc}	δ ($\mu A/mm^2$)	E_{pa} (V)	E_{pc} (V)	ΔE (mV)
Fe-III 1 mM in 0.5 M KCl	C110	12.566	14.2	10.3	1.36	1.114	0.105	0.235	130
	KS540	7.068	6.9	6.2	1.13	0.976	0.115	0.285	170
	TE100	7.068	5.5	3.0	2.5	0.778	0.0	0.33	330

Table 12.2. Current response for a 1 mM AA sample.

Redox sample	Electrode	Area of WE (mm^2)	I_{pa} (μA)	I_{pc} (μA)	I_{pa}/I_{pc}	δ ($\mu A/mm^2$)	E_{pa} (V)	E_{pc} (V)	ΔE (mV)
1 mM AA in 0.1 M PBS	C110	12.566	41.2			3.270	0.26		
	KS540	7.068	21.5			3.041	0.38		
	TE100	7.068	20.5			2.900	0.46		

Table 12.3. Current response for a 1 mM NAD sample.

Redox sample	Electrode	Area of WE (mm^2)	I_{pa} (μA)	I_{pc} (μA)	I_{pa}/I_{pc}	δ (μA/mm^2)	E_{pa} (V)	E_{pc} (V)	ΔE (mV)
1 mM NAD in 0.1 M PBS	C110	12.566	58			4.615	0.55		
	KS540	7.068	24			3.395	0.55		
	TE100	7.068	16			2.263	1.0		

concentration also show the highest current density for both C110 and KS540 at the same redox potential. No traceable anodic peak is present in the NAD redox probe for all three electrodes.

It is very difficult to calculate the working electrode active surface area (S_a) from the nonlinear CV response and by measuring the current peak. The only possible way to estimate the active surface area (S_a) of the electrode is by measuring the anodic current peak (I_{pa}) and with the help of the Randles–Sevcik equation. In table 12.4 the effective active surface area of the three electrodes is compared numerically. A larger active site of the working electrode surface improves the electron transfer capabilities, automatically producing a large current peak. The characteristic plot of the CV response of an 1 mM Fe-III concentration for three different commercial SPEs is shown in figure 12.18, to understand the nature of the redox and oxidation reaction. Figure 12.19 shows an experimental test set-up for multiple SPEs simultaneously using a multi-channel CV instrument.

Table 12.4. Various signal stimuli for three different SPEs.

Redox	Electrode	Geometric area (S_g)	I_{pa} (μA)	Active area (S_a)	S_a/S_g	% Activity
1 mM Fe-III	C110	12.56	14	24.32	1.94	93.63
	KS540	7.07	6.9	13.5	1.91	90.95
	TE100	7.07	5.5	12.4	1.75	75.39
1 mM AA	C110	12.56	41.2	23.4	1.86	86.31
	KS540	7.07	21.5	12.24	1.73	73.13
	TE100	7.07	20.5	11.67	1.65	65.06
1 mM NAD	C110	12.56	58	23.01	1.83	83.20
	KS540	7.07	24	12.6	1.78	78.22
	TE100	7.07	16	10.8	1.53	52.76

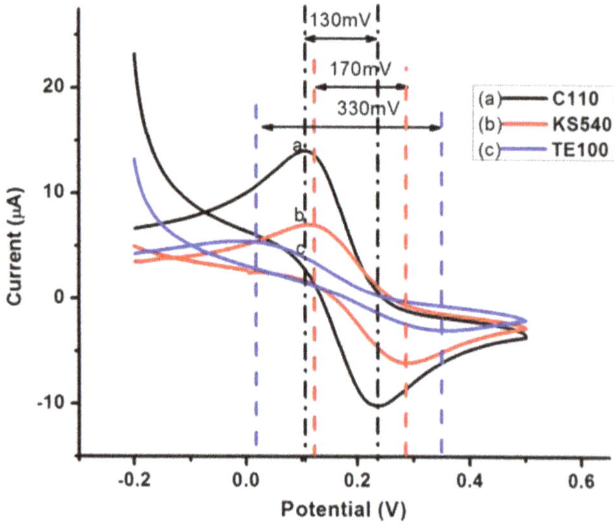

Figure 12.18. Current–voltage characteristics of C110, KS540, and TE100.

12.8 Conclusion

Various types of electrode materials, such as gold, silver, glassy carbon, graphite impregnated with wax, and steel paste, are used widely in various types of electrochemical analysis. However, while solid electrodes provide a very good stable electrochemical response, they are practically used in the laboratory only for research and electroanalysis. This standard solid electrode may not be suitable for the preparation of biosensors or chemical sensors, hence the flexible screen printed electrode is best suited for highly sensitive electrochemical sensors. The SPE based disposable chemosensors and biosensors have emerged quickly in

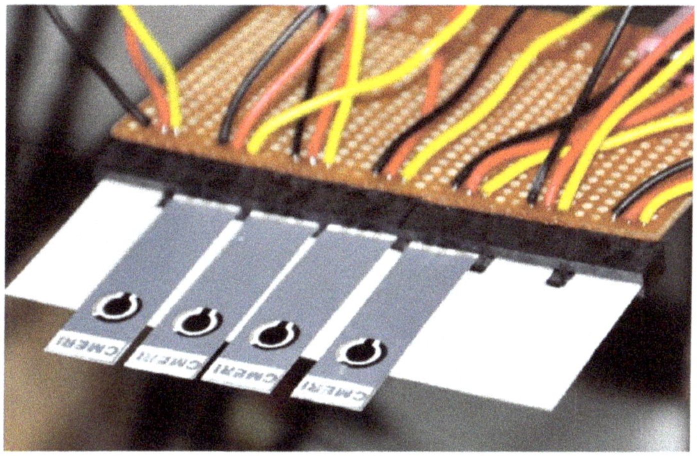

Figure 12.19. A multiple SPE based sensor for detecting biochemical species.

recent years due to their functionality, operating versatility, durability, and flexibility. Developments in the manufacture of SPEs, surface modifications, and immobilization techniques are global research concerns. Flexible SPE based sensing strips have broad commercial appeal to solve the problems of health and environment monitoring. Immunosensors are prepared using such fabricated SPEs, wherein biomolecules are immobilized directly on the electrode surface without any pretreatment or surface modification. This allows low cost and reduces the production time of the device and thus has the potential to be used for practical purposes. The signal amplitude corresponding to the analyte depends on various parameters, such as the design and dimensions, materials and methods, electronics interface, etc.

Acknowledgement

The author thanks Director, CSIR-CMERI, Durgapur for providing support to publish this work. The author wishes to thank Professor H Saha, and Dr C Roychoudhuri for their guidance in this field of work. The author also wishes to thank Gourab Kumar Bagchi and Mr Saugata Dey, Research Fellow for their collective help in preparing this work. The author is also grateful for the completed project 'CSIR 12th Five Year Plan' for rendering financial support.

References

[1] Ping J, Wang Y, Ying Y and Wu J 2012 Application of electrochemically reduced graphene oxide on screenprinted ionselective electrode *Anal. Chem.* **84** 3473–9

[2] Viswanathan S, Rani C, Vijay Anand A and Ho J A 2009 Disposable electrochemical immunosensor for carcinoembryonic antigen using ferrocene liposomes and MWCNT screenprinted electrode *Biosens. Bioelectron.* **24** 1984–9

[3] Qi H, Ling C, Ma Q, Gao Q and Zhang C 2012 Sensitive electrochemical immunosensor array for the simultaneous detection of multiple tumor markers *Analyst* **137** 393–9

[4] Rao V K, Sharma M K, Goel A K, Singh L and Sekhar K 2006 Amperometric immunosensor for the detection of *Vibrio cholerae* O1 using disposable screenprinted electrodes *Anal. Sci.* **22** 1207–11
[5] Darain F, Park D S, Park J S, Chang S C and Shim Y B 2005 A separation free amperometric immunosensor for vitellogenin based on screenprinted carbon arrays modified with a conductive polymer *Biosens. Bioelectron.* **20** 1780–7
[6] Piermarini S, Micheli L, Ammida N H S, Palleschi G and Moscone D 2007 Electrochemical immunosensor array using a 96 well screenprinted microplate for aflatoxin B1 detection *Biosens. Bioelectron.* **22** 1434–40
[7] Grennan K, Killard A J and Smyth M R 2001 Physical characterizations of a screenprinted electrode for use in an amperometric biosensor system *Electroanalysis* **13** 745–50
[8] Mu S, Wang X, Li Y T, Wang Y, Li D W and Long Y T 2012 A novel screenprinted electrode array for rapid highthroughput detection *Analyst* **137** 3220–3
[9] Morrin A, Killard A J and Smyth M R 2003 Electrochemical characterization of commercial and homemade screenprinted carbon electrodes *Anal. Lett.* **36** 2021–39
[10] Metters J P, GomezMingot M, Iniesta J, Kadara R O and Banks C E 2013 The fabrication of novel screen printed singlewalled carbon nanotube electrodes: electroanalytical applications *Sens. Actuators* B **177** 1043–52
[11] Foster C W, Metters J P, Kampouris D K and Banks C E 2014 Ultraflexible screenprinted graphitic electroanalytical sensing platforms *Electroanalysis* **26** 262–74
[12] FanjulBolado P, HernándezSantos D, LamasArdisana P J, MartínPernía A and CostaGarcía A 2008 Electrochemical characterization of screenprinted and conventional carbon paste electrodes *Electrochim. Acta* **53** 3635–42
[13] Pemberton R M, Hart J P, Stoddard P and Foulkes J A 1999 A comparison of 1 naphthyl phosphate and 4 aminophenyl phosphate as enzyme substrates for use with a screenprinted amperometric immunosensor for progesterone in cow's milk *Biosens. Bioelectron.* **14** 495–503
[14] Layek K and Mistry K K 2018 A study of optimization of various parameters in the fabrication of screenprinted electrodes *IEEE Sens. J.* **18** 7917–23
[15] Yu H, Yan F, Dai Z and Ju H 2004 A disposable amperometric immunosensor for α1 fetoprotein based on enzymelabeled antibody/chitosan membrane modified screenprinted carbon electrode *Anal. Biochem.* **331** 98–105
[16] Zhao C, Lin D, Wu J, Ding L, Ju H and Yan F 2013 Nanogold enriched carbon nanohorn label for sensitive electrochemical detection of biomarker on a disposable immunosensor *Electroanalysis* **25** 1044–9
[17] Suresh S, Gupta A K, Rao V K, Om Kumar and Vijayaraghavan R 2010 Amperometric immunosensor for ricin by using on graphite and carbon nanotube paste electrodes *Talanta* **81** 703–8
[18] Xu Q, Yan F, Lei J, Leng C and Ju H 2012 Disposable electrochemical immunosensor by using carbon sphere/gold nanoparticle composites as labels for signal amplification *Chem. Eur. J.* **18** 4994–8
[19] Lai G, Yan F, Wu J, Leng C and Ju H 2011 Ultrasensitive multiplexed immunoassay with electrochemical *Anal. Chem.* **83** 2726–32
[20] Mistry K K, Kashyup S, Roychaudhuri C and Saha H 2014 A comparative study in electrochemical response of some commercial screen-printed electrodes (SPEs) *Sens. Lett.* **12** 1750–9

[21] Ricci F, Amine A, Palleschi G and Moscone D 2002 Prussian blue based screen printed biosensors with improved characteristics of longterm lifetime and pH stability *Biosens. Bioelectron.* **18** 165–74
[22] Hart J P, Abass A K, Cowell D C and Chappell A 1999 Development of a disposable amperometric NH^{+4} biosensor based on a chemically modified screenprinted carbon electrode coated with glutamate dehydrogenase, 2 oxoglutarate, and NADH *Electroanalysis* **11** 406–11
[23] Pundir C S and Phaugat N 2009 Construction of dissolved O_2 metric oxalate biosensor using PVC membrane bound sorghum leaf oxalate oxidase *J. Sci. Ind. Res. (India).* **68** 800–6
[24] Kanyong P, Pemberton R M, Jackson S K and Hart J P 2013 Development of an amperometric screenprinted galactose biosensor for serum analysis *Anal. Biochem.* **435** 114–9
[25] Deepthy S, Layek K, Mukherjee R, Kumar K and Ghosh M 2017 Development of screenprinted electrode based immunosensor for the detection of HER2 antigen in human serum samples *Bioelectrochemistry* **118** 25–30
[26] Mistry K K, Layek K and Mahapatra A 2014 A review on amperometrictype immunosensors based on screenprinted electrodes *Analyst* **139** 2289–311
[27] Wu J, Tang J, Dai Z, Yan F, Ju H and El Murr N 2006 A disposable electrochemical immunosensor for flow injection immunoassay of carcinoembryonic antigen *Biosens. Bioelectron.* **22** 102–8
[28] Taleat Z, Ravalli A, MazloumArdakani M and Marrazza G 2013 CA125 immunosensor based on polyanthranilic acid modified screenprinted electrodes *Electroanalysis* **25** 269–77
[29] Mistry K K, Layek K, Chell T N, Chaudhuri C R and Saha H 2016 Design and development of an amperometric immunosensor based on screenprinted *Anal. Methods* **8** 3096–101
[30] Singh A K 2012 Biosensor and its applications *J. Aquac. Res. Dev.* **1** 1–3
[31] Chen C *et al* 2013 Determination of lead(II) using screenprinted bismuthantimony film electrode *Electroanalysis* **25** 1446–52
[32] Lezi N, Economou A, Dimovasilis P A, Trikalitis P N and Prodromidis M I 2012 Disposable screenprinted sensors modified with bismuth precursor compounds for the rapid voltammetric screening of trace Pb(II) and Cd(II) *Anal. Chim. Acta* **728** 1–8
[33] Murari K, Stanaćević M, Cauwenberghs G and Thakor N V 2004 Integrated potentiostat for neurotransmitter sensing: a high sensitivity, wide range VLSI design and chip *IEEE Eng. Med. Biol. Mag.* **24** 23–9
[34] Mistry K K, Deepthy S, Chaudhuri C R and Saha H 2015 Electrochemical characterization of some commercial screenprinted electrodes in different redox substrates *Curr. Sci.* **109** 1427–36
[35] Honeychurch K C and Hart J P 2012 The chronoamperometric and voltammetric behaviour of glutathione at screenprinted carbon microband electrodes modified with cobalt phthalocyanine *Adv. Anal. Chem.* **2** 46–52

Chapter 13

3D printed enzymatic biofuel cells incorporated with graphene and modified graphite bioelectrodes: a comparative study

Prakash Rewatkar, U S Jayapiriya, Subhas Chandra Mukhopadhyay and Sanket Goel

3D printing, an additive manufacturing technique, is an advanced fabrication method used for developing complex structures with the utmost precision economically. They have a wide variety of applications in the fields of energy, biotechnology, microfluidics, and electronics. Of the various techniques, fused deposition modeling (FDM) has been considered extensively because of its unique process and simplicity. Electrochemical energy storage and harvesting devices will benefit from 3D printing technologies, paving the way for cheaper designs and fabrication in large scale production. These technical features are well suited for biofuel cells for implantable biomedical applications where cost and production have always been issues. Here, a brief discussion of the role of 3D printing in microfluidic enzymatic biofuel cells is provided, along with recent related work in this emerging research area.

13.1 Introduction

The ever increasing demand for clean and efficient energy production techniques has paved way for the use of renewable sources [1]. Rapid population growth and declining environmental health have forced policy makers and researchers to find alternative energy sources which are environmentally friendly and economically viable. Currently, several alternative renewable sources, such as solar energy, biomass, and fuel cells, are promising to generate energy, of which fuel cells have gained considerable attention [2–4]. Fuel cells are electrochemical devices working on the principle of an electrocatalytic redox reaction to generate power from a wide range of resources. They are classified based on the electrolytes and catalysts used [5]. Although these conventional fuel cells possess high energy conversion, efficiency, and

doi:10.1088/978-0-7503-3439-6ch13 © IOP Publishing Ltd 2021

significant power density, they possess major technical challenges such as high cost, bulkiness, and scarcity of noble metals [6].

An alternative for conventional fuel cells can be produced by making use of biological molecules as environmentally friendly catalyst materials. The devices harnessing enzymes or microorganisms as catalysts for energy harvesting by biochemical reactions are called biofuel cells (BFCs) [7]. The main advantages of BFCs are their ability to operate at low temperatures (20 °C–40 °C) and in biological pH conditions, and the option to use different enzymes and microorganisms which offers diversity and specificity [8]. Enzyme based biofuel cells (EBFCs) have gained more attention due to their fuel and electrode flexibility, miniaturization, and wide array of biomedical applications. Further, EBFCs can be classified into two types based on the mediators used for enhancing bioelectrocatalysis [9]. In mediated electron transfer (MET), the presence of mediators helps in faster electron transfer between the electrode and enzyme, whereas in the direct electron transfer (DET) method mediators are not required. Both transfer modes are used widely, although MET suffers from limitations such as toxic mediators and high ohmic losses. Recently, DET has been gaining attention but still has to be improved with advanced electrode materials for enhanced power output [10].

Electrode fabrication and device assembly are the two main components which influence the achievement of successful electrochemical energy storage devices [11]. Among the available scientific technologies, additive manufacturing is considered prominently for advanced fabrication of miniaturized devices. 3D printing is an example of the advanced technologies to realize complex three-dimensional objects [12]. In comparison with traditional methods, 3D printing is a rapid prototyping method which has unique advantages for electrochemical storage devices. These include simple production and modeling of electrodes and other components in specific configurations, direct development of desired architectures with precise chemistry, calculated shapes, [13] and organized porosity, and manufacturing of complete miniaturized devices to allow on-chip optimization of parameters [14].

Several advanced 3D printing technologies have been developed. They can be divided into four main categories based on their mechanism: (i) photo polymerization; (ii) extrusion; (iii) powder based; and (iv) lamination. Fused deposition modeling (FDM) is the most commonly preferred additive manufacturing method as it is capable of producing a large number of functional parts and devices neatly without complex machinery and procedures [15]. A variety of materials are available for FDM processes such as acrylonitrile butadiene styrene (ABS), polycarbonate (PC), and PCABS blends. The concept of developing enzyme based biofuel cells using additive manufacturing technology arose recently and is gaining a lot of attention due to its unique properties and mass production capabilities. Table 13.1 summarizes some of the recent EBFCs developed using various fabrication techniques.

13.2 Experimental details

13.2.1 Materials, reagents, and supplies

a. *3D printable filaments*: The acrylonitrile butadiene styrene (ABS) transparent filament (1.75 mm diameter) for the microchannel fabrication was procured

Table 13.1. Summary of different fabrication technologies.

#	Device fabrication technique	Electrode material	Power density (μW cm^{-2})	Comment	Reference
1	Photolithography	Printed circuit board/Ag	3.6	Time consuming process	[16]
2	Soft lithography	Graphene oxide	2.5	Costly equipment	[17]
3	Paper based	Grade 1 paper	100	Disposable and fragile devices	[18]
4	Xurography	Toray paper	125	Requires mold fabrication	[19]
5	3D printing	3D printed conductive graphene	4.15	Mass production and cost-effective method	[20]

from Matter Hackers, USA. The conductive graphene/polylactic acid (PLA) filament (1.75 mm diameter) for the bioelectrode fabrication was acquired from Black Magic 3D, USA.

b. *Pencil graphite (PG) lead as bioelectrodes*: Commercially available drawing pencils (Apsara brand) of B and 5H grades were procured from a stationary shop.
c. *Enzymes*: Glucose oxidase (GOx) (26 820 units/g solid) from *Aspergillus niger* and laccase from (24 U mg^{-1}) *Trametes versicolor* from Sigma Aldrich, India.
d. *Electron transfer mediator*: The anodic mediator P-benzoquinone (PBQ) and the cathodic mediator 2, 2-azino-bis (3-ethylbenzothiazoline-6- sulphonic acid) (ABTS) from Sigma Aldrich, India.
e. *Anolyte*: D-(+)-glucose from Sigma Aldrich, India
f. *Scientific grade materials and chemicals*: Multiwalled carbon nanotubes (MWCNTs), dimethylformamide (DMF), dimethyl sulfoxide (DMSO), 1-ethyl-3-(3-dimethylaminopropyl) carbodiimide (EDC), N-hydroxy succamide (NHS), monosodium phosphate (NaH_2PO_4), disodium phosphate (Na_2HPO_4), sodium hydroxide (NaOH), phosphoric acid (H_3PO_4), ethanol, and acetone were obtained from SRL, India.
g. *General material, apparatus, and instruments*: 2 ml centrifuge tubes, 100 ml beakers, micropipettes (20 μl and 1000 μl) and their corresponding tips, oxygen and nitrogen gas cylinders, weighing scales, probe sonication, hot air oven, pH meter, 3D printer, and potentiostat/galvanostat (Biologic Instruments).

13.2.2 Preparation of chemicals

1. Ultra-pure MilliQ (18.2 MΩ.cm) water was used for making all the standard chemicals. All the potentiometric electrochemical analyses were accomplished in PBS solution (0.1 M) with optimized pH 7 and 5 for the anode and cathode, respectively.

2. The chemicals Na_2HPO_4 and NaH_2PO_4 were used to prepare 0.1 M PBS solution, where 0.5 M of H_3PO_4 and NaOH solutions were used to adjust the pH to 7 and 5.
3. Homogeneous MWCNT solution was prepared by dispersing MWCNTs (1 mg ml^{-1}) in DMSO solvent via probe sonication for 30 min
4. The linker solution EDC/NHS (1 mg ml^{-1}) was prepared to activate the −COOH groups present in MWCNTs and the interconnection of enzymes on the bioelectrode surface.
5. Enzyme solution from GOx and laccase was prepared by adding a proportion of 5 mg ml^{-1} to 0.1 M PBS at pH 7.0 and pH 5.0, respectively.
6. The glucose stock (0.5 M) was prepared 24 hours before usage to allow D-glucose muta-rotation. All of the solutions were freshly made prior to the experiment, and the analyses were performed at room temperature.
7. The preparation of anolyte was performed in PBS (0.1 M, pH 7), where optimal D-glucose and anodic redox mediator P-benzoquinone (1 mM) were used, which was further subjected to nitrogen gas purging for 10 min to remove the excessive oxygen.
8. In a similar way, the preparation of the catholyte was achieved in 0.1 M PBS with pH 5, where the cathodic redox mediator ABTS (1 mM) was used, which was further subjected to oxygen gas purging for 10 min, appropriate for an efficient laccase mechanism.

13.2.3 Characterization and fabrication equipment

a. *Fused deposition modeling (FDM) based 3D printer.*

   A bench-top FDM technology-based 3D printer (Creator Pro, USA) was used to fabricate the microfluidic device and electrodes, by using ABS and composite graphene/PLA filament. 3D modeling software (Fusion 360) was used to create the basic virtual design of the microchannel and electrodes which was further sliced using 3D printer compatible FlashPrint software prior to final 3D printing.

b. *Electron microscope.*

   Using energy dispersive spectroscopy (EDS) the bioelectrodes were scanned (Apreo Scanning Electron Microscope (SEM) from Thermo Fisher Scientific) to understand the morphological characteristics of the bioelectrodes. Thus the analysis of the structural differences will explain the conducting aspect of the enzyme based biofuel cell.

c. *Galvanostat/potentiostat.*

   The potentiometric electrochemical analyses were accomplished using a potentiostat (SP-150 Bio-Logic France), controlled via EC-Lab software. The standard three-electrode arrangement was used to evaluate the electrochemical analysis while the two-electrode configuration was used to evaluate the polarization performance. The responses of the fabricated 3D printed electrodes and the prepared pencil graphite electrodes (PGEs) were assessed using the three-electrode arrangement, while the complete polarization study of the 3D printed microfluidic EBFC was evaluated using the two-electrode configuration. During the experimentation Ag/AgCl (3 M NaCl) and platinum wire were used as the reference electrode and counter electrode, respectively, while the prepared 3D printed and PGE were used as working electrodes.

13.2.4 3D printed bioelectrode fabrication and preparation

Conductive graphene/PLA filament was used to produce 3D printed graphene (3DPG) bioelectrodes. The detailed fabrication process and chemical preparation steps are explained below:

1. Prior to fabrication, the virtual designs of the bioelectrodes were created in CAD software and then saved in (.stl) format (figure 13.1(a)).
2. Thereafter, the materials were assigned for the design. An appropriate temperature was decided based on the filament material, and also custom based changes could be achieved by varying the temperature of the extruder of the 3D printer.
3. The file was saved in the '.x3g' format which was compatible with the 3D printer (figure 13.1(c)).
4. Once the design was finalized, it was transferred to the 3D printer using the GUI on the printer.
5. The electrode dimensions were selected based on the printer resolution and device requirements. Keeping this in mind, the electrode dimensions of 35 mm × 1.5 mm × 1.5 mm in length, width, and height were selected (figure 13.1(d)).
6. To improve the surface morphology of the electrodes, the 3D printed electrodes were treated with DMF solvent (figure 13.1(e)).

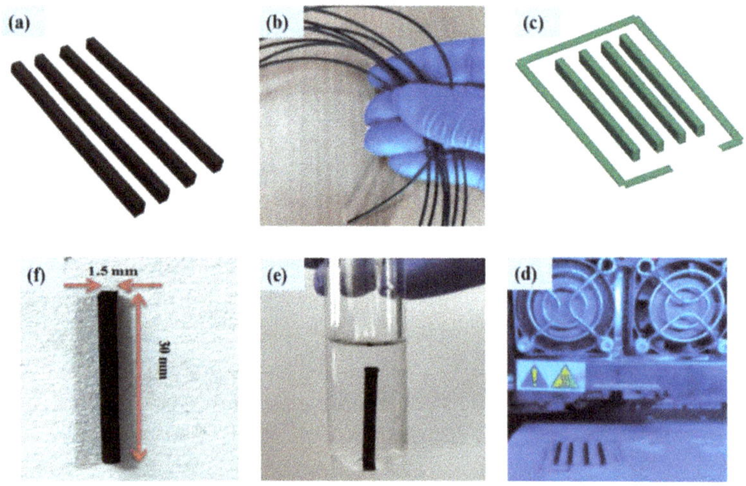

Figure 13.1. Digital images of the 3DPG electrode fabrication to immobilization process: (a) a sketch of the 3DPG electrode, (b) commercially available composite graphene/PLA filament, (c) sliced image of the 3D printed design, (d) fabricated 3DPG electrodes, (e) DMF treatment, and (f) final structure of the 3DPG electrode. (Reproduced with permission from [21]. Copyright 2019 IEEE.)

7. After performing this treatment for 10 min, they were washed with ethanol and left to air dry for 24 h at the ambient temperature [20]
8. Subsequently, the treated electrodes were again submerged in the EDC/NHS solution for better enzyme attachment over the electrode surface.
9. Finally, the DMF treated 3D printed bioelectrodes were placed deep in GOx and laccase enzyme solutions to develop them as bioanodes and biocathodes, respectively.
10. This solution was prepared by dissolving 5 mg of the enzymes (GOx and laccase) in 1 ml of 0.1 M, PBS with pH 7 and 5, and leaving it at ambient temperature for 2 h.

13.2.5 Preparation of pencil graphite bioelectrodes

The use of commercially available PG leads as bioelectrodes for biofuel cell applications has been explored previously by our group [22–25]. In this work, a rigorous study was conducted where almost all grades of pencil leads were tested and the optimal grades of pencil were utilized to prepared the bioanode and biocathode. The preparation steps for PGEs are explained below:

1. Various process steps, from the pencil peeling to cleaning process, have been taken from the previously published work by our group (figures 13.2(a) and (b)) [23].
2. After the cleaning process, the PG was immersed in the MWCNT suspension for molecular bonding on the electrode surface by a dip coating procedure

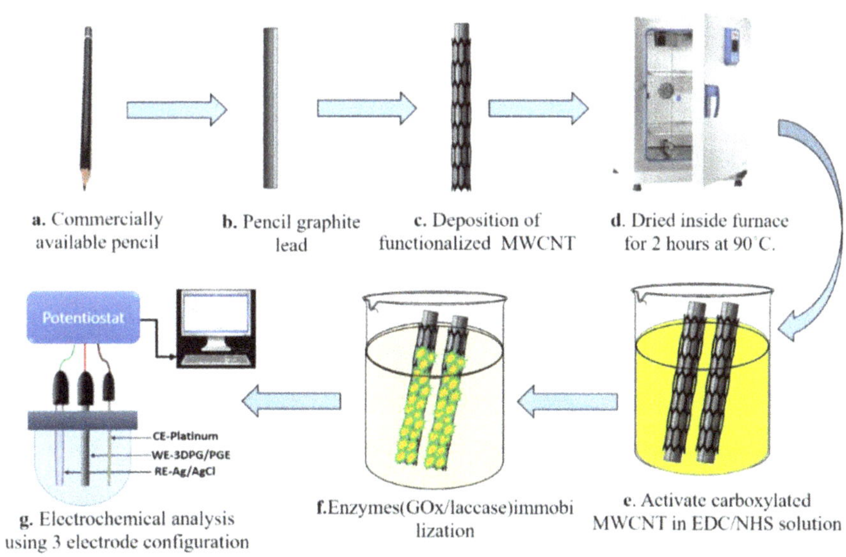

Figure 13.2. Stepwise schematic representation of the fabricated PGE: (a) commercially available drawing pencil, (b) peeling out the PG lead, (c) deposition of MWCNTs over the PGE surface, (d) drying the deposited MWCNTs inside the oven, (e) activation of carboxylic groups using EDC/NHS solution, (f) enzyme immobilization, and (g) an electrochemical study using a three-electrode configuration.

and placed inside the oven for 2 h at 90 °C. This process was repeated four times in order to obtain a uniform MWCNT coating on the PGE surface (figure 13.2(c)).

3. Furthermore, the MWCNT coated PGEs were dipped in an EDC/NHS linker solution for activation of the carboxylic groups to improve the interconnection of enzymes over the PG surface (figure 13.2(e)).
4. In order to realize an enzymatic bioanode, the PGEs was further immobilized with a biocatalyst, GOx enzyme, and left for 2 h to dry at room temperature. In a similar manner, the enzymatic biocathode was prepared by immobilizing laccase enzymes at room temperature (figure 13.2(f)).
5. The GOx and laccase enzyme solutions were prepared with a concentration of 5 mg ml^{-1} in PBS (0.1 M) with pH 7 and 5.
6. Finally, the immobilized bioelectrodes were cleaned with PBS solution to take away unbound enzymes, which helps in providing a better active surface area of the electrode during experimentation (figure 13.2 (g)).

13.2.6 Design and fabrication of a 3D printed microchannel

A novel, cost-effective, and rapidly prototyped plug-and-play type 3D printed microchannel was manufactured using a dual extruder 3D printer. Detailed sequential steps consisting of the CAD design and fabrication images are shown in figure 13.3. The complete design of the Y-shaped microchannel was created in two steps where the electrodes were inserted after the modification (figure 13.3).

1. Each separate part comprised one opening (inlet) for electrolyte streaming on the top side and one small opening on the bottom side for holding the electrodes (figures 13.3(c) and (d)).
2. The bioelectrodes were placed over half of the microchannel to make the process of enzyme immobilization simpler and easier.
3. After joining both parts together by clipping, the complete Y-shaped EBFC was realized with two inlets for providing the anolyte and catholyte, and a single outlet for collecting the consumed electrolyte.

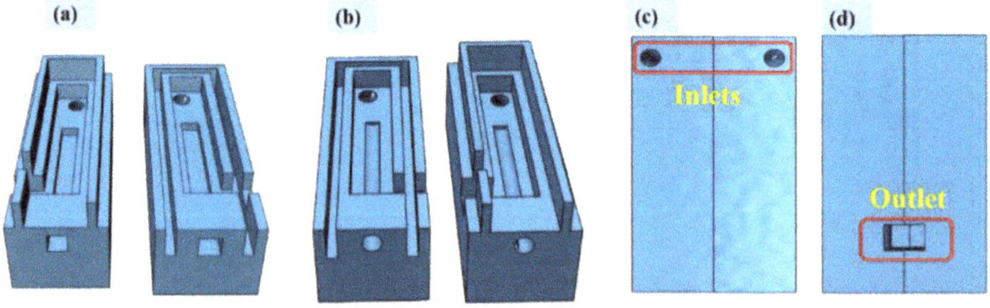

Figure 13.3. The individual halves of the 3D printed microchannel: (a) 3D printed microchannel for 3DPG bioelectrodes, (b) 3D printed microchannel for PG bioelectrodes, (c) top view with two inlets, and (d) bottom view with one outlet.

4. During the fabrication, suitable microchannel dimensions, such as 25 mm (length) × 4 mm (width) × 4 mm (height), were chosen and the transparent ABS material was used for fabrication.

13.2.7 Integration of 3DPG and PGE based 3D printed EBFCs

1. After completing the individual voltammetric studies of the 3D printed and PGE bioelectrodes, the modified 3DPG and PG bioelectrodes were inserted into two individual microchannels.
2. The whole fuel cell was assembled by clipping the two distinct parts together where the two electrodes were 1 mm apart, as shown in figures 13.4(b) and (d).
3. The fully integrated 3DPG and PGE based 3D printed EBFC contained one combined outlet and two inlets, where the fuel and oxidant were injected downstream into the microchannel using two peristaltic pumps.
4. The entire stepwise integration process of the 3DPG and PGE based 3D printed EBFC and the experimental set-up are shown in figures 13.4(b) and (d).

13.2.8 Electrochemical analysis

1. In the potentiometric electrochemical and polarization analysis, electrochemical impedance spectroscopy (EIS), linear sweep voltammetry (LSV), and cyclic voltammetry (CV) were performed.
2. This study evaluates the electrocatalytic activity of enzymes, the reaction kinetics, and the maximum current and power generation ability of the EBFC.
3. These studies were conducted in a conventional three-electrode set-up (5 ml capacity) with the 3D printed and PGEs serve as the working electrodes, while Ag/AgCl and platinum were the reference and counter electrodes, respectively.

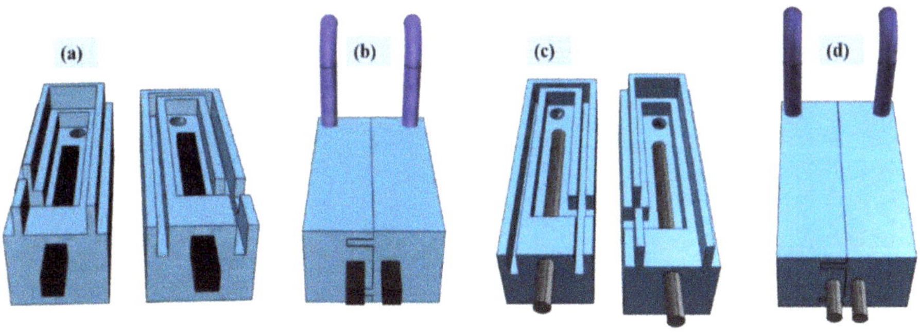

Figure 13.4. 3D schematic illustration of the 3DPG and PGE based EBFC: (a) the individual 3D printed half microchannel with 3DPG bioelectrodes, (b) the integrated 3DPG based 3D printed EBFC, (c) individual 3D printed half microchannels with PGE bioelectrodes, and (d) the PGE based 3D printed EBFC.

4. All potentiometric electrochemical experiments were performed at a scan rate of 10 mV s^{-1} at ambient temperature.
5. Likewise, the polarization and stability studies of the whole integrated 3D printed EBFC were accomplished using the two-electrode configuration.

13.3 Results and discussion

13.3.1 Morphological analysis

Scanning electron microscopy (SEM) was carried out to investigate the morphological appearance of the prepared bioelectrodes. This study helps in understanding the changes occurring on the surface of the electrode due to the modifications performed, which is a significant parameter in the electron–electrode transfer kinetics. Figure 13.5(a) shows a bare PGE with a smaller surface area. The reactive area of the PGE was increased by the surface modification with MWCNTs, which is visible in figure 13.5(b). This modification provided an increased active area which was useful for the immobilization of the biocatalysts for the preparation of the bioelectrodes. It can be understood clearly that there is a uniform distribution of MWCNTs on the PGE, thereby making the modified PGE more porous and electrically conductive. Figures 13.5(c) and (d) show the uniform and distributed

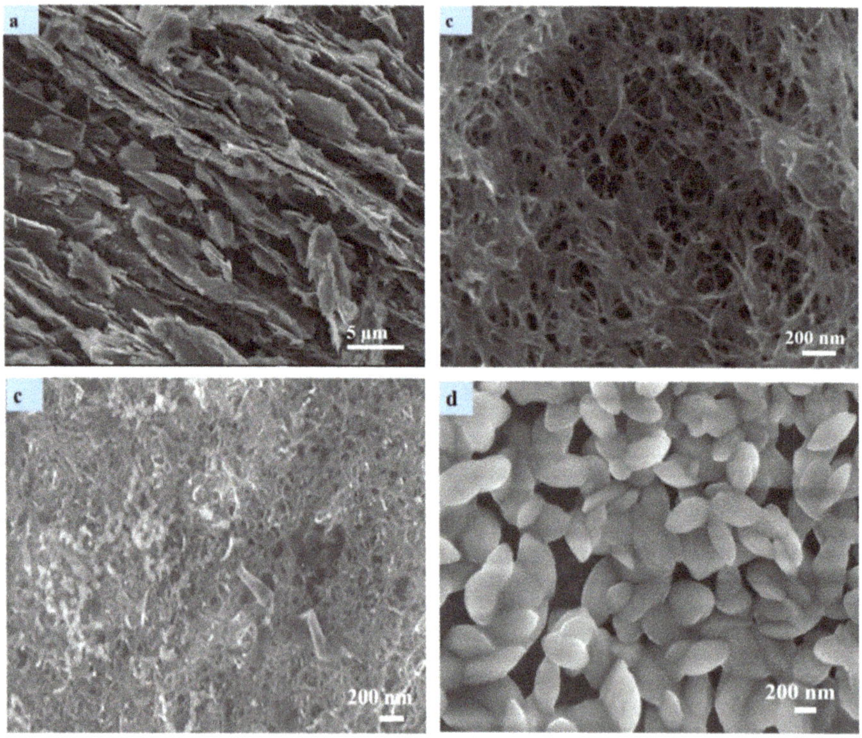

Figure 13.5. SEM images of (a) a bare pencil graphite electrode, (b) a MWCNT modified PGE, (c) a laccase modified PGE, and (d) a GOx modified PGE. Reproduced with permission from [25]. Copyright 2019 IEEE.

layer immobilization of enzymes on the MWCNT modified PGE surface. The greater availability of reactive sites along with the porous MWCNT network created a biocompatible microenvironment for enzyme modification with the probability of enhanced electron transfer with longevity and stability of the bioelectrode.

Similarly, the 3DPG electrodes were also studied for changes in their surface after chemical modification, which is clearly evident from the SEM micrographs at a magnification of 1 μm. Figure 13.6(a) shows the 3DPG electrodes before DMF treatment, where the surface was plain and covered with the PLA material. Figure 13.6(b) presents micrographs after the DMF treatment, where the electrode surface is composed of fibrous materials. Such a porous structure will provide active sites for enzyme immobilization. This proves that the DMF has reacted with the 3DPG and removed the PLA material, thereby making the electrode an electrically conductive substrate. Figures 13.6(c) and (d) show efficient immobilization of the enzymes on the chemically modified 3DPG electrodes. To confirm the elemental composition of the surface of the 3DPG electrodes, EDS analysis was performed, as shown in table 13.2. It was found that there was a visible change in the carbon and oxygen content after the DMF treatment, confirming the increased surface area in the treated electrode compared to the untreated electrode.

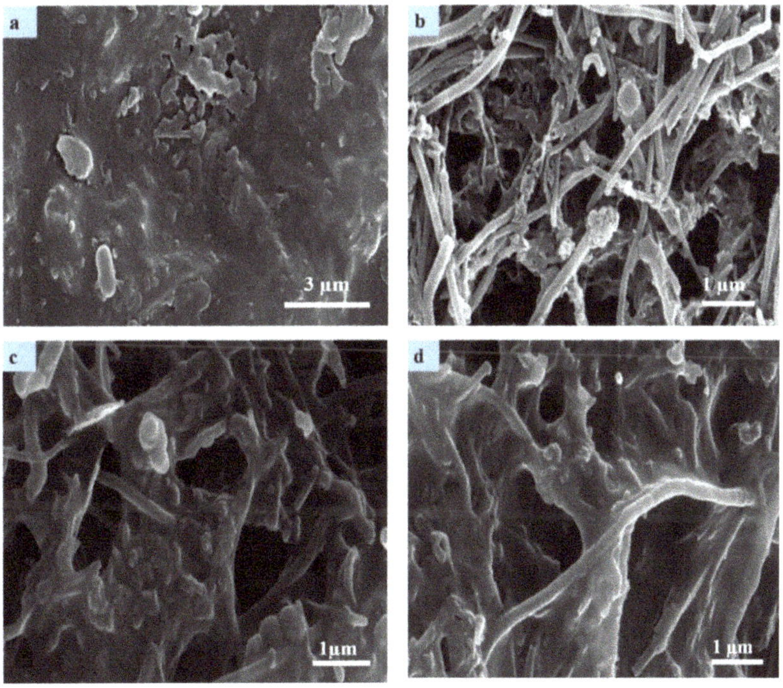

Figure 13.6. SEM images of 3DPG bioelectrodes (a) before DMF treatment, (b) after DMF treatment, (c) modified with GOx, and (d) modified with laccase. (Reproduced with permission from [20]. Copyright 2009 IEEE.)

Table 13.2. Elemental composition of the 3DPG bioelectrodes (3DPG-BE) before and after DMF treatment.

Element	3DPG-BE Wt.%	3DPG-BE/DMF Wt.%	3DPG-BE/DMF/GOx Wt.%	3DPG-BE/DMF/laccase Wt.%
C	65.00	82.14	44	48.52
O	29.99	14.18	34.03	37.99
Ti	1.72	3.14	1.58	1.38
Fe	0.24	0.55	0.23	0.17
Na	0.68		11.07	6.54
P			8.42	4.93
Al	0.24			
Si	0.28			
Cl	0.69			0.46
K	0.57			
Ca	0.59			

13.3.2 Optimization of fuel concentration

The effect of fuel concentration is an important parameter to be studied during the preparation of a bioelectrode. A higher concentration of glucose would lead to internal cell resistance, thereby leading to increased ohmic losses, which has an effect on fuel cell efficiency. Therefore, the optimization of the fuel concentration for the PG and 3DPG electrodes was performed using LSV at a voltage of 1 V and a scan rate of 10 mV s^{-1} for a glucose concentration range of 10 mM–60 mM. From figures 13.7(a) and (b) it was found that the optimum glucose concentrations for the PG and 3DPG electrodes were 100 mM and 40 mM, respectively, which were used in further electrochemical bioelectrode analysis. When the concentration was increased more than the optimum value, there was a decrease in the performance due to the increase in mass transfer resistance [26, 27].

13.3.3 Bioanode characterization

The bio-electrocatalytic behavior of the enzymes deposited on the electrodes in the presence and absence of fuel were studied. CV of the bioelectrodes was performed in the potential range (−1 V, 1 V) at a 10 mV s^{-1} scan rate in 1 mM PBQ solution (0.1 M, pH 7) containing 40 mM glucose as the fuel, with nitrogen saturated solution [28]. PGEs of different grades, such as 5H and B, were studied for their effective bioanode utilization. It can be understood from figure 13.8(a) that the optimized B grade PGE produced a high current density compared to the regular HB grade PG in the presence of glucose. In the case of the B grade PGE, an oxidation peak was found at a voltage of 0.3 V corresponding to the oxidation of glucose with an electric density of 5.1 mA cm^{-2}. The bio-catalytic activity of the 3DPG bioanode was studied before and after DMF treatment in the presence of the fuel. From figure 13.8(b) it is clearly evident that the chemical modification of the 3DPG electrode played a prominent role in the enhancement of the conducting property. In the presence of glucose the 3DPG bioelectrode showed no visible oxidation peaks,

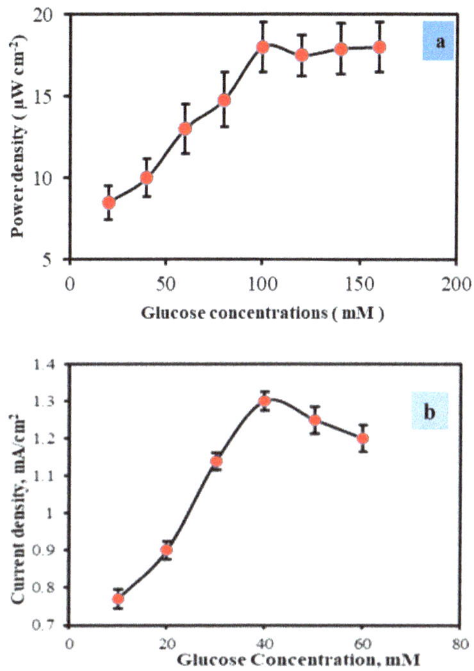

Figure 13.7. Optimization of glucose concentration for (a) PG based bioelectrodes and (b) 3DPG based bioelectrodes. ((a) Reproduced with permission from [15]. Copyright 2019 Elsevier. (b) Reproduced with permission from [20]. Copyright 2019 IEEE.)

whereas the DMF treated 3DPG showed a prominent oxidation peak at 0.48 V corresponding to the glucose oxidation. This confirms that the DMF treatment provided a highly porous surface for entrapment of the enzymes, providing enhanced electrochemical behavior.

13.3.4 Effect of the scan rate

A study was performed to understand the type of reaction occurring in a fuel cell system and the influence of the enzyme mechanism for obtaining optimal performance. A similar CV technique was achieved at various scan rates (10–100 mV s^{-1}) in the presence of 1 mM PBQ as the mediator in the PBS solution (0.1 M, pH 7) with saturated nitrogen and optimum glucose concentrations [18]. As shown in figure 13.9(b), the scan rate results reveal that the maximum current densities of the oxidation reaction of the fabricated PGE based bioanodes increased linearly with scan rate. This confirmed that the obtained oxidation peaks were due to immobilized enzymes on the PG surface of the electrode improving the controlled process at the surface of the electrode. This typical bio-electrochemical redox reaction signifies that the immobilized GOx is exhibiting quasi-reversible electron transfer phenomena on the MWCNT modified PG surface. Similarly, the fabricated 3D printed graphene based bioanodes were also tested for their electrochemical

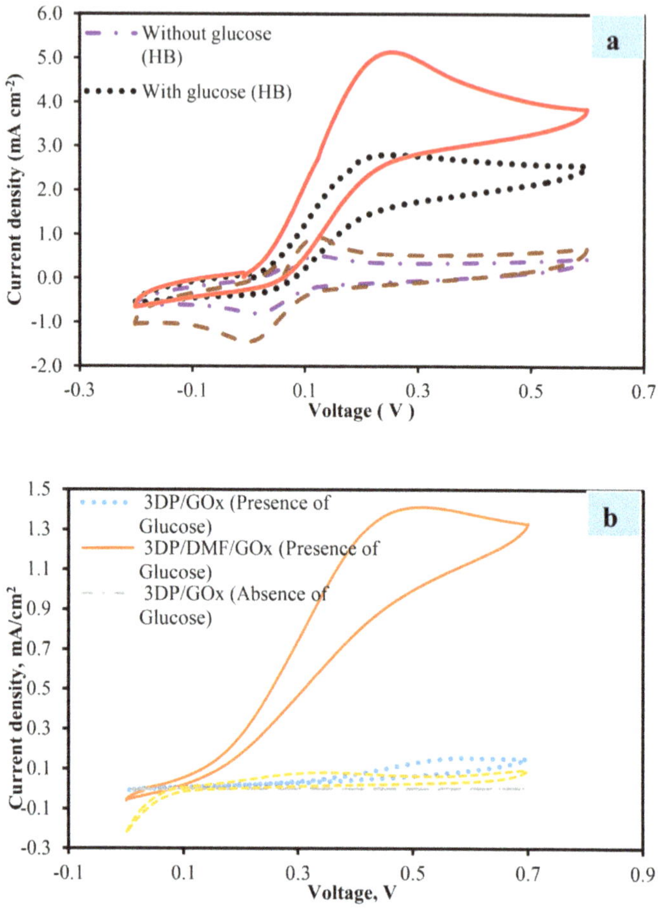

Figure 13.8. Bioanode characterization of (a) PG based bioelectrodes and (b) 3DPG based bioelectrodes with 1 mM PBQ in PBS (0.1 M, pH 7) solution. ((a) Reproduced with permission from [15]. Copyright 2019 Elsevier. (b) Reproduced with permission from [20]. Copyright 2019 IEEE.)

behavior. It was found the peak oxidation current increased with the scan rate, which shows that it is a diffusion-controlled process (as shown in figure 13.9(a)).

13.3.5 Biocathode characterization

The oxygen reduction reaction occurring at the biocathode was studied using the PG and 3DPG electrodes using the CV technique. The CV analysis was carried out in 1 mM ABTS as the mediator in PBS solution (0.1 M, pH 5) with saturated oxygen.

The voltammetric characterization of the regular HB and 5H grade PGE biocathodes were studied to understand the electrochemical reduction reaction. As shown in figure 13.10(a), the optimized 5H grade PGE biocathode showed a higher reduction current density at 0.4 V over the regular HB grade in the presence of oxygen. A visible change in performance was observed in the presence of oxygen,

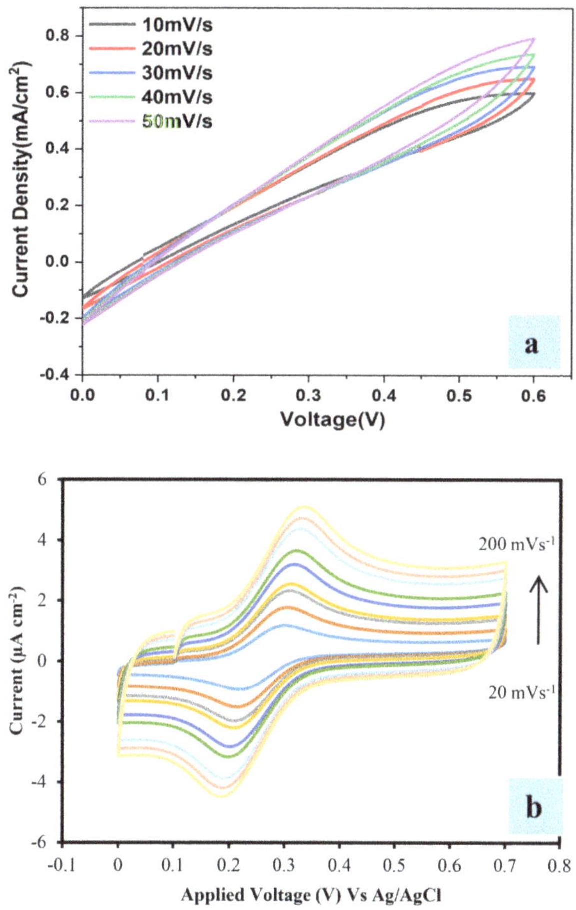

Figure 13.9. Scan rate study of (a) 3DPG based bioelectrodes and (b) PG based bioelectrodes using the CV technique. (Reproduced with permission from [25]. Copyright 2019 IEEE.)

which produced favorable conditions for the laccase biocatalyst. Also, based on the microscopic studies (SEM and EDS), it is clear that the 5H PGE has superior characteristics to regular HB. According to the comprehensive screening analysis of PGE it was found that the optimized grades (B, 5H) for the bioanode and biocathode, respectively, are best suited for miniaturized 3D-EBFCs.

In the case of the DMF treated 3DPG electrodes, the peak reduction current was noted in the presence or absence of oxygen, as shown in figure 13.10(b). In contrast, even after the enzyme immobilization the untreated biocathode did not show any redox peak in the presence of oxygen. CV analysis of the treated 3DPG biocathode showed an exceptionally good redox peak, revealing an efficient oxygen reduction with a current density of 0.216 mA cm^{-2} at 0.42 V. A reduction peak was found at the biocathode with nitrogen saturation, showing the importance of immobilized laccase enzyme for the electrochemical reduction process.

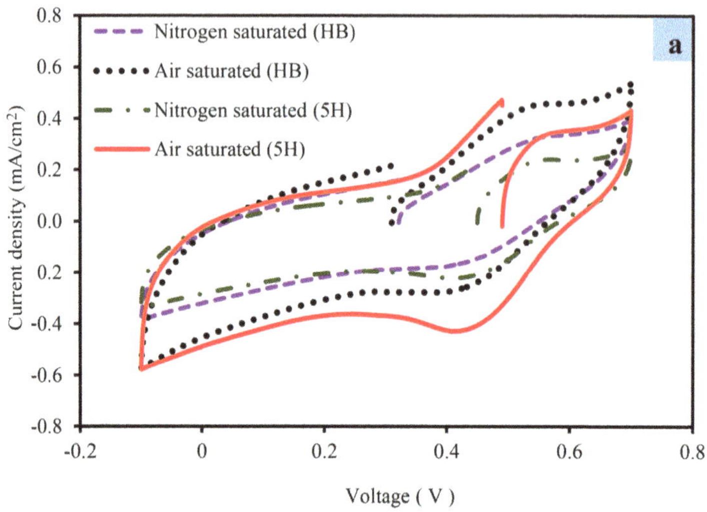

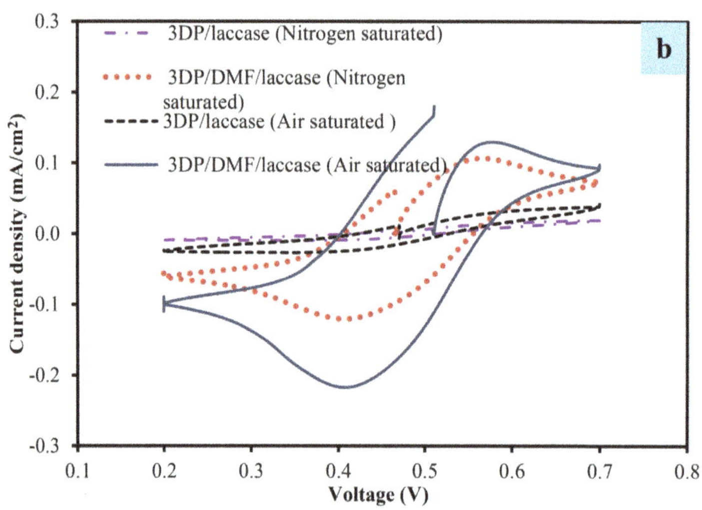

Figure 13.10. Biocathode characterization of (a) PG based bioelectrodes and (b) 3DPG based bioelectrodes with 1 mM ABTS in PBS (0.1 M, pH 5) solution. ((a) Reproduced with permission from [15]. Copyright 2019 Elsevier. (b) Reproduced with permission from [20]. Copyright 2019 IEEE.)

13.3.6 Electrochemical impedance measurements

To analyze the electrode kinetics of bioelectrodes, electrochemical impedance spectroscopy (EIS) was carried out. The electrochemical impedance spectrum consists of a straight line showing the diffusion-controlled phenomena and the curved portion gives the charge transfer resistance (R_{ct}) of the fabricated electrode [29]. Figure 13.11(b) shows the Nyquist plots obtained for PG bioelectrodes at 40 mM fuel concentration in PBS (0.1 M, pH 7.0). It is clearly evident from the EIS results that the unmodified and C-MWCNT modified PGE have undergone

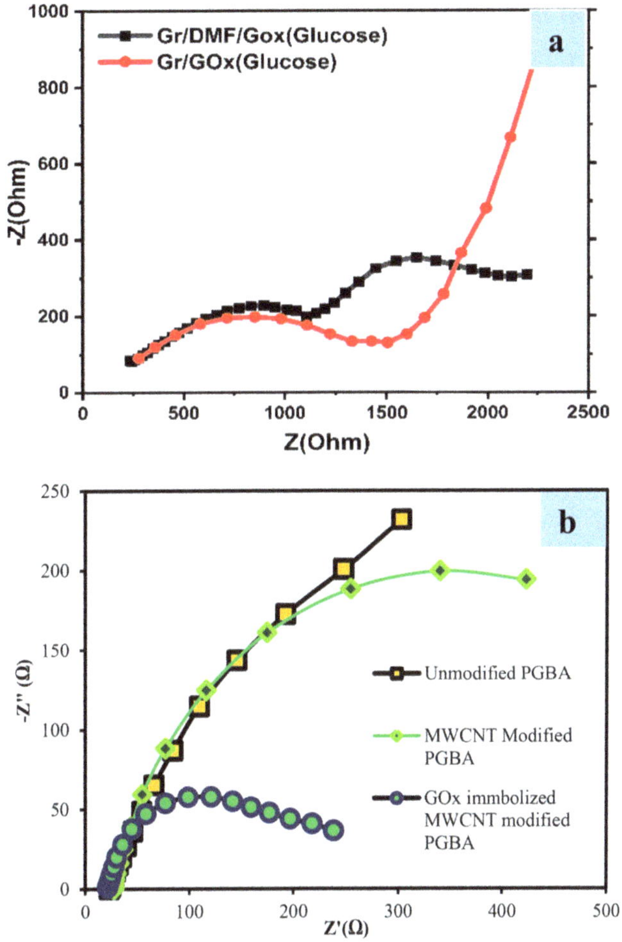

Figure 13.11. EIS study of (a) 3DPG based bioelectrodes and (b) PG based bioelectrodes with 1 mM PBQ in PBS (0.1 M, pH 7) solution. (Reproduced with permission from [25]. Copyright 2019 IEEE.)

diffusion processes, confirming the occurrence of electrochemical reactions [31]. A curve showing an reflecting R_{ct} of 180 Ω (figure 13.11(b)) was observed for the MWCNT modified PGE, indicating the efficient immobilization of GOx on the MWCNT porous network. This also revealed a low charge transfer resistance and uniform dispersion of the conductive MWCNT surface of the modified PGE. Similarly, the 3DPG bioelectrodes were also studied for their charge transfer resistance, as shown in figure 13.11(a). The bioelectrodes before and after treatment were tested in glucose solution (40 mM) containing 1 mM PBQ as the mediator. The R_{ct} was calculated using a Randles circuit and it was found that the untreated electrode delivered high resistance for charge transfer with an R_{ct} value of 7.7 MΩ. In the case of the treated bioelectrode, the electrode–electrolyte resistance was much lower (R_{ct} = 295.5 Ω), proving high conductivity and the significance of activation.

13.3.7 The effect of flow rate

Flow rate in a microfluidic device plays an essential role in device performance, fuel utilization, and ohmic losses. In a microfluidic device, the flow of the two electrolytes (anolyte and catholyte) is streamed based on the principle of co-laminar flow, with a Reynolds number less than 2000 and a very large Peclet number, facilitating the fuel and oxidant flow without mixing in the absence of a proton exchange membrane (PEM). The mixing of the electrolytes can be controlled by varying the rate of their flow into the fabricated device [19]. At higher flow rates, the amount of fuel consumed will be lower but the diffusion region occurs at an accelerated rate. In the case of lower flow rates, cross-over of fuel and bubble generation becomes a major limiting issue, which results in broadening the diffusion zone. Thus it is important to optimize the flow rate for a particular microfluidic device to obtain a high power density [30].

In the 3D printed EBFC device with bioelectrodes, the flow rate was changed in a sequential manner and a time interval of 2 h was allowed for the cell to become stable. The performance of the cell was recorded using the chronoamperometry (CA) technique and the flow rate which delivered a maximum power density was calculated. CA is an electrochemical technique where current is recorded as a function of time for an applied voltage. As shown in figures 13.12(a) and (b), the optimum flow rate was found to be 400 μL min^{-1} and 500 μL min^{-1} for the PG and 3DPG based EBFCs, respectively, which were used in the further study of the full biofuel cell.

13.3.8 Power performance of the biofuel cell

The polarization curves of the completely assembled fuel cell were investigated using the CA technique at optimized flow rates. To plot a polarization curve, the potential was stepped down at intervals from the maximum open circuit potential (OCP) to 0 V and the corresponding current values were measured. The fuel cell performance was continuously recorded by flushing the bioanode side with anolyte and the biocathode side with catholyte at optimized flow rates. The two-electrode configuration, as the reference and counter terminal, were merged together, and the working terminal of the potentiostat was connected to the bioanode and biocathode for recording the performance of the biofuel cell.

1. *PGE based 3D printed biofuel cell*: In this case, two categories of PG bioelectrodes were utilized in the 3D printed EBFC to compare the power output between them with optimized parameters. The first category consisted of both HB grade PG bioelectrodes, while in the second category the optimized (B, 5H) grades were used as the bioelectrodes. The anolyte and catholyte were flushed into the microfluidic device using a peristaltic pump at an improved flow rate of 400 μL min^{-1} at room temperature. A clear difference in power output was found between the two categories of PGE incorporated 3D printed EBFCs, as shown in figure 13.13. The optimized (B, 5H) PGE based 3D printed EBFC presented a high open circuit potential (OCP) of 0.433 V and delivered a peak power density of 18 μW cm^2 at a current density of 60 mA cm^2. In contrast, both bioelectrodes with HB grade PG EBFCs showed a decline in performance with an OCP of 0.210 V and a peak power density of

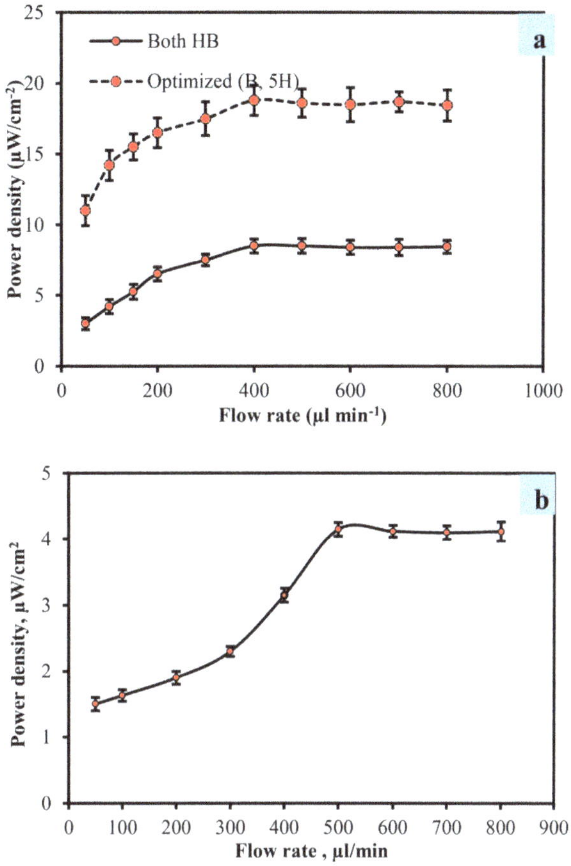

Figure 13.12. Flow rate effect of the (a) PGE based EBFC and (b) 3DPG electrode based EBFC. ((a) Reproduced with permission from [15]. Copyright 2019 Elsevier. (b) Reproduced with permission from [20]. Copyright 2019 IEEE.)

8.5 μW cm^2 at a current density of 56 mA cm^2 compared to the second category. These results prove that the performance was improved by 112% for the optimized (B, 5H) PGE based EBFC with an active surface area of 0.5 cm^2.

2. *3DPG bioelectrode based biofuel cell*: The 3DPG based EBFC was tested for its performance using the optimized parameters. 3DPG bioelectrodes were incorporated into the 3D printed microchannel and clipped together, ensuring an electrode separation of 1 mm. An optimized flow rate of 500 μL min^{-1} was maintained for anolyte and catholyte flow inside the 3D printed microfluidic platform. Figure 13.14 shows the polarization plot for the full cell incorporated with bioelectrodes, recorded at an optimized flow rate. The developed 3D printed EBFC delivered a maximum OCP of 0.315 V with a peak power density of 7.12 μW cm^{-2} with a reactive surface area of 0.4 cm^2. A comparative table indicating all the performance values of the PGE and 3DPG based EBFCs are given in table 13.3.

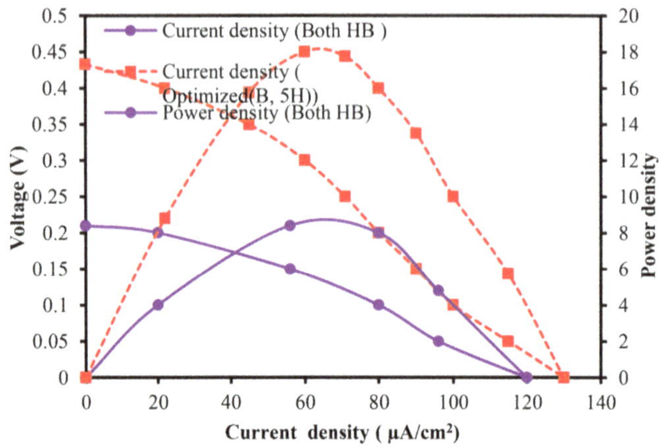

Figure 13.13. Polarization curves of the PGE based EBFC for two categories, HB bioelectrodes (purple dots), and optimized bioelectrodes (B as the bioanode and 5H as the biocathode; red squares). (Reproduced with permission from [15]. Copyright 2019 Elsevier.)

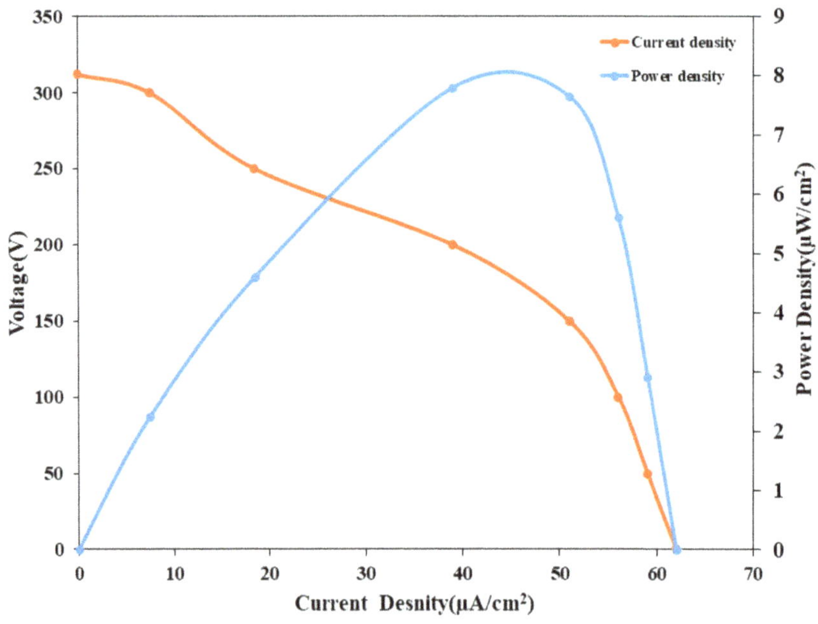

Figure 13.14. Polarization curves for the 3DPG bioelectrode based EBFC.

13.3.9 Stability study

This study will help in evaluating the shelf-life of the fabricated EBFCs under controlled conditions and their ability to recover its performance after storage. The complete biofuel cell was preserved at 4 °C until its use to maintain the activity of the enzymes. The stability of the microfluidic fuel cell was assessed by operating the cell every day for two hours under optimized conditions and the

Table 13.3. Comparison of PGE and 3DPG based EBFC performance.

Performance parameters	PGE based EBFC	3DPG based EBFC
Open circuit potential (V)	0.433	0.315
Maximum current density (mA cm^{-2})	60	62
Maximum power density (μW cm^{-2})	18	7.12

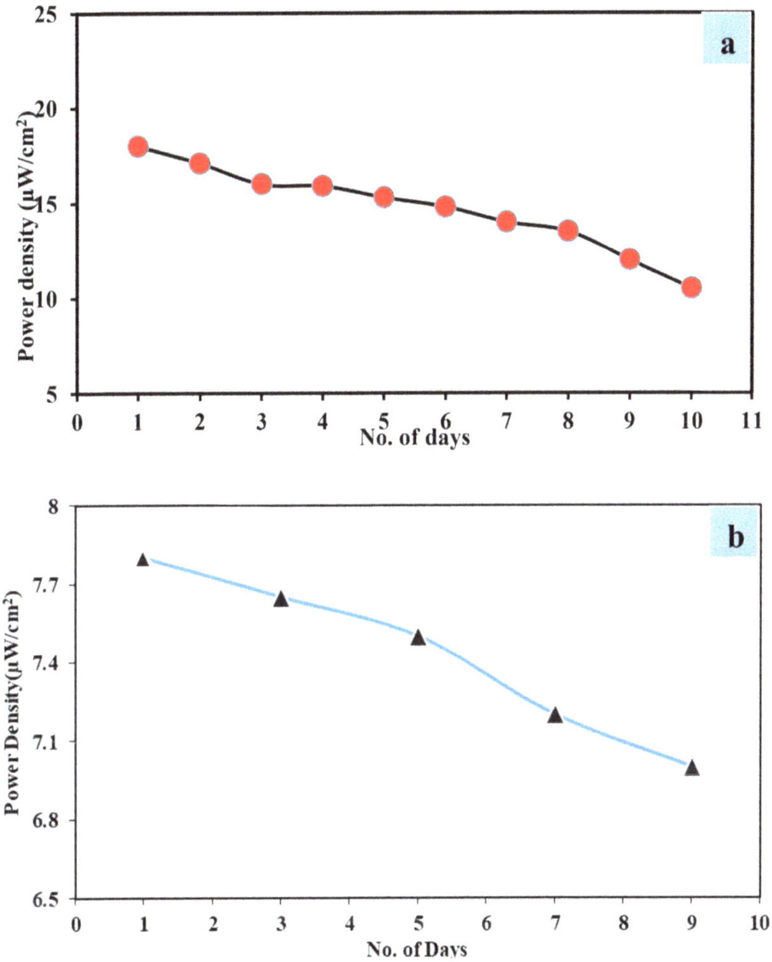

Figure 13.15. Stability plot for the (a) PGE based EBFC (optimized B, 5H) and (b) the 3DPG based EBFC. (Reproduced with permission from [15]. Copyright 2019 Elsevier.)

power output was plotted as a function of time (number of days). From figure 13.15 it can be seen that there was a minimal decrease in power output in the initial days, followed by a linear decrease of the performance in the PGE based EBFC. Similarly, the 3DPG EBFC was evaluated for its performance and it was found that more than half of the power output was recovered up to 10 days, after which it began to decline.

13.4 Conclusions

In this work, the viability and flexibility of fabricating 3D printed microfluidic devices has been discussed. In the case of the PGE based EBFC, it is clear that additive manufacturing can be used to develop a universal microfluidic device which could incorporate a variety of bioelectrodes for various applications. Also, the 3DPG based EBFC reveals that a complete EBFC can be fabricated using 3D printing with a customized design in a cost-effective and time-efficient manner. The stability studies prove that the devices are capable of retaining their performance for a considerably longer period in comparison to conventional electrodes. Further, an outstanding amount of power can be harvested for these 3D printed fuel cells by incorporation of a conductive nanomaterial during the filament manufacturing process itself and during the device fabrication steps. Thus the rapidly evolving 3D printing technology will find its way into biomedical, implantable, and portable applications in the very near future.

Acknowledgement

The authors thank the Central Analytical Lab, Birla Institute of Technology-Pilani, Hyderabad Campus, Hyderabad for their help with various characterizations.

References

[1] Zhu L, Qin C and Peng C 2018 Analysis on energy saving and emission reduction of clean energy technology in ports *IOP Conf. Ser.: Earth Environ. Sci.* **121** 1–5

[2] Li M, Bai Y, Zhang C, Song Y, Jiang S, Grouset D and Zhang M 2019 Review on the research of hydrogen storage system fast refueling in fuel cell vehicle *Int. J. Hydrogen Energy* **44** 10677–93

[3] Rewatkar P, Hitaishi V P, Lojou E and Goel S 2019 Enzymatic fuel cells in a microfluidic environment: status and opportunities. A mini review *Electrochem. Commun.* **107** 106533

[4] Enaganti P K, Dwivedi P K, Sudha R, Srivastava A K and Goel S 2020 Underwater characterization and monitoring of amorphous and monocrystalline solar cells in diverse water settings *IEEE Sensors J.* **20** 2730–7

[5] Abreu C, Nedellec Y, Ondel O, Buret F, Cosnier S, Le Goff A and Holzinger M 2018 Towards eco-friendly power sources: in series connected glucose biofuel cells power a disposable ovulation test *Sens. Actuators* B **277** 360–4

[6] Zebda A, Alcaraz J P, Vadgama P, Shleev S, Minteer S D, Boucher F, Cinquin P and Martin D K 2018 Challenges for successful implantation of biofuel cells *Bioelectrochemistry* **124** 57–72

[7] Yahiro A T, Lee S M and Kimble D O 1963 Bioelectrochemistry: I. Enzyme utilizing bio-fuel cell studies *Biochim. Biophys. Acta* **88** 375–83

[8] Patil M, Pawar P, Kerawall M A K and Goswami P 2016 Enzymatic biofuel cells—a progress review *J. Chem. Biol. Phys. Sci.* **6** 039–58

[9] Moreira F T C, Frasco M F and Peixoto L 2019 Enzymatic self-powered biosensing devices *Bioelectrochemical Interface Engineering* (New York: Wiley) pp 505–19

[10] Gonzalez-Solino C and Lorenzo M 2018 Enzymatic fuel cells: towards self-powered implantable and wearable diagnostics *Biosensors* **8** 11
[11] Foster C W, Down M P, Yan Z, Ji X, Rowley-Neale S J, Smith G C, Kelly P J and Banks C E 2017 3D printed graphene based energy storage devices *Sci. Rep.* **7** 42233
[12] Punpattanakul K, Kraduangdej S, Jiranusornkul N, Chiaranairungroj M, Pimpin A, Palaga T and Srituravanich W 2018 A novel patterning method for three-dimensional paper-based devices by using inkjet-printed water mask *Cellulose* **25** 2659–65
[13] Foo C Y, Lim H N, Mahdi M A and Wahid M H 2018 Three-dimensional printed electrode and its novel applications in electronic devices *Sci. Rep.* **8** 7399
[14] Parandoush P and Lin D 2017 A review on additive manufacturing of polymer-fiber composites *Compos. Struct.* **182** 36–53
[15] Rewatkar P, Bandapati M and Goel S 2019 Miniaturized additively manufactured co-laminar microfluidic glucose biofuel cell with optimized grade pencil bioelectrodes *Int. J. Hydrogen Energy* **44** 31434–44
[16] Mashayekhi Mazar F, Alijanianzadeh M, Molaei Rad A and Heydari P 2020 Power harvesting from physiological serum in microfluidic enzymatic biofuel cell *Microelectron. Eng.* **219** 111159
[17] Jayapriya U S, Rewatkar P and Goel S 2020 Miniaturized polymeric enzymatic biofuel cell with integrated microfluidic device and enhanced laser ablated bioelectrodes *Int. J. Hydrogen Energy* **46** 3183–92
[18] Rewatkar P and Goel S 2018 Paper-based membraneless co-laminar microfluidic glucose biofuel cell with MWCNT-fed bucky paper bioelectrodes *IEEE Trans. Nanobiosci.* **17** 374–9
[19] Renaud L, Selloum D and Tingry S 2015 Xurography for 2D and multi-level glucose/O_2 microfluidic biofuel cell *Microfluid. Nanofluid.* **18** 1407–16
[20] Rewatkar P and Goel S 2019a Next-generation 3D printed microfluidic membraneless enzymatic biofuel cell: cost-effective and rapid approach *IEEE Trans. Electron Devices* **66** 3628–35
[21] Rewatkar P and Goel S 2019b 3D printed bioelectrodes for enzymatic biofuel cell: simple, rapid, optimized and enhanced approach *IEEE Trans. Nanobiosci.* **19** 4–10
[22] Kashyap D, Kim C, Yeol S, Ho Y, Man G, Dwivedi P K, Sharma A and Goel S 2015 Multi walled carbon nanotube and polyaniline coated pencil graphite based bio-cathode for enzymatic biofuel cell *Int. J. Hydrogen Energy* **40** 9515–22
[23] Bandapati M, Dwivedi P K, Krishnamurthy B, Ho Y, Man G and Goel S 2017 Screening various pencil leads coated with MWCNT and PANI as enzymatic biofuel cell biocathode *Int. J. Hydrogen Energy* **42** 27220–9
[24] Bandapati M, Krishnamurthy B and Goel S 2019 Fully assembled membraneless glucose biofuel cell with MWCNT modified pencil graphite leads as novel bioelectrodes *IEEE Trans. Nanobiosci.* **18** 170–5
[25] Bandapati M, Rewatkar P, Krishnamurthy B and Goel S 2019 Functionalized and enhanced HB pencil graphite as bioanode for glucose–O_2 biofuel cell *IEEE Sensors J.* **19** 802–11
[26] Perveen R, Inamuddin U, Haque S, Nasar A, Asiri A M and Ashraf, G. M 2017 Electrocatalytic performance of chemically synthesized PIn–Au–SGO composite toward mediated biofuel cell anode *Sci. Rep.* **7** 1–10
[27] Khan H, Kim C M, Kim S Y, Goel S, Dwivedi P K, Sharma A, Kim Y H and Kim G M 2019 Fabrication of enzymatic biofuel cell with electrodes on both sides of microfluidic channel *Int. J. Precis. Eng. Manuf.-Green Technol.* **6** 511–20

[28] Rewatkar P, Bandapati M and Goel S 2018 Optimized bucky paper-based bioelectrodes for oxygen—glucose fed enzymatic biofuel cells *IEEE Sensors J.* **18** 5395–401

[29] Kashyap D, Dwivedi P K, Pandey J K, Kim Y H, Kim G M, Sharma A and Goel S 2014 Application of electrochemical impedance spectroscopy in bio-fuel cell characterization: a review *Int. J. Hydrogen Energy* **39** 20159–70

[30] González-Guerrero M J, Esquivel J P, Sánchez-Molas D, Godignon P, Muñoz F X, del Campo F J, Giroud F, Minteer S D and Sabaté N 2013 Membraneless glucose/O_2 microfluidic enzymatic biofuel cell using pyrolyzed photoresist film electrodes *Lab Chip* **13** 2972

Chapter 14

Development, simulation and characterization of a novel incontinence sensor system using 2D-printing technology with conductive polymer PEDOT:PSS

Ha-Duong Ngo, Manuel Bäuscher, Michael Schiffer, Dahiana Mojena Medinaa, José Luis Jorcano, Pablo Acedo and Raghied Mohammed Helmy Atta

14.1 Introduction

Today's sensors and microelectronics are bulky and rigid since they are made of semiconductors such as silicon. Unlike conventional microelectronics/nanoelectronics, the next-generation sensors and electronics focus on large area, low-cost applications and new form factors such as flexible and stretchable devices [1–4]. Electronic devices that are fabricated on plastic or rubbery substrates, such as flexible displays, electronic papers, smart packages, skin-like sensors, wearable electronics, implantable medical devices, and many others, could radically change people's perspectives on electronics [1–3].

The potential use of such technology for the development of an easily implementable incontinence sensor is enormous, because of 400 million people worldwide (4%–8% of the world's population) suffer from incontinence [5]. There are approximately 6–8 million people of all genders and ages in Germany alone. Elderly people in particular are affected and mostly women, two to four times more frequently than men, because of childbirth. Urge incontinence is one of the main reasons for referral to a nursing home, because elderly people become weaker and their mobility decreases [6]. The hygienic enhancement provided by a sensor is an advantage for care stuff and the person wearing the diaper, which enhances quality of life. It also increases the effectiveness of employees in nursing homes, because they no longer have to check every single bed in a certain time period, since the collected information could be transferred wirelessly to a monitor or a smartphone [7].

doi:10.1088/978-0-7503-3439-6ch14

© IOP Publishing Ltd 2021

Currently, a few solutions for this problem exist on the market. Pampers® offers a system with color change wetness indicators [8], but it is only available for children's diapers and uses a non-electric method. With the focus on electrical sensing, O'pro9® and Monit® developed promising sensor systems with external attachments to the diaper and a smart way to communicate via a smartphone as an alarm system for diaper change [9, 10]. However, these sensors only detect liquid in a certain area and are not implementable into the diaper manufacturing process. The most promising incontinence sensor system is from the Danish company Abena [11]. They manufacture a built-in diaper incontinence sensor with a wireless connection to a smartphone or computer for the support of caregivers and the monitoring of elderly people. At this point, the project is in the development and testing phase and is not yet on the market. If the tests are successful and market maturity is proven, it could be an alternative to the system developed in this chapter.

The idea in our work is to develop a smart, cheap, fast, and highly sensitive one-way-use sensor platform, easily implementable into the diaper manufacturing process, which is able to detect moisture changes and would serve as sensor platform to detect chemical properties or biomarkers in urine in the future. The most important requirement for the sensor system is the detection of the amount of liquid inside the diaper. The readout should give an electrical signal to a cloud server or caregiver for quick action. The materials inside the diaper must be biocompatible (ISO 10993), flexible, and stretchable. Also, the sensor should be integrated easily into a standard diaper manufacturing process for production at low total cost. The wireless interface must be compatible with standards in medical systems.

For the diaper application, a high realistic analytic model and a precise simulation model are developed to optimize sensor structures and obtain the highest capacitance changes for the later manufactured sensor. For the arrangement of the interdigital capacitor (IDC) sensors, the diaper is divided into different sensing zones, as seen in figure 14.1. Zones 1, 2, 4, and 5 are for detection in the middle part of the diaper, zone 3 for the front, and zone 6 for the rear area [7].

As the electrode material, the novel organic, biocompatible, and conductive screen printing ink PEDOT:PSS (poly(3,4-ethylenedioxythiophene):poly(styrene

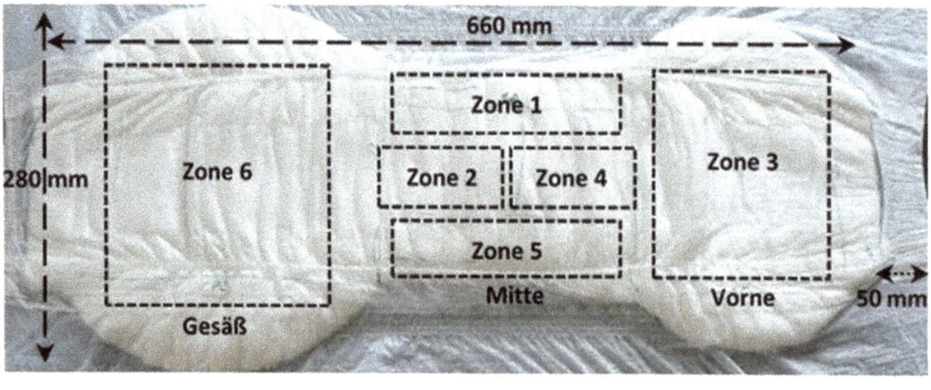

Figure 14.1. Zones for liquid detection in the diaper. Reproduced from [1] CC BY 4.0.

sulfonate)) EL-P5015 (from AGFA ORGACON, viscosity > 100 Pa·s) is used [1, 2, 12]. The specific material parameters for the PEDOT:PSS, the substrate material, and the absorber pad are investigated in the following section.

14.2 Material characterization for the FEM calculation

The determination of the electrical characteristics of a material is necessary, because of the definition of boundary conditions in numerical simulations and the dielectric parameters for an analytic model. The most valuable parameters are the permittivity of the dielectric materials and the electrical conductivity. Thus the measurement set-up and the results will be presented in the following.

14.2.1 Relative permittivity

For the definition of the relative permittivity of the materials used in this work, the measurement instrument Agilent® impedance/material analyzer E4991A, including the compatible Agilent® dielectric material test fixture 16453A, is used. The dielectric constant (complex permittivity) of the material under test is measured by impedance measurement in a parallel plate capacitor set-up.

First the materials need to be prepared for the measurement. It is important to evaluate the material thickness in the tolerance range of the device, with minimum to maximum thicknesses of 0.3 mm $\leqslant t \leqslant$ 3 mm. Also, the diameter of the material under test needs to be less than 15 mm [13].

Four different materials are prepared for the test. Table 14.1 shows the materials with their test thickness and the result of the measurement. In order to reach the minimum thickness for the test fixture, the materials are stacked to a specific thickness with multiple sheets. For every measurement the spring of the test fixture is lifted and lowered. The mean value of every measurement over the frequency is calculated and a mean of all ten measurements is computed. The results are shown in table 14.1. Liquids are not characterized because of no available test fixture [7].

Polypropylene (PP) and polyethylene (PE) are standard substrates for diapers. In this work, polyethylene terephthalate (PET) is used as substrate material, because of its good surface behavior for printing.

14.2.2 Electrical conductivity

The electric conductivity of the conductive material PEDOT:PSS is determined by measuring the resistance of a defined geometric structure of the material under test.

Table 14.1. Materials characterization results. Copyright 2020 IEEE. Reprinted, with permission, from [7].

Material	Thickness (μm)	Measured mean relative permittivity
PP foil	360	2.12
PE foil	420	1.62
PET foil	365	2.86
Absorber material	1150	2.09

The calculated sheet resistance $R_\square$ from the measured mean resistance is 230 $\Omega/\square$. This is comparable to the given sheet resistance of 190 $\Omega/\square$ from the material datasheet of PEDOT:PSS [12]. To obtain the conductivity the resistivity is calculated, with 200 nm as the printed layer thickness [1]. The conductivity is the reciprocal of the resistivity [13, 14]. The measured conductivity is 21.7 $\mathrm{s\ m^{-1}}$ and is suitable for the simulation model [7].

14.3 Experimental models

In this chapter, two different theoretical methods are derived and compared to a practical measurement of manufactured sensor structures at the end of this work. In the following, the two different experimental methods in the form of an analytic model and a simulation model are described.

14.3.1 Analytic model

To achieve a first model for dimensioning purposes of the IDC sensor systems, a deviation from Abu-Abed's [15] approach is made. Figure 14.2(a) shows schematics of an IDC structure from the top view and cross-section with important geometrical parameters.

The capacitances between the electrodes of the IDC structures C_g of printed thin sheets can be neglected, because of the high aspect ratio from the size of the gap to the thickness of the electrode ($w_g \gg t$). Therefore only the fringing capacitance C_f, shown in figure 14.2(b), counts. For the calculation of the fringe field, complex mathematical equations are necessary to obtain an accurate analytic model for interdigital capacitors. Various scientists deal with the accurate modeling of IDC structures. They all have a common approach, Palmer's method using complementary elliptic integrals [16, 17]. The elliptic factor comes from the elliptical field distribution from one electrode to the other, shown in figure 14.2(b). It is said to be complete if the angle of the fringe field is $\varphi = \frac{\pi}{2}$. The value K in the following equation is then the complete elliptic integral of the first order.

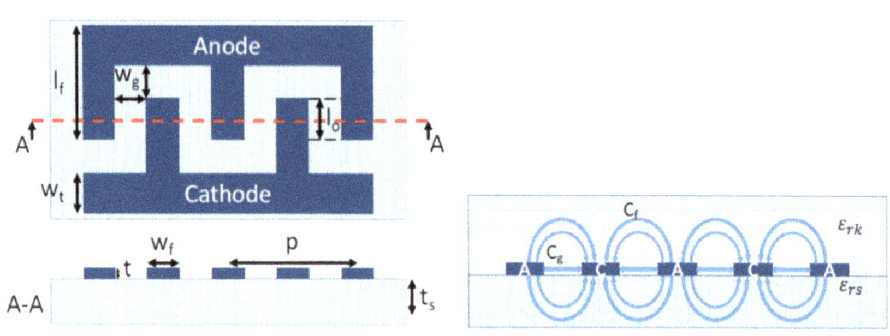

(a) (b)

Figure 14.2. (a) Schematic of an IDC structure. (b) Schematic of the fringe field in the IDC structures. Copyright 2020 IEEE. Reprinted, with permission, from [7].

$$K(k) = \int_0^{\frac{\pi}{2}} \frac{d\theta}{\sqrt{1 - k^2 \sin 2\theta}} = \int_0^1 \frac{dt}{\sqrt{(1 - t^2)(1 - k^2t^2)}}.$$

For simplification it can be expressed as a power series [16].

$$K(k) = \frac{\pi}{2} \sum_{n=0}^{\infty} \left(\frac{(2n)!}{2^{2n}(n!)^2} \right)^2 k^{2n}.$$

The cross-section of the interdigital capacitor seen in figure 14.2(b) shows the interdigital electrodes on a dielectric substrate with its relative permittivity ε_{rs} and the space above the electrodes with the relative permittivity ε_{rk} of the dielectric. Other symbols are taken from figure 14.2(a).

Abu-Abed from the University of Alabama expresses in his publication [15] the capacitance for a unit cell C_{UC} (only for the length of p) with the following equation:

$$C_{UC} = \varepsilon_0(\varepsilon_{rs} + \varepsilon_{rk}) \frac{K\left(\sqrt{1 - k^2}\right)}{K(k)} + 2\varepsilon_0\varepsilon_{rk}\frac{t}{w_g}$$

with $k = \frac{w_g}{p}$.

The total capacitance C_{Total} of a whole IDC structure can be calculated with $N = n - 1$ as the number of unit cells. A disadvantage of this model is the neglected capacitances on the edges of the fingers in this calculation:

$$C_{\text{Total}} = C_{UC}(N - 1)l_f.$$

With this model, the first experiments with different dielectrics or changes in the geometry of interdigitated electrodes can be performed. Figure 14.3 shows these

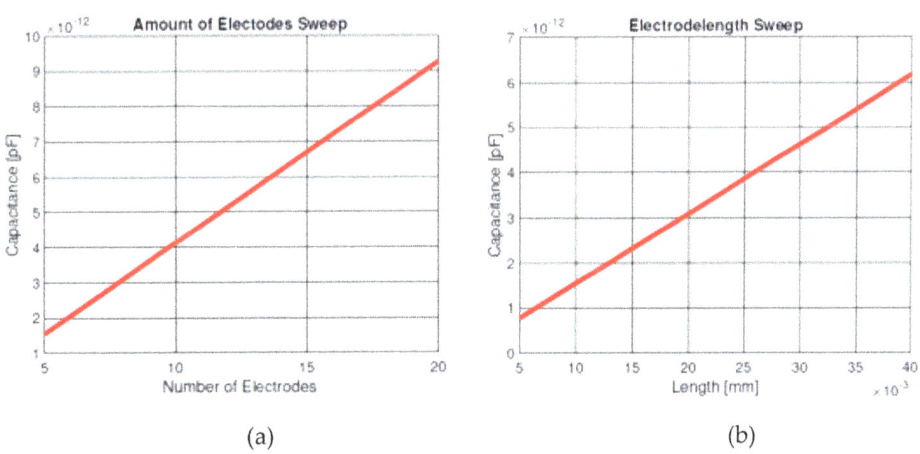

Figure 14.3. Plot for the manipulation of the geometry of an IDC structure with the analytic model. (a) Capacitance over an increasing number of electrode fingers. (b) Capacitance over an increasing length of electrode fingers. Copyright 2020 IEEE. Reprinted, with permission, from [7].

variations in the change of the amount of electrodes and the change of the length of electrodes. A linear increase of the capacitance can be detected [7].

14.3.2 Simulation model

The simulations in this work are carried out with Ansys® Electromagnetics Suite 18.1.0. It enables accurate simulations for electromagnetic, electronic, thermal, and electromechanical systems. The analysis tool used in this work is Ansys® Maxwell3D. The principle is based on numerical simulations and the applied method is finite element analysis (FEA). It is the most dominant numerical simulation method for the calculation of electrical field distributions [18]. It is used to analyze boundary-value problems in an approximate way in order to solve a mathematical, physical problem [19].

Based on the area scheme of the diaper with the geometric dimensions in figure 14.1, a layout for the sensor array is made in figure 14.4. For simulation purposes, to reduce the computation time, and obtain accurate results, the sensor array is divided into its sub-sensors. Figure 14.5 shows the computer aided design (CAD) layouts of the sensors from the inspected sensor system. Table 14.2 shows the geometrical parameters.

The other parameters are equal for all sensors and the symbols are the same as in figure 14.2:

$$t = 200\ \text{nm};\quad t_s = 180\ \mu\text{m};\quad w_f = 1\ \text{mm};\quad w_g = 1\ \text{mm};$$

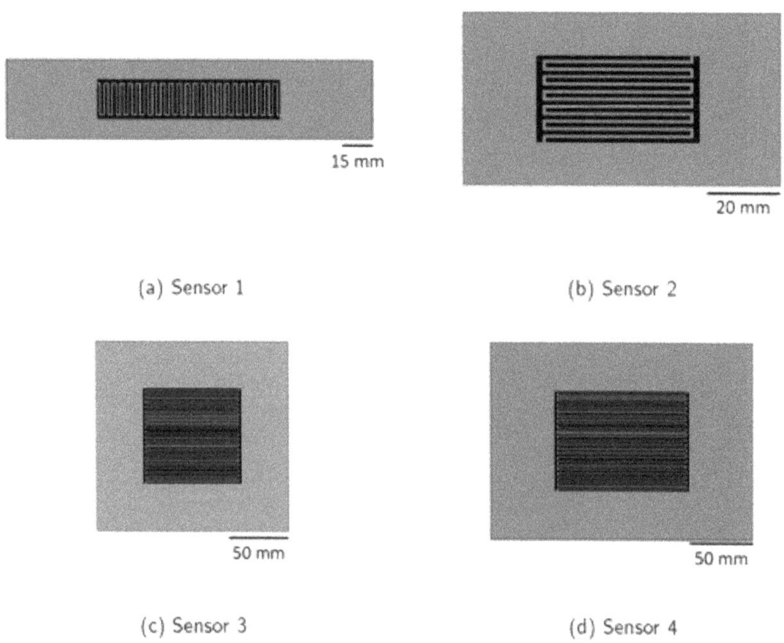

Figure 14.4. Layout of the 2D-printed humidity sensor for diaper application. Copyright 2020 IEEE. Reprinted, with permission, from [7].

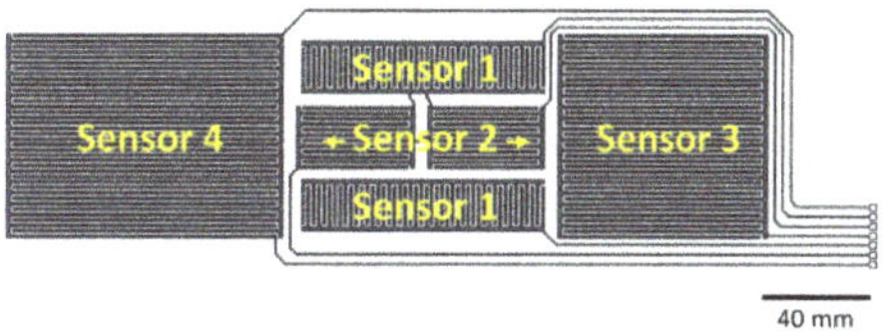

Figure 14.5. CAD layouts for the sub-sensors of the humidity sensor array. Copyright 2020 IEEE. Reprinted, with permission, from [7].

Table 14.2. Geometric parameters of the sensor layouts [7].

Sensor	Number of fingers	Finger lengthl_f (mm0	Substrate geometry $X \times Y$ (mm^2)
Sensor 1	46	15	182 × 38
Sensor 2	12	40	86 × 46
Sensor 3	38	75	158 × 150
Sensor 4	38	100	204 × 150

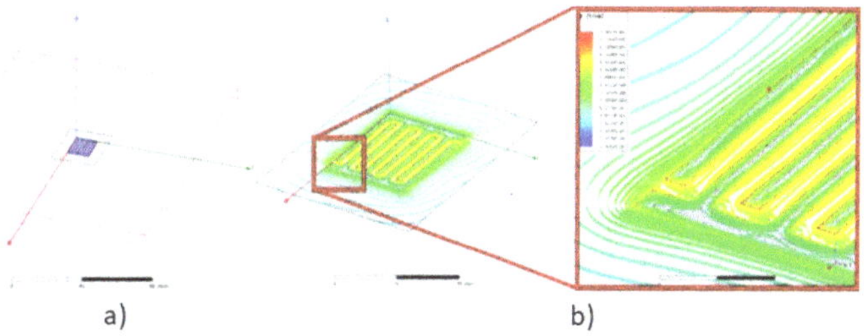

Figure 14.6. (a) Simulation model with a region for an example sensor structure. Copyright 2020 IEEE. Reprinted, with permission, from [7]. (b) Simulation result for the electric field magnitude.

$$p = 4\ \text{mm};\ \varepsilon_0 = 8.854 \times 10^{-12} \frac{As}{Vm}.$$

The CAD models are imported into the simulation model. Afterwards, the discovered material parameters of section 14.2 are added to the bodies and boundary conditions. The voltage potentials are set with 2.5 V for the anode and 0 V for the cathode. To consider the fringe field a region will be created. It is shown by the red cuboid lines around the sensor structure in figure 14.6(a). At the surface of the region, a perfect electrical (infinite conductivity) boundary condition is imposed. In order not to affect the fringing electrical fields of the capacitor, the distance has to be adjusted so that the fields have already decayed. This is one of the most important boundary conditions for the simulation of planar thin sheet IDCs.

Figure 14.6(b) shows the visualized simulation result of the electric field magnitude. All the simulations in this work are carried out with a high performance cluster (HPC) on 12 processor cores. With a standard computer, the computation time would be significantly higher and the random access memory (RAM) would need to be enormous as well [7].

14.4 Sensor design, manufacturing, and communication

In this section, the specifications of the sensor, the design, and the printing process are described. Among the many physical principles (for example optical, chemical, and electrical) which are suitable for measuring the humidity, the resistive and capacitive methods have many advantages, such as easy manufacturing and standard electronics for data acquisition. The resistive and capacitive structures can be integrated directly on flexible substrates. Figure 14.7 shows the sensor concept. The single sensor contains planar interdigital electrodes (IDEs) between the diaper foil and the absorbent pad. The urine absorbed on top of the IDEs changes the sensor resistance and capacitance. The sensor array consists of six sensors for the six areas of the diaper (figures 14.1 and 14.4) and is able to detect the amount of urine in the diaper, and sends the signal to a cloud or to caregivers [1].

Commonly, resistive and capacitive humidity sensors have been fabricated in silicon. But as mentioned above, silicon technology has some limitations, such as low flexibility, low stretchability, and high cost. Another limitation is the high number of silicon humidity sensors which would be necessary in a single diaper. In this work PEDOT:PSS, a conducting polymer [20], was used as the material to form the IDE structures. This material is highly flexible and has a sheet resistivity of 125 Ω/□. The real foil substrate has a size of 280 mm × 660 mm. The process is shown in figure 14.8. For the screen printing process, the screen printer DEK 248 is used. First, a meshed screen is painted with emulsion and covered with a developed photo layer, which has the desired pattern on it. After exposure under ultra-violet light for 3 min, the required mask is hardened. The unexposed areas covered with the emulsion are removed with water spray, and the remaining mask is ready for the screen printing process. Next, the screen is placed onto the substrate and PEDOT: PSS is placed on top of the screen mesh. Then the printing process begins with the uniform motion of a squeegee automatically controlled by the printer machine, which pushes the ink over the whole mask surface smoothly. As the squeegee moves towards the rear of the screen, the ink is distributed onto the defined areas, such that

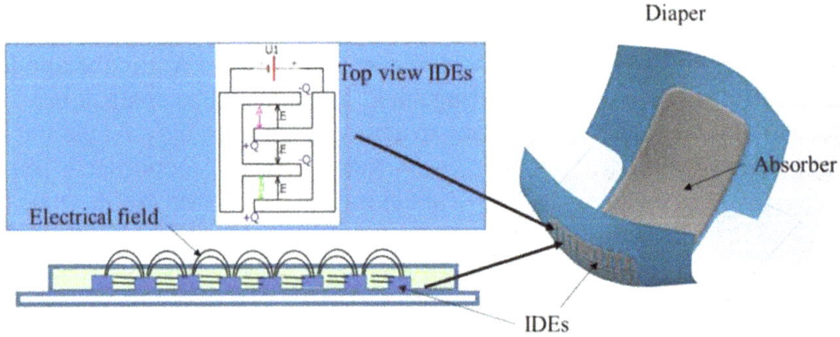

Figure 14.7. Sensor concept for detection of urine in incontinence materials. The sensor system consists of distributed single sensors integrated into the diaper. The resistance and the capacitance of the sensor depend on the amount of urine. Reproduced from [1] CC BY 4.0.

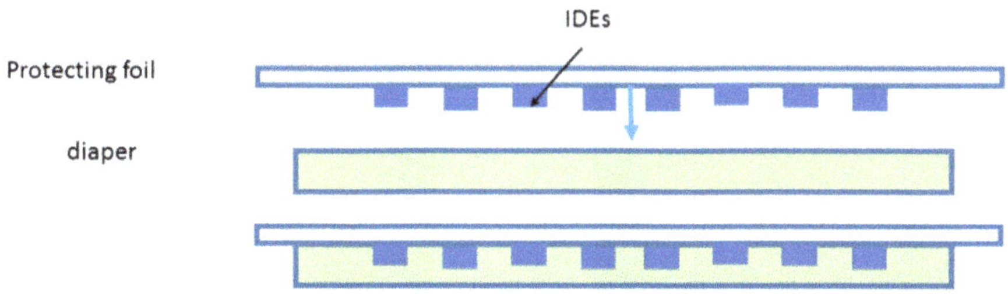

Figure 14.8. Integration of the novel smart incontinence sensor into a standard diaper. The first IDE array is realized on protecting foil using screen printing. In the next step the foil is laminated on the bottom of the diaper. Foil and lamination are standard materials and standard procedures in diaper production. Reproduced from [1] CC BY 4.0.

Table 14.3. Measurement parameters [8].

Parameter	Value
Number of measurements	5
Number of samples per measurement	10
Sampling time	1 s
DC voltage level	2.5 V
Frequency	100 kHz
AC voltage level	30 mV rms^{-1}

when the screen mesh is lifted the designed sensor is printed on the lamination foil. Then the printed sensor structures are cured at 80 °C for 30 min and laminated on PET material.

14.5 Characterization measurement

The characterization measurements of the 2D-printed sensors are performed with the semiconductor characterization system Keithley 4200-SCS and the included capacitance over voltage (CV) measurement card Keithley 4210-CVU. With this set-up the capacitance can be measured precisely.

The following table 14.3 shows the set-up for the capacitance measurement. The parameters are equal to the simulation boundary conditions in section 14.3.

The printed sensor array is cut into its sub-sensors. The measurements are performed for all sensors named in previous sections. Every sensor (device under test (DUT)) is measured initially with air as the dielectric, as figure 14.9(a) indicates.

Then the absorber material is placed on the interdigitated electrodes of each sensor (figure 14.9(b)). No scratches or adhesion problems can be observed. The same measurement is performed for freshwater as the dielectric (figure 14.10(a)). For this measurement a problem occurs. The conductivity of water electrically shortens the electrodes of the sensor. Therefore the capacitance fluctuates from the negative to the positive in each sampling. Another problem is the dissolving of the PEDOT:PSS ink from the substrate. It is not robust enough to resist liquids [7].

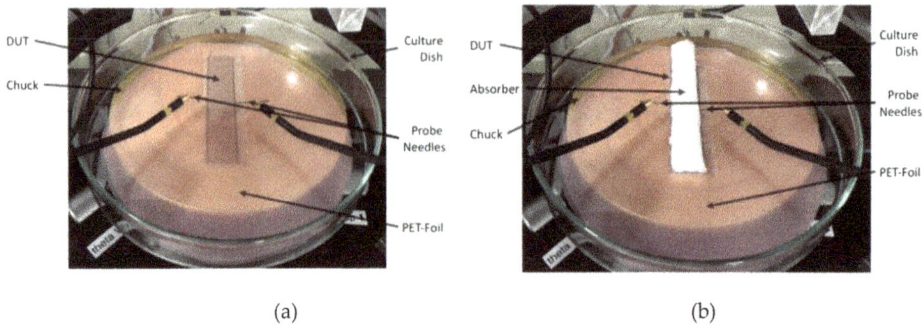

Figure 14.9. (a) Initial measurement set-up of sensor 1. (b) Measurement set-up with the absorber material as the dielectric on top of sensor 1.

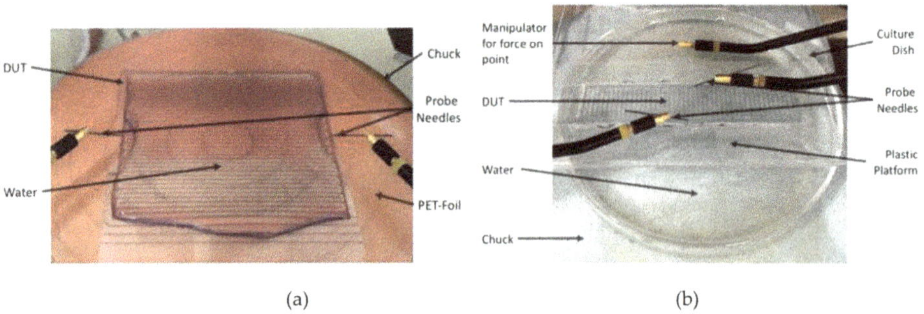

Figure 14.10. (a) Measurement set-up with freshwater as the dielectric on top of sensor 6 (water 1). (b) Measurement set-up with water as the dielectric underneath sensor 1 (water 2).

To solve this problem, another characterization measurement set-up is used, by having the freshwater touching the lower surface of the substrate of the sensor. Figure 14.8(b) shows the new set-up. For this reason, and for reasons of comparison, the other simulation models are created with the dielectric underneath the substrate. The achieved results are discussed with the comparison of all experimental methods in the next section.

14.6 Results and discussion

Figure 14.11 shows the printed sensor array. The foil size is 320 mm × 700 mm.

This sensor is perfectly suitable for wearable and one-way products. The developed system has been tested with synthetic urine (Human-Urin-Imitat 1000-D). Figure 14.11(b) shows the prototype. There are no issues to be observed in terms of flexibility and stretchability in comparison with standard products on the market. In the table on the left-hand side of figure 14.12 the resistance of each sensor is shown without and with urine in contact with the PEDOT:PSS material and the graph shows the capacitive sensor signals over the urine amount in the diaper from sensor 3 and sensor 6 [1].

These first measurement results are comparable to the results of the theoretical methods and the practical characterization measurements of each sub-sensor.

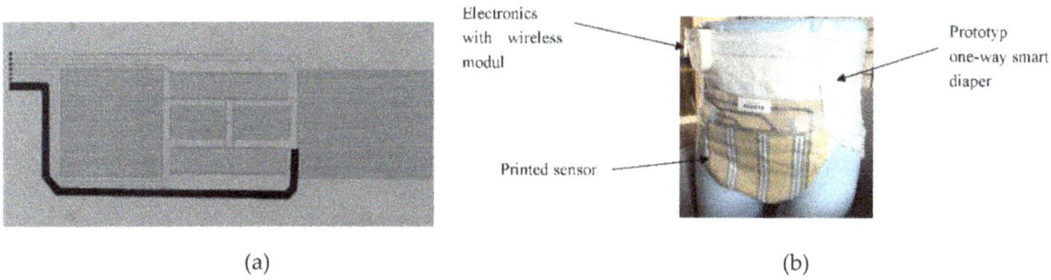

Figure 14.11. (a) A printed sensor array with six PEDOT:PSS ink sensors on a PET substrate. (b) A smart diaper prototype with electronics and a wireless module on a PE diaper substrate. Reproduced from [1] CC BY 4.0.

	Resistance (kΩ) (No Urine)	Resistance (kΩ) (Urine)
Sensor 1	> 1000	140
Sensor 2	> 1000	140
Sensor 3	> 1000	50
Sensor 4	> 1000	135
Sensor 5	> 1000	140
Sensor 6	> 1000	150

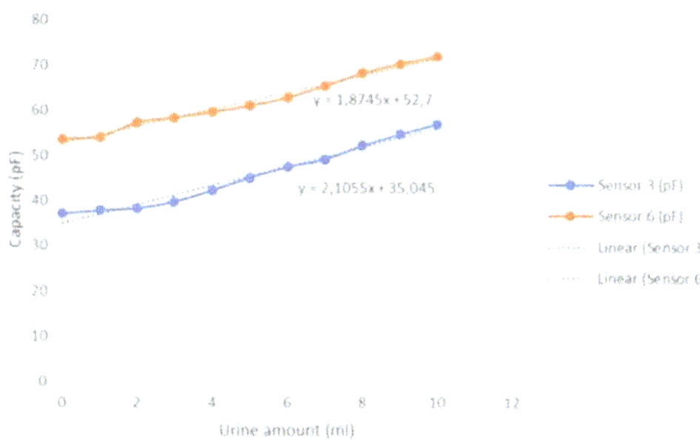

Figure 14.12. Measurement with synthetic urine as a demonstration of proof of concept. Left: table of the resistance of each sub-sensor with urine in contact with printed electrodes. Right: graph of the capacitance signal of sensor 3 and sensor 6 over urine amount. Reproduced from [1] CC BY 4.0.

Figures 14.13 and 14.14 show the behavior of the different experimental models based on the capacitance variation between them, with all three types of dielectrics. Each plot shows on the x-axis the different experimental methods and on the y-axis the capacitance. On top of the graphs are the results of the capacitance deviations with:

- ΔC_{MA} for the deviation between the analytic model and the measurement.
- ΔC_{MS} for the deviation between the simulation and the measurement.
- ΔC_{AS} for the deviation between the analytic model and the simulation.

For better visualization, the dashed line in every plot highlights the base line for the comparisons of the deviations. Every figure represents the results of one sensor type, and in each figure is a plot for every dielectric material used in this work.

The first plot (a) in figure 14.13 shows the results of all experimental methods of sensor 1 with air as dielectric layer. It can be recognized that the capacitance is lower for the characterization measurement of the real sensor sample than in the two other experimental models. The same behavior can be seen in the next plot (b) with the absorber as the dielectric layer on top of the electrodes. The capacitance of the real sensor is weaker than those computed with the simulation or the analytic model [7].

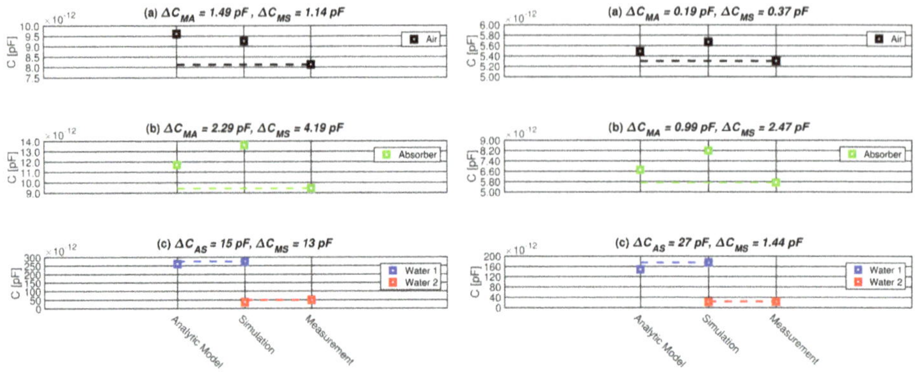

Figure 14.13. Left: Plots for the comparison of the results of sensor 1 with the dielectric materials of (a) air, (b) absorber, and (c) water. Right: Plots for the comparison of the results of sensor 2 with the dielectric materials of (a) air, (b) absorber, and (c) water. Copyright 2020 IEEE. Reprinted, with permission, from [7].

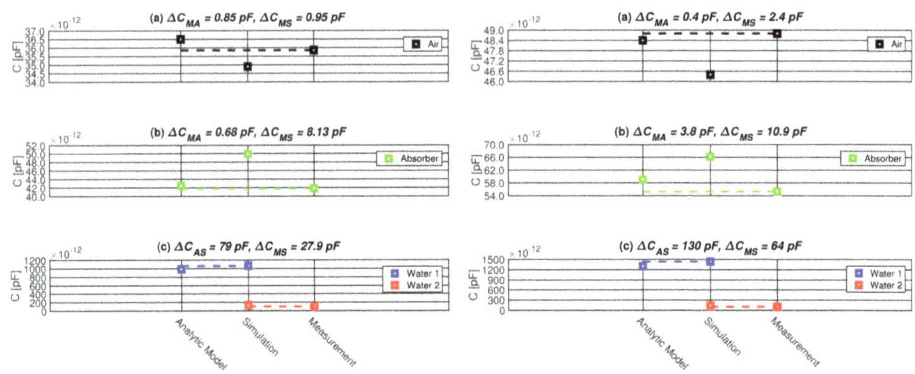

Figure 14.14. Left: Plot for the comparison of the results of sensor 3 with the dielectric materials of (a) air, (b) absorber, and (c) water. Right: Plot for the comparison of the results of sensor 4 with the dielectric materials of (a) air, (b) absorber, and (c) water. Copyright 2020 IEEE. Reprinted, with permission, from [7].

A different kind of plot to (a) and (b) is seem in the last graph (c) of figures 14.13–14.14. As mentioned before, the simulation and characterization with water as the dielectric is divided into two parts because of the shorting of the electric circuit and the damage to the printed electrodes. Therefore the model is changed with water underneath the substrate (figure 14.10(b)), touching the PET foil. Each graph for the model with water as the dielectric underneath the substrate (Water 2, red) shows an enormous shift to smaller values of the capacitance. The analytic model is only derived to have a layer on top of the IDC structures. For this reason it is no longer comparable to the simulation and the measurement of water as the dielectric underneath the substrate. This is why the value for the measurement Water 1 and the value for the analytic model Water 2 in graph (c) of each figure is missing. Nevertheless, two separate comparisons are made in every graph (c) with Water 1 for the comparison between the analytic model and the simulation model, and

Water 2 for the comparison between the characterization measurement and the simulation model [7].

The right-hand side of figure 14.13 describes the results of sensor 2. The geometrical structure of this sensor and thus the area are smaller than for sensor 1, therefore a smaller capacitance is expected. This is seen in the right-hand side of figure 14.13 with capacitances almost half the values of sensor 1 on the left. The same slope behavior for every dielectric can be seen as on the left. Only the results of the analytic model and the simulation model of (a) on the left-hand side of figure 14.13 are closer to the measured sensor sample with only 0.19 pF and 0.37 pF capacitance deviation. In general for sensor 2 all deviations are very small [7].

The left-hand side of figure 14.14 presents the results of sensor 3. The bigger the area of the sensor is (number of electrodes and electrode length), the higher is the capacitance. With growing capacitance, the deviation between each experimental method grows as well. This has an even greater effect on the values of sensor 4 on the right-hand side of figure 14.14, in particular for water as the dielectric. However, it still can be seen clearly that every experimental method is still of the same magnitude.

For more accurate comparisons, the percentage deviations from the experimental methods are shown as graphs in figure 14.15. Three comparisons are made and plotted against sensor area [7]: (a) between the analytic models and the characterization measurements; (b) between the simulation models and the characterization measurements; and (c) between the analytic models and the simulation models.

As mentioned in earlier sections, the analytic model is only derived for the sensor design with the dielectric layer on top of the electrodes. This is possible with the absorber material, but not with water. Therefore, figure 14.15(a) shows only the experimental results for air and the absorber as the dielectric for the comparison between the analytic models and the characterization measurements. The slope behavior is similar, but the percentage deviation of the capacitance shrinks with greater sensor area. The smallest deviation here is shown for sensor 4 in an air environment [7].

The comparison of the characterization measurements and the simulation models (figure 14.15(b)) shows for an approximately similar behavior for the air and

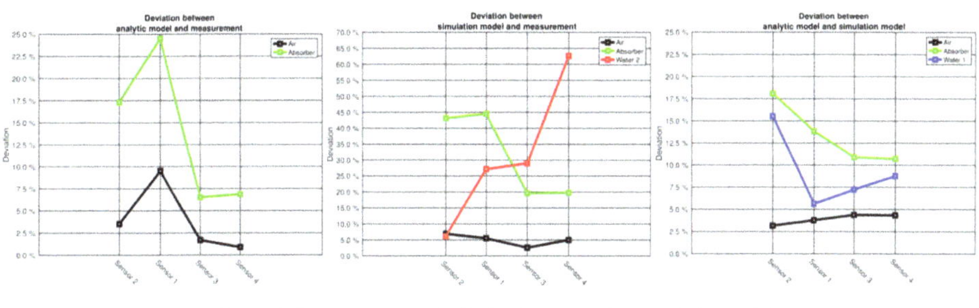

Figure 14.15. (a) Comparison of the percentage deviation between the analytic models and measured samples against sensor area. (b) Comparison of the percentage deviation between the simulation models and measured samples against sensor area. (c) Comparison of the percentage deviation between the analytic models and simulation models against sensor area. Copyright 2020 IEEE. Reprinted, with permission, from [7].

Table 14.4. Comparison of the deviation of the experimental methods by the mean of all sensors to every dielectric material.

Comparison of experimental methods	Air	Absorber	Water
Analytic model with measurement	3.90%	13.79%	
Simulation model with measurement	5.00%	31.70%	31.16%
Analytic model with simulation model	3.92%	13.37%	9.28%

absorber dielectrics as in the previous comparison. For the dielectric layers of air and absorber, the deviation is becomes smaller with greater size of the active sensor area. However, the deviation is shifted to higher levels than with the analytic model in the previous plot. The comparison with water as the dielectric layer is completely the reverse of the expected result. The smaller the area of the sensor is, the lower is the deviation of the capacitance between the simulations and measurements [7].

Figure 14.15(c) shows the comparison of the capacitance deviation of the analytic models and the simulations for all dielectrics above the electrodes of the sensor. It can be seen that the deviation between the two theoretical models is not as high as in the previous comparisons with the practical model. For air, the deviation is low, as in all comparisons. With a different material and thus a different permittivity on top of the sensor, the deviation is smaller with increasing sensor area. This time, the deviation for water is lower than the deviation for the dielectric absorber material.

In general, the deviation with the absorber material is higher than with air as the dielectric for all three comparisons. This is seen in table 14.4 with the values for the deviations averaged for all the sensors.

14.7 Conclusions

The objective of this work was to develop, simulate, and characterize a printed humidity sensor used in a diaper application to detect liquids. The newly developed system has shown very good and repeatable performance, in particular in the change of capacitance with different urine amounts. The specifications were first listed, the theoretical background for the proposed methods was explained, and the materials for this work were characterized. The first theoretical construct, the analytic model, was developed with consideration of the fringe field. The result is a functional, simple calculation, where the first optimization of a humidity sensor, without any simulation model, is possible. For this model the dielectric layer has to be on top of the electrodes. Later a simulation was performed and show a realistic performance. Then the manufacturing process was explained and the whole system design was developed. In further work a complementary capacitive characterization measurement on real sensors was performed. Additional measurements to prove the resistive principle were performed as well and show the resistance change. At the end a comparison was made with the results of working analytic models, with which the optimization of a 2D-printed humidity sensor can be realized. As mentioned, the analytic model provides a simple computation for the first design or for a minor optimization. The simulation had to be performed to give more complex results, with graphical illustration of field distributions as to whether micro-effects, such as pointy edges or orthogonal corners can be seen. These effects need to be considered for an appropriate virtual replication of a sensor. These models can be implemented easily into the development in different application areas, such as silicon technologies for MEMS applications. For this sensor tests are still on-going to improve the mechanical and chemical stability in extreme conditions. The focus of further research will be the implementation of biosensing structures for the detection of biomarkers to find diseases in an early stage. Development work is still necessary to scale up the process for mass production, but with the theoretical models this goal can be achieved. The capacitive principle compared to the resistive one has more advantages, such as the possibility of using a thin flexible passivation layer to protect the PEDOT:PSS structures, because of the material's solubility [1, 2, 7].

References

[1] Ngo H-D *et al* 2018 A novel low cost wireless incontinence sensor system (screen-printed flexible sensor system) for wireless urine detection in incontinence materials *32nd Eurosensors Conf. (Graz)*

[2] Wang B and Ngo H-D *et al* 2020 Development and characterization of a novel low-cost water-level and water quality monitoring sensor by using enhanced screen printing technology with PEDOT:PSS *Micromachines* **11** 474

[3] Someya T 2013 *Stretchable Electronics* (Weinheim: Wiley)

[4] Takei K 2018 *Flexible and Stretchable Medical Devices* (Weinheim: Wiley)

[5] Global Forum on Incontinence 2018 About incontinence http://gfiforum.com/incontinence

[6] APOGEPHA Arzneimittel 2018 Fakten und Zahlen zur Harninkontinenz. Häufigkeit der Harninkontinenz https://harninkontinenz-info.de/harninkontinenz-erkennen/fakten-zahlen.html

[7] Baeuscher M *et al* 2018 Simulation and electrical characterization of a novel 2D-printed incontinence sensor with conductive polymer PEDOT:PSS for medical applications *20th Electronics Packaging Technology Conf. (Singapore)*

[8] Procter and Gamble 2019 Pampers® https://pampers.com/en-us/diapers-wipes/pampers-swaddlers-diapers

[9] © Opro9 Service url: https://www.opro9.com/smart-diaper/

[10] Monit Corp 2020 http://goodmonit.com

[11] Abena 2021 Abena Nova with MediSens http://abenanova.com

[12] AGFA 2018 ORGACONTM conductive polymer screen-printing inks http://agfa.com/specialty-products/wp-content/uploads/sites/8/2018/04/Screen-Printing-inks-2018v2.0_ODE.pdf

[13] Keysight Technologies 2017 Solutions for measuring permittivity and permeability with LCR meters and impedance analyzers *Application Note* http://literature.cdn.keysight.com/litweb/pdf/5980-2862EN.pdf

[14] Heaney M B 2003 Electrical measurement, signal processing, and displays *Electrical Conductivity and Resistivity* ed J G Webster (Boca Raton, FL: CRC Press)

[15] Abu-Abed A S and Lindquist R G 2008 Capacitive interdigital sensor with inhomogeneous nematic liquid crystal film *Prog. Electromagn. Res.* B **7** 75–87

[16] Palmer H B 1937 The capacitance of a parallel-plate capacitor by the Schwartz–Christoffel transformation *Trans. Am. Inst. Electr. Eng.* **56** 363–6
[17] Carlson B C 2010 Elliptic integral *NIST Handbook of Mathematical Functions* ed F W J Olver vol 1 (Cambridge: Cambridge University Press)
[18] Ansys Maxwell 2018 Ansys Electromagnetics Suite 18.1.0, Maxwell 3D
[19] Westermann T 2010 *Modellbildung und Simulation: mit einer Einführung in ANSYS* (Berlin: Springer)
[20] Magdassi S 2017 *Nanomaterials for 2D and 3D Printing* (Weinheim: Wiley)

Chapter 15

Flexible and transparent 'self-adhesive' thin film strain, impact, force, pressure, and temperature sensors: an experimental study

Shreyas P Bhat, B Pavithra and M M Nayak

15.1 Introduction

All over the world, the transparent electronics market has been witnessing remarkable growth in the last couple of decade thanks to technological advancements and continues to evolve to meet the ever-increasing demand for sophisticated products in the consumer, healthcare, and military sectors. Transparent electronics focus on producing 'invisible' integrated circuitry that offers electrical properties like those of conventional devices, depending on the intended functionality. Most existing sensors, however, are opaque due to the optical properties of the raw materials used (substrate, sensing layer, passivation layer, protective coating, etc) for their fabrication process. The choice of appropriate raw materials to produce the desired functionality of sensors while being optically transparent is of interest in camouflaged applications. Hence, the research interest in the development of transparent sensors has soared in recent years to catch up with the futuristic needs of the market [1–4]. Examples are transparent temperature sensors on glass windows, transparent thin film transistors for thin transparent display panels, transparent super capacitors, transparent force sensors, etc. Also a great number of applications can be found in the wearable electronics and military sectors.

Flexible electronics are also introducing revolutionary changes in the design and conceptualization of otherwise conventional electronics [5–7]. They possess the benefits of being inexpensive, robust, lightweight, and more durable compared to their conventional counterparts. The first work toward mechanically unconventional forms of electronics began around 25 years ago with polymer transistors on bendable plastic sheets and paper displays representing the main target applications. Advances in printing and related patterning techniques on flexible substrates were the key for much of the progress in this field [7].

doi:10.1088/978-0-7503-3439-6ch15

© IOP Publishing Ltd 2021

The combination of flexible and transparent electronics to produce transparent flexible sensors opens up avenues for several applications in wearable sensors, healthcare, display technology, defense, etc, where sensors can be employed without being visually obstructive to the surface of the object under test. Various materials can be employed to produce transparent circuitry on transparent flexible substrates depending on the functionalities required and its end application (figure 15.1). For example, a strain, impact, or pressure sensor uses a strain-sensitive resistor or piezoelectric material that changes resistance according to the strain or pressure acting on it. Such sensors can be made transparent by choosing the appropriate raw materials and fabrication equipment for its fabrication process to achieve high optical transparency and the desired electrical conductivity.

A combination of high optical transparency and optimum conductivity for thin films can be achieved in the following two classes:

1. Ultrathin films (<10 nm) of high conductivity metals (Cu, Ag, Au). The optical transmission can go up to 50% and higher. The film resistivity would be a couple of orders higher than the bulk.
2. Wide bandgap semiconductors such as indium tin oxide (ITO) and Al-doped zinc oxide (AZO) exhibit good electrical conductivity and low absorption of light. The optical transmission can go up to 90% and higher. The film resistivity ranges from 10^{-3} to 10^{-4} Ω.cm [8, 9].

In their most common form, thin film strain sensors (TFSSs) are opaque with a sensing layer made of alloy films (a few microns thick) on semi-transparent polymeric substrates [10–12]. The TFSSs are then cemented on the surface of the test object by applying appropriate adhesives and following the curing procedure as recommended by the manufacturer. This cementing step requires consumables and skilled manpower to result in minimum measurement errors (non-uniformity of

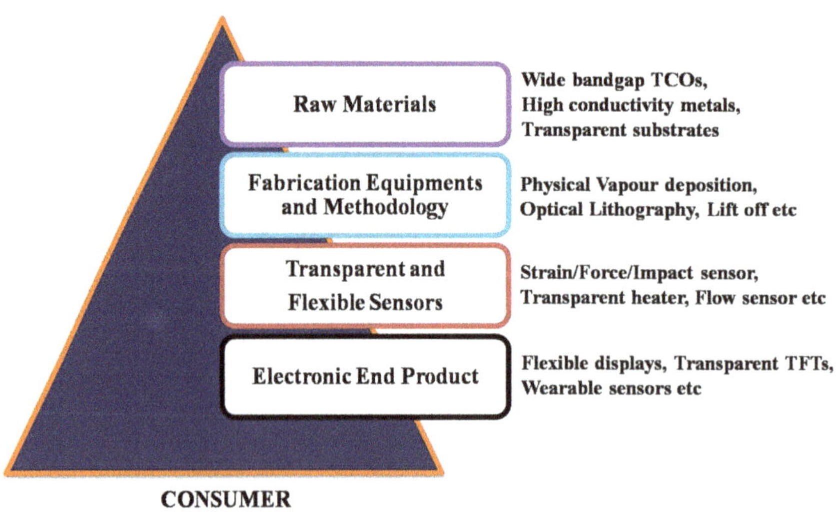

Figure 15.1. The transparent and flexible electronics industry ecosystem.

adhesive spread, particle contamination, positional offset, micro air gaps, etc, compromise measurement accuracy and repeatability). Also, such an installation step may seem tedious and time consuming for someone wanting it for a quick, single time use. The time, cost, and consumables requirements can be minimized if the sensor is made to easily stick on the surface of the test object, without any added steps or materials.

The work presented in this chapter was intended to develop a 'self-adhesive' transparent thin film strain sensor that can be employed for quick installation by anyone without any prior experience of strain sensor mounting. The underlying objectives of this work were to:

1. Develop a high 'figure of merit' transparent thin film (<120 nm thickness on PET).
2. Develop transparent TFSSs with self-adhesive properties.
3. Evaluate the performance of the self-adhesive TFSSs in static and dynamic environments.

15.2 Methodology

This section presents the methodology to choose the appropriate raw materials, and discusses the methods and mechanisms to fabricate a strain sensor to achieve the mentioned objectives. The basic idea of a thin film strain sensor is discussed for the ease of the reader's understanding. Strain sensors are available in various size and shapes. The selection of a strain sensor is application specific, there is no strain sensor which fulfils 'all' requirements [13]. For this reason, huge varieties of strain sensor are available on the market and special types are supplemented when required. The general considerations to prepare a strain sensor are shown in figure 15.2. The simplicity of the raw materials chosen, the ease of fabrication, and the low production costs make the methodology discussed here easy to adopt for beginners in the fabrication process.

15.2.1 Thin film strain sensor philosophy

Strain sensors are identified as transducing elements that convert mechanical stress or physical quantities such as strain, force, load, displacement, and pressure into electrical signals. The physical quantity to be measured causes a change in the geometrical dimensions of the object under measurement. Change in the resistance of the sensor corresponds to the elastic deformation of the test object on which the

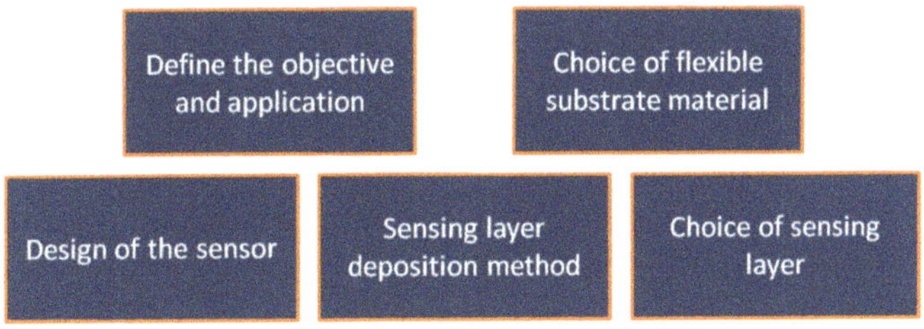

Figure 15.2. Generic considerations to implement a methodology.

strain sensor is mounted. In simple words, strain is the fractional change in length of a body due to the applied stress. Stress may be compressive or tensile. Ideally, the sensor exhibits change in resistance linearly in response to strain. The measurement of strain becomes pivotal in areas requiring critical stress or deformation analysis, such as in the aerospace sector to analyze the structural load on aircraft parts, the seismic testing of bridges and buildings, the monitoring of body parameters by wearable sensors for the healthcare sector, etc.

Every metal and semiconductor has its specific resistance. Suppose the original resistance is R and a strain-initiated change in resistance is given by ΔR, as shown in figure 15.3. The resistance may increase for tensile force whereas it may decrease for compressive force, depending on the sensing material chosen. This simple principle is adopted in this work to develop a transparent thin film strain sensor. Patterned thin film that is transparent in the visible region serves as the resistive path on a transparent polymeric substrate which also acts as a highly insulating mechanical carrier between the test object and sensing layer.

Because of the versatility of thin film technology, it is possible to fabricate devices of the desired electrical specification and optical properties by parametric optimization using vacuum deposition methods. An added advantage of employing thin film technology for strain gauge production is its efficiency in mass production with a considerable cost reduction. Continuous metal and metal alloy films (opaque) have strain sensitivity or gauge factor values around 2–6 [10, 12, 15], whereas ultrathin metal films with high sheet resistance are often discontinuous (semi-transparent), hence they exhibit higher values of gauge factors ranging from a few tens to hundreds, depending on the material(s) used [2, 4]. Discontinuous films, however, are not used extensively because of a lack of repeatability and poor performance stability. The meandering structure of the sensing layer provides maximum gauge resistance while keeping both the length and width of the gauge to a minimum.

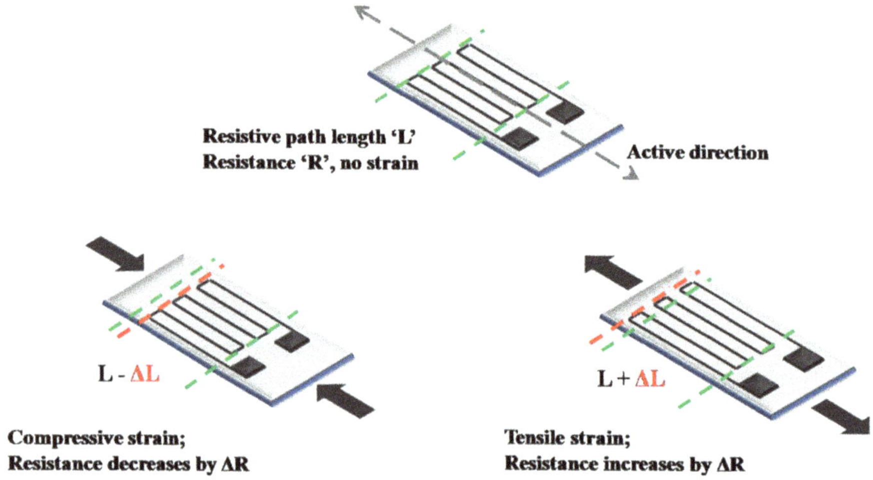

Figure 15.3. Schematic representation of a simple resistive metallic thin film strain sensor.

Strain sensors are bonded onto the test object using specific adhesives followed by a curing procedure, as recommended by the manufacturer. When an object attached to strain sensor undergoes deformation, the strain is transferred from the test object to the adhesive layer, then from the adhesive layer to the flexible substrate, and finally from the substrate to the sensing layer. Figure 15.4 shows a commercially available thin film strain sensor attached to a stainless steel specimen using conventional bonding accessories.

In this work, the objective was to fabricate a simple transparent sensor; hence, the following are few parameters that would help to evaluate the quality of the transparent TFSSs:

1. The strain sensitivity of the sensing layer.
2. The flexibility of the substrate and bonding epoxy/adhesive used.
3. The electrical resistance of the sensor.
4. The optical transparency of the sensor in the visible region (400–800 nm wavelength).

15.2.2 Choice of flexible substrate

To fabricate a transparent TFSS, it is essential to begin by choosing a substrate that exhibits high optical transparency, mechanical flexibility, and good chemical stability. This substrate acts as a mechanical carrier for the sensing layer/circuitry produced on the substrate and hence must possess high electrical insulation resistance between the sensing layer and test object. The substrates used for conventional strain gauge production include polyamide (Kapton), polyethylene terephthalete (PET), polydimethylsiloxane (PDMS), and other polymers of thickness ranging from a few tens to a few hundreds of micrometers.

In this work PET substrate (130 μm thickness) with a pre-applied thin silicone layer (pressure-sensitive adhesive) on one side and a clean surface on the other side was used as the substrate. Both sides were protected by a thin plastic cover.

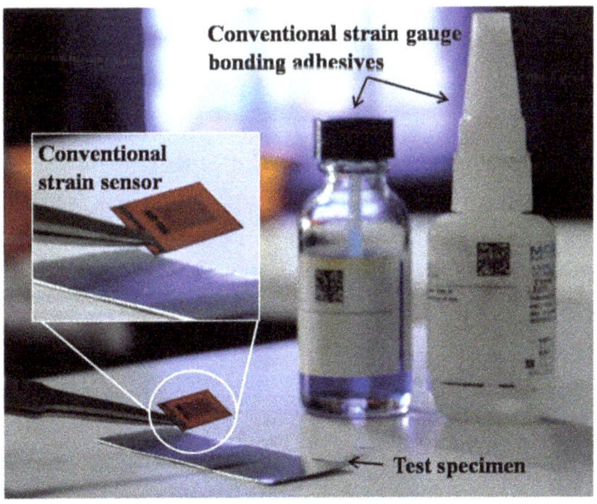

Figure 15.4. A conventional strain sensor placed on a steel specimen using bonding adhesives.

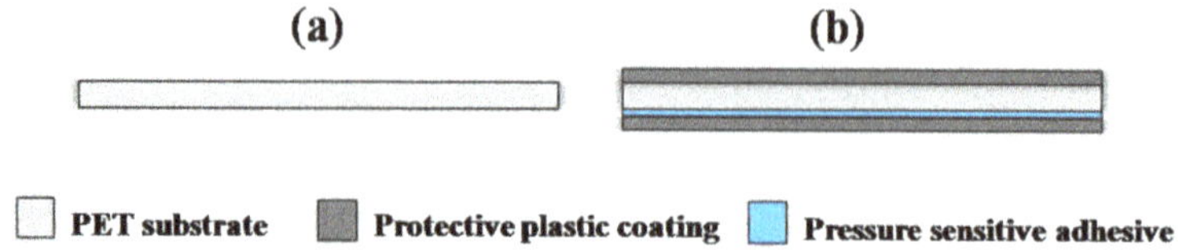

Figure 15.5. Schematic of (a) a conventional PET substrate and (b) a self-adhesive PET substrate.

A schematic representation of substrates is shown in figure 15.5. These substrates are available commercially and are used commonly as screen covers on touch screen displays. The presence of a pressure-sensitive adhesive on the substrate eliminates the need to apply any external adhesives to mount the strain sensor on the test object. The protective plastic coatings are peeled off just before using the respective surfaces; this avoids the accumulation of particle contaminants on the substrate surface efficiently. A few advantages of the self-adhesive PET used as a substrate in this work are:

1. It is inexpensive, costing ~20 times less than the conventional PET sheets available on the market.
2. It has a smooth substrate surface (<40 nm roughness) with no surface irregularities.
3. It has a high specific electrical resistance contributing to good insulation.
4. It is easy to handle, can be repositioned, and is non-invasive.
5. It is vacuum compatible and chemically inert to acetone and IPA solvents.

A poor interface may lead to poor adhesion causing film delamination and surface irregularities on the substrate. Hence an important consideration is the interface between the thin film and polymeric substrate, as it dictates the film adhesion, residual stress, and binding energy during film condensation. The quality of the interfacial region typically depends on the substrate morphology, surface contamination, chemical interactions, energy available for the surface diffusivity of adsorbed atoms, internal stress due to thermal expansion coefficient mismatch between the film and the substrate, etc. Optical micrographs of a few common types of irregularities observed due to a poor substrate–film interface are shown in figure 15.6.

The mechanical bonding of this interface depends on the interlocking mechanism of the cluster of atoms being deposited on the surface of the substrate. In contrast the chemical bonding depends on the formation of chelates of polymer oxygen atoms with those of the atoms deposited on the substrate, as seen in figure 15.7.

15.2.3 Thin film deposition method

A thin film can be typically defined as a layer of any material ranging from monolayer (a few fractions of a nanometer) to several nanometers in thickness on a physically thick substrate (a few hundred micrometers). They exhibit interesting characteristics as their material properties depend on the inter-related deposition

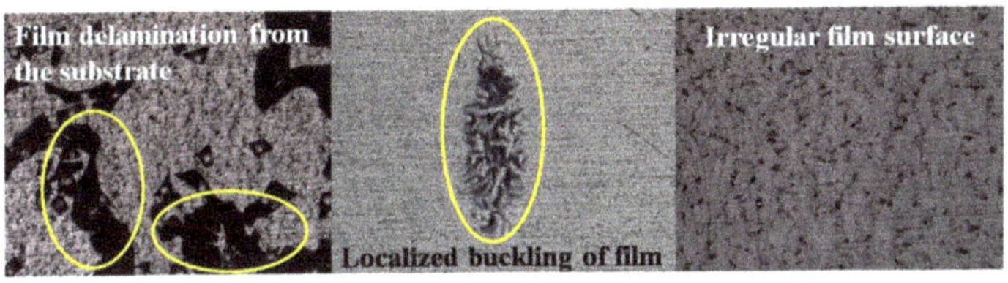

Figure 15.6. Micrographs of surface irregularities due to a poor film–substrate interface.

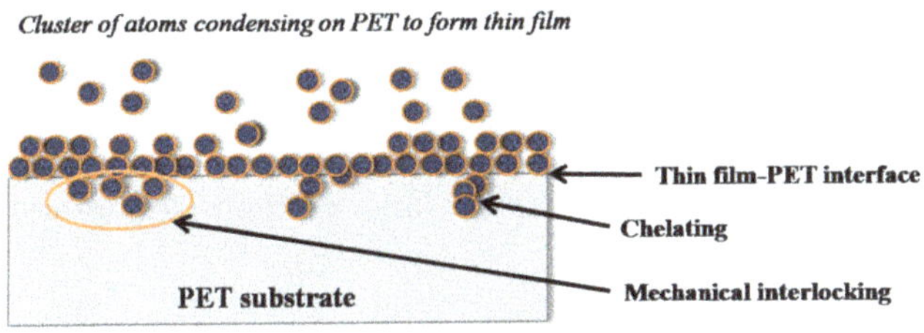

Figure 15.7. Thin film–PET substrate interface mechanism.

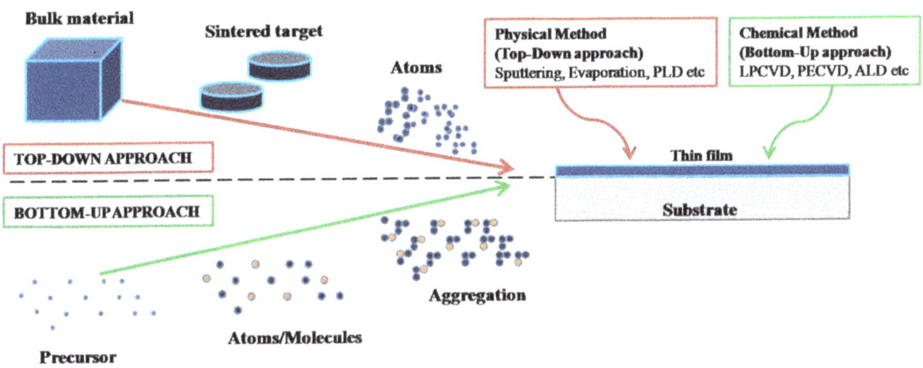

Figure 15.8. The two common types of thin film deposition approaches.

parameters and method of deposition. The process parameters during deposition can be optimized to produce thin films with similar or distinct properties compared to their bulk counterparts. In addition to that, thin film may not be as dense as bulk, exhibits stress states, contains several types of defects, is influenced strongly by the substrate surface (substrate–film interface), exhibits different composition and microstructure depending on deposition conditions, etc.

Thin films can be produced by several methods of which two broad categories are:

1. Chemical vapor deposition (CVD).
2. Physical vapor deposition (PVD).

Schematic representations of the two approaches are shown in figure 15.8. CVD techniques, in general, employ a bottom-up approach which involves atom by atom assembly (from precursor or chemical gases) to form a layer of nanometer scale on the substrate. Reactant gases are introduced in the chamber, chemical reactions occur on the wafer surface leading to the growth or deposition of a solid film, which generally takes place at temperatures higher than 500 °C, hence providing great control over the layer being formed.

In contrast, PVD techniques employ a top-down approach in which the bulk material is broken down (by electrical or thermal energy) to produce thin films at the nanometer scale on a given substrate. Vapors of the constituent materials are created inside the chamber and condensation occurs on the substrate surface leading to film deposition. PVD does not require a high substrate temperature, and hence can take place even at room temperature. Due to the ease of processing, low cost, and feasibility of accommodating the low thermal budget substrate used in this work (PET), the PVD method was employed. We will not attempt to describe all PVD techniques, but focus on the two common and inexpensive methods of deposition used in this work.

1. *Magnetron sputtering*: In a sputtering technique, a solid target (source) material is sublimated using a DC or RF power supply to create a cloud of vapor in a vacuum chamber back-filled with inert process gas (Ar or He). A glow discharge (plasma) is needed to make the process gas conductive and generate positively charged ions that then accelerate to strike the target placed on a magnetron acting as a cathode. The cathode is bombarded by positive ions to sustain the plasma through secondary electron emission and ejects atoms from the cathode. These ejected species will land everywhere in the vicinity of the cathode including the substrate depending on chamber geometry, mean free path, and power applied to the target. The target surface undergoes sputter removal and the substrate surface undergoes sputter deposition. A generic sputtering chamber set-up is shown in figure 15.9.
2. *Electron beam evaporation*: In the electron beam evaporation technique, the tungsten filament is fed with a high voltage DC power supply to generate a cloud of free electrons in a vacuum, like in an incandescent light bulb. The filament is placed at a high negative potential, around −10 kV,

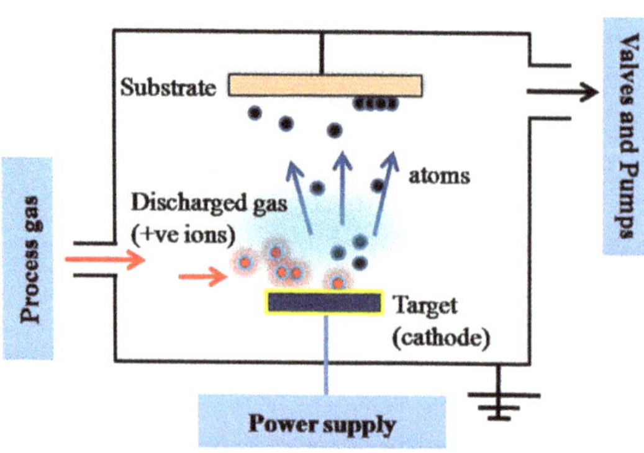

Figure 15.9. Generic schematic of the sputtering chamber.

resulting in the electrons being accelerated away from the filament, as shown in figure 15.10. These electrons generated by 'thermionic emission' are ultimately delivered to the surface of the target material placed inside a water cooled crucible using a combination of permanent magnets and electromagnets to deflect the beam by an angle of 270 °C. The atomic flux from the source material travels in the chamber defined by the mean free path and then the hot atoms condense on the substrate placed on the substrate holder.

Thin films deposited by the PVD technique can be patterned using two easy methods:

1. Hard masking.
2. Soft masking.

The precision and repeatability of pattern transfer in a PVD process depends on the ability of vapors to distribute in well-defined openings and then condense to form the intended design/structures. The physical barrier above the substrate surface prevents deposition on the covered regions of the substrate. The shadow masks with patterned openings used for hard masking are usually made of stainless steel using laser beam or electro-discharge machining. However, the usage of shadow masking is limited to critical widths larger than 100 μm. This is due to the scattering of deposited particles around the mask edges causing a shadowing effect. Also, the creation of micro air gaps between the substrate and hard mask may compromise pattern repeatability and thus produces poor film edge sharpness.

To avoid these problems, soft masking or lithography is employed for more reliable and repeatable transfer of patterns. It is a process by which a light sensitive polymer (photoresist) is exposed and developed to form a three-dimensional relief image on the substrate. The ideal photoresist image has the exact shape and

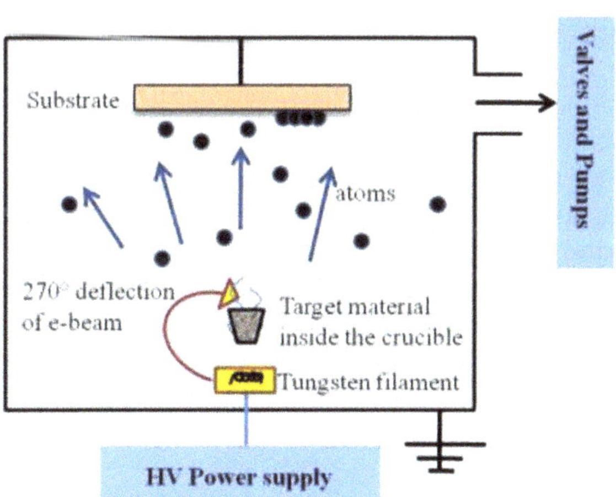

Figure 15.10. Generic electron beam evaporation chamber representation.

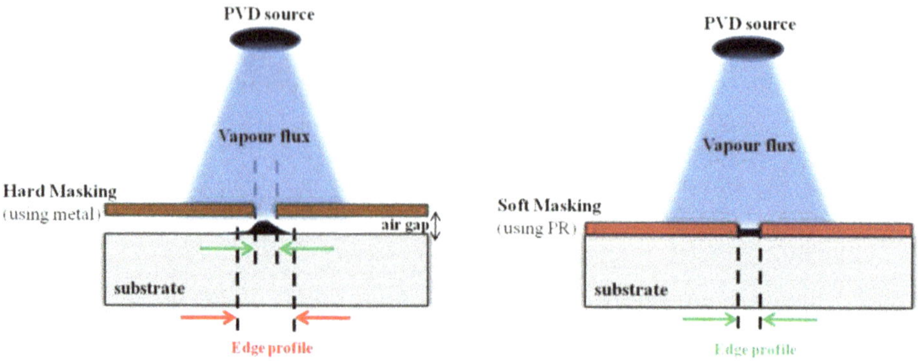

Figure 15.11. Representation of hard and soft masking for pattern transfer on the substrate.

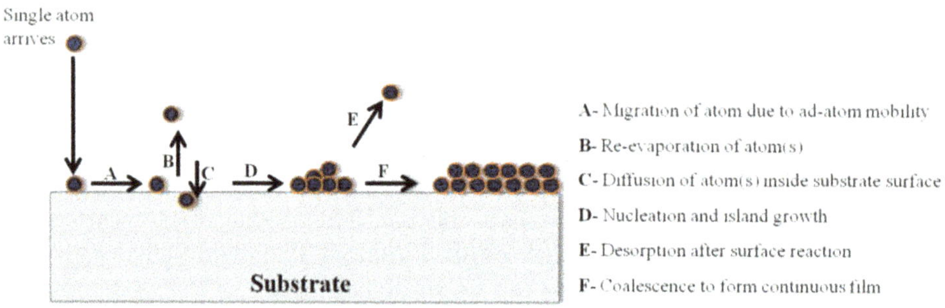

Figure 15.12. Schematic diagram of the generic film formation mechanism by PVD.

dimensions of the designed pattern in the plane of the substrate, with vertical walls through the thickness of photoresist. Thus the final resist pattern is binary, i.e. specific regions on the substrate are covered with photoresist while other regions are not, hence producing sharp and repeatable patterns on the substrate. A schematic representation of the effect of hard and soft masking is seen in figure 15.11.

The evolution of microstructures of thin film is highly dependent on the deposition conditions and relevant process parameters; however, the mechanism behind the morphological evolution of thin films by PVD techniques is similar. Thin film formation by PVD at room temperature is summarized in figure 15.12. Atoms that are deposited on the substrate diffuse around until they meet other atoms, nucleate, and form islands. The islands grow with further deposition and eventually grow to come in contact with other islands. When islands touch they begin coalescing to minimize their surface area. Island coalescence will drastically slow down or stop when a continuous film has formed. The film structure will continue to evolve as the film thickness increases. The microstructural evolution of film growth to a large extent is set by the island nucleation, growth, and coalescence processes.

More interesting details about the thin film formation mechanism can be read in [16].

15.2.4 Choice of the transparent sensing layer

Any material in appropriately thin form acts flexible mechanically with respect to the bending strains, which decrease with material thickness [7]. For example, a silicon wafer of 500 μm is brittle, but thin membranes of silicon are relatively flexible. In this work, the desired sensing functionality was achieved by transparent film of thickness 100–110 nm on a flexible polymeric substrate. As discussed in the introduction, there are two popular methods to produce transparent films. This work focuses on the use of transparent conductive oxides (TCOs) to produce transparent films with high optical transparency and the lowest possible electrical resistivity by parametric optimization of deposition. Let us examine briefly what TCOs are.

15.2.4.1 Transparent conductive oxides

TCOs are generally n-type wide bandgap semiconductors with a relatively high concentration of free electrons in the conduction band. The wide bandgap is responsible for high optical transmittance and free electrons which increase electrical conductivity. The conductivity is due to doping either by oxygen vacancies or by extrinsic dopants (such as zinc, tin, gold, silver, etc). In the absence of doping these oxides become very good insulators. TCOs can thus have high optical transmission (>70%) at visible wavelengths and electrical conductivity (ranges from 10^{-3} to 10^{-4} Ω.cm) close to that of metals.

Examples of TCOs are indium tin oxide (ITO), zinc oxide (ZnO), tin oxide (SnO_2), aluminum doped zinc oxide (AZO), indium oxide (In_2O_3), cadmium oxide (CdO), etc. They are used in a wide range of applications, such as transparent electrical contacts in photovoltaic cells, de-foggers in aircraft windows, liquid crystal devices, touch screens panels, thin film transistors, etc. The interesting question of how materials can be both transparent and conductive is answered well in [17–19].

15.2.4.2 Figure of merit for transparent films

An ideal TCO film must exhibit optical transmission as high as possible and electrical resistivity as low as possible. Amongst all the TCOs available on the market, ITO exhibits superior properties and thus holds a huge market space in the transparent electronics sector. The question of qualitatively evaluating and comparing the transparent films that exhibit conductivity is answered well by a metric called the figure of merit (FOM), developed by Haacke [20]. It helps to evaluate a transparent film's properties from its fundamental electrical sheet resistance and optical transmission over required wavelengths. The FOM formulated by Haacke is

$$\text{FOM} = \frac{(\text{Optical transmission of film over certain wavelengths})^{10}}{\text{Sheet resistance of transparent film}}. \qquad (15.1)$$

The unit of the FOM is expressed in $10^{-3}\ \Omega^{-1}$ or m.Ω^{-1}.

An FOM comparison of ITO and AZO thin films is shown in figure 15.13, deposited by sputtering as a function of increasing RF power. It is evident that the ITO film exhibits an almost ten times higher FOM than the AZO film, for the same

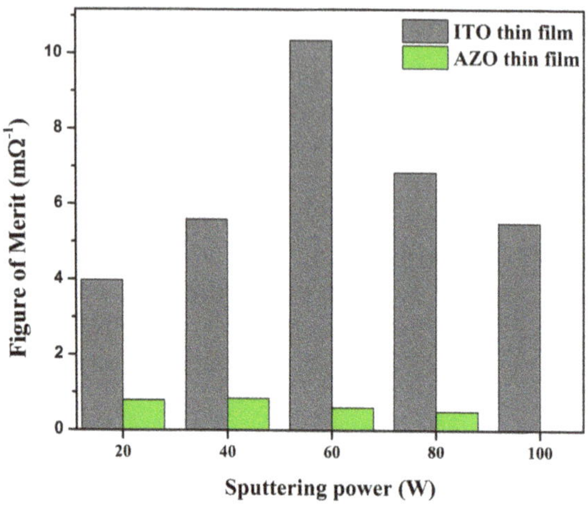

Figure 15.13. FOM comparison of ITO and AZO thin films.

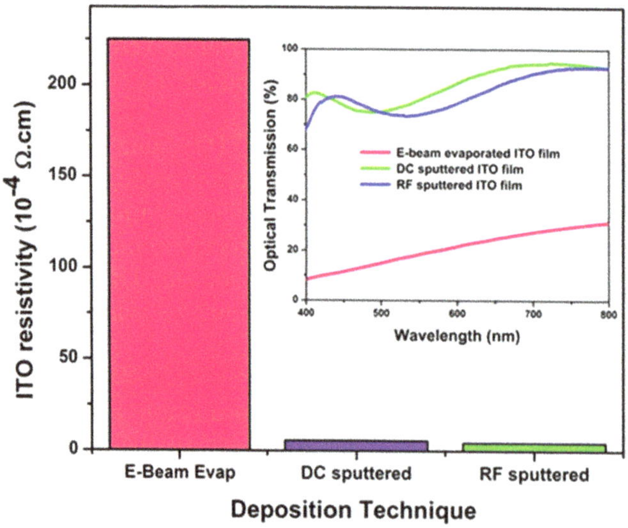

Figure 15.14. Film resistivity and optical transmission for different methods of ITO film deposition.

thickness. The higher the FOM value, the better is the quality of the thin transparent film. Thus ITO was employed for further processes in this work.

15.2.4.3 Choice of film deposition technique

In this segment an appropriate method to deposit ITO film is discussed. Figure 15.14 shows a comparison of ITO thin film resistivity and optical transmission (the plot in the inset of figure 15.14) deposited by electron beam evaporation and sputtering at room temperature. Different methods of deposition yield different microstructures and material compositions leading to changes in its sheet resistance and optical absorption for certain wavelength ranges.

ITO is a ternary compound that contains indium, tin, and oxygen. Decomposition of the ITO source by electron beam evaporation (EBE) at room temperature produces a thin film with the lowest optical transparency (almost opaque) and very high resistivity in the order of 10^{-2} Ω.cm. When the ITO compound is evaporated, it decomposes into sub-oxides and free oxygen. The oxygen release disturbs the stoichiometry in the films which depends significantly on the deposition conditions and chamber geometry. High quality ITO thin film was achieved by sputtering using an RF power supply at room temperature producing superior optical transparency and film resistivity in the order of 10^{-4} Ω. cm. The sputtered particles have kinetic energy up to one order of magnitude higher than the evaporated particles. The kinetic energies of the ejected atoms thus tend to enhance the surface mobility resulting in better film packing density and improved crystallinity at considerably low temperatures [21]. Thus the RF sputtering technique was employed for further experiments to produce the thin ITO films in this work.

15.2.4.4 Trilayer structure

Single layered conductive or semi-conductive thin film cannot exhibit specific resistivity lower than its bulk value. To further improve the FOM of transparent thin films, bilayer and trilayer film structures have been studied in the last few years [4, 14, 22, 23], wherein the multilayer structure provides superior electrical conductivity (lower sheet resistance) than a single layer TCO film while maintaining high optical transparency. A very thin layer of high conductivity metal such as Ag, Cu, Au, or Al can be used to produce a bilayer or trilayer film structure.

To improve the conductivity of the ITO thin film, a thin layer (~10 nm) of silver was sputtered and sandwiched between the ITO layers (i.e. ITO/Ag/ITO). The structure of ITO and Ag as shown in figure 15.15 is termed trilayer transparent film. Since the layers are in parallel, the effective sheet resistance would be lower than that of single layer ITO film and is given by

$$\frac{1}{R_{s(\mathrm{Eff})}} \approx \frac{1}{R_{s(\mathrm{Ag})}} + \frac{2}{R_{s(\mathrm{ITO})}}, \tag{15.2}$$

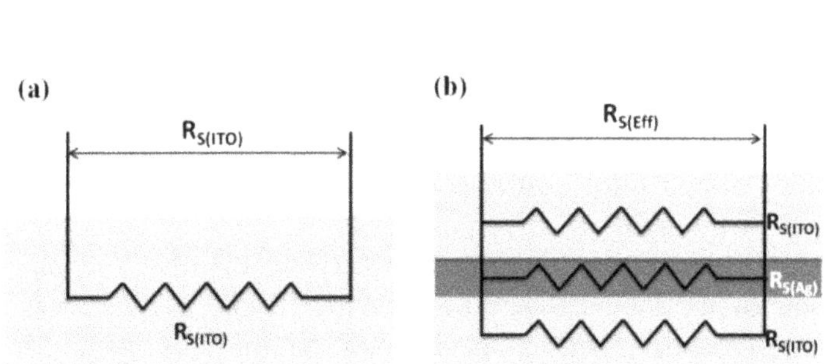

Figure 15.15. Schematic of (a) single layer ITO film and (b) trilayer IAI film representations.

Table 15.1. Comparison of the FOMs achieved by different PVD methods.

Name	Method	R_s (Ω/□)	Resistivity (Ω.cm)	$\%T_{avg}$ (400–800 nm)	FOM (10^{-3} Ω^{-1})	Remarks
a	EBE	2251	225×10^{-4}	0.21	0	Single layer
b	RFMS	32	4.4×10^{-4}	89.53	10.3	Single layer
c	RFMS	5.5	6×10^{-5}	85.34	37.1	Trilayer

where $R_{s(\mathrm{Eff})}$ is the effective sheet resistance of the Ag-inserted ITO film, $R_{s(\mathrm{Ag})}$ is the sheet resistance of the silver layer, and $R_{s(\mathrm{ITO})}$ is the sheet resistance of the ITO layer

A quick summary of the FOM values achieved by different techniques of deposition to produce high quality transparent thin film is shown in table 15.1. The trilayer structure (ITO/Ag/ITO) with 50/10/50 nm thickness yields a remarkably low sheet resistance (5.5 Ω/□) while producing high optical transmission (~85.34%) in the visible wavelengths. Thus this trilayer structure was employed as a sensing layer to fabricate the transparent thin film strain sensor.

15.3 Self-adhesive thin film strain sensor

15.3.1 Fabrication process

Step 1: Substrate preparation

1. The front protective cover is peeled off the self-adhesive PET substrate (figure 15.16). Since the PET surface was always protected, substrate cleaning may be optional.
2. The use of a carrier substrate to mount the flexible substrate is optional.
3. De-hydration baking for a few minutes is used to minimize the moisture adsorbed on the substrate.

Step 2: Photoresist coating

1. A thick uniform coating of a positive photoresist of specific thickness (1.2 μm) is accomplished by a well-controlled and simple process of spin coating.
2. Pre-exposure 'soft-baking' is used to reduce the residual solvent concentration in the photoresist for its stabilization and to improve the adhesion with the substrate.

Step 3: Exposure and development

1. The desired pattern is exposed using 390 nm wavelength optical sources. The solubility of the photoresist in the exposed region changes due to cross-linkages of the light sensitive polymer.
2. Aqueous bases called developers selectively dissolve the photoresist as a function of its chemical composition (it reacts and removes the exposed regions of positive resist), leaving behind the soft mask for the pattern transfer process. The photoresist–developer interaction time defines the shape and the profile of the photoresist.

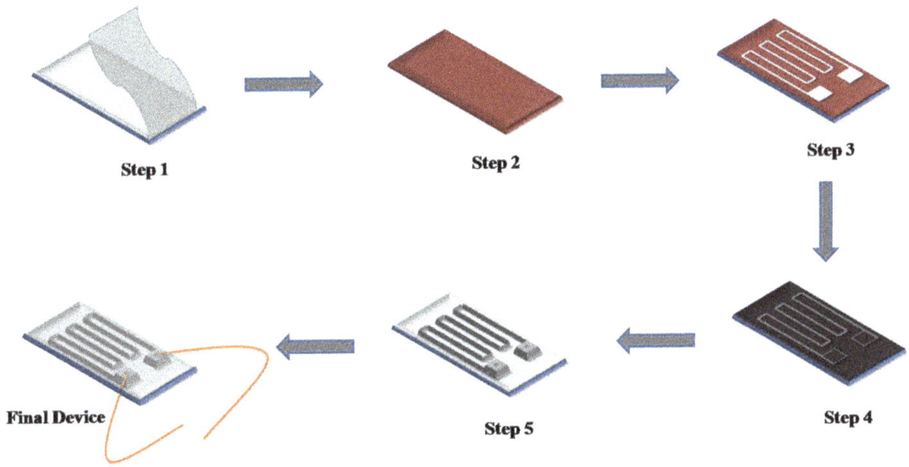

Figure 15.16. Pictorial representation of the TFSS fabrication process flow.

Step 4: Thin film deposition

1. A thin transparent sensing layer is deposited by the RF sputtering method at room temperature.
2. Parametric process optimization is carried out to achieve the best FOM for the thickness of ITO/Ag/ITO (50/10/50 nm).
3. The deposited film does not exhibit any surface defect and is continuous.

Step 5: Lift-off

The entire substrate surface is coated with the film after sputtering deposition. Treating the substrate in acetone dissolves the photoresist and the material on top of the photoresist is removed, hence transferring the meandering structures onto the substrate.

Step 6: Final device

1. Thin enameled copper wires (enamel removed at the ends of the wire) are bonded to the contact pads using silver epoxy.
2. The device surface is coated with parylene to protect from environmental contaminants.

15.3.2 Transfer of the self-adhesive sensor

The measurements using the strain sensor assume that the strain acting on the test object is transferred faithfully to the sensing material without any loss. This requires good bonding between the strain sensor and the object under test. This bonding is best provided by a standard resin and hardener, followed by a curing procedure as recommended by the strain gauge manufacturers. When an object attached to a strain sensor undergoes deformation, the strain is transferred from the test object to the adhesive layer, then from the adhesive layer to the substrate, and finally from the substrate to the sensing layer, as shown in figure 15.17.

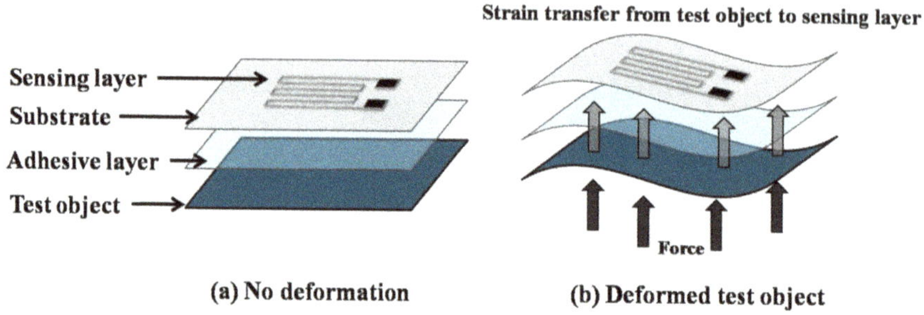

Figure 15.17. Cross sectional exploded view of cantilever beam bending measurement set-up.

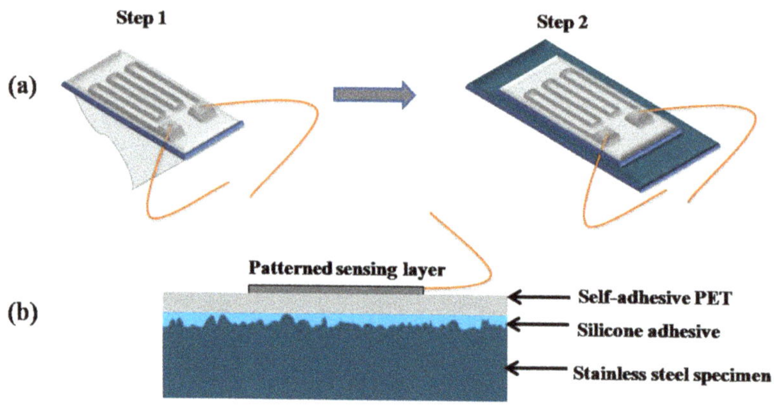

Figure 15.18. (a) The two-step mounting procedure for self-adhesive TFSSs. (b) A cross sectional schematic of a TFSS with silicone self-adhesive on a rough SS surface.

Typical applications of conventional strain sensors are thus labor intensive. For any commercial or academic purpose, the requirement for applying and curing adhesives can be eliminated by employing self-adhesive strain sensors, where a thin uniform layer of pressure-sensitive adhesive is already present on one side of the substrate. It can be attached to the test object surface with light pressure (finger pressed) for quick installation.

After the fabrication of TFSSs as discussed in a previous section, the sensor is ready to use for measurements. The two-step mounting procedure is shown in figure 15.18. The protective cover on the adhesive side of the PET is removed and can be placed easily on the test surface; in this case a stainless steel (SS) specimen of thickness 500 μm was used as the test specimen.

The silicone or polysiloxanes layer acting as the adhesion layer possesses the advantage of sticking uniformly on low energy surfaces (such as plastics) and also high energy surfaces (such as metals). The adhesive layer is transparent in the visible wavelength. The ability to reposition the TFSS with clean removability, feasibility to remove trapped air bubbles between the substrate and test object, good resistance to solvents, etc, are a few of the advantages of the substrate.

15.3.3 Testing of the transparent TFSSs

The fabricated sensor was tested for its optical transparency in the visible region and its resistance before evaluating its electromechanical response. The TFSSs' resistances were in the range of 3–4 KΩ. The insulation resistance between the TFSS and SS specimen was greater than 9999 MΩ at 50 V DC. The self-adhesive thin film strain sensor was subjected to:

1. A static strain test to evaluate its resistive response and strain sensitivity.
2. An impact force response test was carried out by a ball drop test.
3. Shock-tube tests were used to investigate the dynamic strain response.
4. A low pressure pneumatic calibrator was used to evaluate the response due to pressure.
5. The effect of temperature on the resistance of the sensor is also measured.

The observations and results of the tests carried out are discussed in the following sections.

15.3.3.1 Optical transparency of the sensor

The optical transmission spectra were recorded by a UV–visible spectrophotometer in the wavelength range of 400–800 nm, as shown in figure 15.19. The optical transmittance of single layered ITO thin film deposited by RF sputtering was the highest. A higher carrier density and the presence of silver particles between the ITO layers increases the optical scattering and absorption in the visible region, hence the transmission exhibited by trilayer film was marginally less than that of single layer

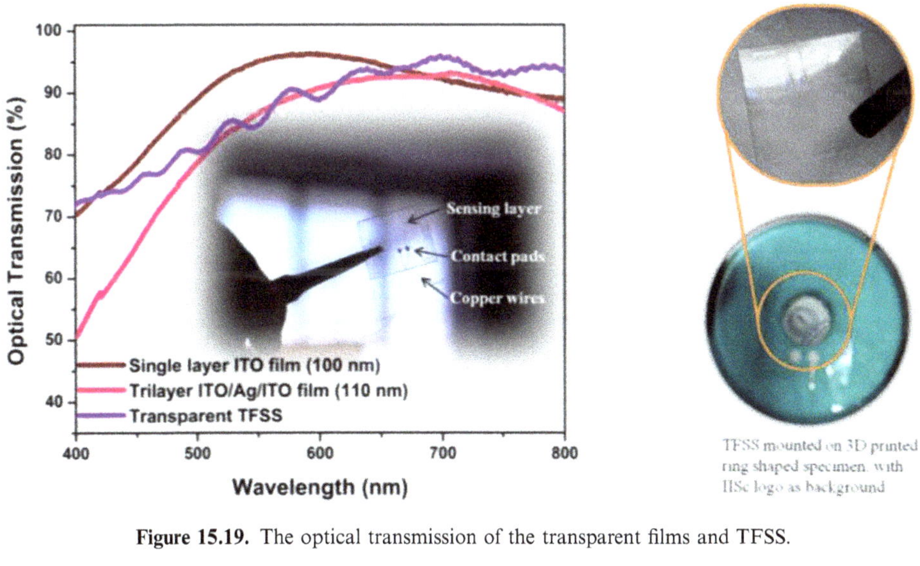

Figure 15.19. The optical transmission of the transparent films and TFSS.

ITO film. The average transmission of the TFSS improved compared to the blanket trilayer films due to the meandering structures causing relatively low optical absorption. A transparent TFSS on PET is camouflaged when attached to the test specimen, as shown in figure 15.19 where the institute logo placed behind the TFSS is visible, thus can be employed in sensing applications needing high optical visibility through the sensor and test objects.

15.3.3.2 Static strain response

The term 'static measurement' includes all the measurements of strain components which are constant in time. The most common techniques to evaluate the static strain response are

1. Cantilever beam bending.
2. Three point bending.
3. Four point bending.

In this work the cantilever beam bending technique was employed. The TFSS was attached onto an SS specimen of known dimensions by a two-step transfer method, as shown in figure 15.18. To measure the response to longitudinal stress, the SS cantilever beam was fixed at one end and a bending moment was applied on the free end using a height gauge to a maximum deflection of 1000 μm corresponding to a strain value of 1333 $\times 10^{-6}$. A pictorial representation of the cantilever bending method is shown in figure 15.20.

Beam deflection causes a change in the length (Δl) of the specimen attached to the TFSS. Tensile strain causes the resistance to increase, as shown in the plot in figure 15.20(b). The change in resistance of the TFSS at each deflection was recorded and the hysteresis observed was marginal. The applied strain was well within the fracture limit depicted by the linear relation of stress and strain in the inset of figure 15.20.

To arrive at the strain values from deflection of the cantilever, the applied force at any point on the beam was found using

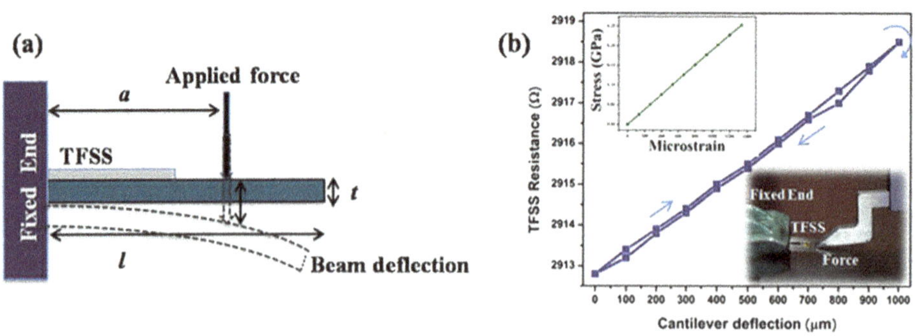

Figure 15.20. (a) Cross sectional view of the cantilever beam bending measurement set-up. (b) Change in TFSS resistance as a function of cantilever deflection.

$$F = \frac{6Ebt^3\,(\text{Deflection})}{12a^2(3l - a)}. \tag{15.3}$$

The following equation (15.4) converts the force into appropriate stress and equation (15.5) computes the strain value calculated from the stress using Hooke's law and the Young's modulus of SS (190 GPa):

$$\sigma = \frac{6Fa}{bt^2} \tag{15.4}$$

$$\varepsilon = \frac{\sigma}{E}, \tag{15.5}$$

where σ = stress on the beam, $\varepsilon = (\Delta l/l)$ strain on the beam, E = Young's modulus of the SS specimen, F = force applied, l = length of the beam, a = length of the beam from the fixed end to the point of force applied, t = beam thickness, and b = beam width.

The strain sensitivity of the TFSS can be calculated by measuring the relative change in resistance for a known amount of strain on the specimen under test. The measure of sensitivity of the strain sensor is called the gauge factor, given by

$$\text{gauge factor} = \frac{(\Delta R/R)}{\varepsilon}. \tag{15.6}$$

The higher the value of the gauge factor, the more sensitive is the sensor, thus the greater is the electrical output. The gauge factors reported in the literature not only depend on the type of sensing material used, but also on the conditions of deposition, its resistivity, substrate imperfections, and the magnitude of strain transfer. TFSSs prepared by sandwiching Ag in ITO films resulted in a longitudinal gauge factor in the range of 1.4 ± 0.15, as seen in figure 15.21. The relevant sensor design and dimension details can be found in the inset of the figure 15.21.

The strain sensitivity exhibited by the self-adhesive TFSS was lower than its bonded counterpart [24]. This behavior can be attributed to the less efficient strain transfer from the SS test specimen to the sensing layer due the migratory nature of the silicone particles that act as the adhesive layer of the TFSS. The silicone migration is due to the low surface energy substance (silicone) in contact with a higher energy surface (stainless steel specimen) [25].

15.3.3.3 Impact force response

An impact is a force or shock applied over a short period of time. Impact sensors are pivotal for structural health monitoring and failure analysis in a multitude of fields such as the aerospace, automotive, and process control industries. They also give information about the energy absorbed or energy required to fracture the mechanical components under test.

An experimental set-up for a 'ball drop test' was designed and developed at the Centre for Nano Science and Engineering, IISc, as elaborated in [26], to study the impact force response on test samples. The transparent TFSS was attached to a

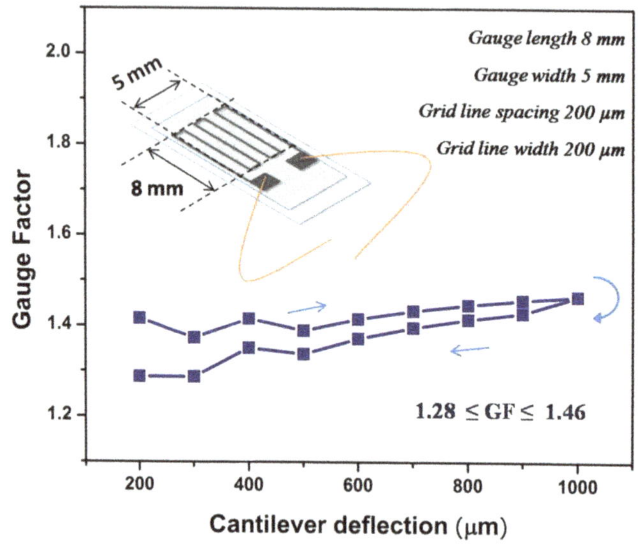

Figure 15.21. TFSS gauge factor versus cantilever deflection.

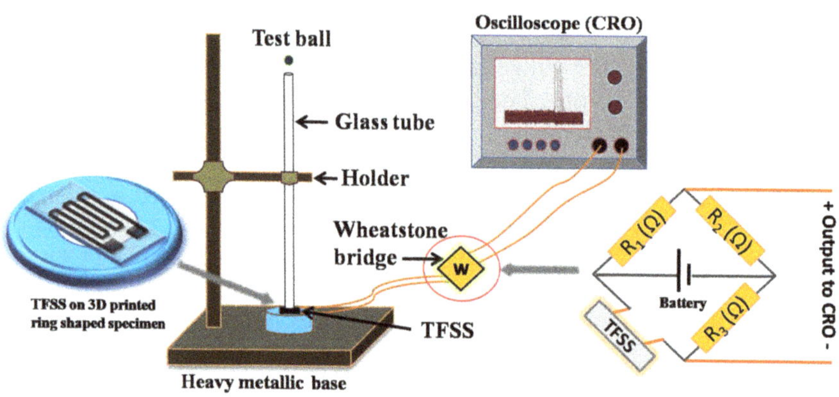

Figure 15.22. Schematic representation of the ball drop test experimental set-up.

circular 3D printed specimen, as shown in figure 15.19. The schematic of the test set-up is shown in figure 15.22. A glass tube holder with a heavy metal base served as a support structure for the glass tube placed vertically. The height of the glass tube was typically the drop height of the test ball that was dropped from the top side of the glass tube while the TFSS was placed at the bottom end of the tube. The base of the tube holder was made of heavy metal such that the stiffness and mass of the base was higher than that of the test balls.

This experimental set-up is a specific implementation of the law of the conservation of energy. With the assumption that the air drag in the glass tube and its side walls are negligible and the acceleration due to gravity is equal to 9.81 m s^{-2}, the testing condition of the experiment can simplify the energy relationship before and after the impact of the ball. The assumptions deduce that the potential energy before impact is equal to the kinetic energy after the impact. Hence, the conservation of energy equation is

$$mgh = \frac{1}{2}mv^2, \tag{15.7}$$

where m = mass of the test ball, h = drop height, g = acceleration due to gravity, and v = impact velocity. For a simpler understanding, the drag forces caused by the air resistance and side walls of glass tubes during the ball drop are neglected here. Considering the conservation of energy equation, the impact velocity of the test balls can be calculated using the relation

$$v_{\text{impact}} = \sqrt{(2gh)}. \tag{15.8}$$

The impact velocity is independent of the mass of the test balls. By employing the work energy principle to estimate the impact force due to ball drop, the net work done (W) during an impact event is given by

$$W_{\text{net}} = \frac{1}{2}\, mv_{\text{final}}^2 - \frac{1}{2}\, mv_{\text{initial}}^2. \tag{15.9}$$

The initial velocity (v_{initial}) in this experiment is zero. Hence, the net work done (W_{net}) during the impact event will be same as the net kinetic energy. The magnitude of the impact force exerted by the test balls can be calculated by determining the height of balls after the impact event. If the rebound height of the test ball after impact event is d, then the impact force (F_{impact}) is given by

$$F_{\text{impact}} = \frac{W_{\text{net}}}{d}. \tag{15.10}$$

The assessment of the rebound height (d) after the impact event depends largely on the type of collision. A significant rebound height after impact corresponds to much lower energy absorption by the material, hence a nearly elastic collision as in this case with the TFSS made on a PET substrate.

The impact by a test ball on the specimen causes a change in the resistance of the TFSS due to the induced strain on the transparent sensing layer, which is converted to a proportional voltage value using the Wheatstone bridge connected along with a storage oscilloscope, as seen in figure 15.22. The output response recorded for the ball dropped is seen in figure 15.23(a). The voltage peak difference was due to the ball bouncing back on the TFSS, hence the amplitude of the voltage decreases due to the lower impact forces after the first collision. The duration for the voltage to rise to its peak value after the first collision is seen in figure 15.23(b). This response is attributed to the impact force corresponding to the material density of the test ball, thus finding application in material discrimination studies.

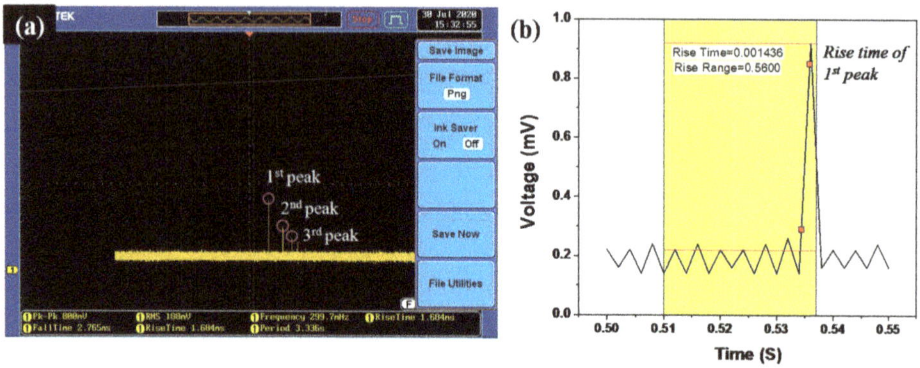

Figure 15.23. (a) The peak voltages recorded on the CRO for the dropped ball. (b) The time taken to rise to the maximum voltage value for the first collision.

15.3.3.4 Dynamic response evaluation

The measurements of time varying strain are defined as a dynamic response where only the dynamic component is determined, e.g. the amplitude of an oscillation. Strain gages are also well suited to the measurement of dynamic strain conditions. On account of their low mass there is no noticeable influence on the vibration characteristics of the test object [13].

A simple compression driven shock-tube set-up was employed to evaluate the dynamic response of the transparent TFSS mounted on an end flange. A compressed gas driven shock-tube consists of a driver and the driven section is shown in figure 15.24. The low pressure gas is referred to as the driven gas and is subjected to the shockwave. The high pressure gas is known as the driver gas and is monitored by a pressure monitor. A diaphragm controlled by a solenoid valve separates the two sections. The driven section contains commercial piezoresistive sensors to measure the shock speed as soon as the diaphragm bursts, producing a series of pressure waves that compress into a shock propagating through the driven gas. The distance between the commercial piezoresistive sensors is known and the voltage rise time of the two sensors was recorded on an oscilloscope. The time taken by the shockwave to travel the known distance is provided by the difference in the time of rise of each sensor. The speed of the shockwave can hence be calculated [27].

The primary shockwave that is generated due to the diaphragm burst travels through the driven section to strike the TFSS attached to the end flange, as shown in figure 15.24. The shockwaves travel through the driven section in microseconds and impacts the end flange before being reflected inside the driven section. The reflected waves lose energy and dampen out. The primary shockwave impacting the TFSS was recorded using the instrumentation set-up and connected to the digital storage oscilloscope which displays the output. The variation of the electrical resistance of the TFSS due to the shockwave impact is converted into voltage using a Wheatstone bridge network, thus measuring the instantaneous response in the time scale of a few microseconds.

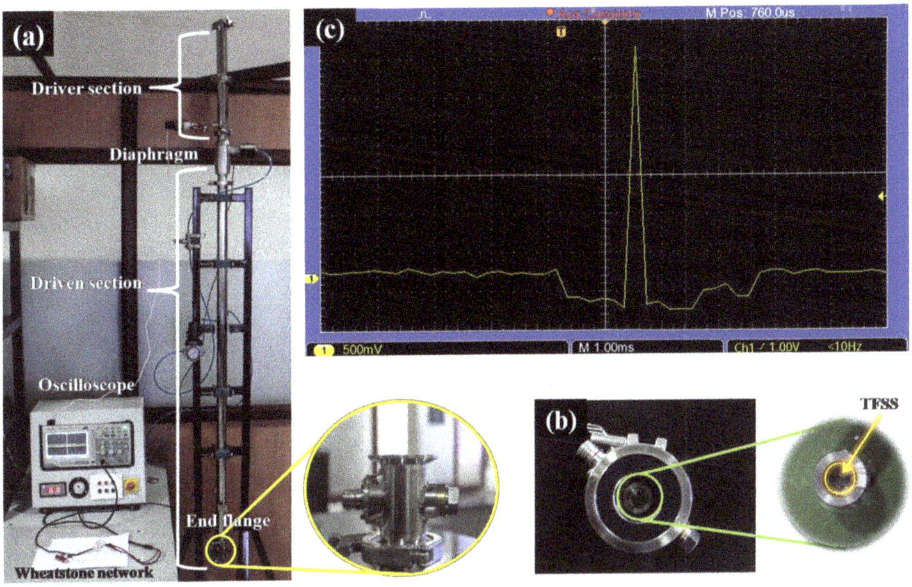

Figure 15.24. The experimental set-up for the dynamic response using shock-tube testing: (a) magnified side-view of the end flange connected to the shock-tube set-up; (b) top view of the end flange with the TFSS mounted region magnified; and (c) voltage response of the TFSS as observed on an oscilloscope.

If d is the distance (in millimeters) between the two commercial piezoresistive sensors and t (in seconds) is the time taken by the shockwave to travel the distance between the two sensors, then the velocity of the shockwave, V_{shock} (in m s^{-1}) is given by

$$V_{\text{shock}} = \frac{d}{t}. \tag{15.11}$$

The Mach number (M) is calculated as a ratio of the velocity of the shockwave to the velocity of sound in the same medium (air, in the present case) which is given by

$$M = \frac{V_{\text{shock}}}{a}. \tag{15.12}$$

Here a is the speed of sound in air (m s^{-1}) given as

$$a = \sqrt{(\gamma_1 RT)}, \tag{15.13}$$

where γ_1 (= 1.4) is the specific heat ratio, R (= 287 J kg^{-1}) is the specific gas constant, and T (= 300 K) is the air temperature.

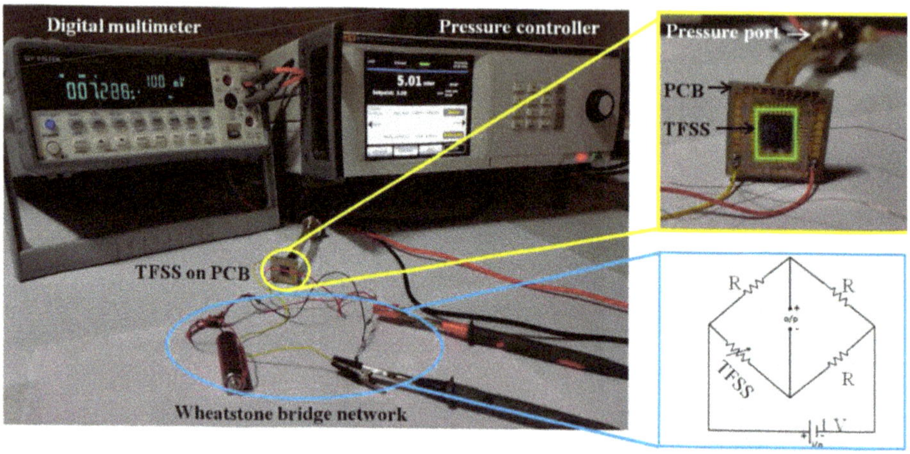

Figure 15.25. Experimental set-up to evaluate the pressure response of the TFSS mounted on PCB.

In this case the burst of the diaphragm results in the shockwave being generated with a Mach number of 1.16 and the response time of the TFSS was observed to be 138.2 μs.

15.3.3.5 Pressure sensing using TFSS

Pressure is defined as the force per unit area. A strain sensor based pressure transducer converts pressure into a measurable electrical quantity. Their function is based on the piezoresistive effect of the TFSS to change its output resistance in response to the physical deformation of the sensing element caused by the input pressure. By connecting the strain gauges in a Wheatstone bridge configuration in the experimental set-up, the small changes in resistance can be measured judiciously to give rise to a precise electrical signal proportional to the applied pressure. In this case the pneumatic pressure (in units of mBar) acting on the transparent TFSS changes the gauge length proportionally from L to $(L + \Delta L)$, which is directly proportional to the applied pressure.

Experimental set-up to evaluate the pressure response of the transparent TFSS is shown in figure 15.25. The TFSS was bonded directly on the PCB while connected in a Wheatstone bridge configuration to the digital multimeter, as seen in the magnified view in the figure. The bottom side of the PCB was connected to a plastic pressure port so that a low pressure pneumatic calibrator could be attached through a flexible tube to apply the input pressure to the TFSS. The maximum pressure was limited to 20 mBar with the excitation voltage of 1 V. The TFSS calibration was carried out by applying pressure in equal steps of ascending and descending values, and the graph of change in output resistance and voltage as a function of applied pressure is shown in figure 15.26.

Table 15.2 shows the measured quantities as a function of input pressure and summarizes the characteristics of the TFSS, i.e. sensitivity in units of Ω mBar^{-1} and mV mBar^{-1}, termed as sensitivity 1, and also the sensitivity with respect to the measured span in units of Ω Ω^{-1} mbar^{-1} and mV V^{-1} mBar^{-1}, termed as sensitivity 2,

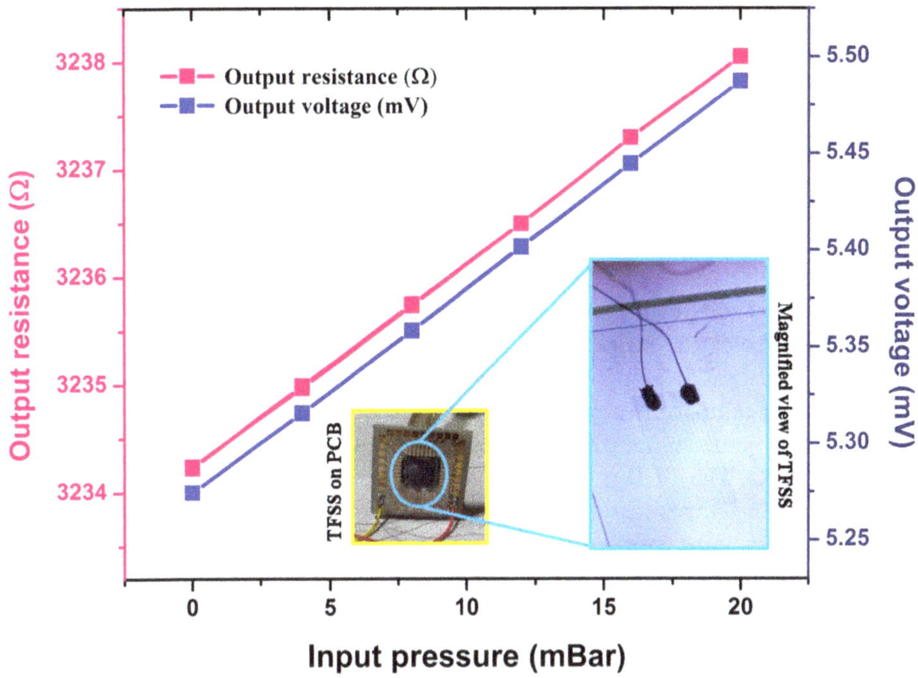

Figure 15.26. Output resistance and voltage as a function of applied pressure.

Table 15.2. The calibration of the TFSS (output response in ohms and millivolts).

Pressure (mBar)	Output (Ω)	Output (mV)
0	3234.23	5.2746
4	3234.99	5.3156
8	3235.75	5.3583
12	3236.50	5.4015
16	3237.30	5.4447
20	3238.05	5.4873
16	3237.30	5.4448
12	3236.50	5.4018
8	3235.74	5.3585
4	3234.97	5.3157
0	3234.24	5.2746
Sensitivity 1	0.191 Ω $mBar^{-1}$	0.011 mV $mBar^{-1}$
Span	3.820 Ω	0.2127 mV
Sensitivity 2	0.050 Ω $Ω^{-1}$ $mBar^{-1}$	0.0106 mV V^{-1} $mBar^{-1}$
Max. non-linearity and hysteresis	0.481%FSO	0.424%FSO

for ease of understanding. Maximum non-linearity and hysteresis %full scale output (%FSO) were calculated using the best fit straight line method.

15.3.3.6 Temperature response of the TFSS

An ideal strain sensor would change resistance only due to the deformations of the test surface to which the sensor is attached. However, in real applications the temperature, the material properties, the adhesive that bonds the gauge to the surface, and the stability of the metal all affect the detected resistance. The behavior of the electrical resistance of the transparent TFSS as a function of temperature is evaluated in this section to gauge its usefulness as a temperature sensor.

Ohm's law states the proportionality between voltage and current ($V = I \times R$), assuming a constant R of the resistor. In reality, ohmic resistors exhibit reversible changes of their temperature values when cooled or heated by changing the ambient temperature. These changes can be quantified by the temperature coefficient of resistance (TCR). The TCR is a measure of change in the resistance of a material per degree change in temperature, as given

$$TCR = \frac{(R - R_O)}{R_O(T - T_O)} \times 10^6 \tag{15.14}$$

in where R = electrical resistance at given temperature, R_O = electrical resistance at room temperature, T = any given temperature to measure TCR, and T_O = room temperature. The resulting value of the TCR is described in parts per million per degree celsius (ppm °C^{-1}).

The TFSS was placed inside a hot and cold chamber and the required connections were taken through the electrical feed-through. The sensors were heated to the desired temperature using internally mounted radiant cooler and heater set-ups. The transparent TFSS with a trilayer structure exhibited a positive temperature coefficient of resistance. The plot of change in resistance as a function of temperature is shown in figure 15.27, along with the TCR values at given temperatures.

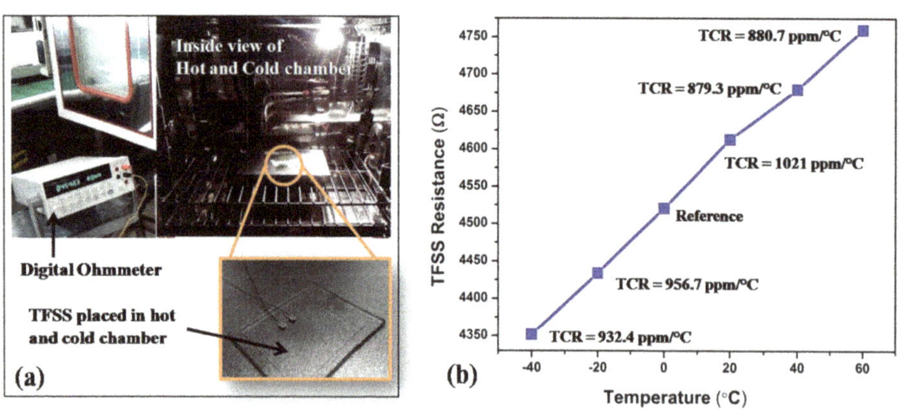

Figure 15.27. (a) Experimental set-up for the temperature response measurement in a hot and cold chamber. (b) Plot of change in resistance as a function of temperature with TCR.

15.4 Conclusion

In this chapter an experimental study carried out to develop inexpensive and non-intrusive transparent thin film strain sensors on self-adhesive substrates is presented. The methodology employed to choose appropriate raw materials for sensor fabrication and the methods to achieve high quality transparent films to serve as the strain sensing layer were elaborated. The fabrication process flow and performance evaluation of the TFSSs were discussed. Insights from the testing of TFSSs to evaluate their static and dynamic strain responses were discussed along with the description of the relevant experimental set-up.

References

[1] Kim K *et al* 2019 Recent advances in transparent electronics with stretchable forms *Adv. Mater.* **31** 1804690

[2] Lee T *et al* 2016 Transparent ITO mechanical crack-based pressure and strain sensor *J. Mater. Chem.* C **4** 9947–53

[3] Chen J *et al* 2020 Nano-engineered materials for transparent supercapacitors *ChemNanoMat* **6** 42–52

[4] Wu C-C 2018 Highly flexible touch screen panel fabricated with silver-inserted transparent ITO triple-layer structures. *RSC Adv.* **8** 11862–70

[5] Han S-T *et al* 2017 An overview of the development of flexible sensors *Adv. Mater.* **29** 1700375

[6] Zhou J *et al* 2008 Flexible piezotronic strain sensor *Nano Lett.* **8** 3035–40

[7] Rogers J A, Someya T and Huang Y 2010 Materials and mechanics for stretchable electronics *Science* **327** 1603–7

[8] Afre R A *et al* 2018 Transparent conducting oxide films for various applications: a review *Rev. Adv. Mater. Sci.* **53** 79–89

[9] Dixit R *et al* 2017 Preparation and properties of transparent conducting oxide (TCOs) thin films: a review *Int. J. Innov. Res. Sci.* **6** 1944–59

[10] Kalpana H M and Siddeswara Prasad V 2014 Development of the invar36 thin film strain gauge sensor for strain measurement *Meas. Sci. Technol.* **25** 065102

[11] Fiorillo A S, Critello C D and Pullano S A 2018 Theory, technology and applications of piezoresistive sensors: a review *Sens. Actuators* A **281** 156–75

[12] Kazi I H *et al* 2003 The electromechanical behavior of nichrome (80/20 wt.%) film *Thin Solid Films* **433** 337–43

[13] Hoffmann K 2012 *An Introduction to Stress Analysis and Transducer Design Using Strain Gauges* (Millbrook: HBM) pp 126–209

[14] Puyoo E *et al* 2018 Ultrahigh sensitivity to strain of cracked thin films based on metallic nanoparticles in a dielectric matrix *IEEE Sens. Lett.* **2** 1–4

[15] Parker R L and Krinsky A 1963 Electrical resistance–strain characteristics of thin evaporated metal films *J. Appl. Phys.* **34** 2700–8

[16] Magnfält D 2014 Fundamental processes in thin film growth: the origin of compressive stress and the dynamics of the early growth stages *Dissertation* Linköping University

[17] Razeghi M 2002 *Fundamentals of Solid State Engineering* (Berlin: Springer) p 410

[18] Sze S M and Kwok K N 2006 *Physics of Semiconductor Devices* (New York: Wiley)

[19] Materion 2015 Transparent conductive oxide thin films *Technical paper*

[20] Haacke G 1976 New figure of merit for transparent conductors *J. Appl. Phys.* **47** 4086–89
[21] Qiao Z 2003 Fabrication and study of ITO thin films prepared by magnetron sputtering *Dissertation* University of Duisburg-Essen
[22] Bingel A *et al* 2016 AZO/Ag/AZO transparent conductive films: correlation between the structural, electrical, and optical properties and development of an optical model *Opt. Mater. Express* **6** 3217–32
[23] Ren N, Zhu J and Ban S 2017 Highly transparent conductive ITO/Ag/ITO trilayer films deposited by RF sputtering at room temperature *AIP Adv.* **7** 055009
[24] Bhat S P, Bharathi S H and Nayak M M 2020 Strain sensing capabilities of Ag-sandwiched ITO as transparent thin film resistorIOP Conf. Ser.: Mater. Sci. Eng. **872** 012164
[25] Meyer J and Smith C A 2012 Effect of silicone contamination on assembly processes *IPC Apex Expo (Palm Bay, FL)* pp 1622–42
[26] Joshi S *et al* 2013 A novel piezoelectric thin film impact sensor: application in non-destructive material discrimination. *Sens. Actuators* A **199** 272–82
[27] Srivatsa S *et al* 2020 Dynamic response study of Ti_3C_2-MXene films to shockwave and impact forces *RSC Adv.* **10** 29147–55

IOP Publishing

Printed and Flexible Sensor Technology
Fabrication and applications
Subhas Chandra Mukhopadhyay and Anindya Nag

Chapter 16

Fabrication and characterization of CeO_2 thin film based oxygen sensors

N Ramshanker, Kolla Lakshmi Ganapathi, M S Bhat and S Mohan

A stable, scalable metal oxide oxygen sensor is a pre-requisite to develop miniaturized oxygen sensors that could find numerous applications in various fields such as the automotive industry, medical facilities, leak detection, and industrial safety. In this chapter the authors report on the fabrication of high-performance radio frequency (RF) magnetron sputtered cerium oxide (CeO_2) thin film based oxygen sensors by optimizing the film thickness. CeO_2 films of various thicknesses in the range of 90–340 nm were deposited on aluminum oxide (Al_2O_3) substrates at 400 °C using RF magnetron sputtering. The elemental composition and structural properties of the CeO_2 thin films were characterized by x-ray photoelectron spectroscopy (XPS) and x-ray diffraction (XRD); these are the key parameters to understand the gas sensitivity characteristics. The gas sensors were fabricated and tested with oxygen gas. The films of thickness 260 nm showed high sensitivity (~12.6) towards oxygen gas along with fast response and recovery times ($\leqslant$10 and 15 s, respectively) at an operating temperature of 400 °C. The developed CeO_2 thin film based oxygen sensors operated at 400 °C and are highly sensitive, stable, and scalable with fast response and recovery times.

16.1 Introduction

At present, the use of gas sensors has been extended to almost all fields, such as environmental monitoring, industry, transportation, medicine, and agriculture [1]. Currently, reliable air quality monitoring is a major concern worldwide due to worsening environmental pollution [2]. Harmful gases such as hydrocarbons, carbon monoxide, hydrogen sulfide, nitrogen oxides, ammonia, hydrogen, methane, etc, are released from industrial and agricultural processes and also from vehicle exhausts [3]. To overcome this problem the scientific and technological community has developed a

 © IOP Publishing Ltd 2021

wide range of gas sensors and many innovations have been achieved in gas sensor technology during the last 50 years [4].

One such sensor, namely the oxygen sensor, has found various applications in the automotive industry, medical facilities, industrial safety, modified atmospheric packaging, and many other fields [5]. Most oxygen sensors are designed to measure the quantity of oxygen in the air or in a given environment. Fresh air contains 20.9% oxygen by volume. One of the most common and dangerous atmospheric hazards is oxygen deficiency and it is very difficult for humans to detect this deficiency or depletion of oxygen in the atmosphere. Also, it is necessary to monitor leaks in oxygen cylinders in the medical sector and related fields using very precise instruments/detectors [6]. The types of oxygen sensors that have been developed are electrochemical, zirconia, optical, Clark, infrared, electro galvanic, ultrasonic, laser, and paramagnetic sensors. The type of sensor is chosen depending on the application and operating environment [5, 7].

Oxygen sensors were first developed by the company Bosch during the late 1960s, in which zirconium dioxide was used as a sensing element. Zirconia based oxygen sensors have been used extensively in automobiles to measure the oxygen concentration in vehicle exhaust. They work on solid-state electrochemical fuel cells. These sensors do not measure the oxygen concentration, but estimate the difference between the amount of oxygen in exhaust gases and the amount of oxygen in the air. The failure of these oxygen sensors could be due to ageing and wear and tear [8, 9]. Later, yttria (8% mol Y_2O_3) stabilized zirconia (YSZ) has attracted a lot of interest and is being used as a sensing material since it provides better stability in harsh environments due to its ionic conductive nature [10–13].

Another sensor that has attracted much attention is the semiconductor metal oxide (SMO) based resistive type gas sensor due to its high efficiency (in terms of response time and high sensitivity), low power consumption, ease of fabrication, simplicity of use, and the fact that it can be integrated with the existing complementary metal oxide semiconductor (CMOS) technology to make portable devices [2, 3, 14–16]. This sensor works on the principle of the detection of the change in the conductivity of a sensing layer upon adsorption and reaction with the analyte gas molecules. Therefore the selection of a sensing material and its structure is highly critical as it decides the sensing performance (sensitivity and selectivity). This type of sensor is called a chemiresistive or resistive gas sensor [4, 14, 17, 18].

Tin oxide (SnO_2) has dominated as a sensing material and has been extensively studied for gas sensing applications for a wide range of gases including hydrocarbons, NO_x, CO, NH_3, H_2S, NO_2, CO_2, SO_2, H_2, CH_4, etc, and many more [19–27]. However, SnO_2 based sensors are not stable at elevated temperatures and have poor selectivity [28, 29]. Titanium dioxide (TiO_2) is also one of the common materials used in oxygen sensing that does not require a reference sample, unlike the zirconia based oxygen sensors [30]. However, titania sensors need a complex compensatory electric circuit and cannot be used at oxygen partial pressures greater than 10 Pa [31–33].

In this scenario, ceria based materials have emerged as promising sensing elements because of their good dielectric properties, chemical and thermodynamic stability, high thermal shock resistance, and good ionic and electronic conductivity [19, 34–38]. Cerium oxide (CeO_2) is an n-type oxide semiconductor with a wide bandgap of 3–3.6 eV. The unique advantage of CeO_2 is the co-existence of trivalent (Ce^{+3}) and tetravalent (Ce^{+4}) states of the Ce atom and the ease of transition from one to the other [38–41]. The ratio of Ce^{+3}/ Ce^{+4} states, the oxygen vacancies, and the orientation of the crystal planes play a major role in the oxygen reactivity on the CeO_2 surface [42–44]. Extensive research has been carried out towards the development of CeO_2 as a sensing material for oxygen sensor applications. However, all these studies have been limited to thick films operated at high temperatures (>600 °C).

In 1991 Beie *et al* [45] developed a prototype CeO_2 based oxygen sensor in the form of both thick and thin films. The thicknesses of the thick films prepared by spin coating and screen printing were in the ranges of 5–50 and 5–30 μm, respectively. The ceria thin films were deposited by the RF sputtering technique and the thickness was between 0.5 and 1.5 μm. The authors reported that the films with a particle size of 2–3 μm showed a good response. However, the operating temperature was between 800 °C and 1000 °C. Later, Izu *et al* [46] prepared CeO_2 powder by mist pyrolysis and mixed it with an organic binder to form a paste which was screen printed. The response times obtained for particle sizes of 1 μm or less were 35, 22, 15, 11, and 8 s when operated at 1081 K, 1130 K, 1178 K, 1226 K, and 1275 K, respectively. Further, Izu *et al* [47] prepared a CeO_2 fine powder with an average particle size of 200 nm and fabricated thick (20–30 μm) oxygen sensors by screen printing. These sensors showed a response time of 10 s when operated at 712 °C. Again, Izu *et al* [48] reported oxygen sensors based on porous CeO_2 thick films (20–30 μm) with an average particle size of 100 nm. The reported response time was 11 s while operated at 615 °C. Izu *et al* [49] evaluated the performance of porous CeO_2 thick film based oxygen sensors with respect to the particle size and firing temperature. They found that the response time was not affected by particle size but depends on the firing temperature. They further observed that the films with particle sizes between 100–500 nm showed a good sensitivity to oxygen gas.

Most of the reported literature on CeO_2-based oxygen sensors involves very high temperatures (>600 °C). Moreover, the CeO_2 films have been prepared by chemical routes such as spray pyrolysis, mist pyrolysis, screen printing, and the sol–gel process [45–48, 50, 51]. The thicknesses of these CeO_2 films reported for gas sensing applications are in the range of micrometers to millimeters [45–54]. RF sputtered CeO_2 films are very rarely used for oxygen gas sensing applications [54, 55]. In addition, no systematic investigations have been carried out to understand the effect of CeO_2 film thickness on oxygen sensing.

In the above context, the authors report the fabrication and characterization of a CeO_2 thin film based oxygen sensor that is highly stable, scalable, and sensitive, with fast response and recovery times. This has been achieved by optimizing the CeO_2 film thickness in RF magnetron sputtering. Systematic investigations were carried out to understand the effect of thickness on the composition and structural properties of the deposited CeO_2 films using XRD and XPS, which have been used to understand the sensing properties. The deposited CeO_2 films with an optimum thickness of 260 nm showed high sensitivity (~12.6) towards oxygen gas along with fast response and recovery times (⩽10 and 15 s, respectively) at an operating temperature of 400 °C.

16.2 Experimental details

The selection of the film deposition technique plays a crucial role in determining the properties of the film. The most commonly used processing methods for thin film growth are physical vapor deposition (PVD) and chemical vapor deposition (CVD). The latter method is used to grow pretty films using specialized precursor materials and the process is carried out at higher temperatures, usually between 300 °C and 900 °C [56]. Some precursors and their by-products are toxic and corrosive and, hence, they need to be handled carefully [57]. PVD is a line-of-sight coating process that allows thin coatings and sharp edges [58]. It is divided broadly into evaporation and sputtering.

Sputtering has the following advantages compared to other PVD methods [59]:

- Wafer-scale thin films can be deposited with very good uniformity in thickness (1%–2%).
- The thin film composition and other properties including grain size can be controlled very precisely by regulating the various process parameters comprising deposition rate, deposition temperature, and pressure.
- Multi-component thin films can be deposited easily.

Sputtering is a thin film deposition technique in which the atoms or molecules are ejected from a target material by high-energy particle (ions, neutrals, etc) bombardment. The ejected atoms or molecules or ions, called unit species, will condense on the substrate to form a thin film. This method can be classified broadly into direct current (DC) and radio frequency (RF) sputtering. DC sputtering has become one of the most widely used techniques for depositing various metallic or conductive films, including gold, platinum, titanium, and aluminum. On the other hand, RF sputtering or reactive sputtering is frequently used to deposit oxides/nitrides such as SiO_2, CuO, ZnO, SnO_2, Si_3N_4, TiN, etc. The commonly used frequency for sputtering applications is 13.56 MHz since this is the RF spectrum that has been allocated to industrial applications [60–62].

The RF magnetron sputtering technique is the most preferred mode used in the deposition of oxide thin films. The desired composition and structural properties can be obtained by fine tuning the process parameters [62].

Figure 16.1 shows a picture of the RF sputtering unit that was developed in-house. The sputtering chamber houses the cathode, substrate holder/heater, and shutter assembly. The cathode is powered by an RF generator (13.56 MHz from AE-Dressler Cessar-1310). The material to be sputtered is taken in the form of a circular disc and it is mounted onto a copper backing plate in the sputter-up configuration mode. Here, the chosen target is CeO_2 of 99.99% purity with 3 inch diameter and 0.12 inch thickness. A shutter assembly will protect the sample during

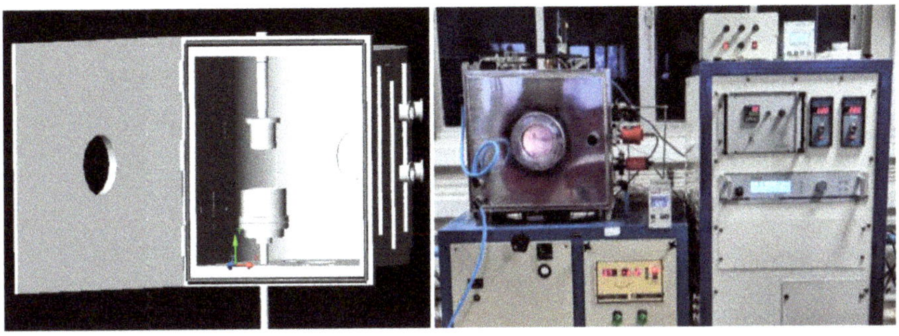

Figure 16.1. (a) AutoCAD drawing of the vacuum chamber. (b) In-house fabricated RF sputtering unit.

pre-conditioning of the target. A substrate rotation mechanism is incorporated to control the uniformity of the deposited films.

16.2.1 Sensor film (CeO_2) deposition and characterization

The initial step is cleaning of the substrate—in this case, it is sapphire (Al_2O_3)—to remove organic and inorganic contaminants. The Al_2O_3 substrates were cleaned ultrasonically in the organic solvents acetone and isopropyl alcohol, each for about 5 min. Then they were dried using high purity nitrogen gas. The cleaned Al_2O_3 substrates of size 1 cm × 1 cm were loaded into the chamber. The sputtering procedure was commenced by evacuating the chamber to a pressure lower than 3×10^{-6} mbar by a set of backing (rotary) and high vacuum (turbo molecular) pumps. Then Ar gas was introduced into the chamber using a mass flow controller (MFC) at a specified pressure. In the case of reactive sputtering, this was followed by allowing O_2 into the chamber at a set rate.

The important process parameters in RF sputtering are base pressure, working pressure, target power, reactive gas pressure, substrate to target distance, and substrate temperature. These parameters need to be optimized with respect to the structural, compositional, and electrical properties of the films to be deposited. The base pressure and working pressure determine the purity of the sputtering environment. The target power and the sputtering gas (Ar) influence the deposition rate and structure of the deposited film. The structural and electrical properties of the deposited CeO_2 film change significantly on varying the process parameters, such as working pressure, power, and substrate temperature. A uniform coating can be obtained by varying the distance between the substrate and target. The reactive gases, such as oxygen or nitrogen, are used to change the composition of the deposited films. The deposited samples are annealed for a certain duration to obtain the required stoichiometry. All the above process parameters need to be optimized for any specific new material depending on the required properties.

The main aim of the present work was to study the effect on thickness by varying the duration of deposition. Although the sputtering parameters (such as RF power and Ar:O_2 ratio) were variable, the RF power was fixed at 75 W and the Ar:O_2 flow was kept at 25:2 sccm. The deposition parameters are listed in table 16.1 and the

Table 16.1. Sputtering parameters used for deposition of CeO_2.

Technique	RF sputtering
Target	CeO_2 (99.99%)
Base pressure	2×10^{-6} mbar
Ar	25 sccm
O_2	2 sccm
Sputtering gas pressure	2.5×10^{-3} mbar
Power	75 W
Substrate to target distance	8.0 cm
Substrate temperature	400 °C
Deposition time	30–315 min
Annealing temperature	500 °C
Annealing duration	2 h

Table 16.2. The various techniques used to characterize the deposited CeO_2 thin films.

Description	Tool
Thickness	Ellipsometry, Dektak surface profiler, and cross-sectional scanning electron microscopy
Structural analysis with phase identification	X-ray diffraction
Chemical composition and core level binding energies	X-ray photoelectron spectroscopy

various techniques used to characterize the deposited CeO_2 thin films are shown in table 16.2.

16.2.2 Sensor fabrication and characterization

To measure the changes in electrical conductivity (change in resistance), aluminum electrodes of 100 nm thickness were deposited using the thermal evaporation technique. These electrodes were patterned by placing a shadow mask underneath the substrate. The figure 16.2 shows the schematic cross section of the basic oxygen sensor.

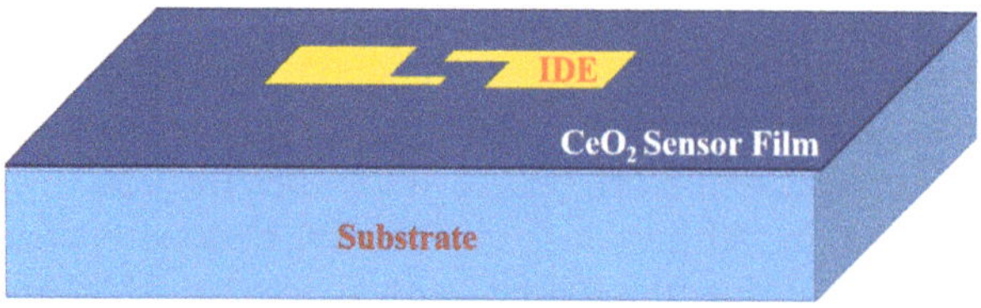

Figure 16.2. Schematic view of the sensor.

16.2.3 Electrical sensitivity measurement

The gas sensing set-up that was developed in-house is shown in figure 16.3. The sensors were mounted on a hot plate that was coupled with a ceramic base stand. The film (placed above the heater) was connected to two probes that were further connected in series with a Keithley Source-Measure Unit (Model number: 2405). The electrical connections for th egas sensing measurements, thermocouple, and temperature variation were made using instrumentation feed through.

The temperature of the hot plate was set to a desired value and the sensor was exposed to N_2 gas until a constant baseline resistance was reached. Then the analyte gas (in this case oxygen) was introduced into the chamber and the change in resistance of the sensor was recorded. During this time a steady resistance was achieved. Subsequently, the sensing chamber was flushed with pure N_2 to allow the surface of the sensitive film to regain the atmospheric condition and the resistance reached was steady and remained stable.

16.3 Results and discussion

The main objective of the study was to find the minimum thickness of CeO_2 thin films that would achieve a high performance (in terms of response time, recovery time, and sensitivity) in oxygen sensing. The CeO_2 thin films were deposited on Al_2O_3 substrates by varying the duration of deposition and keeping all other parameters (RF power, Ar:O_2 flow, and working pressure) constant (as shown in table 16.1).

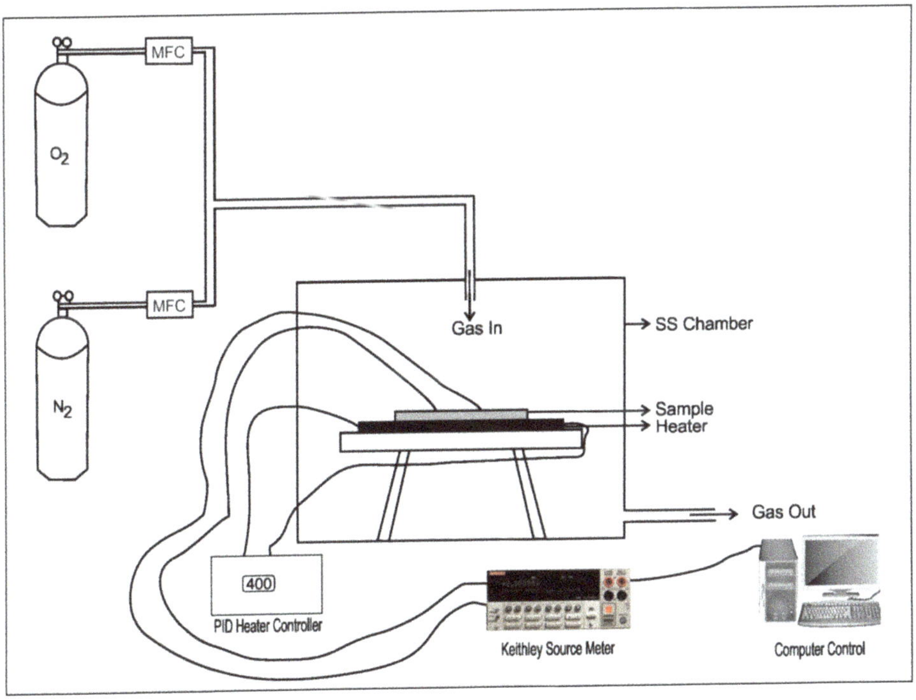

Figure 16.3. Schematic diagram of the gas sensing unit used for measuring sensor characteristics.

16.3.1 Physical properties

An Ellipsometer was used to measure and calculate the thickness of the film; the measured thickness was also validated using a Stylus profilometer and cross-sectional SEM. The figure 16.4 shows the variation of thickness with deposition time. It was observed that the thickness varies linearly with time, which indicates that the process is stable. The measured thickness was found to be in the range of 32 to 340 nm at a deposition rate of ~1.08 nm min^{-1}. From the ellipsometry data, the refractive index was calculated and found to be ~2.4. This is in good agreement with reported values of bulk CeO_2 in literature [63]. Figure 16.5 shows the cross-sectional scanning electron microscopy image of the CeO_2 film deposited for 240 min and the measured thickness was around 260 nm ± 15 nm.

16.3.2 Gas sensing characteristics

The change in resistance depends on nature of the sensor material (n-type or p-type) and the type of gas (reducing or oxidizing). For n-type semiconductors materials, the resistance increases for oxidizing gases and decreases for reducing gases [19, 64–66]. Hence the sensitivity (S) of CeO_2 (n-type) for the oxidizing gas (oxygen) is defined as

$$\text{Sensitivity } (S) = \frac{R_g}{R_a}, \tag{16.1}$$

where R_g is the resistance of the sensing film upon exposure to the target gas and R_a is the initial film resistance. It would be vice versa for p-type materials.

The performance of the gas sensor is determined by the response and recovery times; these are defined as the time taken to achieve 90% of the final change in conductance following a step change in gas concentration at the sensor, which should be as minimal as possible for faster detection. The response time and recovery time can range between seconds to minutes depending on the toxicity of the

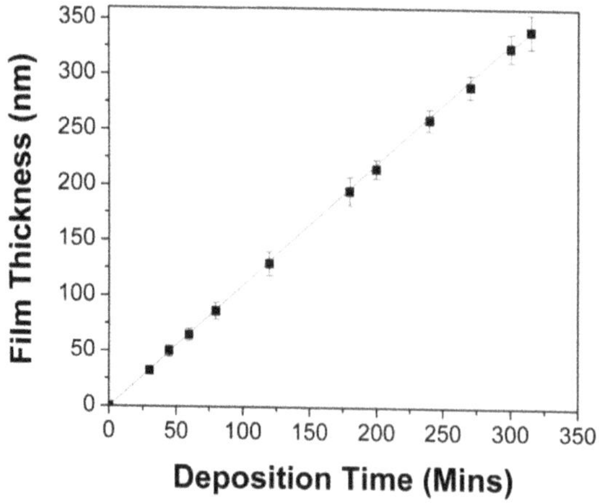

Figure 16.4. Variation of thickness w.r.t deposition time.

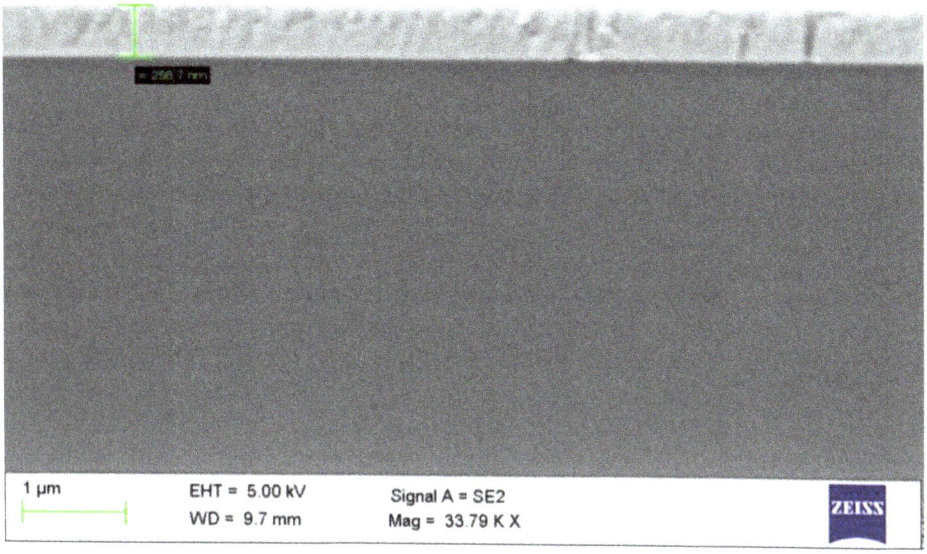

Figure 16.5. Cross-sectional scanning electron microscopy image of the CeO_2 film.

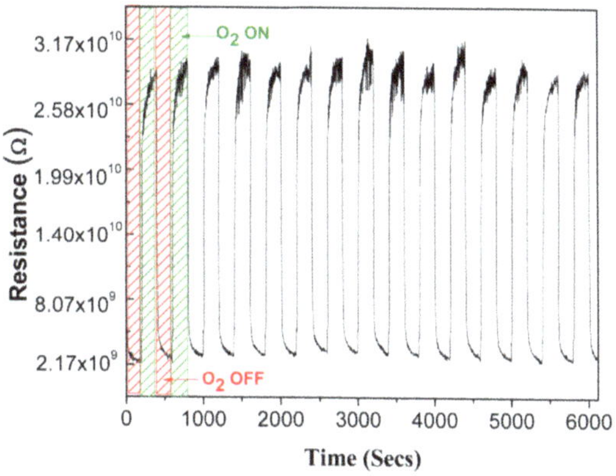

Figure 16.6. Typical response curve of a CeO_2 sensing film for oxidizing gas (260 nm CeO_2 thin film).

gas. In addition to this, there should not be any drift in the baseline resistance of the sensor even over long operating times—this ensures stability [18, 64, 65].

All the deposited CeO_2 thin films were subjected to oxygen gas sensing at various temperatures ranging between 50 °C and 400 °C. The figure 16.6 corresponds to the typical response curve of an oxygen sensor. From the response curve it is observed that a constant baseline resistance was achieved in a N_2 environment and upon exposure to the oxidizing (oxygen) gas (in the absence of N_2) the resistance increased from its baseline and became saturated. At this point, oxygen flow was stopped and N_2 was introduced. A drop in the resistance was observed and this became further saturated at its original baseline.

Figure 16.6 shows the sensing characteristics of the 260 nm film at an operating temperature of 400 °C for 100% O_2 gas. The measured response time and recovery time were found to be ⩽10 s and 15 s, respectively, with a very high sensitivity of 12.6 at 100% O_2 gas.

The CeO_2 films of thickness up to 125 nm did not reveal any change in resistance after exposure to oxygen gas. This might be due to very high resistance. The films of thickness around 130 nm showed sensing, but the response and recovery times were observed to be higher (29 and 31 s, respectively). Moreover, the sensitivity was also very poor. The performance of the sensor (sensitivity, response, and recovery times) was observed to improve with an increase in thickness from 130 to 260 nm. However, a decreasing trend was again observed in the sensor performance above a certain optimum thickness of 260 nm. The operating temperature during sensing was also varied between room temperature (RTP) to 400 °C and the best sensing results were found at 400 °C. The sensing performance of different films at 400 °C is listed in table 16.3, which shows the variation of response and recovery along with the electrical resistivity of CeO_2 thin films at an operating temperature of 400 °C

The table 16.3 clearly indicates that the oxygen sensitivity increased up to a film thickness of 260 nm; however, it decreased in films of thicknesses above 260 nm.

To ascertain the minimum oxygen detection limit, the oxygen concentration was varied between 1% and 100%. Figure 16.7 shows the sensitivity of the 260 nm CeO_2 thin film sensor when exposed to different oxygen levels (100%, 50%, 35%, 20%, 10%, and 1%) at 400 °C. The sensor also responds to a very low level of oxygen concentration of 1%.

Overall the performance of the 260 nm CeO_2 film with respect to oxygen sensing was found to be good in terms of response time, recovery time, and sensitivity. The reasons for this high sensing performance needed to be ascertained and hence further characterization was carried out to study the variation of film sensitivity with respect to thickness.

16.3.3 Compositional properties of CeO_2 thin films

The elemental composition of the deposited CeO_2 thin films was calculated from XPS wide or survey spectra (figure 16.8). These clearly indicated that the deposited films were pure (single phase) in nature. In other words, no other impurity peaks were observed except Ce and O.

Table 16.3. Sensor performance of CeO_2 thin films of various thicknesses.

Thickness (nm)	Response time (s)	Recovery time (s)	Sensitivity	Resistance (Ω)
130	29	31	1.87	8.0×10^{-9}
195	25	30	4.12	5.5 × 10–9
220	15	19	5.42	4.0×10^{-9}
260	**10**	**15**	**12.93**	$\mathbf{2.5 \times 10^{-9}}$
320	15	18	5.12	2.0×10^{-9}
340	20	22	2.12	3.2×10^{-9}

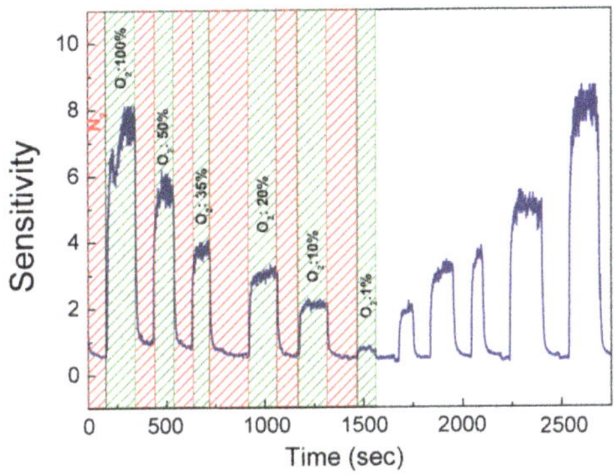

Figure 16.7. Sensitivity towards a varied concentration of O_2.

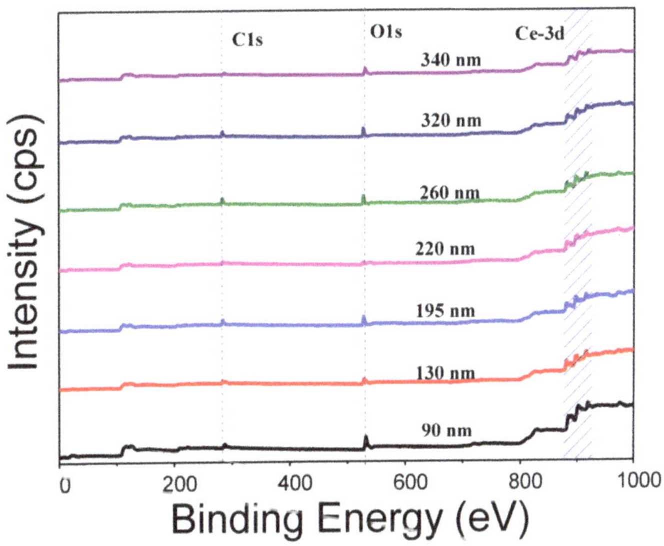

Figure 16.8. XPS wide spectra of CeO_2 films of various thicknesses.

By using the relative sensitivity factors (RSFs) of oxygen and cerium, the ratio of O to Ce atomic concentrations was extracted. The RSFs are used to scale the measured peak areas so that variations in the peak areas are representative of the amount of material on the surface of the sample. The RSF values were obtained from the library of elements for XPS transitions. The RSF of Ce-3d is 51.62, Ce-3d$_{3/2}$ is 21.12 and oxygen is 2.93. For an increase in thickness up to 195 nm, the ratio of O to Ce increased from 1.4 ± 0.1 to 2 ± 0.1 and for a thickness >195 nm it became saturated at 2 ± 0.1. This gives a clear indication that the deposited films are stoichiometric (CeO_2). A similar observation was also made by Sunkara *et al* [51].

However, the structural properties of the films including phase, orientation, grain size, and surface energy, and the diffusion coefficient also influences the sensing properties, apart from the composition [4, 18, 66–68]. Due to the strong hybridization of the Ce-4f and O-2p states in CeO_2, the Ce-3d core level would have a complex structure (Ce^{+3}or Ce^{+4}) as shown in figure 16.9.

The core level spectra of Ce were deconvoluted using the CasaXPS software with a Shirley background correction with 30% Gaussian and 70% Lorentizian peak fitting. Charging effects were corrected by setting the C1s peak for adventitious carbon at 284.6 eV for the surface and by flood gun after Ar^+ ion etching and shifting the entire spectrum correspondingly. The peak positions at the binding energies (BEs) of 882.0, 884.9, 888.5, and 897.9 eV correspond to the components of Ce-$3d_{5/2}$. Similarly, the peaks at 900.6, 903.6, 907.2, and 916.2 eV are attributed to the components of Ce-$3d_{3/2}$. In brief, the major component (peaks at the BEs of 882.5, 888.2, 898, 901.1, 907.2, and 916 eV) correspond to Ce^{+4} cations (as marked in figure 16.9) in the CeO_2 lattice and the minor component (peaks at the BEs of 881.1, 885.83, 899.65, and 904.3 eV) corresponds to Ce^{+3} in the amorphous Ce_2O_3 [69–71]. The total fraction of Ce_2O_3 (Ce^{+3}) was calculated using the following equation [71]:

$$\%Ce^{+3} = \frac{Ce^{+3}}{Ce^{+3} + Ce^{+4}}. \tag{16.2}$$

The fraction of Ce^{+4} (CeO_2) can be calculated from the following equation as no other Ce oxidation states except Ce^{+4} and Ce^{+3} were found in the CeO_2 films:

$$\%Ce^{+4} = 1 - \%Ce^{+3}. \tag{16.3}$$

Figure 16.10 shows the variation of the % Ce^{+4} and % Ce^{+3} components with respect to CeO_2 film thickness. It was observed that the % Ce^{+4} component increased with CeO_2 film thickness, which indicates that the defects and oxygen vacancies decreased [72–74]. This indicates clearly that the stoichiometry and quality of the films improve with an increase in thickness (from 90 to 340 nm). However, the stoichiometry of films that are ⩽90 nm can be improved by tuning the deposition parameters including the flow rate and deposition pressure; this needs to be investigated.

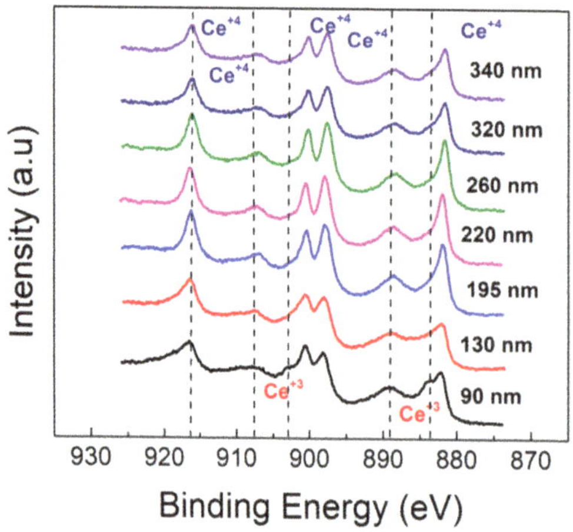

Figure 16.9. High resolution XPS spectra (Ce-3d level) of CeO_2 films of different thicknesses. Copyright 2019 IEEE. Reprinted, with permission, from [44].

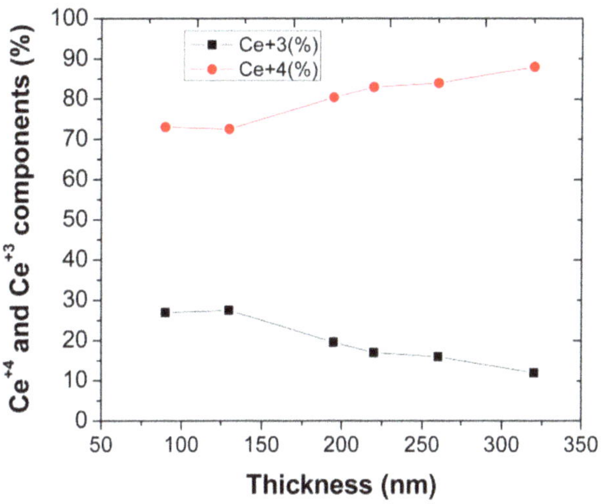

Figure 16.10. Variation in the oxidation states of Ce (Ce^{+4} and Ce^{+3}) components w.r.t CeO_2 film thickness.

Table 16.4. The atomic concentration values of lattice oxygen (O_I) and adsorbed oxygen (O_{II}).

Thickness (nm)	O_I (529.3 eV)	O_{II} (531 eV)	$O_{II}/(O_I+O_{II})$
90	51.9	48.1	0.48
130	33.9	66.1	0.66
195	53.0	47.0	0.47
220	58.3	41.7	0.41
260	76.0	24.0	0.24
300	84.6	15.4	0.15

The quantitative analysis of oxygen vacancies gives insight into the high resistance of the 300 nm CeO_2 film. Oxygen vacancies can be obtained from the relative percentage of adsorbed oxygen. The calculated atomic concentration of the lattice oxygen (O_I) and adsorbed oxygen (O_{II}) are summarized in table 16.4. The obtained proportion of $O_{II}/(O_I + O_{II})$ is plotted with respect to thickness in figure 16.11 as an inset. It clearly shows that the proportion of $O_{II}/(O_I + O_{II})$ decreases with an increase in thickness; this indicates that the oxygen vacancy concentration decreases with an increase in thickness, which is in good agreement with the Ce^{+3} concentration. Hence, the high resistance in 300 nm films can be attributed to a low oxygen vacancy concentration that gives a low carrier concentration.

16.3.4 Structural properties

X-ray diffraction was performed for phase identification of the crystalline material using the Rigaku SmartLab system. Figure 16.12 shows the XRD patterns of the CeO_2 films of various thicknesses after annealing at 500 °C for 2 h. The films clearly showed an improvement in crystallinity, which indicates the presence of more-

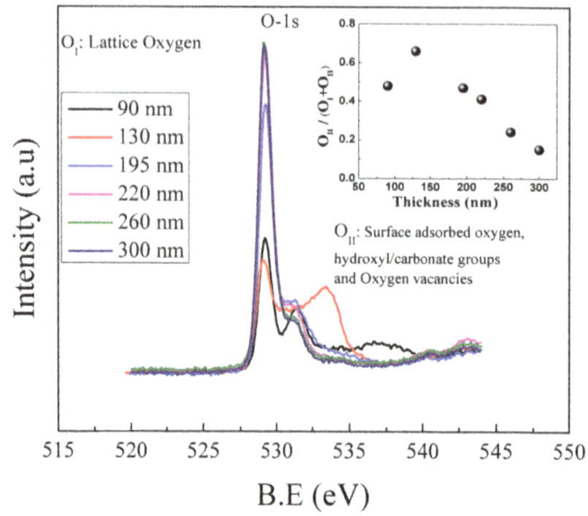

Figure 16.11. High resolution XPS of O-1s in CeO_2 films and inset shows the content of oxygen vacancies obtained from the XPS relative percentage of adsorbed oxygen.

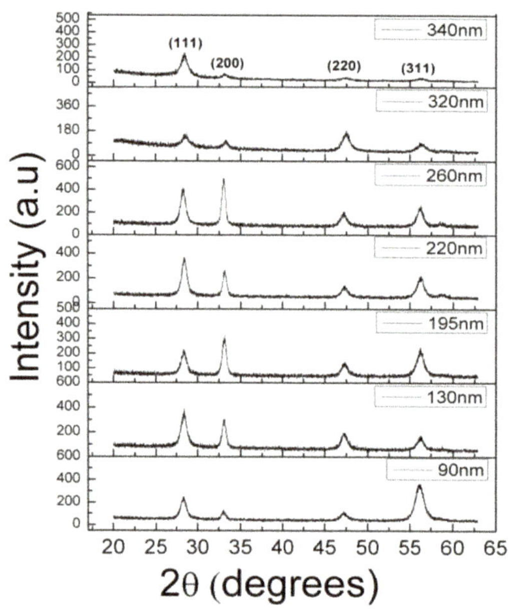

Figure 16.12. XRD patterns of CeO_2 films of various thicknesses. Copyright 2019 IEEE. Reprinted, with permission, from [44].

oriented CeO_2 of phases (111), (200), (220), and (311). It was also noted that the XRD peaks became sharper after annealing, which may indicate an improvement in the grain size with annealing. The characteristic peaks of the XRD patterns matched well with that of CeO_2 (JCPDS no. 043–1002) having a fluorite cubic structure with the F*m*-3*m* space group. The XRD patterns were composed of four strong and broad peaks centered around 28.4°, 33.1°, 47.3°, and 56.2° that correspond to the

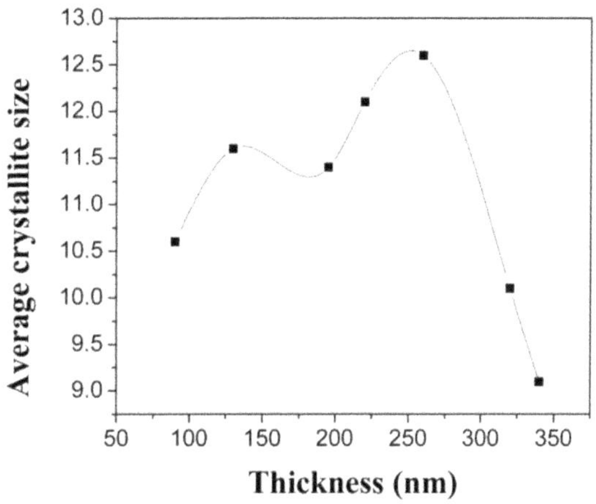

Figure 16.13. Average crystal size of the deposited CeO_2 thin films.

(111), (200), (220), and (311) reflections of CeO_2, respectively. This indicates that all the films are polycrystalline in nature. No other phases of CeO_2 (including Ce_2O_3 and CeO_{2-x}) were observed. However, Ce_2O_3 could be present in the amorphous form, which may be below the detection limit of the XRD.

The major factor that affects the sensing properties of gas sensors is the microstructure of the polycrystalline material [75, 76]. Hence, an understanding of the crystallite size is very important since it provides information of the morphology of the sensing material, which can be obtained by the shape, position, and width of the peaks.

The crystallite size can also cause peak broadening. The well-known Scherrer's equation was used to calculated the crystallite size for all the reflections (111), (200), (220), and (311) [77]:

$$D = \frac{K\lambda}{\beta \text{ Cos } \theta}, \tag{16.4}$$

where D is the crystallite size, β is the full width half maximum (FWHM), K is a constant close to unity (i.e. 0.94) for spherical particles, λ is the incident x-ray radiation wavelength (i.e. $\lambda_{\text{Cu-K}\alpha}$ 1.52 Å), and θ is the Bragg reflection angle.

From figure 16.13 it is seen that the crystallite size increases with an increase in thickness up to 260 nm and decreases for films above 260 nm. This increase in crystallite size for films up to 260 nm can be attributed to an improvement in grain growth. Above 260 nm the mobility of the adatoms may be low as the film grows beyond a certain critical thickness which may affect the grain growth and film orientation. This needs a detailed investigation to obtain insight into the nucleation process.

In general, the (111) orientation would have the lowest surface energy due to the highest packing density; this favors the growth of CeO_2 films in the (111) orientation [46, 48]. In contrast to this, it was observed that the intensity of the (200) plane increases with an increase in the CeO_2 film thickness.

In order to quantify the preferred orientation and surface energy of the deposited CeO_2 films, the texture coefficient (TC) was calculated using [78]

$$TC(hkl) = \frac{\frac{I(hkl)}{Io(hkl)}}{\frac{1}{N}\sum N\frac{I(hkl)}{Io(hkl)}} \tag{16.5}$$

where N is the number of diffraction peaks, and $I(hkl)$ and $I_o(hkl)$ are the measured and corresponding recorded intensities, respectively, according to the JCPDS no. 043–1002 card.

The texture coefficients for the (111) and (200) planes are plotted in figure 16.14. The TC for the (200) plane is maximum for the 260 nm film, which indicates that the film has the highest surface energy compared to other films with required stoichiometry due to the low packing density along this plane [46, 48].

From the structural (XRD), compositional (XPS), and electrical results, it was observed that the 260 nm films show better crystallinity and high surface energy with low resistivity and the required stoichiometry compared to other films.

From the sensing data it was observed that the 260 nm film has maximum sensitivity to gas sensing, which might be due to the (200) plane as its preferred orientation with highest surface energy; due to this, it has high reactivity compared to other planes because of defects or oxygen vacancies [79, 80]. The rest of the planes (111), (220), and (311) are also equally important since they indicate the nature of the film; here, it is polycrystalline. In addition to this, the low sensitivity in films < 260 nm may be due to high resistance.

From this, it can be concluded that the 260 nm films, with the lowest resistance due to better crystallinity, the required stoichiometry, and an orientation in the (200) direction which has the highest surface energy, seems to have the optimum thickness for CeO_2 based oxygen sensors.

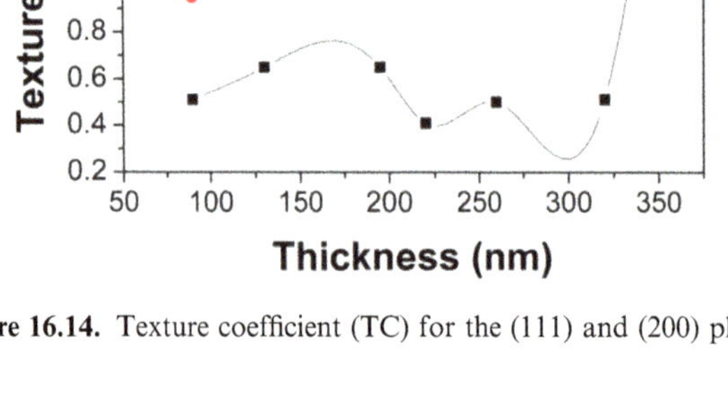

Figure 16.14. Texture coefficient (TC) for the (111) and (200) plane.

16.4 Conclusion

The CeO_2 films of various thicknesses ranging between 90 and 340 nm were deposited on Al_2O_3 substrates at 400 °C using RF magnetron sputtering. Systematic investigations were carried out to understand the effect of thickness on the composition and structural orientation of the deposited CeO_2 films by using various characterization techniques, such as ellipsometry, XPS, and XRD. The gas sensors were fabricated with CeO_2 film as a sensing element and tested with oxygen gas. It was observed that the sensitivity increases with increasing CeO_2 film thickness from 130 nm to 260 nm and then decreases for films above 260 nm. A CeO_2 film thickness of 260 nm showed high sensitivity (12.6) towards oxygen gas along with fast response and recovery times of $\leqslant$10 sand 15 s, respectively, at an operating temperature of 400 °C. For pure CeO_2 thin film based sensors, these are the best values reported to date.

The superior performance of 260 nm film towards oxygen sensing can be attributed to low resistance due to its crystalline nature and the presence of a highly reactive surface due to the (200) preferred orientation which is a highly reactive plane compared to other planes such as (111), (220), and (311).

References

[1] Hu Y, Tan O K and Zhu W 2006 Nanosized metal-oxide semiconducting $SrTi_{1\pm x}O_{3-\delta}$ oxygen gas sensors for low-temperature application *IEEE Sensors J.* **6** 1389–94

[2] Gulial S, Shiva Nagendra S M, Khare M and Khanna I 2015 Urban air quality management —a review *Atmos. Pollut. Res.* **6** 286–304

[3] Li Z *et al* 2019 Advances in designs and mechanisms of semiconducting metal oxide nanostructures for high-precision gas sensors operated at room temperature *Mater. Horiz.* **6** 470–506

[4] Neri G 2015 First fifty years of chemoresistive gas sensors *Chemosensors* **3** 1–20

[5] Gaslab 2021 How does an oxygen sensor work?

[6] Arrieta A, Brickell P, Camparada V, Fry C, Gachot R, Lewandowski J, Nielsen A, Patel M and Ritlop D 2009 Hazards of inert gases and oxygen depletion *European Industrial Gases Association IGC Document 44/09/E* http://www.linde-gas.com.ph/en/images/Hazards%20of%20inert%20gases%20and%20oxygen%20depletion_tcm30-13909.pdf

[7] AZO Sensors 2019 How oxygen sensors work and their applications https://azosensors.com/article.aspx?ArticleID=1792 (Accessed: 19 Sept. 2020)

[8] Wikipedia Oxygen sensors http://frankshospitalworkshop.com/equipment/documents/oxygen_concentrators/wikipedia/Oxygen%20sensor.pdf

[9] Freudenburger B 2020 Oxygen sensors 101 *Report* Walker Products http://walkerproducts.com/wp-content/uploads/2020/05/OXYGEN-SENSORS-101-INFORMATION-BOOKLET-Europe.pdf

[10] Tsampas M N *et al* 2015 Applications of yttria stabilized zirconia (YSZ) in catalysis *Catal. Sci. Technol.* **5** 4884–900

[11] López-Gándara C *et al* 2009 SZ-based oxygen sensors and the use of nanomaterials: a review from classical models to current trends *J. Sens.* **2009** 258489

[12] Sillassen M *et al* 2009 Ionic conductivity and thermal stability of magnetron-sputtered nanocrystalline yttria-stabilized zirconia *J. Appl. Phys.* **105** 104907

[13] Yu S *et al* 2002 Development of a silicon-based yttria-stabilized-zirconia (YSZ), amperometric oxygen sensor *Sen. Actuators* B **85** 212–8

[14] Barsan N *et al* 2007 Metal oxide-based gas sensor research: how to? *Sens. Actuators* B **121** 18–35

[15] Zhou Q *et al* 2015 Fast response integrated MEMS microheaters for ultra low power gas detection *Sens. Actuators* A **223** 67–75

[16] Pearton S J *et al* 2010 Recent advances in wide bandgap semiconductor biological and gas sensors *Prog. Mater. Sci.* **55** 1–59

[17] Mirzaei A *et al* 2019 Resistive gas sensors based on metal-oxide nanowires *J. Appl. Phys.* **126** 241102

[18] Wang C *et al* 2010 Metal oxide gas sensors: sensitivity and influencing factors *Sensors* **10** 2088–106

[19] Eranna G *et al* 2004 Oxide materials for development of integrated gas sensors—a comprehensive review *Crit. Rev. Solid State Mater. Sci.* **29** 111–88

[20] Windischmann H and Mark P 1979 A model for the operation of a thin-film SnO_x conductance-modulation carbon monoxide sensor *J. Electrochem. Soc.* **126** 627

[21] Tang Z *et al* 1998 An integrated gas sensor based on tin oxide thin-film and improved micro-hotplate *Sens. Actuators* B **46** 174–9

[22] Yannopoulos L N 1987 A p-type semiconductor thick film gas sensor *Sens. Actuators* **126** 263–73

[23] Fujimoto C *et al* 1996 Evaluation of the efficiency of deodorants by semiconductor gas sensors *Sens. Actuators* B **32** 191–4

[24] Martin M A *et al* 1999 Study of the interferences of NO_2 and CO in solid state commercial sensors *Sens. Actuators* B **58** 469–73

[25] Seiyama T and Kagawa S 1966 Study on a detector for gaseous components using semiconductive thin films *Anal. Chem.* **38** 1069–73

[26] Girardin D *et al* 1997 Modelling of SO_2 detection by tin dioxide gas sensors *Sens. Actuators* B **43** 147–53

[27] Yamazoe N *et al* 1979 Interactions of tin oxide surface with O_2, H_2O and H_2 *Surf. Sci.* **86** 335–44

[28] Radecka M *et al* 1998 SnO_2–TiO_2 solid solutions for gas sensors *Sens. Actuators* B **47** 194–204

[29] Liu X *et al* 2012 A survey on gas sensing technology *Sensors* **12** 9635–65

[30] Li Z *et al* 2018 Resistive-type hydrogen gas sensor based on TiO_2: a review *Int. J. Hydrogen Energy* **43** 21114–32

[31] Zheng L *et al* 2000 TiO_{2-x} thin films as oxygen sensor *Sens. Actuators* B **66** 28–30

[32] Kaiser W J and Logothetis E J 1983 Exhaust gas oxygen sensors based on TiO_2 films *SAE Technical Paper* 830167

[33] Esper M J *et al* 1979 Titania exhaust gas sensor for automotive applications *SAE Technical Paper 790140* pp 518–26

[34] Liao L *et al* 2008 Single CeO_2 nanowire gas sensor supported with Pt nanocrystals: gas sensitivity, surface bond states, and chemical mechanism *J. Phys. Chem.* C **112** 9061–5

[35] Naqvi S T R *et al* 2018 Synthesis, design and sensing applications of nanostructured ceria-based materials *Analyst* **143** 5610–28

[36] Fronzi M *et al* 2009 Stability and morphology of cerium oxide surfaces in an oxidizing environment: a first-principles investigation *J. Chem. Phys.* **131** 104701

[37] Zamiri R *et al* 2015 Dielectrical properties of CeO_2 nanoparticles at different temperatures *PLoS One* **10** e0122989
[38] Vangelista S *et al* 2017 Structural, chemical and optical properties of cerium dioxide film prepared by atomic layer deposition on TiN and Si substrates *Thin Solid Films* **636** 78–84
[39] Guo S *et al* 1995 A spectroscopic ellipsometry study of cerium dioxide thin films grown on sapphire by RF magnetron sputtering *J. Appl. Phys.* **77** 5369–76
[40] Trovarelli A and Llorca J 2017 Ceria catalysts at nanoscale: how do crystal shapes shape catalysis? *ACS Catal.* **7** 4716–35
[41] Liu Y, Ding Y, Zhang L, Gao P X and Lei Y 2012 CeO_2 nanofibers for *in situ* O_2 and CO sensing in harsh environments *RSC Adv.* **2** 5193–8
[42] Majumder D and Roy S 2018 Development of low-ppm CO sensors using pristine CeO_2 nanospheres with high surface area *ACS Omega* **3** 4433–40
[43] KasÏpar J, Fornasiero P and Graziani M 1999 Use of CeO_2-based oxides in the three-way catalysis *Catal. Today* **50** 285–98
[44] Ramshanker N *et al* 2019 RF sputtered CeO_2 thin films-based oxygen sensors *IEEE Sensors J.* **19** 10821–8
[45] Beie H J and Gnorich A 1991 Oxygen gas sensors based on CeO_2 thick and thin films *Sens. Actuators* B **4** 393–9
[46] Izu N *et al* 2002 Numerical analysis of response time for resistive oxygen gas sensors *Sens. Actuators* B **87** 99–104
[47] Izu N *et al* 2002 Resistive oxygen gas sensors based on CeO_2 fine powder prepared using mist pyrolysis *Sens. Actuators* B **87** 95–8
[48] Izu N *et al* 2003 Fast response of resistive-type oxygen gas sensors based on nano-sized ceria powder *Sens. Actuators* B **93** 449–53
[49] Izu N *et al* 2003 The effects of the particle size and crystallite size on the response time for resistive oxygen gas sensor using cerium oxide thick film *Sens. Actuators* B **94** 222–7
[50] Shin W *et al* 2004 Millisecond-order response measurement for fast oxygen gas sensors *Sens. Actuators* B **100** 395–400
[51] Manorama S V *et al* 2003 On the platinum sensitization of nanosized cerium dioxide oxygen sensors *Sens. Actuators* B **89** 299–304
[52] Jasinski P, Suzuki T and Anderson H U 2003 Nanocrystalline undoped ceria oxygen sensor *Sens. Actuators* B **95** 73–7
[53] Chen C-Y and Chang K-H 2012 Temperature independent resistive oxygen sensor prepared using zirconia-doped ceria powders *Sens. Actuators* B **162** 68–75
[54] Gerblinger J *et al* 1995 High temperature oxygen sensor based on sputtered cerium oxide *Sens. Actuators* B **26** 93–6
[55] Izu N *et al* 2004 Resistive oxygen sensors using cerium oxide thin films prepared by metal organic chemical vapor deposition and sputtering *Japan. J. Appl. Phys.* **43** 6920–24
[56] Reed M L and Fedder G K 1998 Photolithographic microfabrication *Handbook of Sensors and Actuators* vol 6 (Amsterdam: Elsevier) pp 13–61
[57] Barron A R 2011 Chemical vapor deposition *Chemistry of Electronic Materials* (Houston, TX: Connexions) https://cnx.org/contents/EJYWe4UY@9.4:m7vjnKhA@2/Chemical-Vapor-Deposition
[58] TechSolve 2021 PVD vs CVD—how to choose the right tool coating https://techsolve.org/pvd-vs-cvd-how-to-choose-the-right-tool-coating/

[59] Gould R D, Kasap S and Ray A K 2006 Thin films *Springer Handbook of Electronic and Photonic Materials* (Berlin: Springer)

[60] Baklanov M *et al* 2007 *Dielectric Films for Advanced Microelectronics* vol 12 (New York: Wiley)

[61] Martin P M 2009 *Handbook of Deposition Technologies for Films and Coatings: Science, Applications and Technology* (Amsterdam: Elsevier)

[62] Angstrom Sciences Magnetron sputtering https://angstromsciences.com/magnetron-sputtering-deposition

[63] Singh P and Srivatsa K M K 2016 On the wettability and optical properties of nanocrystalline CeO_2 thin films *Indian J. Pure Appl. Phys.* **54** 359–64

[64] Kanan S M *et al* 2009 Semiconducting metal oxide based sensors for selective gas pollutant detection *Sensors* **9** 8158–96

[65] Arafat M M *et al* 2012 Gas sensors based on one dimensional nanostructured metal-oxides: a review *Sensors* **12** 7207–58

[66] Hunter G W *et al* 2006 Factors influencing the gas sensitivity of metal oxide materials *Encyclopedia of Sensors* vol 3 (Valencia, CA: American Scientific) pp 413–47

[67] Ahlers S *et al* 2020 A critical review of solid state gas sensors *J. Electrochem. Soc.* **167** 037570

[68] Lin T *et al* 2017 The morphologies of the semiconductor oxides and their gas-sensing properties *Sensors* **17** 2779

[69] Zhang Y *et al* 2016 Oxygen vacancies on nanosized ceria govern the NO_x storage capacity of NSR catalysts *Catalysis Sci. Technol.* **6** 3950–62

[70] Holgado J P *et al* 2000 Study of CeO_2 XPS spectra by factor analysis: reduction of CeO_2 *Appl. Surf. Sci.* **161** 301–15

[71] Zhu Y *et al* 2014 X-ray photoelectron spectroscopy analysis and band offset determination of CeO_2 deposited on epitaxial (100), (110), and (111) Ge *J. Vacuum Sci. Technol.* B **32** 011217

[72] Zhao X *et al* 2013 Sensitive determination of thymol based on CeO_2 nanoparticle-decorated graphene hybrid film *New J. Chem.* **37** 4045–51

[73] Ujjain S K *et al* 2014 Nanoceria based electrochemical sensor for hydrogen peroxide detection *Biointerphases* **9** 031011

[74] Sudarsanam P *et al* 2014 Structural evaluation and catalytic performance of nano-Au supported on nanocrystalline $Ce_{0.9}Fe_{0.1}O_{2-d}$ solid solution for oxidation of carbon monoxide and benzylamine *RSC Adv.* **4** 43460–9

[75] Ji H *et al* 2019 Gas sensing mechanisms of metal oxide semiconductors: a focus review *Nanoscale* **11** 22664–84

[76] Zhang J *et al* 2017 Metal-oxide-semiconductor based gas sensors: screening, preparation, and integration *Phys. Chem. Chem. Phys.* **19** 6313–29

[77] Cullity B D 1978 *Elements of X-ray Diffraction* (New York: Addison-Wesley)

[78] Agashe C *et al* 1988 Structural properties of SnO_2:F films deposited by spray pyrolysis *Sol. Energy Mater.* **17** 99–117

[79] Younis A *et al* 2016 Tuning the surface oxygen concentration of {111} surrounded ceria nanocrystals for enhanced photocatalytic activities *Nanoscale* **8** 378–3987

[80] Hu Z *et al* 2016 Effect of ceria crystal plane on the physicochemical and catalytic properties of Pd/Ceria for CO and propane oxidation *ACS Catal.* **6** 2265–79

Chapter 17

Humidity detection in low ppm using printable flexible sensors

Md Rahat Mahboob, Shakeb A Khan and Tarikul Islam

17.1 Introduction

Sensors are used to collect all sorts of information from different sources for the measurement, monitoring, and control of various physical and chemical parameters. They play very important roles in enhancing the performance of equipment and improving the quality of human life. The advancements in the field of materials science have revolutionized the physical and technical aspects of the sensors. They can be fabricated either on rigid substrates or on the flexible/printable substrates [1, 2]. The flexibility of a sensor is an important property that allows us to fabricate sensors of different shapes which can be mounted easily on non-planar surfaces [2]. Flexible sensors and electronics are compatible with movable parts and arbitrarily curved surfaces, which ensure their application in wearable sensors and large-area electronics. Thus flexible sensors hold great promise in the fields of medicine, healthcare, environment, food technology, biotechnology, etc. They normally detect various stimuli based on their fabrication and working principles and respond with an electrical signal. A sensor must have desirable static and dynamic characteristics, including small size, low power consumption, excellent interchangeability, and flexibility [3]. Non-flexible sensors also have wide-ranging applications and are fabricated mainly on silicon, alumina, glass, and quartz substrates. These substrates are rigid, brittle in nature, and refuse to change their shape. In contrast, flexible sensors utilize polyimide, polyetheretherketone, polyethersulfone, polycarbonate, poly(ethylene naphthalate), and polyethylene terephthalate (PET) as substrates. Flexible sensors offer various advantages such as low fabrication cost, light weight, foldability, ease of handling, better thermal and chemical properties, integration of sensors and electronics, and low power consumption [2].

Flexible substrates are also suitable for measuring humidity over a wide range. The relative humidity, which indicates the presence of water vapour in the air, is

doi:10.1088/978-0-7503-3439-6ch17 © IOP Publishing Ltd 2021

defined as the ratio of partial water vapour pressure to the saturation vapour pressure and affects everything from human comfort to critical fields of technologies. Thus it becomes essential to measure humidity under different scenarios. The dynamic range of humidity generally extends from percentage relative humidity (%RH) to moisture in parts per million/billion/trillion (ppm/ppb/ppt). The %RH measurement covers higher humidity ranges whereas ppm/ppb cover lower humidity ranges. Further, the temperature and harsh environmental conditions introduce measurement errors by affecting the sensing parameters. Because of the wide dynamic range of water vapour, it is very difficult to measure the entire range with a single sensor. Low humidity measurements in ppm/ppb or at high levels above 95% RH are challenging.

The literature reveals that different technologies are used to fabricate humidity sensors for various ranges and applications [4, 5]. Most of the humidity sensors are of capacitive, resistive, or impedance type. Capacitive sensors are used extensively because such sensors are well suited to modern automation systems, with the possibility of a wide range, low temperature error, and low power consumption [1–6]. The materials for sensing are selected on the basis of the working environment, measurement range, and some critical sensing parameters such as sensitivity, response time, drift due to aging, interchangeability, and hysteresis. The fabrication method should be simple, using easily available chip materials and should be suitable for batch production.

17.2 Background

There are many industrial applications where moisture measurement in the ppm range using a flexible substrate is useful. Important applications include the moisture content of SF_6 gas in gas insulated electrical power substations, plutonium glove boxes in nuclear power plants, hydrocarbon gas pipe lines, transformer oil, the fabrication of electronic components and their failure analysis, hydrogen gas cooled alternators, the fabrication of organic LEDs, etc [4, 5].

The materials used for humidity detection are polymers, ceramics, porous silicon, electrolytic materials (P_2O_5), carbon nanomaterials, and some composite materials. A flexible humidity sensor generally utilizes polymers in the form of liquid crystal polymer (LCP), PET, polydimethylsiloxane (PDMS), polytetrafluoroethylene (PTFE), and polyimide as a flexible substrate.

The unique structure of LCP enables it to exhibit excellent chemical, fire, and moisture resistance properties in addition to mechanical flexibility. It is utilized widely as the substrate and packaging material in MEMS. A flexible humidity sensor has been fabricated on LCP substrate using carbon nanocoil sensing material to detect humidity from 4%RH–95%RH [6]. PDMS is a hydrophobic material known for its optically clear, inert, non-toxic, and non-flammable nature, and has also been exploited for the fabrication of a flexible humidity sensor that can detect humidity in the range of 30%RH–95%RH [7]. A polymer-based flexible capacitive humidity sensor on a PET substrate was able to detect humidity in the range of 45% RH–95%RH wirelessly [8]. A flexible resistive sensor that could detect humidity in the range of 20%RH–80%RH utilizes a micropore PTFE substrate [9]. However, the unique properties of polyimide have also attracted many researchers. Flexible humidity sensors in the %RH range have been developed that use polyimide as substrate [2, 3, 10, 11]. A hybrid system-in-foil (HySiF) also utilized a polyimide substrate due to its flexibility and humidity sensing properties at high humidity levels. The system contains microcontroller chips, temperature sensors, and capacitive humidity sensors [2, 3].

Polyimide is one of the popular materials for humidity sensing in the %RH range [2, 3, 10–13]. It is flexible and has stable mechanical, chemical, and thermal properties for sensing applications [10–13]. In addition, polymer films are well suited to standard integrated circuit (IC) processing techniques, which enable them to integrate the humidity sensor with other sensors as well as signal conditioning units on Si/SiO_2/plastic substrate [2, 3]. This material is common in the IC fabrication industry for various other products such as photoresists, interlayer dielectrics, coating/passivation materials, and mechanical stress relief, etc [3]. It is also suitable for fabricating flexible, light, and low-cost sensors for RFID tags for non-contact humidity measurement in sealed packages [14, 15]. In the normal relative humidity range polyimide shows drift due to aging and ambient temperature variation. Thus it requires a compensation scheme. Also, it is not suitable for very high and low ppm humidity measurement. This is because a polymer sensor has poor durability at high levels and poor sensitivity at low humidity [3, 16]. Polyimide of different thicknesses can be fabricated in a lab or procured directly from a manufacturer. Polyimide sheets of different thicknesses and varieties are available in the market. Some of the polyimides that are normally reported in the literature for sensor fabrication are from Kapton [17] and Upilex [18].

However, when thermal, mechanical, and chemical stability are concerned, ceramic sensors are more popular [19]. Efforts have been made to produce a metal oxide (MOX) gas sensor on a polyimide substrate for a fully flexible multiparametric gas sensor array [13, 20]. Among the different forms of ceramics, porous alumina has been studied extensively for humidity sensing over a wide range. Some important electrical properties of Al_2O_3 are its dielectric constant (9.0–10.1), dielectric strength (10–35 kV mm^{-1}), and volume resistivity (at 20 °C > 10^{14} Ω-cm) [19]. Among the different phases of Al_2O_3 only two are commonly used for humidity sensing: γ-Al_2O_3 (amorphous) and α-Al_2O_3 (corundum). γ-Al_2O_3 is more sensitive for humidity sensing than α-Al_2O_3 due to its high porosity and effective surface area [21]. The *α phase* is most thermodynamically stable but has reduced surface area, which is not suitable for ppm level moisture detection [21]. The formation of the γ phase takes place from 450 °C to 550 °C whereas the α phase occurs at 1200 °C [21, 22]. However, the γ-Al_2O_3 is also stable thermally and chemically in comparison to other humidity sensing materials, such as polymers, porous silicon, etc [13, 21, 23]. Among the various techniques employed for commercial trace moisture measurement, parallel plate capacitive anodized nanostructured Al_2O_3 hyper thin film sensors offer low-cost technology with acceptable accuracy. However, the cost of such a capacitive moisture sensor is high (~2500 UKP) [24–26].

Aluminum oxide thin film sensors are most commonly reported for absolute humidity measurement.

This chapter presents the fabrication of a flexible metal oxide thin film capacitive sensor to measure moisture at the ppm level. The sensor is fabricated by combining the useful properties of flexible polyimide substrate and porous ceramic thin film to fabricate an inexpensive, linear sensor without any shorting problems. A thin film of polyimide of 12.5 μm thickness is used as the flexible substrate. A thin hydrophobic metal oxide film of alumina is deposited on each side of the substrate. To form the capacitive structure, silver electrodes are placed on the sensing films. The fabricated sensor has hybrid dielectric layers of ceramic and polyimide between the metal plates. The response characteristics of the sensor are determined in the presence of moist nitrogen gas. The moisture in the gas varies from 3.6 to 770 ppm.

17.3 Working of the sensor

The structure of the sensor consists of hybrid dielectric layers of metal oxide (γ-Al_2O_3) and polyimide between two metal plates. Three types of sensors designs and their electrical equivalents are highlighted to understand the effect of different dielectric materials with respect to humidity.

17.3.1 A parallel plate capacitor with polyimide as the dielectric sensing film

In the first capacitive sensor, two silver electrodes are placed simply on each side of the polyamide substrate. Figure 17.1 shows a representation of the sensor.

The capacitor is represented by a parallel form of C_p and $R_{p.}$ The contact terminals have negligible resistance. The fringing electric field is neglected. Since the area of the electrode (A) is greater than the thickness of the dielectric (t_p) film ($A >> t_p$) [27], the humidity sensing capacitance C_p can be written as

$$C_p = \frac{\varepsilon_0 \varepsilon_{rp} A}{t_p}, \tag{17.1}$$

where ε_{rp} is the effective dielectric constant which can be given by Looyenga's empirical equation [5] as

$$\varepsilon_{rp} = \left\{\gamma\left(\varepsilon_{\omega}^{1/3} - \varepsilon_p^{1/3}\right) + \varepsilon_p^{1/3}\right\}^3, \tag{17.2}$$

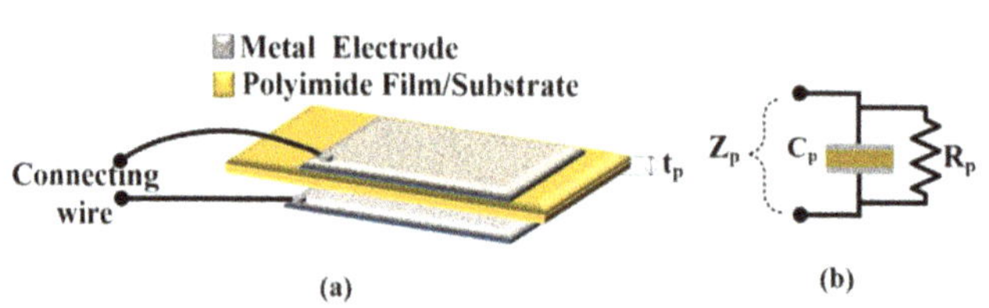

Figure 17.1. (a) Schematic of a parallel plate capacitor with polyimide as the dielectric film. (b) The equivalent circuit of the sensor.

where ε_ω and ε_p are the dielectric constants of water and polyimide, respectively, and γ is the volume fraction of water adsorbed in the sensing film. The dielectric constant of the polyimide varies from 2 to 4 [12].

The equivalent impedance (Z_p) as seen from the terminals of the sensor is written as

$$Z_p = \frac{R_p}{1 + \omega^2 C_p^2 R_p^2} - j\omega \frac{C_p R_p}{1 + \omega^2 C_p^2 R_p^2}, \tag{17.3}$$

where ω is the excitation signal frequency. The resistance R_p of (17.3) at any humidity (%RH) can be represented by the empirical relation of [16]

$$\log\frac{R_p}{R_\mathrm{d}} = (\log c - \log\Phi)/(1 + \frac{d}{\Phi}), \tag{17.4}$$

where R_d is the resistance at dry conditions. The factor $\Phi = \mathrm{RH}^k$, where k is the correction index, and c and d are the constants depending on the composition of the film and its pore morphology. The resistance changes due to changes in the electronic and ionic conductivity of the film in the presence of water molecules [16].

17.3.2 A parallel plate capacitor with metal oxide (γ-Al_2O_3) as a dielectric sensing film

This structure contains only γ-Al_2O_3 as the dielectric sensing film. The sensing film is sandwiched between two electrodes in order to form a capacitive structure, as shown in figure 17.2.

The humidity sensing capacitance C_a for a metal oxide thickness of t_a can be written as

$$C_a = \frac{\varepsilon_0 \varepsilon_a A}{t_a}, \tag{17.5}$$

where ε_a is the effective dielectric constant which can be given by Looyenga's empirical equation [5, 12] as

$$\varepsilon_a = \left\{\gamma(\varepsilon_\omega^{1/3} - \varepsilon_{al}^{1/3}) + \varepsilon_{al}^{1/3}\right\}^3, \tag{17.6}$$

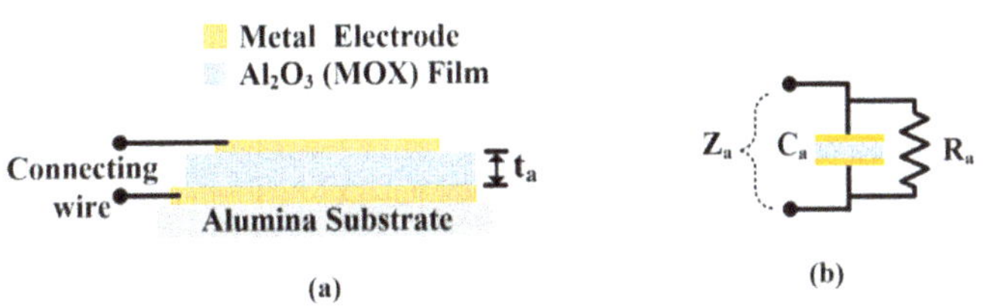

Figure 17.2. (a) The schematic of a parallel plate capacitor with only γ-Al_2O_3 as the dielectric film. (b) The equivalent circuit of the sensor.

where ε_{al} is the dielectric constant of the metal oxide (γ-Al_2O_3) without water molecules ($\varepsilon_{al} \approx 9 - 11$) [5]. The resistance R_a at any humidity (%RH) is similar to equation (17.4).

Similar to the earlier derivation, the equivalent impedance Z_a of the pure γ-Al_2O_3 film is

$$Z_a = \frac{R_a}{1 + \omega^2 C_a^2 R_a^2} - j\omega \frac{C_a R_a}{1 + \omega^2 C_a^2 R_a^2}. \tag{17.7}$$

However, this type of sensor is commonly fabricated on non-flexible substrates using glass, alumina, or FTO as the substrate [21, 23, 28]

17.3.3 A parallel plate capacitor with hybrid dielectric layers of γ-Al_2O_3 and polyimide films

The second sensor utilizes the polyimide as well as metal oxide film. In this capacitive structure, the sensor has γ-Al_2O_3 film deposited on each side of the pure polyimide substrate. Three parallel plate capacitors with two different dielectrics are formed [27]. Figure 17.3(a) shows a schematic diagram of the sensor. Two capacitors are formed due to the oxide film on each side of the substrate and another capacitor due to the inner polyimide substrate. The equivalent circuit of the sensor, which consists of three capacitors in series, is shown in figure 17.3(b). The two capacitors due to the oxide films are identical. The capacitor for the MOX film is represented by the parallel form of C_a and R_a, and for the polyimide substrate by C_p and R_p. The most contributing factor for humidity sensing is C_a.

The impedance (Z_p) of the capacitor due to polyimide is given by (17.3), where Z_a is the impedance due to metal oxide film (equation (17.7)) and Z_p is the impedance due to polyimide film (equation (17.6)). The impedance (Z_{eq}) of the three series combination of capacitors and resistors is given by [27]

$$Z = 2Z_a + Z_p, \text{ or}$$

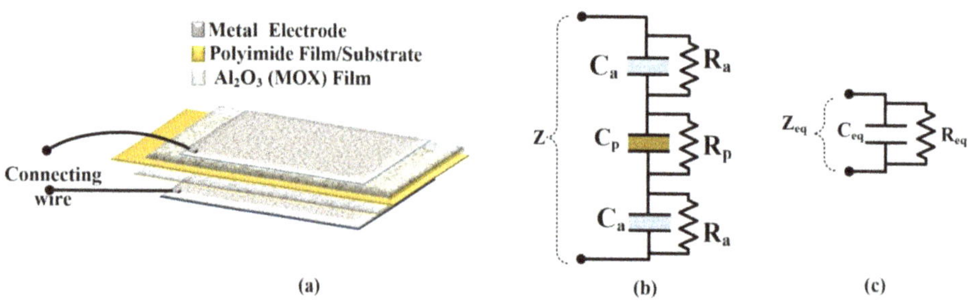

Figure 17.3. (a) A schematic of the proposed sensor with γ-Al2O3 dielectric sensing film on both sides of the polyimide substrate. (b) The equivalent circuit representation of the sensor. (c) The simplified form of the equivalent circuit of the sensor.

$$Z = \left(\frac{2R_a}{1 + \omega^2 C_a^2 R_a^2} + \frac{R_p}{1 + \omega^2 C_p^2 R_p^2}\right) - j\omega\left(\frac{2R_a^2 C_a}{1 + \omega^2 C_a^2 R_a^2} + \frac{R_p^2 C_p}{1 + \omega^2 C_p^2 R_p^2}\right). \tag{17.8}$$

Three capacitors are further represented in a simplified form by C_{eq} and R_{eq} in figure 17.3(c). The equivalent impedance (Z_{eq}) which is equal to Z of the sensor, as shown in figure 17.3(c), is represented by

$$Z_{\text{eq}} = \frac{R_{\text{eq}}}{1 + \omega^2 C_{\text{eq}}^2 R_{\text{eq}}^2} - j\omega\frac{C_{\text{eq}} R_{\text{eq}}}{1 + \omega^2 C_{\text{eq}}^2 R_{\text{eq}}^2}. \tag{17.9}$$

At low frequency, the terms $(1 + \omega^2 C_a^2 R_a^2)$, $\left(1 + \omega^2 C_p^2 R_p^2\right)$, and $(1 + \omega^2 C_{\text{eq}}^2 R_{\text{eq}}^2)$ are approximated as $\omega^2 C_a^2 R_a^2$, $\omega^2 C_p^2 R_p^2$, and $\omega^2 C_{\text{eq}}^2 R_{\text{eq}}^2$, respectively. Now comparing the real and imaginary parts of Z and Z_{eq} of (17.8) and (17.9), the capacitance C_{eq} is written as

$$1/C_{\text{eq}} = 2/C_a + 1/C_p \tag{17.10}$$

and the resistance R_{eq} as

$$R_{\text{eq}} = \frac{(2C_p + C_a)^2 R_a R_p}{2C_p^2 R_p + C_a^2 R_a}. \tag{17.11}$$

17.4 Sensing material and pore morphology

The fabrication process of a humidity sensor starts with the identification of the sensing material and its compatibility with the substrate. The sensing material must possess the required features to attract the target sensing stimuli. For humidity sensing applications, the target is to estimate or quantify the moisture content. The humidity or moisture is the amount of water present in an atmosphere or material. Thus the sensing film must be able to attract the water molecules present in an environment. Pore morphology helps us to understand the potential of a sensing material to attract the target stimuli. The nanostructure of the films is generally characterized with the help of x-ray diffraction (XRD), the Brunauer–Emmett–Teller (BET) method, Barett–Jayner–Halenda (BJH) analysis, atomic force microscopy (AFM), and field emission scanning electron microscopy (FESEM). The literature suggests that the sol–gel method proposed by Yoldas yields promising material with suitable pore morphology for sensing humidity.

17.4.1 Preparation of γ-Al_2O_3 film using the sol–gel method

The sol solution for the fabrication of MOX film was prepared using the Yoldas method [22]. An organometallic compound of aluminium sec. butoxide ($C_{12}H_{27}AlO_3$) was hydrolyzed into an excess amount of water with a molar ratio of ~1:100. The metal alkoxide was added gradually into the warm water and then stirred on a magnetic hot plate at 90 °C for 1 h. This resulted in the formation of milky colloidal AlO(OH) boehmite. The solution was then peptized by adding concentrated hydrochloric (HCl) acid with a molar ratio 0.07 mol per mol of hydroxide. Finally, the peptized solution was refluxed initially for about 18 h at 80 °C and then for about 24 h at 25 °C room temperature. After refluxing, a fixed amount of 0.5 gm polyvinyl alcohol was added in the prepared solution to improve the adherence of the sensing film on the polyimide. The film obtained from the prepared sol solution is highly porous.

17.4.2 Pore morphology studies of the porous γ-Al_2O_3 film

Pore morphology plays an important role in the sensing parameters of the sensor. It is defined by the average dimension of the pores, the distribution of the pores, the thickness of the porous layer, and the stability of the sensing film [29]. The morphology of the γ-Al_2O_3 film, including the specific surface area, is normally determined by BET, BJH, and AFM. For BET, a powder sample of γ-Al_2O_3 was prepared at 450° for 1 h. The specific surface area from the BET results was found to be approximately 200 m^2 g^{-1} [22]. The BET surface area was determined using the NOVA 2000e Model. The distribution of the pores was determined from the BJH adsorption graph. The distribution of the pores in the film was found to be 1.5–64 Å with a BJH pore volume ~0.21 cm^3 gm^{-1}. Most of the pores have pore radii less than 10 Å. The average pore radius obtained from the Dubinin–Astakov (DA) plot is ~9.5 Å. The DA plot is shown in figure 17.4. Atomic force micrographs of the bare polyimide and metal oxide film coated polyimide films are shown in figure 17.5.

The roughness of the film was studied using AFM for both γ-Al_2O_3 and polyimide films.

Figure 17.5(a) shows an AFM image of the top surface of pure polyimide film. The top surface of the polyimide film is fine, indicating an almost smooth surface. Figure 17.5(b) shows an AFM image of γ-Al_2O_3 film on polyimide. However, the polyimide film with metal oxide film is very rough. It indicates a very rough surface

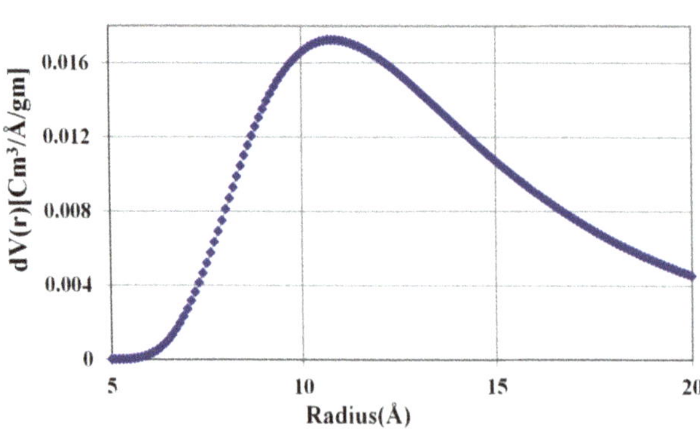

Figure 17.4. DA plot of the γ-Al_2O_3 determining the average pore size and its distribution.

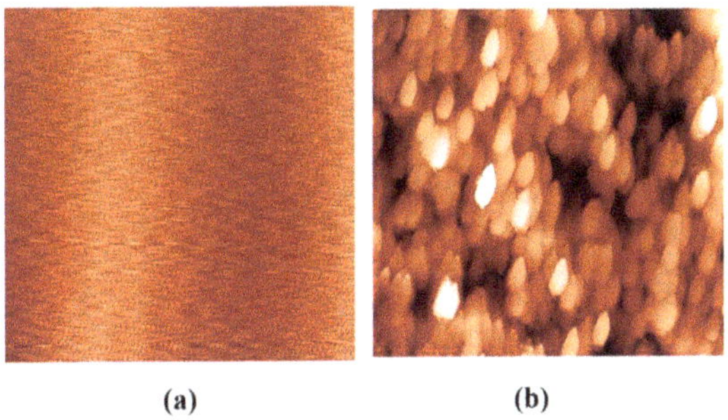

Figure 17.5. AFM of the film: (a) polyimide only and (b) γ-Al_2O_3 film on polyimide substrate.

with the presence of Al_2O_3 nanocrystals. The formation of thin film is clear from the image. The effective surface area of the polyimide film is much less than the surface area of the porous Al_2O_3 film. The rms value of the roughness of the γ-Al_2O_3 film is 55.5 nm.

17.5 Fabrication of the sensor

Two types of capacitive sensors were fabricated for the measurement of moisture in the ppm range. These sensors are hereafter called:

Sensor A: Pure polyimide film with metal electrodes on both sides

Sensor B: Polyimide film coated with MOX (γ-Al_2O_3) with electrodes on both sides

17.5.1 Fabrication procedure of sensor A

Sensor A has only polyimide as the sensing film. The thickness of the polyimide dielectric film procured from Upilex (Japan) is 12.5 μm. Two identical polyimide flexible substrates of size 22 mm × 22 mm were taken and treated with a mixture of H_2SO_4 and hydrogen peroxide for 30 s at 45 °C to remove dust and other impurities. The treated polyimides were then rinsed thoroughly with deionized water (DI). The substrates were then dried in a flow of dry nitrogen gas. Two identical silver parallel electrodes of size 20 mm × 20 mm were pasted on each side of one of the polyimide substrates. The deposited silver electrode was like a mesh with pore dimensions in the order of ~0.5–4 μm so that water molecules could penetrate the porous electrode and be adsorbed in the polyimide (sensing) film. The electrodes were heated at 200 °C for 1 h to dry the silver paste. The other polyimide substrate was used for the fabrication of sensor B.

17.5.2 Fabrication procedure of sensor B

Sensor B has hybrid dielectric films of polyamide and γ-Al_2O_3 sensing film. In the second sensor, a thin film of binder mixed sol of Al_2O_3, as described in section 4.1, was coated on each side of the treated polyimide film by a coating method. The film was kept in the solution for a few seconds and then pulled up at the speed of 20 mm min^{-1}. Each time after film deposition, the thin film was dried at a temperature of

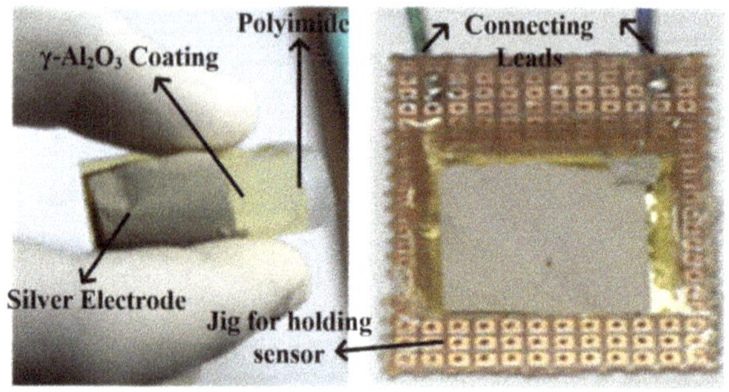

Figure 17.6. Photograph of sensor B with combined polyimide–Al_2O_3 dielectric films.

80 °C for 5 m. Several layers of metal oxide films were deposited on the polyimide substrate. The dried film was then sintered at 450 °C for 1 h. 450° is the minimum temperature required for the formation of amorphous gamma alumina oxide film. This oxide film is extremely hydrophilic and is suitable for water vapour adsorption. The size of the electrode for both the capacitors is the same. Silver electrodes were then printed on the oxide films on both sides and heated again for 1 h at 200 °C. Finally, the terminals were connected by soldering silver wires on each electrode. A photograph of sensor B is shown in figure 17.6.

However, the polyimide substrate for sensor B was heat treated at 450 °C for 1 h to improve the mechanical stability and adherence of the oxide film on the polyimide [14]. The dielectric constant of the polyimide varies from 2 to 4 [2], and for the γ-Al_2O_3 it varies from 9 to 11 [5].

17.6 Electrical characteristics of the sensor

Some important characteristics of the sensor which were investigated with the variation of moist nitrogen gas were the sensitivity, response, and recovery time, reproducibility, and hysteresis. These characteristics are selected on the basis of their usefulness for practical applications. These parameters are studied and often listed in the data sheet of humidity sensors available on the market [24–26]. The determination of the electrical characteristics of the sensor required an experimental set-up, and the schematic of the set-up is shown in figure 17.7. The system mixes the dry nitrogen gas with wet nitrogen gas to generate a particular moisture level.

The moisture level of the dry N_2 gas was around 3–4 ppm. The concentration of the moist N_2 gas was measured by a commercial dew point meter (Model SADP {R}, SHAW, UK). The reference dew point meter has an absolute accuracy of ±1 ppm [25]. The electrical characteristics were determined by an impedance analyzer (4294 A) at a room temperature of 25 °C. The sensors were excited by an ac signal of amplitude 500 mV (rms) and frequency 1 kHz. The fabricated sensors were tested in the moisture range of 3.5 to 770 ppm with the help of the experimental set-up

The fraction of water adsorbed on both sensors A and B plotted with the variation of moisture in the range of 3.5–762 ppm is shown in figure 17.8(a). The volume fraction of water adsorbed on the sensors is determined using (17.1) and (17.2) from the capacitive responses shown in figure 17.8(b).

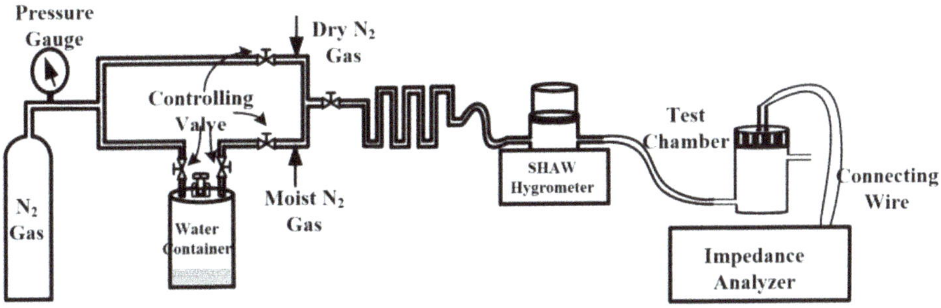

Figure 17.7. Schematic of the experimental set-up.

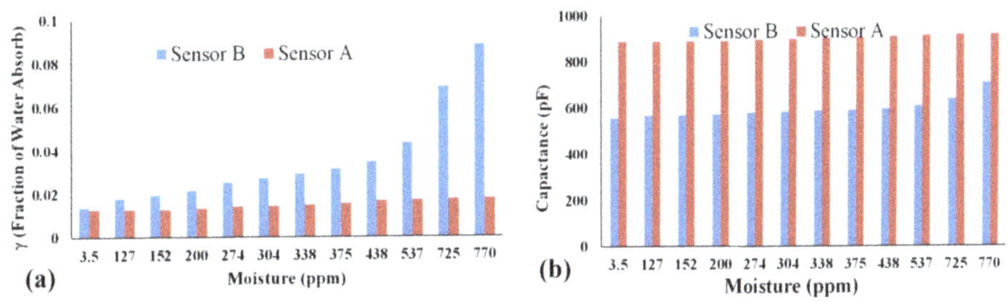

Figure 17.8. (a) The fraction of water adsorbed and (b) the capacitive response by sensors A and B with variation of moisture from 3.5 to 762 ppm. Copyright 2015 IEEE. Reprinted, with permission, from [13].

The capacitive response curve of sensor B, as shown in figure 17.8(b), indicates that the capacitance value increases (approximately) linearly up to 700 ppm with an increasing moisture level. The capacitance value after 700 ppm increases sharply due to water condensation in the polyimide film through the MOX film [12]. However, the capacitance variation for the desired measurement range for sensor A is very small. It is shown in figure 17.8(a) that up to 200 ppm the polyimide does not adsorb moisture, and later the adsorption is very low compared to the metal oxide film coated sensor. The fraction of moisture adsorbed by the film approximately follows Dubinin's equation, as reported in [12]. The Dubinin's equation is given by

$$\gamma = \gamma^0 \exp[-(RT \ln {}^{x}/_{E})^n - \alpha(T - T_0)] \tag{17.12}$$

where, γ^0 is the maximum fractional volume of water adsorption at the temperature T^0, R is the universal gas constant, T is absolute temperature, and α is the thermal coefficients of limiting adsorption. The variable E is the free energy of adsorption and n is an empirical factor, x is the humidity level. Thus, it was obtained that sensor A sensitivity is negligible as compared to sensor B. Therefore, the electrical response characteristics of sensor B were investigated in details.

For sensor B the capacitance value C_{eq} increases almost linearly up to 700 ppm and then there is a nonlinear change in capacitance value with an increase in moisture in ppm. Below 700 ppm, the change in capacitance value (C_p) for sensor A is very small in comparison to C_{eq} due to the MOX film, and since C_p is in series with 2 C_a in (17.10)

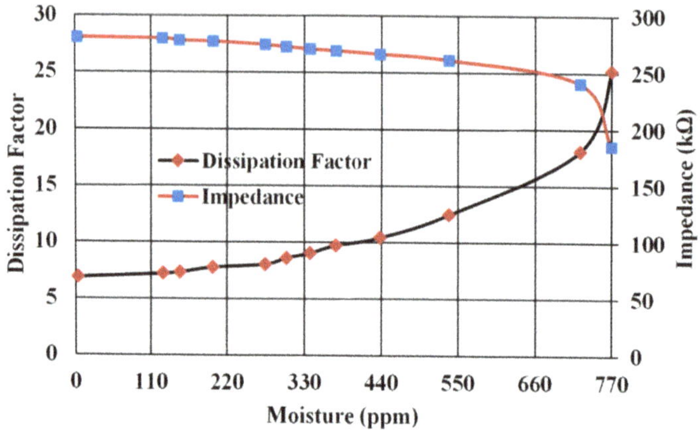

Figure 17.9. Dissipation factor and impedance plot of sensor B for different moisture concentrations from 3.5 to 770 ppm. Copyright 2015 IEEE. Reprinted, with permission, from [13].

causing an almost linear response. When the sensor is exposed to moist N_2 gas with a certain concentration, the change in the effective dielectric constant of the sensor depends on the ambient vapor concentration and the effective surface area of the porous film, which in turn depends on the pore morphology. After adsorption, water molecules diffuse in the porous bulk and an equilibrium condition is reached. The effective surface area of sensor B is much higher than that of sensor A. A change in dissipation factor (DF) with moisture at 1 kHz signal frequency is shown in figure 17.9. The dissipation factor is given by $DF = 1/\omega C_{eq} R_{eq}$, where ω is the signal frequency, C_{eq} is the capacitance, and R_{eq} is the resistance of sensor B. It indicates that the dissipation factor is low until 700 ppm after which it increases rapidly. A high dissipation factor indicates a rapid change in the conductivity of the sensor. The change in impedance with moisture is also plotted in figure 17.9. The impedance value of sensor B decreases with an increase in moisture as it occurs due to an increase in capacitance and a decrease in resistance.

The rise and recovery times of the sensor are determined from the step change in moisture from 565 to 762 ppm and are shown in figure 17.10. The plot has two graphs, one is for the input moisture and the other is for the output sensor response. Although the moisture concentration changes suddenly, the capacitance value takes some time to reach the steady state value. These parameters should be as small as possible. The response and recovery times of the sensor are approximately 158 and 216 s, respectively. This parameter was determined by the time taken by the sensor for a 10%–90% change in capacitance value for the step change in moisture. Water molecules adsorb and desorb in the porous layer at different rates. It has been observed that the sensor takes more time during the desorption of water molecules from nano order pores. In particular, the desorption of water molecules from the porous layer takes a long time. Difficulty in desorption results in longer recovery times of the sensor.

17.7 Results and discussion

Some important features of the fabricated sensors are:

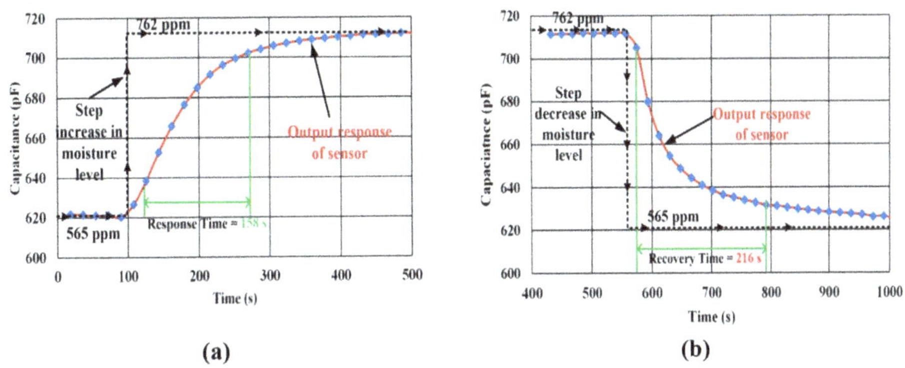

Figure 17.10. Transient response of sensor B for 565 to 762 ppm for a step change of the moisture (a) response curve and (b) recovery curve.

i. Ease of fabrication.
ii. No shorting problem.
iii. High flexibility.
iv. Bulk producible.
v. Low cost and low power consumption.

Experiments were conducted with care to investigate the response parameters as normally found in the data sheet of a commercial dew point meter for ppm range moisture measurement. When the sensor is exposed to moisture the water molecules first penetrate in the pores of the γ-Al_2O_3. With the progress of time the water molecules accumulated between the two films. However, some moisture also condenses in the polyimide film. At the time of condensation the capacitance increases abruptly due to a sudden increase in dielectric constant. At low humidity water molecules are chemisorbed on the oxide surface and are dissociated into hydroxyl ions (OH^-) and protons (H^+). When the concentration increases, water molecules will be physisorbed on a chemisorbed layer and subsequently multilayer physisorbed water molecules will be formed. The physisorbed water molecules will condense in the nano order pores of the film with a radius less than the Kelvin radius [30], causing a large change in effective dielectric parameter as well as the conductivity of the sensor. The large change in dielectric constant leads to an abrupt increase in capacitance value as well as dissipation loss [13, 21, 23, 30]. The capacitance changes by 93 pF for a change in moisture of 762 ppm and the capacitance value changes by 80 pF for a change in moisture from 762 to 770 ppm. Since the moisture adsorption in the polyimide film is very small, the change in dielectric constant of sensor A is small, hence the change in capacitance value is very small. The maximum capacitance change for sensor A is only 25 pF, whereas for sensor B the capacitance change is 93 pF for the same moisture range. The capacitive sensitivity of sensor B was found to be 130 fF ppm^{-1} ($\Delta C/\Delta ppm$). The high sensitivity is due to the availability of ample spaces in the nanoporous oxide film for the condensation of water vapour at low concentration.

Table 17.1. Hysteresis data for sensor B.

Increasing moisture (ppm)	Sensor capacitance (pF)	Decreasing moisture (ppm)	Sensor capacitance (pF)
3.5	559.23	770	710.26
127	568.56	725	626.23
152	572.47	537	594.35
200	574.15	485	587.13
274	580.37	356	580.19
304	583.94	246	575.55
338	587.63	187	572.05
375	591.27	113	566.6
438	597.21	71	564.05
537	609.61	32	562.52
725	639.36	10	560.02
770	710.26	6.5	556.25

In sensor B three capacitors are formed, two are formed due to the oxide dielectric layers and the other is formed due to the polyimide film. The effective capacitance across the terminals of the capacitance is the series combination of the three capacitors. The polyimide capacitance is just like an offset capacitance. Because of the series formation of the capacitances, overall sensitivity is reduced, but it also improves the linearity of the response. Because of this the response is linear. The maximum nonlinearity is only 4%. Normal humidity sensors working on the adsorption and desorption principle shows nonlinearity because of unequal rates of water vapour condensation. The sensor requires an additional circuit for nonlinearity compensation. At higher concentrations water molecules may condense on the polyimide substrate through the MOX film, causing a great change in the effective dielectric constant of the hybrid structure, as is observed in figure 17.8(b). The high response and recovery times are due to the large overall thickness of the sensor. The total thickness of sensor B including the MOX and polyimide is ~22 μm (MOX film $\approx$ 10 μm, polyimide film = 12.5 μm). In addition, the silver electrode on both sides for capacitance formation has a macro porous layer with thickness of the order of ~10 μm. The organic solvent in the silver paste used for making the electrode may also affect the response time to a certain extent. Therefore the sensitivity can be further improved by reducing the thickness of the oxide and polyimide substrates. Because the adsorption is a bulk phenomenon and depends on the film thickness, the thinner the films are, the larger will be the capacitance and thus the larger will be the change in capacitance with respect to humidity, as reported in [8, 12, 13]. Hysteresis is another important phenomenon which occurs due to unequal adsorption and desorption of water molecules. The hysteresis for the sensor B was obtained on plotting the data obtained as presented in table 17.1 while increasing and decreasing the moisture level. It shows quite significant hysteresis at higher humidity but towards lower ppm it is negligible. This phenomenon is observed for most sensors working on the adsorption and desorption principle, but in the case of very small thicknesses of Al_2O_3 film, the hysteresis error is found to be negligible [23]. The hysteresis error of sensor B was found to have a maximum of ±5.04% using

$$\text{Hysteresis error (\%)} = \frac{\pm\max\left(C_p^{in} - C_p^{de}\right)}{2 \times \Delta C} {*}100. \tag{17.13}$$

The hysteresis effect can be compensated by refreshing the sensing layer with a micro heater [2, 3]. There may be a change in the capacitance of the sensor due to a change in the excitation frequency. Therefore experiments were also performed to determine the capacitance with a change in excitation frequency at fixed humidity. The normalized capacitance of sensor B with a change in humidity from 3 to 725 ppm is shown in figure 17.11. It is observed that the capacitance value at fixed humidity decreases with increase in signal frequency. The sensitivity at 50 kHz frequency is lower compared to 1 kHz. At high frequency the relative permittivity of water reduces to some extent. Also, the water molecules which are polar in nature can easily follow a slow change in the electric field, but due to their relatively large mass water molecules find it difficult to follow the rapid change in electric field. Thus at high frequency, the effective dielectric constant change is comparatively lower than the low frequency change. At low frequency, the capacitance change may have some contribution due to the double layer capacitance, which is negligible at high frequency. Some minor reduction of the sensitivity can be addressed using a suitably designed amplifier circuit. Due to the advancement of IC technology, a sensitive electronic circuit can be designed with inexpensive amplifier ICs [2, 3]. Some long-term tests need to be performed to study the effect of aging but the literature on pure polyimide for RH sensing shows small drift [12].

The long-term calibration drift is a major drawback of the porous Al_2O_3 sensor fabricated using the electrochemical anodization method [21]. This drift is due to the formation of the hexagonal cylindrical pore structure of the γ-Al_2O_3 phase by

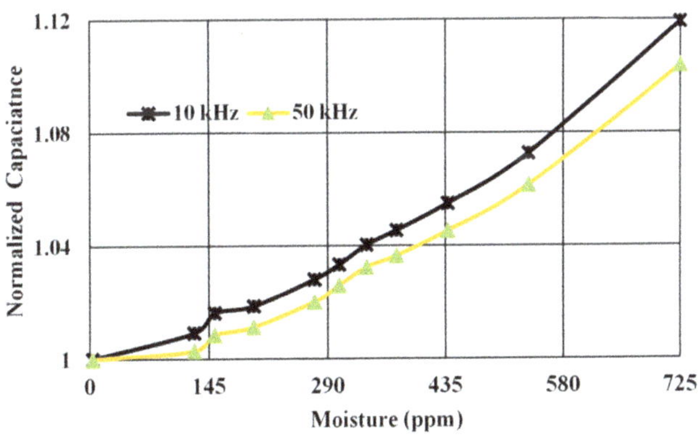

Figure 17.11. Normalized capacitive response of sensor B in ppm moisture concentration at different signal frequencies.

anodization. When the structure is exposed to humidity there is an irreversible decrease in the surface area and porosity [29]. This is even a major issue for commercial Al_2O_3 based dew point meters. Drift is also observed for the metal oxide sensor formed by the sol–gel method. It may be possible to reduce the drift in the capacitance value by putting the sensor in a drying agent such as a molecular sieve after its use. Drift can also be minimized by heating the sensor for some time. By heating, temperature unstable chemical bonds can be removed. Refreshing by heating can be done easily for the oxide sensor fabricated using the sol–gel method. This is because sensor B is fabricated by heating the oxide film at 450 °C for 1 h. Further, note that due to the large electrode size, the offset capacitance for both sensors is quite large. Although the capacitive response of sensor A due to pure polyimide film towards moisture sensing is small, its effect can be nullified by making a differential read-out circuit with a second capacitive sensor B with γ-Al_2O_3 film fabricated on the same polyimide substrate.

If the voltage output of the detection circuit is proportional to $\left(\frac{1}{C_{\text{eq}}} - \frac{1}{C_p}\right)$, the effect of C_p capacitance in (17.10) can be minimized.

This is because $\left(\frac{1}{C_{\text{eq}}} - \frac{1}{C_p}\right) = \left(\frac{2}{Ca} + \frac{1}{C_p} - \frac{1}{C_p}\right) = \left(\frac{2}{Ca}\right)$, where C_a is the capacitance due to the MOX sensing film.

17.8 Conclusion

Nanostructured metal oxide is a promising material for humidity sensing applications over a wide dynamic range. In recent times, major efforts have been ongoing on to fabricate inexpensive, flexible sensors on a printable flexible substrate. However, it is difficult to deposit oxide sensing film on a polyimide substrate. In this chapter two parallel plate capacitive sensors have been fabricated on a flexible polyimide substrate for measuring humidity in the range 3.5–770 ppm. One sensor is fabricated by depositing oxide film on the polyimide substrate. Fabrication of the sensor is simple, based on inexpensive materials, and suitable for large scale production. The sensor with oxide film shows a maximum sensitivity of 130 fF ppm^{-1} with a response time of 158 s. The output of the device is highly reproducible. The flexibility of the device ensures its use in inaccessible areas, for example it can be placed inside a sealed package to monitor its moisture content. The sensor can be made in different shapes according to the application and packaging needs. Due to the presence of the polyimide film beneath the MOX film, the electrodes of the capacitor do not short, hence ensuring a high yield ratio. This particular design is suitable for low power consumption integrated circuit applications with integration of the sensor, electronics, and read-out on the same device. The high offset capacitance due to the large electrode size and small thickness and the drift in output can be compensated for by integrating another capacitor with pure polyimide film on the same platform. A differential capacitive interfacing circuit can then be integrated on the same substrate to have a final output which is drift and offset compensated. The device provides a low-cost solution for ppm moisture sensing on plastic foil.

References

[1] Kapic A, Tsirou A, Verdini P G and Carrara S 2020 Humidity sensors for high energy physics applications: a review *IEEE Sensors J.* **20** 10335–44

[2] Khan S, Lorenzelli L and Dahiya R S 2015 Technologies for printing sensors and electronics over large flexible substrate: a review *IEEE Sensors J.* **15** 3164–81

[3] Elsobky M *et al* 2020 characterization of on-foil sensors and ultra-thin chips for hysif integration *IEEE Sensors J.* **20** 7595–604

[4] McMillan G K and Considine D M 2009 Humidity and moisture systems *Process/Industrial Instruments and Controls Handbook* 5th edn (New Delhi: Tata/McGraw-Hill) 4.194–4.247

[5] Islam T, Khan A U, Akhtar J and Rahman M Z 2014 A digital hygrometer for trace moisture measurement *IEEE Trans. Ind. Electron.* **61** 5599–605

[6] Wu J *et al* 2019 Carbon nanocoil-based fast-response and flexible humidity sensor for multifunctional applications *ACS Appl. Mater. Interfaces* **11** 4242–51

[7] Zhou C *et al* 2020 Rapid response flexible humidity sensor for respiration monitoring using nano-confined strategy *Nanotechnology* **31** 125302

[8] Voutilainen J, Happonen T, Häkkinen J and Fabritius T 2016 Screen-printed remotely readable environmental sensor pair *IEEE Sensors J.* **16** 2523–3531

[9] Chen W, Wei Z, Huang J, Yi Z and Huang Q 2019 A fast response flexible humidity sensor based on PTFE micropore substrate *IEEE Sensors* (Piscataway, NJ: IEEE) pp 1–4

[10] Hajian S *et al* 2020 Development of a fluorinated graphene-based resistive humidity sensor *IEEE Sensors J.* **20** 7517–24

[11] Xinglin Y *et al* 2019 Flexible wearable humidity sensor based on nanodiamond with fast response *IEEE Trans. Electron Devices* **66** 1911–16

[12] Schubert P J and Nevin J H 1985 A polyimide based capacitive humidity sensor *IEEE Trans. on Electron Devices* **32** 1220–23

[13] Islam T, Mahboob M R and Khan S A 2015 A simple MOX vapor sensor on polyimide substrate for measuring humidity in ppm level *IEEE Sensors J.* **15** 3004–13

[14] Deng W-J *et al* 2018 Experimental study of the bending effect on *LC* wireless humidity sensors fabricated on flexible PET substrates *J. Microelectromech. Syst.* **27** 761–63

[15] Fathi P, Karmakar N C, Bhattacharya M and Bhattacharya S 2020 Potential chipless RFID sensors for food packaging applications: a review *IEEE Sensors J.* **20** 9618–36

[16] [29]; Traversa E 1995 Ceramic sensors for humidity detection: the state of art and future development *Sens. Actuators* B **23** 135–56

[17] DUPONT 2020 KAPTON https://www.dupont.com/content/dam/dupont/amer/us/en/products/ei-transformation/documents/EI-10144-Kapton-MT-Data-Sheet.pdf

[18] UBE 2015 Upilex https://ube.com/upilex/catalog/pdf/upilex_s_e.pdf

[19] Goodfellow Metals Ltd 1995 *Metals, Alloys, Compounds, Ceramics, Polymers, Composites* (Cambridge: Goodfellow Metals)

[20] Courbat J, Briand D, Yue L, Raible S and Rooij N F 2012 Drop coated metal oxide gas sensor on polyimide foil with reduced power consumption for wireless application *Sens. Actuators* B **161** 862–68

[21] Kumar L, Saha D D, Khan S A, Sengupta K and Islam T 2012 A medium range hygrometer using nano-porous thin film of γ-Al_2O_3 with electronic phase detection *IEEE Sensor J.* **12** 1625–32

[22] Yoldas B E 1975 Alumina gel that form porous transparent Al_2O_3 *J. Mater. Sci.* **10** 1856–61

[23] Mahboob M R, Khan A U, Kumar L and Islam T 2020 Investigation of chip temperature on response characteristics of the humidity sensor from ppm to %RH *IEEE Trans. Device Mater. Reliab.* **20** 576–83

[24] Vaisala 2019 Vaisala HUMICAP sensor *Vaisala Technology Description* www.vaisala.com

[25] Shaw Moisture Meters Ltd 2018 www.shawmeters.com

[26] GE Panametrics 2014 Aluminium oxide moisture probe www.industrial.ai/measurement-sensing

[27] Golding E W and Widdis F C 2009 Capacitors, capacitance, dielectrics *Electrical Measurement and Measuring Instruments* 5th edn (New Delhi: Reem) pp 135–79

[28] Mahboob M R, Zargar Z H and Islam T 2016 A sensitive and highly linear capacitive thin film sensor for trace moisture measurement in gases *Sens. Actuator* B **264** 658–64

[29] Banerjee G and Sengupta K 2002 Pore size optimization of humidity sensor—a probabilistic approach *Sens. Actuators* B **86** 34–41

[30] Mahboob M R 2017 Development of integrated sensor for wireless humidity detection *PhD Thesis* 2017 (New Delhi: Jamia Millia Islamia) https://shodhganga.inflibnet.ac.in/handle/10603/209186

Chapter 18

Printable and flexible sensor for droplet detection

Zubair Hassan Zargar and Tarikul Islam

Sensors are considered as the backbone of all measurement systems. Accurate and stable sensors are required for reliable measurement of any physical quantities. This chapter discusses the importance of droplet detection and the need for the improvement in the sensor geometry thereof, for precise and accurate detection of micro-droplets. A new type of printable and flexible sensor is introduced for droplet detection. The sensor utilizes the capacitive transduction principle and the detection is contactless, thus increasing the life of the detection system. The capacitance value depends only on the length of the electrode shell and is independent of the other geometrical properties, such as the shape and area of the electrodes. The single length dependence increases the accuracy of the sensor compared to the other conventional electrode structures employed, such as interdigitated or parallel plate capacitive sensors for droplet detection. The sensor is fabricated with a flexible copper cladded substrate and tested to determine the response parameters. The output of the sensor is sensitive, repeatable, and drift free for micro-liquid droplets.

18.1 Introduction

For the last decade, the evolution from classical rigid silicon electronics to flexible electronics has progressed quickly. The motivation in this transition is the reduction in production costs, and in the weight and thickness of the final products. The other impetus is the fabrication of lightweight and very thin devices and sensors, with some unique properties such as transparency and stretchability, in particular for wearable electronics for health monitoring, implantable electronics, etc [1–3]. Flexible wearable electronics can enable the continuous monitoring of patients for medical care and avoid more serious health problems.

The materials play a vital role in the transition from rigid systems to flexible system design. Compared to the other various schemes developed to fabricate

doi:10.1088/978-0-7503-3439-6ch18

© IOP Publishing Ltd 2021

flexible sensors, printing technology has assured simplified processing steps, high flexibility, simple patterning techniques, customization, bulk fabrication, and reduced material wastage, resulting in cost-effective manufacturing [4, 5]. The largest printed electronics market constitutes printed and flexible sensors. It is estimated that the market for printed electronics/sensors will reach $4.5 billion by 2030. With the development in printing technology it is even possible to design sensor systems on irregular non-planar surfaces, which is otherwise not possible with conventional, i.e. wafer based, fabrication technology [6]. Printing technology also reduces the initial investment costs for the manufacturing of micro- and nano-scale flexible components. It has allowed the development of products and sensors on thin plastic films and rubber-like substrates at low temperatures to be incorporated within flexible electronics. The most widely used polymer foils are polyethylene terephthalate (PET), polyethylene naphthalate (PEN), polyimide (PI), polydimethylsiloxane (PDMS), polyurethane (PU), etc. These polymer substrates have opened new possibilities for the low-cost fabrication of electronics by printing electronics on areas larger than the standard commercially available wafers. Printed sensors on flexible and stretchable substrates span a diverse range of technologies and applications, ranging from large area printed image and pressure sensors, radio frequency identification tags (RFID), tactile force, humidity and gas sensing devices, solar cells, light emitting diodes (LED) and transistors [7–11], and so on. These kinds of sensors were developed with the help of polymers, which have high dielectric strength with less dielectric loss.

The overall classification of printing/coating systems can be divided into two broad categories: the non-contact and contact-based patterning. As the name suggests, in contact printing there is direct physical contact between the substrate and the modules with the required inked patterns implanted on them. In a non-contact process the solution is dispensed via openings and nozzles of suitable dimensions onto the substrate. The structure is then developed on the substrate by moving the substrate in a pre-programmed manner. Screen printing, gravure printing, offset printing, and flexography are the classical methods of printing, while in recent years digital direct writing methods such as inkjet printing and aerosol jet printing have bee gaining popularity. The mass production of printed electronics and structures with a minimum size of 10 –30 μm can be obtained by these processes. Non-contact printing techniques are generally preferred owing to their advantages, such as flexibility to the fabrication process, reduced material wastage, high quality, high resolution of patterns, and easy control by adjusting only a few process parameters. Recently, newly developed printing methods, such as nano-imprint, micro-contact printing, and transfer printing, have also attracted significant interest, in particular for inorganic monocrystalline semiconductor based flexible electronics. Nano-imprinting lithography has the capability of producing structures on substrate surfaces in the nano-scale range, usually less than 20 nm. To develop flexible sensors and electronics, the choice of the base material, i.e. the substrate, is essential. Typically, there are three types of substrates that can be employed for this use: thin glass, metal foils, and plastic/polymer substrates. Although thin glass is flexible its brittle property limits its use, and metal foils involve high material cost. On the other hand, polymer substrates can be used for low-cost batch fabrication of flexible devices using the various printing technologies discussed above. Plastic materials are preferred for applications requiring a high degree of bendability, transparency, and emissive properties.

In the last decade, a lot of attention has been given to micro-droplet based microfluidic systems, due to their superior analytical performance, low cost, and high throughput. Droplet based microfluidic devices rely on the use of individual fluid compartments, micro-droplets, rather than a continuous fluid stream. The detection of droplet content is one of the fundamental requirements for most of the micro-droplet based microfluidic systems. Thus micro-droplet detection systems have significant importance in the field of biomedicine, pharmacy, and medical research, where small amounts of selective liquids need to be delivered in a controlled way with a high degree of precision and reproducibility [12–22]. Demand for a controlled fluid dispensing system is high in certain fully automated industrial processes, where different types and amounts of adhesives, lubricants, cooling agents, etc, must be dispensed accurately. A highly precise dispensing system is needed in inkjet 3D printing to obtain the required quality assurance and the resolution in printing. The control of the dispensing event is essential for fault detection, such as missing droplets, clogged nozzles, or incorrect liquid volume or type. Although microfluidic systems can be as small as a coin, the detection systems, pumps, and microscopes that are needed for experimentation are quite bulky for realizing a true lab-on-a-chip system.

This chapter discusses the use of a four-electrode capacitive structure developed on a flexible polyimide substrate working on the basis of the Thompson–Lampard theorem for contactless micro-droplet detection. The capacitive sensor for micro-droplet detection was designed using finite element software. The structure was then fabricated and tested for the detection of droplets of different liquid samples.

18.2 Droplet detection methodologies

Several technologies have been developed for the controlled and flexible detection of microfluidic droplets. Process control systems are often required in dispensing devices to monitor the dispensing process continuously in order to avoid malfunctions due to clogging or nozzle wetting. This control is usually achieved by monitoring the flow or pressure inside the dispensing device or using optical cameras to image droplets in flight. Various methods to monitor the dispensing process and to measure droplet volumes have been reported in the literature [14–22]. Most droplet based microfluidic studies rely on high-speed imaging to provide details of droplet generation and transport. These systems require expensive, bulky, high-speed cameras, and exhaustive post-image analysis.

18.2.1 Droplet detection by optical methods

The detection of droplets by optical techniques can be divided broadly into four categories: (i) absorption spectroscopy, (ii) vibrational spectroscopy, (iii) evanescent field sensing, and (iv) image processing based sensing. Droplet detection using optical methods is performed by integrating a light source and a detector aligned with the dispensing channels. The detection of droplets by optical methods is a non-contact measurement technique and it requires minimal changes to the fluidic system. In optical detection the dispensed droplet causes a change in the intensity of the light by absorption, reflection, and diffraction. This change in light intensity can then be calibrated to determine the droplet size, velocity, and shape [16]. For determination of velocity of a free-flying droplet, a stroboscopic camera is needed, which increases the size as well as the cost of the detection method. An optical flow cell [17] can be used to measure the size and velocity of the droplets. The reported cell is capable of measuring the velocities of droplets flowing through a micro-channel in the range of 0.2–5.45 mm s^{-1}, and sizes of droplets above 0.8 mm. The optical absorption principle can be utilized to determine the composition of a liquid droplet as well. Although the detection is non-contact, the liquid droplets are passed through the microchannel. Thus there is a fair chance of the contamination of the channels by different liquids. The continuous monitoring of droplet size and velocity in real time can be achieved by imaging the center line of the microchannel [18]. However, this technique requires a bulky and high-speed camera. Also, there can be wastage and contamination of liquid when passed through the capillary.

18.2.2 Droplet detection by electrical methods

Due to greater and greater miniaturization of industrial products, the use of optical techniques for droplet detection is limited by the need for bulky and expensive cameras, other off-chip components, and the requirement for skilled manpower. Electrical sensing, on the other hand, is a very efficient, simple, low power, and low-cost technique, and it can be automated for industrial applications. The electrical method provides a scalable and label-free alternative for droplet detection that allows multiple sensors in a small footprint at low cost. Both contacting and non-contacting electrical methods can be applied for droplet detection [19–22]. The measured parameters, such as limit of detection, sensitivity, and dynamic range, are primarily affected by the signal-to-noise ratio and sampling rate. The transduction principle can be impedimetric, electrochemical, pulse resistive, or microwave sensing. The first two transduction principles are most widely used for droplet detection. The presence of the droplet causes a corresponding change in resistance or capacitance of the structure, which yields information about the size and the type of the dispensed liquid. However, the change in resistance is too low to be detected with conventional detection circuits when the droplets are non-wetting. Also, a continuous conducting phase is required to reflect the change in resistance in the sensing area, which gives corresponding information about the droplet in use. There is always a possibility of loss or contamination of the liquid and electrode structure when the system requires contact with the dispensed liquid for measurement. Capacitive sensors are popular in sensing applications. The advantage of droplet detection by the capacitive technique is that droplet characteristics can be detected without any contact between the electrode surface and the droplet, thus avoiding loss and contamination of the sample liquid and eventually increasing the accuracy and throughput of the detection system. The presence of the droplet within the electrodes causes a change in the dielectric constant, which causes a change in capacitance between the electrodes. That change in the capacitance can be calibrated in terms of the volume and the dielectric constant of the dispensed droplet.

A parallel plate capacitive design can detect droplets with volumes in the range of 20 to 65 nL with around 2 nL resolution [19]. However, such types of sensors suffer from drift due to physical deformation of electrodes. Although the detection is non-contact, there is no protective layer over the electrodes. Thus there may be contamination of electrodes due to dispensed droplets. The lateral movement of droplets away from the central axis may also cause error in the measurement.

The conventional interdigitated capacitive sensor shown in figure 18.1(a) can also be used for the determination of the velocity and volume of freely falling droplets [22]. The sensing is based on the difference in the dielectric property of the dispersed and continuous phases. The droplets are formed in a microchannel using a T-junction generator and pressure controller. A thin passivation layer of PDMS is used between the electrodes and liquid to prevent cross contamination. The sensor electrodes are fabricated using a lift off technique and soft lithography is used to develop PDMS microchannels. The complete chip is formed by bonding the microscope slide containing the electrodes on its surface with the PDMS substrate containing the microchannel. The droplets are formed in a microchannel using a T-junction generator and pressure controller. For a given geometry, the size and speed of the droplets can be controlled by adjusting the pressure to the two inlets. There is a change in the capacitance of the structure as the water droplet passes through the sensor surface, shown in figure 18.1(b). The capacitance is measured with the help of the dedicated capacitance-to-digital converter AD7746. The capacitance of the sensor begins to increase as the droplet enters the sensing region at position 2. At position 3, the maximum value of the capacitance (ΔC) is obtained as the droplet is fully inside the sensing region. The capacitance value of

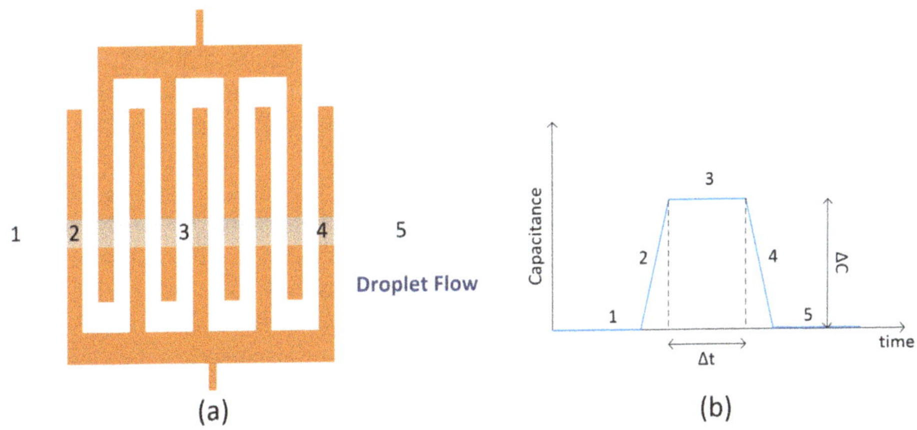

Figure 18.1. (a) Schematics of an interdigitated sensor for droplet detection. (b) Capacitance signal obtained with respect to various droplet positions. Reprinted from [22] Copyright (2011), with permission from Elsevier.

the sensor decreases back to its base value as the droplet exits the sensing electrodes. The size of the droplet can be measured by the saturation signal ΔC. The speed of the passing droplet can be detected using the duration of the saturation state Δt and the distance (Δx) traveled by the droplet when it is completely inside the sensing region. Droplet sizes of 250, 480, and 700 μm were successfully detected by this interdigitated sensor. However, a limitation of this design is that droplets must be less than the length of the sensor or the signal will be completely saturated. Also, the output capacitance and sensitivity of the sensor depend on several geometrical parameters of the electrodes, such as the length of electrodes, number of fingers, electrode width, and electrode gaps. Also, the external field should be properly shielded. The dependence of the sensor characteristics on so many parameters may in turn decrease its accuracy and reliability. The thinner the passivation layer is, the higher the sensor sensitivity. However, this must be balanced by the need to keep droplets from contacting electrodes. The thickness of the passivation layer needs to be optimized.

18.3 Limitations of current electrical sensors and the need for cross-capacitive sensors for droplet detection

In electrical sensing, a proper electrode system is required which can give highly accurate and reliable information about the type and size of the dispensed micro-droplet in use. As already discussed in the above section, electrode systems such as the inter-digital type and parallel plate capacitive type sensors are used for this purpose. The accuracy and stability of these sensors are influenced mainly by the parasitic capacitances, and the variation of the geometrical properties. Small parasitic capacitances, which are caused by the lead wires, and fringing fields change the actual value of the capacitance. To eliminate these unwanted capacitances, and hence to increase the accuracy, one can use a guard ring to minimize the fringing field. The sensors should be placed within a grounded metallic shield. Geometrical properties such as electrode distance, electrode area, bending, and tilt also greatly influence the accuracy and stability of the capacitors. A simple parallel plate or interdigitated capacitor if used for droplet detection is highly sensitive to bending and variation in electrode gap. When low-cost materials are used, these deformations are likely to occur with time due to differences in the thermal coefficients of expansion. However, using high quality materials such as Zerodur glass with gold plated electrodes, the error due to thermal expansion can be minimized. However, the cost of the resulting sensor would be too high to be used by a consumer, and low-cost industrial products thus require improvement in the design of the capacitive droplet sensor. The instability and inaccuracy due to changes in geometrical parameters can be solved by using the cross-capacitor proposed by Thompson and Lampard [23, 24]. A cross-capacitor has four symmetrically placed cylindrical conducting electrodes separated by very small insulating gaps. In such a structure, the cross-capacitance value between the opposite electrodes depends only on one length measurement. This structure thus markedly improves the precision of its calculated value.

18.4 The Thompson–Lampard theorem

In 1956 the two metrologists A M Thompson and D G Lampard came up with a curious and useful theorem in electrostatics, popularly known as the Thompson–Lampard (TL) theorem [23]. They established a new theorem in electrostatics that provides calculable capacitors with special types of electrode structures possessing certain symmetry. The need for precise and accurate measurement of capacitors led to the origin of this theorem. The Thompson–Lampard theorem is little known, apart from in metrological circles and among those concerned with electrical standards. Thus for the benefit of the reader, the TL theorem is discussed in this section, including its origin and some of the proof. Capacitors of various forms such as parallel plates, concentric cylinders, and modifications of these structures are used as capacitance standards. However, their values usually depend on two or more precision length measurements to achieve a high order of accuracy. When the capacitance value of the electrical standard depends on a greater number of physical parameters, the accuracy and stability of the capacitor are reduced. To achieve the highest accuracy, the capacitance of the structure should depend on the least physical dimensions of the capacitor. This led Thompson and Lampard to investigate a new form of cylindrical cross-capacitor, the value of which depends on the single length measurement of the electrodes.

Consider a cylindrical conducting shell whose cross-section is represented by the closed curve P shown in figure 18.2. The cross-section has one axis of symmetry ac, but it is otherwise arbitrary. Further, if the two right angle planes divide this shell into four parts, the line of intersection of these planes is parallel to the length of the cylinder and one of the planes containing the symmetry axis ac. Then the direct capacitance (C_0), per unit length of the cylinder in a vacuum, between the opposite parts of the shell (ab to cd or ad to bc), due to the field inside (or outside) the shell, is a constant:

$$c_0 = \frac{\ln 2}{4\pi^2}\text{e.s.u} = 0.017557623\ \text{e.s.cm}^{-1}. \tag{18.1}$$

If the velocity of light is taken as 2.997 924 60 × 10^8 m s^{-1}, this direct capacitance per unit length (C_0) in a vacuum is equivalent to 0.019 535 485 pF cm^{-1}. Thompson and Lampard gave this direct capacitance per unit length and called it the cross-

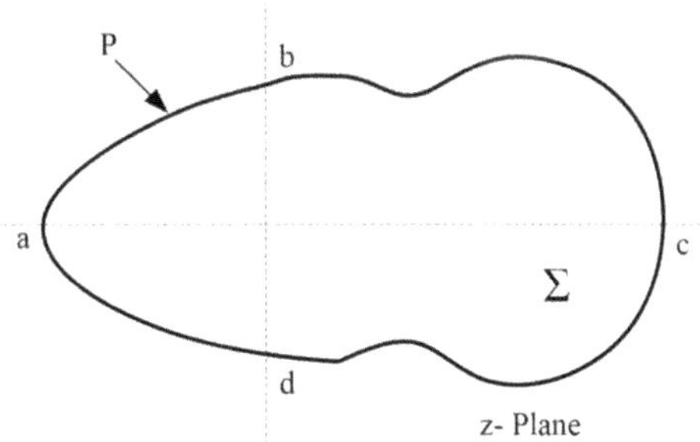

Figure 18.2. Cross-section of a cylindrical shell. Reprinted with permission from [24]. Copyright (1957), the IET.

capacitance. The cross-capacitance depends only on the length of the shell. It is independent of the shape and the area of the cross-section, provided the one-line symmetry is maintained. To determine the cross-capacitance value between the opposite parts of the shell, say *ab* and *cd*, the electrostatic conditions used for establishing this theorem are such that the surface *ab* is held at unit potential, while the rest of the cylindrical surfaces are kept at zero potential. The direct capacitance per unit length (cross-capacitance) between faces *ab* and *cd* is then simply the negative of total charge per unit length that appears on the face *cd* due to the unit potential applied to the face *ab*. The charge on the inner surface of section *cd* is only considered in this theorem. The charge on the outside of the surface is nullified by another continuous grounded cylinder at a very minimum distance from the main cylinder.

The verification of the TL theorem requires a lot of theoretical calculations and knowledge of mathematical transformations in detail. However, some parts of the proof of the theorem are discussed here. The cross-section of the cylindrical shell shown in figure 18.2 can be transformed into the circle shown in figure 18.3 using Riemann's theorem. The axis of symmetry *ac* of figure 18.2 maps to *a′c′* by preserving the symmetry. The direct capacitance per unit length between the opposite electrodes of the mapped circle, say *a′b′* and *c′d′*, due to the field inside the shell is to be computed. If all the faces of the cylinder are kept at ground potential except the face *a′b′*, which is kept at unit potential, then this direct capacitance per unit length, called the cross-capacitance, is the negative of the total charge per unit length that appears on the surface *c′d′* due to the unit potential applied to the surface *a′b′*.

If $V(x,\alpha)$ is the function of the potential inside the cylinder having a circular cross-section of radius r, then the boundary conditions for the potential function must be

$$V\,(r,\,\alpha) = 1,\ \ \frac{-\theta}{2} \leqslant \alpha \leqslant \frac{+\theta}{2} \tag{18.2}$$

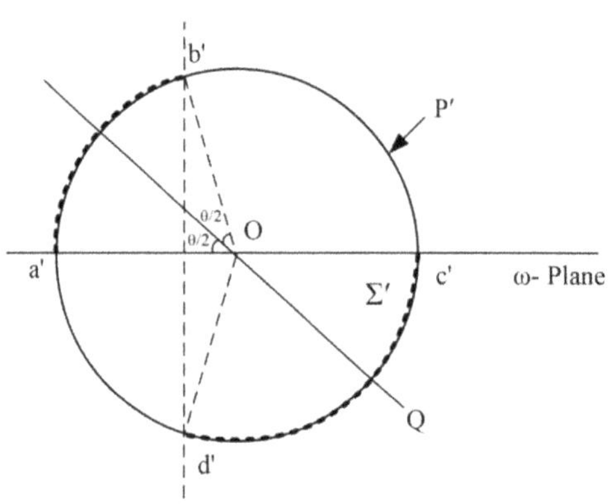

Figure 18.3. Cross-section of a cylindrical shell after conformal transformation. Reprinted with permission from [24]. Copyright (1957), the IET.

$$= 0, \ \frac{\theta}{2} < \alpha < 2\pi - \frac{\theta}{2}. \tag{18.3}$$

Also, Laplace's equation must be satisfied by $V(x,\alpha)$ at all the points inside the shell, i.e.

$$\frac{\partial^2 V}{\partial x^2} + \frac{1}{x}\frac{\partial V}{\partial x} + \frac{1}{x^2}\frac{\partial^2 V}{\partial \alpha^2} = 0 \tag{18.4}$$

Further, $V\ (x,\alpha)$ must satisfy the following two conditions

i) it is even, periodic function of α, as OQ is the origin of the angular coordinates.

ii) it has no singularity at $x = 0$.

Satisfying these two requirements, equation (18.4) has the most general solution of the form

$$V\ (x,\ \alpha) = \sum_{n=0}^{\infty} A(n)x^n \cos n\alpha. \tag{18.5}$$

When $x = r$ (the radius of the circular cross-section of the cylinder), equation (18.5) can be rewritten as

$$V\ (r,\ \alpha) = \sum_{n=0}^{\infty} A(n)r^n \cos n\alpha. \tag{18.6}$$

It is Fourier the cosine series in α. Applying the boundary condition of equations (18.2) and (18.3), one can obtain

$$A(n)r^n = \frac{1}{\pi}\int_{-\theta/2}^{+\theta/2} \cos n\alpha d\alpha; \ \ n = 1, 2, 3 \tag{18.7}$$

$$= \frac{2}{n\pi} \sin \tfrac{n\theta}{2}; \ \ n = 1, 2, 3, \tag{18.8}$$

when $n = 0$,

$$A(0) = \lim_{n\to 0} \frac{2}{n\pi} \sin\tfrac{n\theta}{2} = \frac{\theta}{\pi}\lim_{n\to 0} \frac{\sin\frac{n\theta}{2}}{\frac{n\theta}{2}} \tag{18.9}$$

$$A(0) = \frac{\theta}{\pi}. \tag{18.10}$$

Finally, we have

$$V\ (x,\ \alpha) = \frac{\theta}{\pi} + \sum_{n=1}^{\infty} \frac{2}{n\pi}\left(\frac{x}{r}\right)^n \sin\frac{n\theta}{2} \cos n\alpha. \tag{18.11}$$

The charge per unit length $\sigma(\alpha)$ is obtained using Gauss's theorem and is given by

$$\sigma(\alpha) = -\frac{1}{4\pi}\frac{\partial V}{\partial x}\text{, at } x = r \tag{18.12}$$

$$= -\frac{1}{2\pi^2 r}\sum_{n=1}^{\infty} \sin\frac{n\theta}{2}\cos n\alpha \tag{18.13}$$

Equation (18.13) can be written into a geometric series of complex exponentials given by

$$\sigma(\alpha) = -\frac{1}{8\pi^2 r}\left[\frac{\sin(\theta/2 + \alpha)}{1 - \cos(\theta/2 + \alpha)} + \frac{\sin(\theta/2 - \alpha)}{1 - \cos(\theta/2 - \alpha)}\right]. \tag{18.14}$$

The direct capacitance per unit length (C_0) is obtained by integrating $\sigma(\alpha)$ over the face $c'd'$:

$$C_0 = \int_{-\theta/2}^{2\pi-3\theta/2} \sigma(\alpha) r\, d\alpha = -\frac{1}{8\pi^2}[\ln(1 - \cos\theta) - \ln(1 - \cos 2\theta) - \ln 2 + \ln(1 + \cos\theta)] \tag{18.15}$$

$$= \frac{1}{4\pi^2}\ln 2 \text{ e. s. u. cm}^{-1}. \tag{18.16}$$

Thus the direct capacitance per unit length of the cylinder between faces $a'b'$ and $c'd'$ is always constant if the cross-section of the cylinder has one axis of symmetry and can be taken as a closed curve. The theorem is established by keeping the face $a'b'$ at unit potential and the other faces at zero potential. Similar result will follow if the direct capacitance per unit length is calculated between the other two opposites, i.e. the $a'd'$ and $b'c'$ faces, and applying the potentials to the faces accordingly.

The equation (18.16) can be rewritten as the following equation, where ϵ_0 is the electrical permittivity of the free space within the electrodes and l is the length of the electrode shell:

$$C_{\text{cross}} = l\frac{\epsilon_0}{\pi}\ln 2 = 0.019535485 \text{ pF cm}^{-1}. \tag{18.17}$$

If the space between the electrodes is filled with the dielectric material of permittivity ϵ, equation (18.17) is modified to

$$C_{\text{cross}} = l\frac{\epsilon}{\pi}\ln 2 = l\frac{\epsilon_r\epsilon_0}{\pi}\ln 2. \tag{18.18}$$

Here ϵ_r is the dielectric constant of the material within the electrodes.

Thus the cross-capacitance depends on the length of the electrodes only and not on the other physical dimensions of the electrodes, thus single dimensional high accuracy can be ensured. If the length of the cylindrical structure is known, then the capacitance value can be related directly to the properties of the droplets. Therefore, it decreases the sources of error due to the variations of the shapes and so on. However, for the cross-capacitive sensor the gap between the electrodes should be small, as a large gap distorts the electric field and therefore affects the capacitance value. But the asymmetry between the electrodes, up to some limit, does not cause so much inaccuracy. Excellent shapes of the electrodes based on the TL theorem can be developed for precise and accurate measurement of physical quantities [25–27].

18.5 The fabrication of printable and flexible cross-capacitive sensors for droplet detection

The aim of this section is to discuss the fabrication of a printable and flexible sensor for non-contact droplet detection [27–29]. The sensor utilizes the cross-capacitive transduction principle discussed in the previous section.

18.5.1 Detection principle

The schematic arrangement of the cross-capacitive sensor for micro-droplet detection is illustrated in figure 18.4(a). The sensor consists of four identical symmetrically placed flexible electrodes having a circular cross-section and separated by insulating gaps. This capacitive structure is placed just below the liquid dispensing nozzle. The excitation signal is applied to electrodes 1 and 3, and the other two cross-electrodes 2 and 4 are grounded. The electrical equivalent circuit, when the electrodes 2 and 4 are grounded, is shown in figure 18.4(b). In the equivalent circuit, C_{cross} is the actual measurable capacitance between the electrodes 1 and 3. Two grounded capacitances are formed at the terminals 2 and 4, respectively. The presence of free-flying micro-droplets causes the change in the permittivity ($\in_r$) of the region within the electrodes. This results in a change in its cross-capacitance

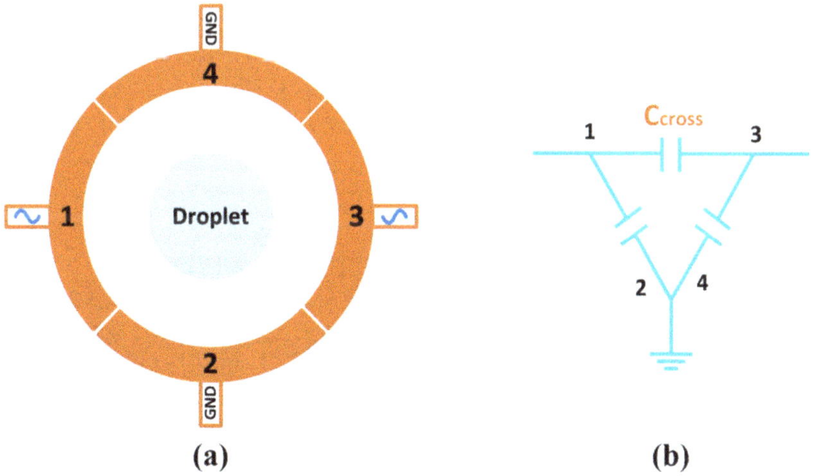

Figure 18.4. (a) Schematic diagram of a cross-capacitive sensor for micro-droplet detection. (b) Electrical equivalent circuit. Copyright 2019 IEEE. Reprinted, with permission, from [27].

value, given by equation (18.18), and provides corresponding information about the droplet. Such a change in permittivity is caused momentarily when the droplet is present within the electrodes. The change in cross-capacitance of the sensor can thus be correlated with the volume and type of the micro-droplet being dispensed along the length of the sensor.

18.5.2 Sensor fabrication

The sensor is fabricated using a double-sided copper cladded flexible polyimide substrate (Upilex-50 s, Japan). This is a material which allows low-cost fabrication of the whole sensor unit. First a mask of four identical electrodes (length = 8 mm, width = 5.8 mm) with identical gaps of 0.5 mm is printed on one side of the copper cladded substrate, shown in figure 18.5. The mask is printed on the substrate by an inkjet printing process. To provide insulation between the electrodes, the copper film in the gaps is removed by a chemical etching process.

The flexible structure shown in figure 18.5 was then placed around a circular hollow shell of pure Teflon with an outer diameter of 8 mm and and inner diameter of 6 mm with the four electrodes facing inwards. The single outer copper electrode serves the purpose of a guard electrode for the electrostatic shielding of the inner electrodes. A schematic diagram of the fabricated micro-droplet sensor is given in figure 18.6.

The Teflon shell on which the electrodes are placed provides the mechanical support to the electrodes as well as protection of the electrodes from liquid contamination. However, by placing the electrodes between the droplet and the electrodes, the effective capacitance value is modified. The modified capacitance value is given by the following equation, where the Teflon shell of internal radius R_1 and outer radius R_2 is coaxially placed inside a circular cross-capacitor of radius R_3, and the dielectric constant of the Teflon cell is ϵ_T:

$$C_c = \left(\frac{\epsilon \ln 2}{\pi}\right) l . \left(1 + \frac{\epsilon_T^2 - 1}{2\epsilon_T \ln 2}\right) \ln\left[\frac{\left\{1 + \left(\frac{R_2}{R_3}\right)^2\right\}^2 \left\{1 + \left(\frac{R_1}{R_3}\right)^4\right\}}{\left\{1 + \left(\frac{R_1}{R_3}\right)^2\right\}^2 \left\{1 + \left(\frac{R_2}{R_3}\right)^4\right\}}\right]. \tag{18.19}$$

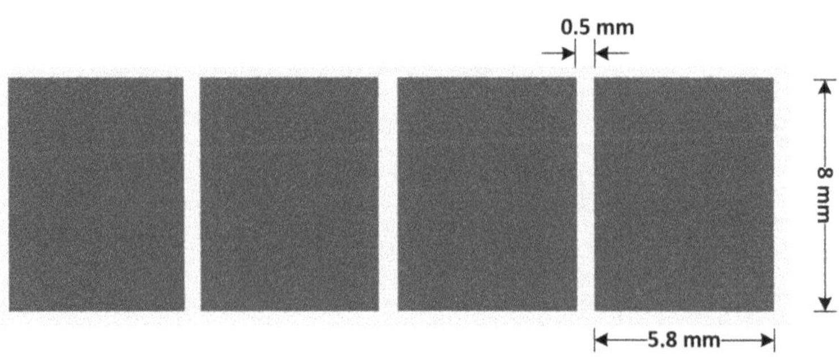

Figure 18.5. Mask of printed electrodes.

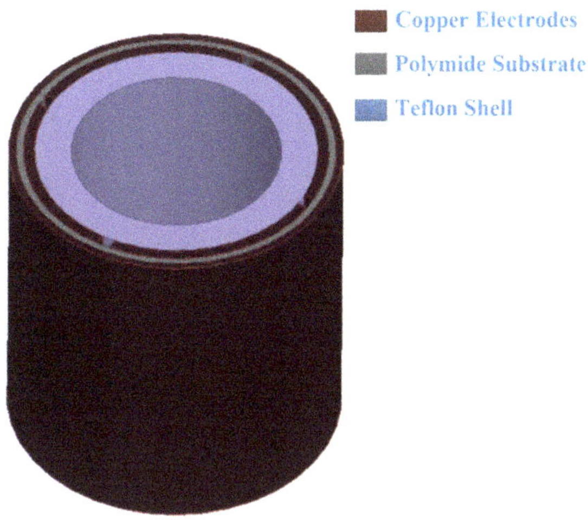

Figure 18.6. Schematic diagram of the micro-droplet sensor. Copyright 2019 IEEE. Reprinted, with permission, from [27].

In case of the present sensor, $\frac{R_2}{R_3} = 1$, $\frac{R_1}{R_3} = 3/4$, and $\in_T = 2$ for pure Teflon, the factor by which the cross-capacitance value given by equation (18.18) is multiplied comes out to be equal to 1.080. An experiment is conducted to determine the actual value of the free space capacitance and this was compared with the theoretical value determined using equation (18.19). The maximum deviation of the experimental value from its theoretical value of the sensor was only 0.8 fF.

18.5.3 Determination of the response characteristics of the sensor

A schematic of the experimental set-up used for determining the capacitance variation for different droplets is shown in figure 18.7. It consists of an automatic dispenser system, the sensor prototype, adaptable sensor holder, a gravimetric digital balance, and an AD7150 Evaluation board [30]. The sensor is placed exactly under the automatic dispenser, so that the droplets fall along the inner space of the sensor without touching the wall of the Teflon tube. The liquid droplets of different sizes are dispensed using micro-needles of different radii through the sensor and then fall on the highly precise electronic gravimetric balance (AY220 SHIMADZU, 0.1 mg resolution). This balance is used to measure the mass and then to determine the volume of the droplets. Thus with this simple set-up, micro-droplets of different volumes of different liquids can be dispensed.

18.5.4 Interfacing of the sensor

The sensor is interfaced to a PC using the AD7150 Evaluation board shown in figure 18.8. The board is a high-performance capacitance-to-digital converter (CDC) with a sensitivity of 1 fF. The sensor terminals 1 and 3 are connected to the input capacitance terminals Cin pin and the Exc pin of the AD7150 CDC board. The terminals 2 and 4 are maintained at ground potential. The single outer copper electrode is also grounded to provide an electrostatic shielding to the structure.

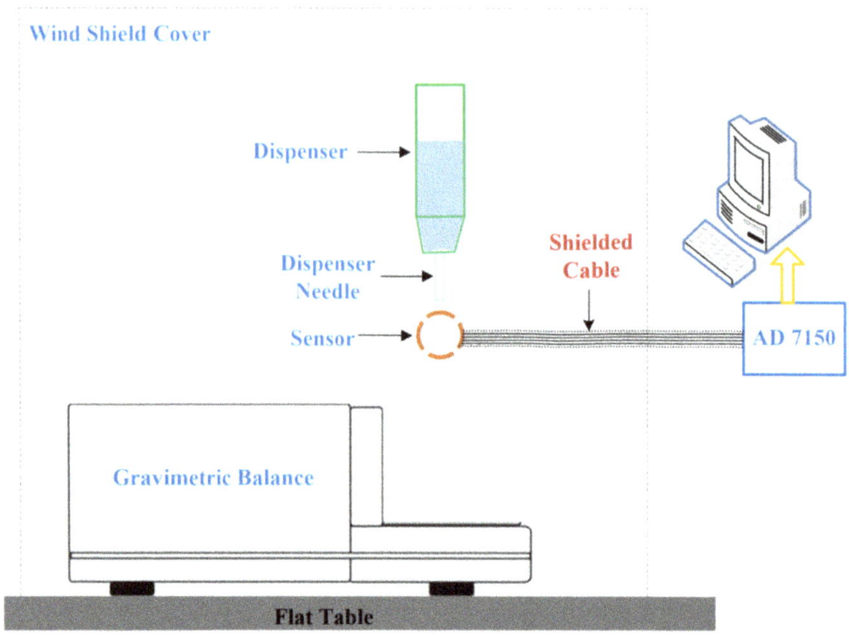

Figure 18.7. Schematics of the complete experimental set-up.

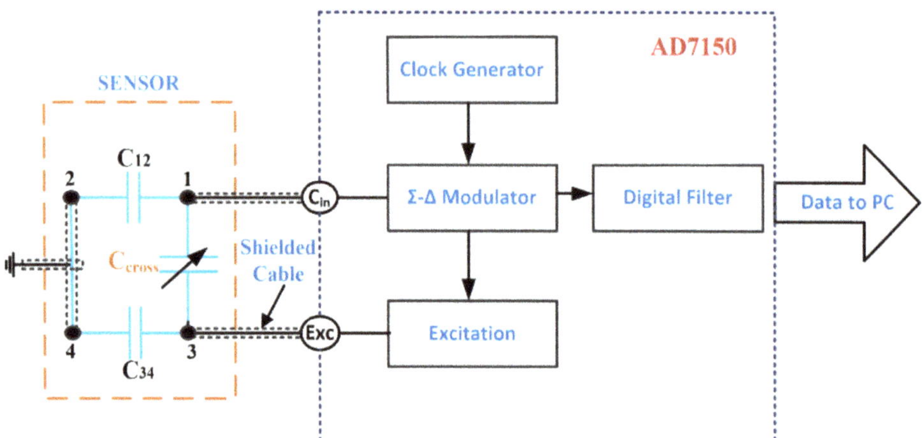

Figure 18.8. Schematic of the sensor interfacing to the CDC board. Copyright 2019 IEEE. Reprinted, with permission, from [27].

The terminals of the sensor are connected by a shielded cable nearly 2 cm in length to the electronic circuit through surface-mount device (SMD) connectors. Necessary precautions are taken to minimize the effect of parasitic capacitances.

C_{13} is the cross-capacitance between 1 and 3 electrodes, when 2 and 4 electrodes are grounded. C_{12} and C_{34} are the grounded capacitances at terminals 1 and 3. These grounded capacitances are measured by an impedance analyzer (4294A). Due to the symmetrical structure, these capacitances are almost equal, and the measured values were found to be ~1.23 pF. However, the effect of these ground capacitances on the output of the CDC was nullified by the internal architecture of the board. To minimize these capacitances, their value should be below 100 pF, and this condition was fulfilled for the present sensor.

The digital data were acquired continuously by the AD7150 CDC and stored on the PC after every 10 ms. The automatic dispenser dispensed liquid droplets at a regular interval of 2 s. The digital data (D) were then converted into a corresponding capacitance value C (pF) using the formula:

$$C\ (\text{pF}) = \frac{D_0 - 12288}{40944} \times C_i, \tag{18.20}$$

where C_i is the input capacitance range (0–0.5 pF). The distance between the dispenser nozzle and the sensor tube is kept minimum as far as possible. The AD7150 measures the capacitance at input terminals and compares its measurement result with a threshold, which can be either fixed or dynamically adjusted by the on-chip adaptive threshold algorithm. A positive change in the capacitance value is observed when a droplet falls through the sensor.

In the case of the present application, the droplet remains inside the volume of the sensor for a very short time ($\approx$ 23 ms). The output should follow the fast change in sensor capacitance without any loss of data.

The response of the adaptive threshold algorithm to an input capacitance step change is an exponential settling curve characterized by the equation

$$\text{Average}(N) = \text{Average}(0) + \text{change}\ (1 - e^{N/\text{TimeConstant}}), \tag{18.21}$$

where Average(N) is the value of the average N complete CDC conversion cycles, after a step change in the input. Average(0) is the value before the step change, and TimeConstant = $2^{(\text{ThrSettling}+1)}$.

ThrSettling is of four-bit value, so it can take any binary value from 0 to 15 and TimeConstant can take any value from $2^{(0+1)}$ to $2^{(15+1)}$. TimeConstant can be selected in the range between 2 and 65,536 conversion cycles, in steps of power of 2, by programming the ThrSettling bits.

Care must be taken when choosing the average TimeConstant for the final application. For maximum sensitivity and reliable detection the settling TimeConstant is kept at a minimum. That is two conversion cycles or 20 ms. This is achieved by setting ThrSettling bits to 0 value in the set-up register.

18.6 Experimental results

18.6.1 Detection of droplet volume and speed

The experiments were conducted at a room temperature of 25 °C and relative humidity of 40%RH. Four different droplet sizes, 9.3, 8, 5, and 3 μL, were dispensed along the length of the sensor using micro-needles of different radii and the change in the cross-capacitance value of the sensor for each drop was recorded. The peak capacitance values for different droplets are shown in figure 18.9. It shows that there

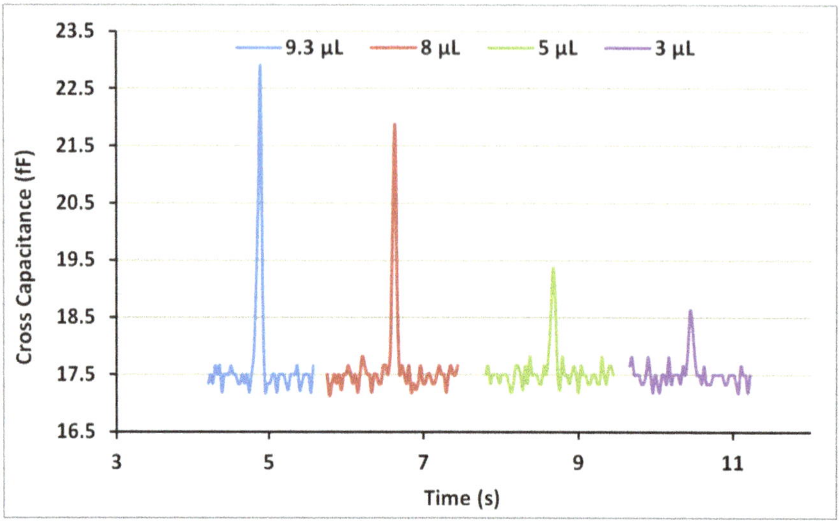

Figure 18.9. Capacitive response of the sensor for droplets of different volume.

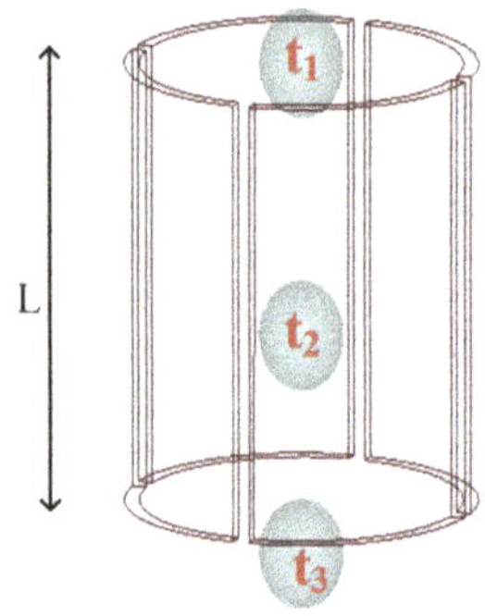

Figure 18.10. Determination of droplet velocity. Copyright 2019 IEEE. Reprinted, with permission, from [27].

is a sharp change in the capacitance value for each droplet. The capacitance value reaches an initial offset capacitance when there is no droplet in the sensor. However, the height of the peaks of the capacitance values increase with an increase in droplet volume because the free-flying droplets cause a momentary change in the dielectric constant of the space within the electrodes according to their volume.

When the droplet of 9.3 μL from the dispenser enters the sensor at any instant t_1, the capacitance begins to increase very fast. The maximum capacitance is recorded when the droplet is fully within the electrodes at the instant $t_{2,}$ and there is again a decrease in capacitance value once the droplet exits the sensing region at the instant t_3. The overall droplet movement along the length of the sensor is illustrated in figure 18.10.

The capacitance values were measured after every 10 ms for the time of flight of the micro-droplet, as indicated by the dots. Since at t_1 and t_3, the droplet is close to the end of the sensing tube, some part of the droplet is not in the sensing range of the sensor. Thus the capacitance at these times will be lower compared to the instants

Table 18.1. Capacitance values at different instants of time.

Position	Time (ms)	Cross-capacitance (fF)
t_1	40	18.9
t_2	60	22.9
t_3	70	18.7

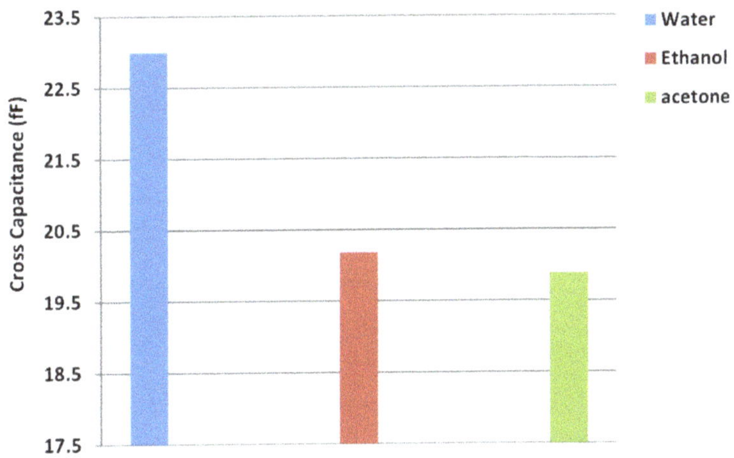

Figure 18.11. Response of the sensor for droplets of the same volume with different dielectric constants.

when the droplet is fully inside the volume at t_1. The capacitance values at these three instants are shown in table 18.1. By knowing the length (L) of the sensor which is 8 mm, the exit velocity of the droplet can be estimated by $L/(t_3 - t_1)$, that comes out to be around 26.6 cm s^{-1}.

18.6.2 Determination of the dielectric constant and conductivity

The sensor was also tested for droplets having different dielectric constants. Droplets from three liquids such as water with $\varepsilon_r = 80$, ethanol with $\varepsilon_r = 24.3$, and acetone with $\varepsilon_r = 20.7$ of the same volume were used for this purpose. Droplets from each of these liquids were dispensed through the sensor and the response of the sensor was recorded and is plotted in figure 18.11. Due to the difference in their dielectric constants, the droplets of these liquids provide different and distinct capacitance peaks. The value of the capacitance peak decreases with a decrease in the dielectric constant of the dispensed droplet. Thus the sensor can detect different types of liquid droplets.

The sensor was also used to measure the conductivity of the liquid droplets. Different salt solutions having different molar concentrations (M) from 1 to 4 M were prepared by mixing fixed amounts of salt to deionized (DI) water. The conductivity of the deionized water was almost zero ($\approx$ 0.06 μS cm^{-1}). Droplets

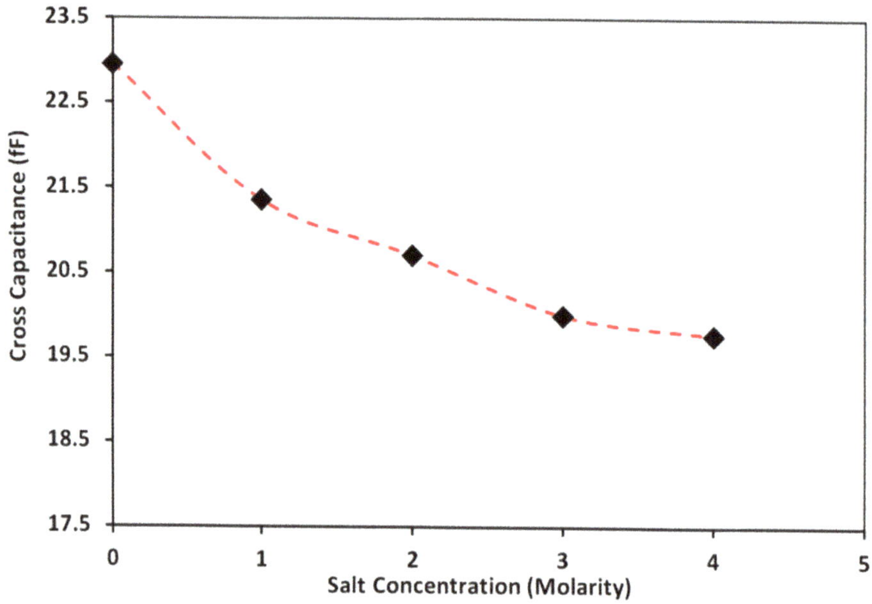

Figure 18.12. Response of the sensor for droplets having the same volume but different salt concentrations.

from each salt solution were passed through the sensor and the capacitance value corresponding to each type of liquid drop was noted. The variation of the capacitance peak with the variation of water conductivity is shown in figure 18.12. It is observed that with an increase in salt concentration (M), the capacitance peak decreases. The reason is that the dielectric constant of water decreases with an increase in salt concentration, as does the corresponding change in cross-capacitance value.

18.6.3 Repeatability and drift study

Repeatability is another important parameter of the sensor. It is the ability of the sensor to give the same output response again and again under similar conditions, and in a short period of time. The repeatability should be very good for any sensor. The repeatability test was conducted by passing identical water droplets through the sensor again and again, and the capacitance peak for each drop was noted. The maximum deviation in successive capacitance peaks was 0.13%, so the output of the sensor was highly repeatable

Over a period, the sensor output for the same quantity of sample drifts from its initial value. This is due to wear and tear to the sensor, changes in material properties due to the effect of the environment. Most commercial sensors suffer from this disadvantage, and the sensors need to be recalibrated after every 6 to 12 months for accurate and reliable measurements. Calibration is a costly and time-consuming affair. Efforts should be made to minimize the drift in sensor output. An experiment was conducted to study the drift of the sensor output by observing the capacitance value for an identical water droplet for several months. It was found that the capacitance value of this cylindrical cross-capacitive sensor drifted by a negligible amount. This is due to the non-contact detection and the absence of any thin sensing layer.

The contactless micro-droplet detection by the cross-capacitive sensor is both volume-dependent as well as medium sensitive. The sensitivity increases as the droplet volume increases. The maximum sensitivity of the sensor was found to be 0.6 fF μL^{-1}. The calibration curve of the developed sensor is of second order given in:

$$C_c\,(\mathrm{pF}) = 0.0533V^2 + 0.0843V + 17.5, \tag{18.22}$$

where V is the droplet volume in μL.

Also, for measuring very small capacitance (in the range of a few femtofarads), the electrostatic shielding of the structure is very important. The lateral displacement of the droplet position with respect to the center of the sensor may cause some error. According to Maxwell's first law, when a dielectric (ε_r) is placed into an electric field (E), the change in electric flux density (D) is proportional to the electric field. Thus the droplets which pass through the high electric field will cause a larger change in flux density and eventually a larger change in capacitance value. The effect of change in the lateral position of the droplet along the sensor tube on the sensor's response is analyzed by finite element analysis. The maximum change in cross-capacitance, when the position of the droplet is changed from the center position to the extreme right or left was found to be only 0.5 fF. The error caused due to the lateral displacement of the droplet in the case of the parallel plate capacitor used for droplet detection is higher due to the inhomogeneous field within the electrodes [14]. The error caused due to the change in the position of the droplet, however, is much lower in the case of the cross-capacitive sensor. This is due to the symmetry of the proposed structure.

The low cost, ease of fabrication, lack of shorting problems, bulk producibility, highly repeatability, negligible drift, and superior response characteristics are some of the important features of the fabricated sensor. The non-contact detection of droplets ensures its use in many applications such as chemical, biological, medical, and automatic industrial dispensing systems. The sensor can also be used for the precise and accurate measurement of the dielectric constant of liquids.

18.7 Conclusion

This chapter discusses a printable and flexible micro-droplet sensor, whose design is based on the Thomson–Lampard theorem. Such a calculable capacitor is very accurate and has a stable capacitance value. The base capacitance or the offset capacitance depends only on the length of the electrodes, and not on any other physical dimensions such as radius. The sensor electrodes were not in contact with the dispensed liquid droplets. The droplets passed through a hollow Teflon shell and the electrodes were placed on the outer periphery of the shell. This arrangement avoided contamination of the copper electrodes by different liquid samples. In addition, the thick solid Teflon shell provided mechanical strength as well as proper symmetry to the structure, which can otherwise cause some measurement error.

The sensor is of non-contact type as the electrodes are placed on the outer side and since Teflon is highly hydrophobic. However, if somehow due to a dispensing error the liquid droplet comes into contact with the Teflon, the inner wall may become wet which can increase the base cross-capacitance of the sensor and result in a higher capacitance peak for that droplet. If the inner wall is wetted, it can be refreshed by ethanol solvent using cotton buds. As a very good solvent, ethanol removes the unwanted film from the Teflon surface and evaporates very quickly. The sensor shows excellent response characteristics to detect droplets of different sizes and different types. The output of the capacitance is highly repeatable, free of drift, and highly reliable.

References

[1] Nathan A *et al* 2012 Flexible electronics: the next ubiquitous platform *Proc. IEEE* **100** 1486–517

[2] Liu Y *et al* 2018 Flexible, stretchable sensors for wearable health monitoring: sensing mechanisms, materials, fabrication strategies and features *Sensors* **18** 645

[3] Souri H and Bhattacharyya D 2019 Highly stretchable and wearable strain sensors using conductive wool yarns with controllable sensitivity *Sens. Actuators* A **285** 142–8

[4] Khan S *et al* 2015 Technologies for printing sensors and electronics over large flexible substrates: a review *IEEE Sens. J.* **15** 2015

[5] Pease R F and Chou S Y 2008 Lithography and other patterning techniques for future electronics *Proc. IEEE* **96** 248–70

[6] Yang H, Leow W R and Chen X 2018 3D printing of flexible electronic devices *Small Methods* **2** 1700259

[7] Dahiya R S, Metta G, Valle M and Sandini G 2010 Tactile sensing—from humans to humanoids *IEEE Trans. Robot.* **26** 1–20

[8] Jung M *et al* 2010 All-printed and roll-to-roll-printable 13.56-MHz operated 1-bit RF tag on plastic foils *IEEE Trans. Electron Devices* **57** 571–80

[9] Sandström A, Dam H F, Krebs F C and Edman L 2012 Ambient fabrication of flexible and large-area organic light-emitting devices using slot-die coating *Nat. Commun.* **3** 1002

[10] Kim T I *et al* 2013 Deterministic assembly of releasable single crystal silicon-metal oxide field-effect devices formed from bulk wafers *Appl. Phys. Lett.* **102** 182104

[11] Krebs F C 2009 Fabrication and processing of polymer solar cells: a review of printing and coating techniques *Solar Energy Mater. Solar Cells* **93** 394–412

[12] Niu X and DeMello A J 2012 Building droplet-based microfluidic systems for biological analysis *Bio. Society Trans.* **40** 615–23

[13] Ali K *et al* 2018 Label-free sensing in microdroplet-based microfluidic systems *Chemosensors* **6** 23

[14] Ernst A *et al* 2011 Non-contact determination of velocity and volume of nanoliter droplets on the fly *IEEE Sens. J.* **11** 1736–42

[15] Comley 2004 Continued miniaturisation of assay technologies drives market for nanolitre dispensing *Drug Discovery World* Summer 1–8

[16] Tröndle J, Ernst A, Streule W and Zengerle R 2010 Non-contact optical sensor to detect free flying droplets in the nanoliter range *Sens. Actuators* A **158** 254–62

[17] Hassan S *et al* 2017 Optical flow cell for measuring size, velocity and composition of flowing droplets *Micromachines* **8** 58

[18] Jakiela S, Makulska S, Korczyk P M and Garstecki P 2011 Speed of flow of individual droplets in microfluidic channels as a function of the capillary number, volume of droplets and contrast of viscosities *Lab Chip* **21** 3603–8

[19] Ernst A, Streule W, Schmitt N, Zengerle R and Koltay P 2009 A capacitive sensor for non-contact nanoliter droplet detection *Sens. Actuators* A **153** 57–63

[20] Cole M C and Kenis P J A 2009 Multiplexed electrical sensor arrays in micro fluidic networks *Sens. Actuators* B **136** 350–8

[21] Isgor P K, Marcali M, Keser M and Elbuken C 2015 Microfluidic droplet content detection using integrated capacitive sensors *Sens. Actuators* B **210** 2015

[22] Elbuken C, Glawdel T, Chan D and Ren C L 2011 Detection of microdroplet size and speed using capacitive sensors *Sens. Actuators* A **171** 55–62

[23] Thompson A M and Lampard D G 1956 A new theorem in electrostatics and its application to calculable standards of capacitance *Nature* **177** 888

[24] Lampard D G 1957 A new theorem in electrostatics with applications to calculable standards of capacitance *Proc. Inst. Elect. Eng.* **104** 271–80

[25] Zargar Z H and Islam T 2020 A thin film porous alumina based cross-capacitive humidity sensor *IEEE Trans. Instrum. Meas.* **69** 2269–76

[26] Lee R D, Kim H J and Semenov P Y 2001 Precise measurement of the dielectric constants of liquids using the principle of cross capacitance *IEEE Trans. Instrum. Meas.* **50** 298–301

[27] Zargar Z H and Islam T 2019 A novel cross-capacitive sensor for non-contact micro droplet detection *IEEE Trans. Indus. Electron.* **66** 4759–66

[28] Islam T and Zargar Z H 2017 A novel capacitive sensor for non-contact micro droplet detection *Indian Patent, Appl.* No. 201711024806 A (Published), J. No. 31/2017

[29] Zargar Z H, Islam T and Akram K J 2018 Design and development of a non-contact cross-capacitive micro droplet detector *IEEE Sensors (New Delhi)* (Piscataway, NJ: IEEE) pp 1–4

[30] Analog Devices AD7150 *Datasheet* https://www.analog.com/media/en/technical-documentation/data-sheets/AD7150.pdf

Chapter 19

Flexible and conformal antennas for wireless communication and sensing applications

Raheel M Hashmi, B Mohamadzade and A S M Sayem

19.1 Introduction

Recent years have demonstrated tremendous growth in flexible and conformal electronics, particularly in wireless wearable devices for medicine, wellbeing, and healthcare. All such devices rely on electromagnetic waves and radiofrequency (RF) antennas to communicate information. This information can range from routine monitoring of data such as ambient readings from sensing devices, to time-critical information, such as data during medical treatments, diagnostic procedures, clinical research, and healthcare delivery. All such wearable electronic devices require compact high-performance antennas, which act as their eyes and ears to connect with the myriad of wireless systems all around us. With new developments in wearable technologies, significant interest in antenna research and development has bee sparked. This has led to the development of flexible and conformal antennas.

The most common applications of such antennas are found in wireless medical devices. Wireless medical devices are independent of any physical wires that tether a patient to a medical bed, creating an obstructive and unwelcome mesh of several sensors around the body to facilitate monitoring, diagnosis, and therapy. These provide real-time data connectivity for healthcare providers (doctors, nurses, carers, etc) while increasing the wearer's comfort. The data can be transmitted or received around the clock, while the wearer enjoys seamless mobility. Another significant advantage from a consumer's perspective is that antennas make many modern medical devices portable in contrast to their conventional bulky counterparts. This has fuelled the development and provisioning of significantly compact, wearable wireless medical devices, with decreasing costs and increasing affordability.

Antennas for wearable wireless devices are categorized with respect to the device type: (i) ingestible, implantable, and injectable antennas that can suitably be added

doi:10.1088/978-0-7503-3439-6ch19 © IOP Publishing Ltd 2021

to micro-electronic devices; (ii) wearable antennas deployed temporarily or permanently on the exterior of the body or embedded within clothing materials, and (iii) antennas mounted on control stations, which connect through to a wireless network gateway, collecting data and issuing commands to the wearable device, where necessary. The design of wearable antennas is challenging due to the requirements of compactness and small space availability in the device platform. The main challenges revolve around aesthetics, the ability to disguise the device, and increased unobtrusiveness, while ensuring moderate-to-good electrical efficiency, and high mechanical robustness and resilience. The performance of antennas at their operating frequency is critical and, in turn, the operating frequency bands experience the effect of the expected loss that can occur due to operation in close proximity to the body (absorption, reflection, etc), communication efficiency (data transfer speed), and device miniaturization. WiFi, Bluetooth, and near-field communications (NFC) are commonly used network options for wireless medical devices and, thus, methods of near-field (inductive coupling) and far-field (RF radiated energy) communication are equally interesting and effective, depending on the application. Table 19.1 lists the frequency bands listed by spectrum regulatory authorities for use in wearable device networks, often called wireless body area networks (WBANs).

In wearable applications, important attributes to consider include the antenna's form factor, its electrical stability when placed on the body, the selection of fabrication processes that are robust, and its mechanical resilience to withstand changes in posture, motion artefacts, subject specificity, and moisture (humidity/perspiration). Increasing demands for unobtrusiveness and blended integration have

Table 19.1. Global spectrum definitions for wearable wireless communication systems. (Reproduced from [1]. CC BY 4.0.)

Frequency	Application
401–402 MHz 402–405 MHz 403.5–403.8 MHz 405–406 MHz	Medical Implant Communication Services (MICS)
413–457 MHz	Medradio Micropower Networks (MMN), transmit and relay data for implanted and body-worn medical devices for diagnostic and therapeutic functions
2.4 and 5 GHz	WiFi, smart hospital beds, mobile nursing stations
2.404–2.478 GHz	Bluetooth, indoor navigation for patients, connectivity between device and smartphone for health-data monitoring
2.36–2.4 GHz	New Medical BAN (MBAN)
3.1–10.76 GHz	Ultrawideband (UWB)
57–64 GHz 59–66 GHz	The band plans and rules at mmW-60 WBAN

led to the development of antennas that are washable, deformable, and can be embedded within the clothing of the wearer, or be attached to the body, so they are not interrupting the usual routine of the individuals. These antennas are largely based on polymer materials, where the polymer material serves as the dominant material used for construction and even encapsulation of the overall device.

Figure 19.1 shows some examples of flexible and conformal antennas, including a fabric-only based antenna and fabric–polymer composite antennas. The latter are more interesting and relevant in this chapter. The non-conductive parts of the antennas are made out of pure-polymers or polymer–ceramic composites, to tailor the requisite electrical properties, i.e. their relative permittivity and loss factors. The conductive parts of the antennas are made of conductive threads embroidered on polymer sheets, nanoparticles embedded within the polymer, or conductive-fabric encapsulated within thin layers of polymer material. Table 19.2 summarizes important material characteristics in PDMS-based antennas with respect to operating frequency. The encapsulated antennas further offer the advantage of embedding any associated RF switching circuits within the same protective polymer casings to prevent oxidation and corrosion that may result from perspiration, making them suitable for a wide range of environmental conditions. Polymer-based antennas are

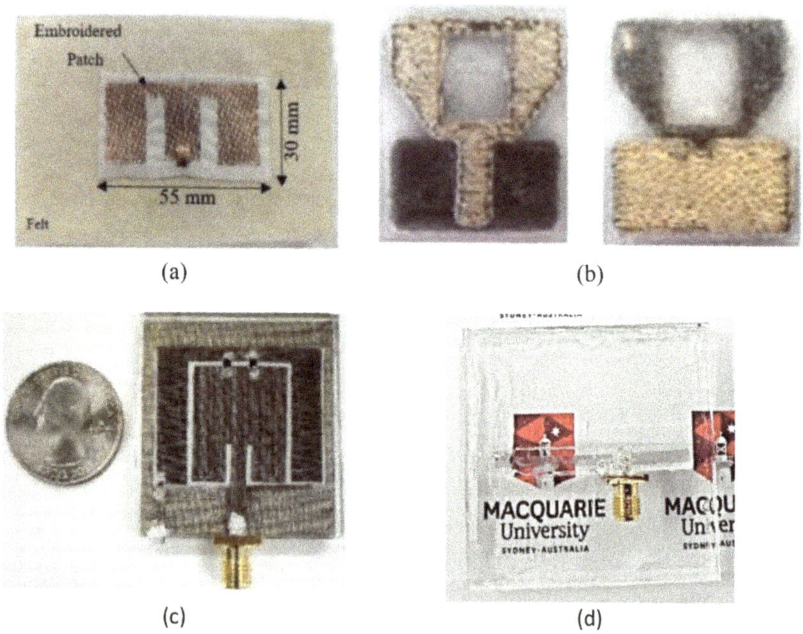

Figure 19.1. Examples of flexible and conformal antennas: (a) fabric-based embroidered E-shape microstrip antenna, these were the pre-cursor to polymer-based wearable antennas. (Reproduced from [1]. CC BY 4.0.) (b) Water resistant antenna with an embroidered conductive thread pattern on a PDMS substrate. (Reproduced from [2]. CC BY 4.0.) (c) Water-proof flexible antenna made out of embedding conductive fabric within polymer. (Reproduced with permission from [3]. Copyright 2018 IEEE.) (d) Transparent flexible antenna made out of pure water and polymer. (Reproduced with permission from [4]. Copyright 2020 IEEE.)

Table 19.2. Key material characteristics in PDMS-based antennas with respect to operating frequency. (Reproduced from [1]. CC BY 4.0.)

	Conductive parts		Non-conductive part (substrate)			
Ref.	Material	Conductivity or resistivity	Material	Relative permittivity ε_r	Loss factor $\tan\delta$	Frequency (MHz)
[5]	Carbon nanotube	$0.9\ \Omega\ \mathrm{sq}^{-1}$	PDMS	3.8	0.015	2250
[6]	Silver nanowires	$8.13 \times 10^3\ \mathrm{s\ m}^{-1}$	PDMS	2.67	0.01	2920
[7]	Amberstrand fibres	$1.6–2.6\ \Omega\ \mathrm{m}^{-1}$	PDMS	3	0.02	915
[8]	Copper foil	NA	PDMS	2.7	0.013	5800
[9]	Nickel–copper–silver-coated nylon	$0.01\ \Omega\ \mathrm{sq}^{-1}$	PDMS	2.8	0.02, 0.04	24505800
[10]	Copper micromesh	$3.28 \times 10^4\ \mathrm{s\ m}^{-1}$	PDMS	2.8	0.02	2460 to 2940

also noted for their excellent resilience to mechanical stress and, thus, are suitable for sustained bending and deformation.

To design wearable antennas, RF modelling is first carried out using full-wave 3D numerical electromagnetic simulation software. Examples of such software include ANSYS HFSS, Altair FEKO, and CST Microwave Studio. For polymer-based wearable antennas, the characterization of polymers, conductive threads, nanoparticles, and conductive fabrics is carried out even before the RF modelling stage to establish an understanding of the electrical properties that must be used during the modelling process. Since these are all non-conventional materials, their electrical properties are often not well-known. The modelling of these antennas must also employ analytical models of the human body to quantify the signal attenuation and loss due to the close proximity of lossy biological tissues [11] and detuning effects that result from the protective encapsulation layers made of polymers or biocompatible materials. Once modelling is completed, prototypes are fabricated using customized layer-by-layer processes (detailed in later sections in this chapter), and experimental validation is performed *in vitro* using phantoms and tissue-mimicking gels. These phantoms and gels emulate the behaviour of biological tissues during the validation stages. Some examples of biologically stable polymers for use in wearable antennas are polydimethylsiloxane (PDMS) from Dow Corning, MED-1134 from NuSil, and Parylene-C polymer.

In the forthcoming sections, we will cover the manufacturing process for polymer-based antennas, including material characterization and morphology of PDMS-embedded conductive-fabric composites. Then some example antennas will be discussed, covering flexible antennas for ultrawideband (UWB) digital impulse radio communications, semi-transparent, fully transparent, flexible, and conformal antennas. The chapter will conclude with a perspective on future wearable networks and considerations for high-frequency operating bands a.k.a 'millimetre-wave' communications, envisioned in the near future.

19.2 Realizing flexible and conformal antennas

19.2.1 Characterization of materials

As stated earlier, the structure of a flexible antenna consists of two parts, the conducting and non-conducting parts. Conducting parts are usually made out of conductive threads or conductive fabrics that are, in turn, made out of woven micro-threads. Conductive fabrics are composed of metal-plated multifilament/monofilament threads that are intertwined with each other in, ideally, a one-to-one ratio between the vertical and horizontal directions. In order to form meaningful conductive sheets of fabric, the distance between adjacent groups of threads is kept very small, often less than 0.04 mm, leading to properties similar to a solid metallic plate [12]. In our studies, we have used several conductive fabrics provided by the manufacturers, Marktek Inc and Less EMF Inc. Various fabrics differ from each other in terms of their thicknesses, the material of plating applied to the threads, and the thread count or 'density' per unit area. The result is changes in the effective conductivity of the thread or resulting fabric, which is directly related to the efficiency and performance of the fabricated antenna. A simple rule of thumb is that the higher the conductivity of the conductive parts, the better the efficiency and the lower the resistive losses within the antenna structure. The non-conducting parts, commonly called the substrate, can be composed of polymer materials. Our studies involved the use of pure PDMS as the substrate as well as a PDMS–ceramic composite polymer, which was made by doping the pure polymer with ceramic powder. The advantage of using the composite substrate is to control the electrical properties and tailor them for desired RF performance. Figures 19.2(a) and (b) show the measured relative permittivity and loss factor of the pure PDMS and PDMS–ceramic composite samples, as a function of frequency. These properties are

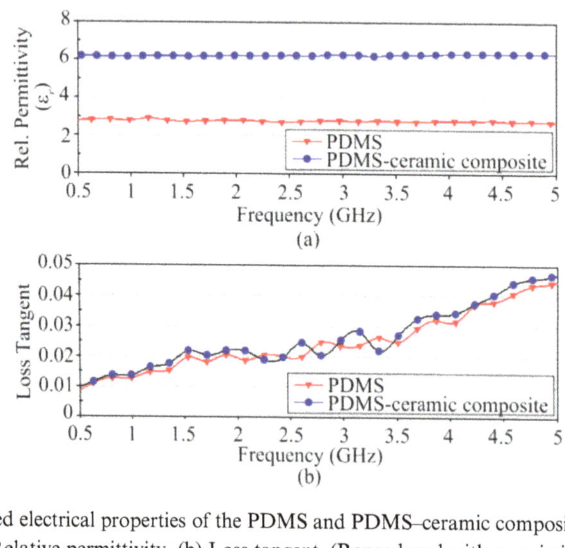

Figure 19.2. Measured electrical properties of the PDMS and PDMS–ceramic composite with 20% volume of $SrTiO_3$ loading. (a) Relative permittivity, (b) Loss tangent. (Reproduced with permission from [3]. Copyright 2018 IEEE.)

measured using an Agilent 85070E Dielectric Probe Kit. The PDMS–ceramic composite sample in this case is made by doping 20% volume of strontium titanate ($SrTiO_3$) in pure PDMS. In itself, $SrTiO_3$ has a very high relative permittivity of 270 and a loss factor of <0.001 at microwave frequencies [13, 14]. In figure 19.2 it can be noted that the PDMS–ceramic composite demonstrates twice the permittivity compared to pure PDMS with a comparable loss characteristic. Such properties are useful for the miniaturization of antennas, as antennas designed using substrates with higher permittivity will have smaller sizes. Nonetheless, it should be pointed out that the flexibility of PDMS–ceramic composites decreases with an increase in the volume of the ceramic used for doping, thus the trade-off between flexibility, miniaturization, and performance.

The use of PDMS polymer to develop antennas and RF components has attracted increasing attention in recent years. It has a remarkably low Young's modulus (<3 MPa), thus presenting extremely high flexibility/conformality compared to any other polymer [15, 16] and also has an acceptable loss in the microwave region. The next section shows its versatile fabrication process, allowing it to be shaped into various form factors due to its initial liquid state [17]. The molecular structure of PDMS gives it unique properties of water-resistance, heat-resistance (up to 400 °C), resilience to UV exposure, and chemical stability [18, 19]. However, PDMS has inherently weak metal–polymer adhesion [20], rendering common metal deposition techniques or direct attachment of metal sheets to form antennas very challenging. Complete encapsulation of the conductive fabrics inside PDMS is an effective alternative for forming strong integration between the conductive fabrics and PDMS [3]. Figure 19.3 presents a scanning electron microscope (SEM) image of such a conductive-fabric–PDMS composite [3]. Figure 19.4 shows the results of various mechanical stress and deformation tests (figure 19.4(a)) applied to conductive-fabric–PDMS samples. The tests are applied to samples of four different conductive fabrics when embedded within PDMS through a layer-by-layer construction (figure 19.4(b)) and when simply attached to the top of a PDMS layer (figure 19.4(c)). Note the detachment of the conductive-fabric–PDMS bonding of the former case, which is minimal, when compared with the latter case, where the

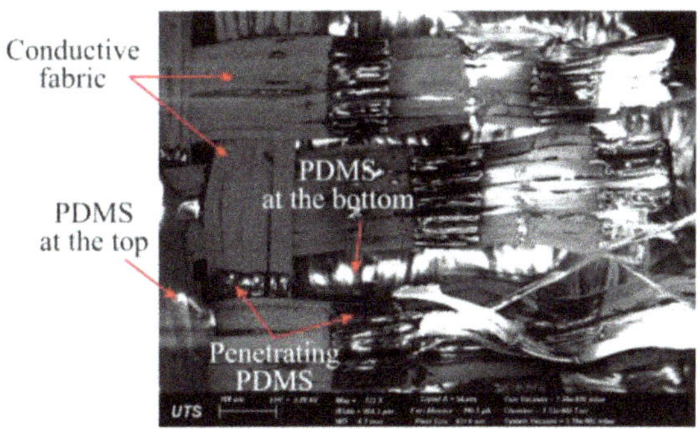

Figure 19.3. Scanning electron microscope image of a PDMS-embedded conductive fabric. (Reproduced with permission from [3]. Copyright 2018 IEEE.)

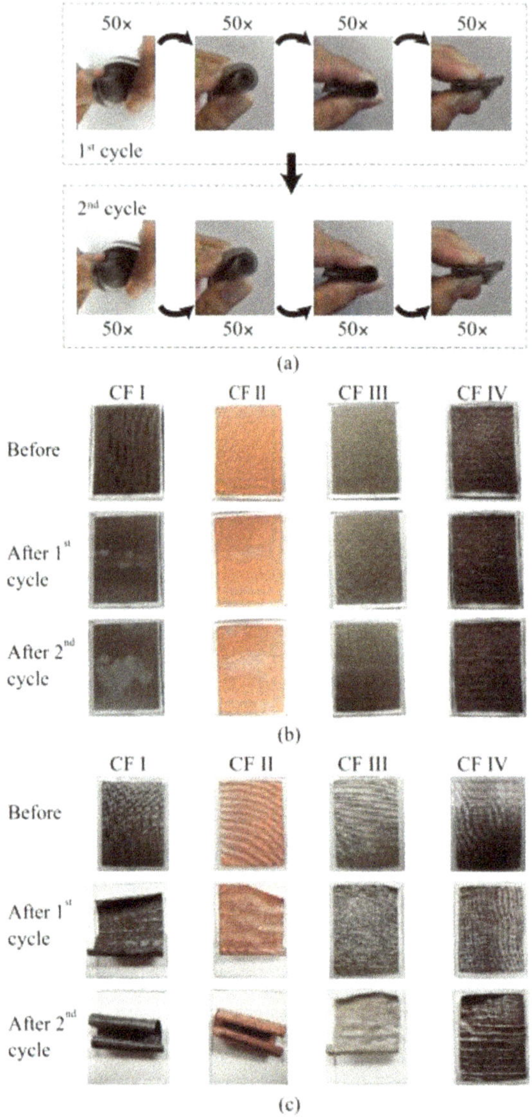

Figure 19.4. (a) Illustration of mechanical testing scenarios. Photograph of the samples before and after tests, when: (b) the conductive fabric is embedded inside the PDMS layer and (c) the conductive fabric is simply attached on top of the PDMS layer. (Reproduced with permission from [3]. Copyright 2018 IEEE.)

conductive fabrics have been detached from the PDMS after deformations. Another interesting point to observe in figure 19.4 is the variation in detachment between different conductive fabrics, e.g. conductive fabric (CF) I detaches the most whereas CF IV detaches the least. This variation is due to the morphology of the conductive fabrics and the way in which PDMS penetrates through the space available between the multifilament conductive yarns intertwining with each other that form the fabrics (figure 19.3). The greater the available space, the more PDMS percolates through the fabric, forming strong PDMS–PDMS adhesion and sandwiching the fabric between the layers. For the fabrics shown in figure 19.4, CF I has the least spacing available between the thread clusters whereas CF IV has the most, and thus the least detachment (figures 19.4(b) and (c)).

19.2.2 Fabrication process

Once the raw materials are identified and characterized, the stage of physically realizing the antenna is next. Traditional fabrication processes are not well suited to fabricating non-conventional material-based flexible antennas, therefore alternative manufacturing processes have emerged. For example, a layer-by-layer fabrication process is used in fabricating polymer-based antennas, mimicking the construction of a printed-circuit-board (PCB) antenna. Let us explain the fabrication process of an example antenna that is made using the PDMS and conductive fabric. Figure 19.5 shows the steps undertaken in the fabrication of two types of flexible antennas, i.e. a microstrip patch antenna (PA) made using pure PDMS, and a resonance-frequency tuned microstrip patch antenna (RFPA) made using a PDMS–ceramic composite.

1. At the beginning of the fabrication process, four ring-shaped moulds and a base were made. Then the PDMS solutions (pure and ceramic mixed) were prepared and the conductive fabric was cut by a sharp razor pursuing the dimensions of the antenna's patch and ground plane.
2. In the next step, the first mould was attached to the base by using silicone sealant and PDMS solution was poured into the mould. The PDMS solution was then degassed in a vacuum desiccator to remove the air bubbles. The PDMS solution was then cured in an oven at 65 °C for about 2 h, thus the bottom encapsulation layer was formed.

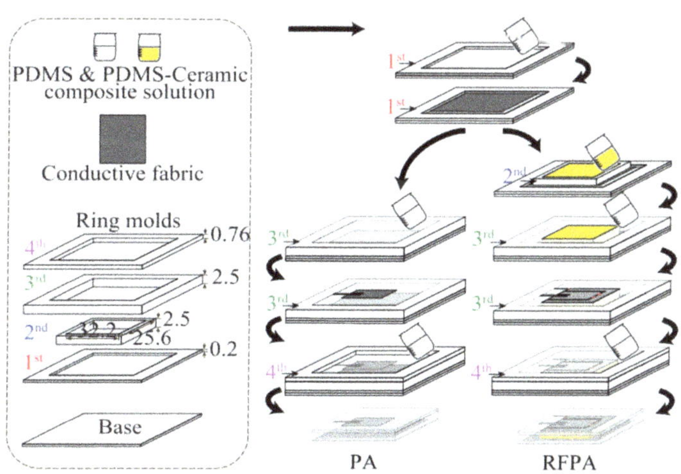

Figure 19.5. Schematic illustration of the fabrication process of PDMS-embedded conductive-fabric antennas. (Reproduced with permission from [3]. Copyright 2018 IEEE.)

3. The ground plane was attached on top of the bottom encapsulation layer with a small amount of uncured PDMS solution and then cured in the oven at 75 °C for 20 min.
4. For RFPA the second mould was used to pour the PDMS–ceramic composite substrate which was cured in an oven at 65 °C for about 2 h. Then the second mould was removed and the third mould was attached to the first mould using silicone sealant. Pure PDMS solution was poured into it, after which it was degassed in a vacuum desiccator and then cured in an oven at 65 °C for about 2 h. Thus the substrate layer was formed. For PA, the second mould was not required, and the third mould was attached directly to the first mould and the substrate layer was formed with pure PDMS solution through the process of desiccation and curing.
5. In the next stage, the main patch was attached on top of the substrate layer by following the same procedure of step 3. For the RFPA, varactor diodes were attached in antenna geometry using silver conductive epoxy which was cured in the oven at 65 °C for 2 h.
6. The fourth mould was attached on top of the third mould using silicone sealant and step 2 was repeated to form the top encapsulation layer.
7. After finishing the antenna fabrication, the prototype was peeled from the mould. For measurement purposes, a small portion of PDMS was peeled off from the top and bottom encapsulation and an SMA connector was connected to the antenna by using silver conductive epoxy which was cured in the oven at 65 °C for 2 h.

It is worth mentioning that the flexible antennas developed from the conductive-fabric–polymer composite have comparatively lower gain and efficiency than traditional rigid antennas. The low conductivity of the conductive fabrics and the high loss of the polymers are the major factors in performance degradation. To improve the performance, in [21] some methods have been suggested, such as applying a coating on the fabric of some coating materials with high electrical conductivity and mixing the polymer with less lossy materials to alleviate the dielectric loss.

19.2.3 Test and validation

Wearable antennas are operated near the environment of the human body whose unique characteristics affect antenna performance. The presence of lossy human body tissue near the antenna affects the resonance frequency, impedance matching, bandwidth, gain, and efficiency [22]. Moreover, the absorption of electromagnetic radiation in human body tissues has hazardous effects on the body. The adverse effects of the antenna–body electromagnetic coupling should be considered in the design and experimental validation stages of the antenna.

In the design stages of the simulation, the effects of the human body are analysed by placing the antenna in close proximity to human body tissues. Different models of tissues are available in electromagnetic computational software. For experimental validation, wearable antennas are tested on fabricated phantoms. Extensive research can be found in the literature which has explored the fabrication processes of human-tissue equivalent phantoms mimicking the electrical properties of actual biological tissues in a wide range of frequencies [23–26]. Based on the physical state, human-body-equivalent phantoms can be classified into four principal groups, i.e. liquid, semi-liquid/gel, semi-solid/jelly, and solid. Consistent electrical properties, manufacturing cost, and durability are the major determining factors for the selection of particular phantoms. The manufacturing process of liquid phantoms is relatively simple [27], but they require container to hold them and suffer from inconsistent electrical properties due to dehydration and fungal contamination. Semi-liquid phantoms are slightly stiffer than liquid phantoms but still need a container to retain their shape. The other type of phantoms are semi-solid phantoms which have a more solid composition than semi-liquid phantoms. These types of phantoms can be customized easily to any shape and thus are useful in a wide range of applications. Moreover, semi-solid phantoms have consistent electrical properties.

Figure 19.6 shows an antenna measurement set-up with a muscle-tissue-equivalent semi-solid phantom [3] made by pursuing the procedure described in [25]. This semi-solid phantom [3] was fabricated with deionized water (85.15%), agar (2.64%), polyethylene powder (PEP) (10.52%), sodium chloride (0.18%), TX-151 (1.46%), and sodium dehydroacetate (0.05%). Figure 19.6(a) shows the phantom in flat form and figures 19.6(b) and 6(c) show a male mannequin filled with the semi-solid phantom.

The fourth type of phantom is solid or dry phantoms [24], generally made with ceramic materials. In solid phantoms it is quite difficult to adjust the electrical properties to emulate actual human body tissues.

19.3 Examples of flexible and conformal antennas

The state-of-the-art research approaches towards the development of flexible and conformal antennas are listed in table 19.3. In this table the properties of the key conductive and dielectric materials used in wearable antenna fabrications are

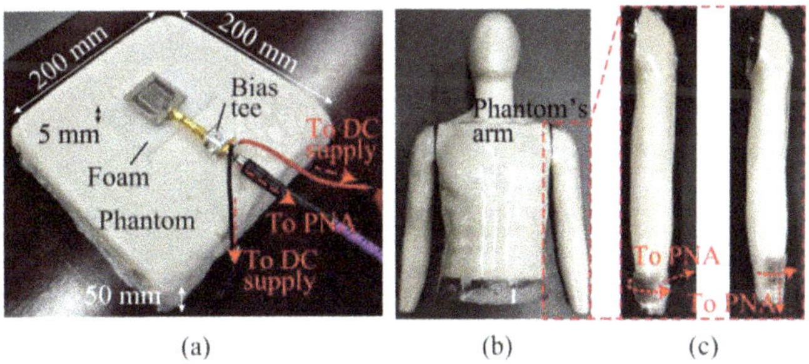

Figure 19.6. Experimental set-up for testing and measurements on phantoms. (a) RFPA mounted on the flat phantom. (b) Torso phantom for bending tests. (c) PA wrapped on the phantom's wrist with E-plane bending (left) and H-plane bending (right). (Reproduced with permission from [3]. Copyright 2018 IEEE.)

Table 19.3. State-of-the-art research approaches towards the development of flexible and conformal antennas. Reproduced from [28]. CC BY 4.0.

Ref.	Material name	Material type	Dielectric constant (relative permittivity)		Conductivity ($S\ m^{-1}$) or resistivity (ohms sq^{-1})	Suitable for folding/bending	Suitable for sustained folding/bending
			ε_r	**tanδ**			
[29]	ShieldIt Super	Textile			1.18×10^5 $S\ m^{-1}$	Yes	No
	Felt	Textile	1.22	0.016			
[30]	Graphene	Conductive ink			0.37×10^5 $S\ m^{-1}$	Yes	No
	Polyvinyl chloride foam, polyimide film	Polymer, foam	1.5, 3.5				
[31]	Copper tape	Copper				Yes	No
	Denim	Fabric	1.68	0.03			
[32]	Nylon ripstop fabric	Fabric			0.01 $\Omega\ sq^{-1}$	Yes	No
	PDMS	Polymer	2.82	0.023			
[33]	Shieldex	Fabric			0.009 $\Omega\ sq^{-1}$	Yes	No
	C-Foam PF-4	Foam	1.06	0.0001			
[34]	EGaIn	Liquid metal			3.4×10^6 $S\ m^{-1}$	Yes	Yes
	S-glass	Fibres and textile	3.4				
[35]	EGaIn	Liquid metal			3.4×10^4 $S\ m^{-1}$	Yes	Yes
	PDMS	Polymer	2.67	0.0375			
[36]	Galinstan	Liquid metal			3.46×10^6 $S\ m^{-1}$	Yes	Yes
	TC5005	Silicone	2.8–3.1				
[37]	EGaIn	Liquid metal			3.4×10^4 $S\ m^{-1}$	Yes	Yes
	Platinum-cured thermoset silicone	Polymer	2.2	0.08			
[38]	E-threads	Threads			1.9 $\Omega\ m^{-1}$	Yes	Yes
	Organza	Fabric	(~1)				
[39]	Copper film	Metal			4.4×10^5 $S\ m^{-1}$	Yes	No
	Paper	Paper	2.2	0.04			
[40]	Copper sheets	Metal				Yes	Yes
	Verowhite	Polymer	2.8	0.01			
[41]	Copper	Metal			5.96×10^5 $S\ m^{-1}$	Yes	Yes
	FR4 substrate	Composite	4.4				

summarized. It is clear that there is a variety of non-conventional materials used in the realization of flexible antennas. The commonly used flexible conductors are conductive textiles, copper tape, and liquid metals, and widely used substrates are textiles, foam, paper, and polymers.

19.3.1 Ultrawideband flexible antennas

Ultrawideband (UWB) technology is gaining significant importance in modern wireless communications due to its prominent features of fast speed, large bandwidth, low power consumption, and less sensitivity to multipath effects [42]. UWB antennas have numerous applications in a variety of fields, such as wireless body area networks (WBANs), unmanned aerial vehicles (UAVs), synthetic aperture radar (SAR), self-controlled ground sensor networks, vehicular communications, and network surveillance [43, 44]. Some of these applications, e.g. wireless body area networks (WBANs), require antennas with a conformal physical structures. Efforts have been devoted by researchers to obtain UWB features in wearable antennas while preserving the other features of wearable devices. Two recent notable research approaches towards the development of flexible UWB antennas are demonstrated below.

A flexible and compact UWB antenna with a monopole-like radiation pattern [45] is illustrated in figure 19.7. The prototype was fabricated by following the same layer-by-layer formation process as described in section 2.2. This antenna was fabricated with PDMS and conductive fabrics. Two different conductive fabrics were used in the patch and ground plane. Nickel–copper–silver-coated nylon ripstop from Marktek Inc, with a thickness of 0.13 mm, was used for realizing the patch and nickel–copper-coated ripstop from Less EMF Inc, with a thickness of 0.08 mm, was used for realizing the ground plane. The first fabric was selected for the patch due to its higher conductivity which incorporates less loss and the second fabric was selected for the ground due to its higher porosity which allows more PDMS percolation, thus maintaining sturdy attachment and becoming suitable for the ground plane which consists of a large surface area.

The performance of this flexible UWB antenna was tested in a bent state to evaluate its feasibility in conformal operations. For this purpose, the prototype was wrapped over a hollow plastic tube around the *x*- and *yyyyy*-axis with a radius (r_b)

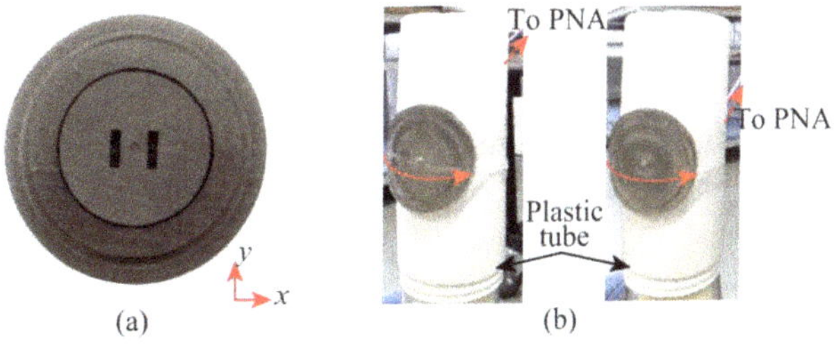

Figure 19.7. (a) Photograph of the fabricated UWB antenna. (b) Bending experimental set-ups over a plastic tube with an outer radius of 40 mm. Case 1 (left): the antenna was bent around its *xxxxx*-axis. Case 2 (right): the antenna was bent around its *yyyyy*-axis. (Reproduced with permission from [45]. Copyright 2020 IEEE.)

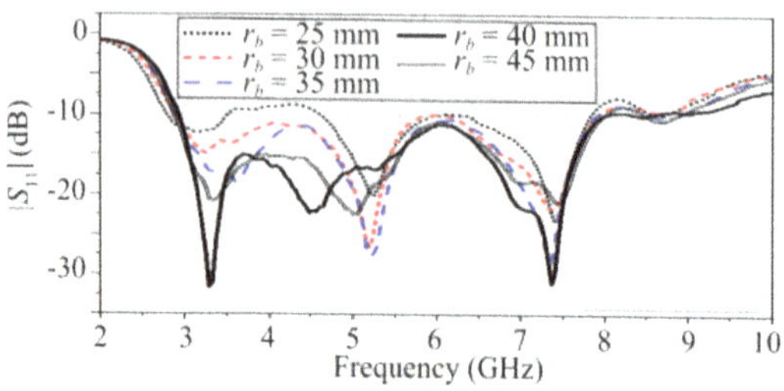

Figure 19.8. Measured $|S_{11}|$ of the flexible UWB antenna bent with different radii. (Reproduced with permission from [45]. Copyright 2018 IEEE.)

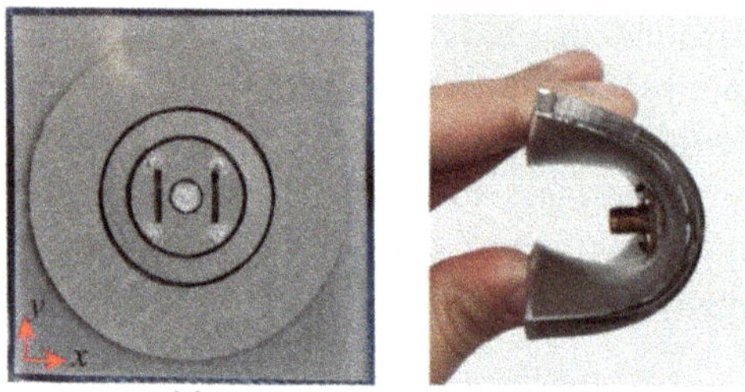

Figure 19.9. Fabricated band-notched UWB antenna. (a) View from the top. (b) View while bent. (Reproduced with permission from [47]. Copyright 2020 IEEE.)

varied from 25 to 45 mm and the corresponding magnitude of input reflection coefficient, $|S_{11}|$, was measured. In most wireless applications, an antenna is deemed to be operational when the value of $|S_{11}|$ falls below a reference level of -10 dB. The bending test set-up is shown in figure 19.7(b). Figure 19.8 exhibits the measured $|S_{11}|$ results, which reveal that the antenna maintains a stable 10 dB return loss across the operating band in bent states up to 30 mm radius of bending. In addition to the bending test, the system fidelity factor (SFF) was also calculated to investigate its feasibility in UWB pulse transmission [46, 47]. It was found that the antenna achieved nearly 86% SFF, which is well above the marginal SFF of 50% [46], demonstrating its compatibility in preserving the pulse shape.

In wideband antennas, the interference from the existing wireless network technologies, e.g. IEEE 802.11a and HIPERLAN/2 in the 5.15–5.825 GHz band, affects the communication channel adversely. In this regard introduction of a band notch (i.e. a portion of bandwidth that is scraped out of the operating band such that the antenna rejects signals within that bandwidth) can be an effective technique for avoiding the interference problem. An example of such a band-notched flexible UWB antenna [47] is shown in figure 19.9. This antenna is also fabricated with conductive fabric and PDMS. To evaluate its performance in conformal operations, it was wrapped over cylindrical plastic tubes with varying radii (r_b), from 20 to 35 mm, in both the x- and y-axes, as shown in the inset of figure 19.10. The corresponding voltage standing wave ratios (VSWRs) for the corresponding bending radii were measured, which are shown in figure 19.10. A VSWR close to a reference

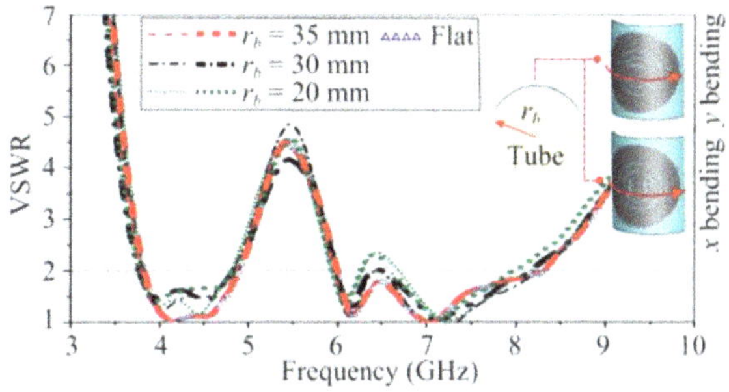

Figure 19.10. Measured VSWR of the antenna under various *xxxxx*-axis (thin lines) and *yyyyy*-axis (thick lines) bending radii compared to that of the flat case (triangles). The inset illustrates the bending test scenarios over a plastic tube. (Reprinted with permission from [47]. Copyright 2020 IEEE.)

level of 2 suggests that the antenna is reasonably accepting the RF signals, whereas higher values, such as those between 5 and 6 GHz, suggest an antenna rejecting the signals. From the measured VSWR results, it is ascertained that the antenna maintains stable performance up to a bending radius of 30 mm.

19.3.2 Transparent flexible antennas

In addition to flexibility and robustness, unobtrusiveness is gaining significant attention in wearable devices to enhance aesthetics and wearers' comfort. Unobtrusive wearable devices are vitally important for seamless monitoring of patients' real-time health status without interrupting routine activities. It has been ascertained that optical transparency is the most relevant way of achieving unobtrusiveness in body-worn devices. Moreover, optically transparent antennas can be merged with solar cells, displays, automobile windshields, and windows, thus their application spectrum is broadening tremendously. However, it is quite challenging to realize transparent antennas that are concurrently flexible and robust. Transparent antennas are generally developed from conductive thin films or meshed conductors. However, most of these classes of transparent conductors are not flexible and have poor adhesion to flexible substrates. Moreover, there is a trade-off between the optical and electrical properties of the realized antennas utilizing these traditional materials. Apart from these extensively used thin films and meshed conductors, transparent-conductive-mesh–PDMS composite is demonstrated as an effective alternative for realizing robust, flexible, and transparent antennas [48]. A dual-band antenna operating at the 2.5 and 5 GHz Industrial, Scientific, and Medical (ISM) band was designed in [48]. The conductive parts (patch and ground plane) of the demonstrated antenna in [48] were made from transparent conductive mesh, VeilShield supplied from Less EMF Inc, USA. VeilShield is a mesh structured shielding fabric, consisting of woven mesh polyester threads coated with zinc (Zn)-blackened nickel (Ni) over copper. The monofilament threads entangled with each other in a 1:1 ratio with a ±0.2 mm spacing. VeilShield is 57 μm thick, has nearly 72% optical transparency, and 0.1 Ω sq^{-1} sheet resistance. In the antenna, PDMS was used as the substrate as well as the protective coating. The unique feature of this

Figure 19.11. Photographs of the fabricated flexible transparent antenna. The Macquarie University logo and name, printed on paper, can be seen through the antenna (left). (Reproduced with permission from [48]. Copyright 2019 IEEE.)

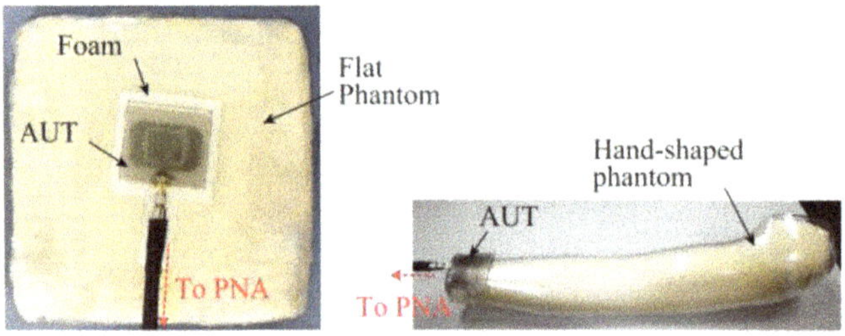

Figure 19.12. Antenna measurement set-up. (a) Antenna on the flat phantom. (b) Antenna on the hand-shaped phantom. (Reproduced with permission from [48]. Copyright 2019 IEEE.)

explored antenna is its excellent robustness in harsh physical deformations. The mesh structured topology of VeilShield allows uniform PDMS penetration through it and after curing a strong attachment forms between the VeilShield and PDMS, which remains stable in extreme bending scenarios [49–51]. It can be noted that VeilShield forms stronger integration with polymers than other opaque conductive fabrics which allow less penetration of polymer though their geometries. It is worth mentioning here that the demonstrated fabrication method in [48] is simple and less costly than the traditional fabrication methods of transparent antennas. Figure 19.11 shows the fabricated prototype of the transparent antenna. The clear visibility of the printed texts though the prototype in left photo exhibits the high transparency of the prototype and the bent view of the prototype in right photo exhibits its excellent flexibility.

The performance of this dual-band flexible transparent antenna was examined in human body environments and compared with free space performance to ascertain its compatibility with wearable applications. Figure 19.12 shows the measurement set-up in human body environments. A human-muscle-equivalent UWB semi-solid flat phantom was used to assess the performance near biological tissue and the hand-shaped phantom was used to assess the performance under bending, mimicking real life applications. Figure 19.13 illustrates the predicted and measured $|S_{11}|$ results. Here, the predicted results are the results computed numerically in CST Microwave Studio 2017. Figure 19.13 reveals strong consistency between the predicted and measured performance, thus validating the design. The small discrepancy emerges

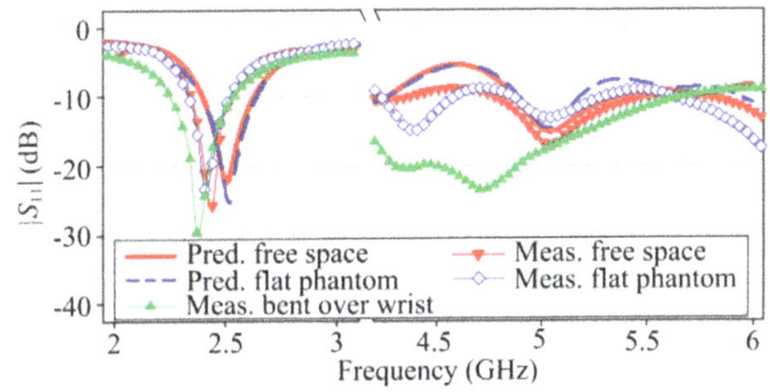

Figure 19.13. Predicted and measured $|S_{11}|$ results of the antenna in free space and on phantoms. (Reproduced with permission from [48]. Copyright 2019 IEEE.)

from the inaccuracy of the manual fabrication of the antenna. It can be observed that the $|S_{11}|$ performance on the phantom still remains in the desired operating bands, which is possibly due to the utilization of the full ground plane which acts as a shield between the antenna and phantom. The measurement results show that in free space a 200 MHz (from 2.33 to 2.53 GHz) 10 dB return-loss bandwidth was obtained in the lower band and 900 MHz (from 4.7 to 5.6 GHz) was obtained in the higher band. In the operation on the flat phantom, the bandwidths were 170 MHz (from 2.34 to 2.51 GHz) and 700 MHz (from 4.8 to 5.5 GHz) in the lower and higher bands, respectively. Using the conductive-mesh–polymer composite, a new flexible and transparent ultra-high-frequency (UHF) circularly polarized (CP) radiofrequency identification (RFID) tag antenna was further developed in [52]. The antenna was tested under bending states in different orientations and it was found that the antenna demonstrated an excellent read range and CP performance when subjected to sustained bending.

Another prominent research investigation towards the development of flexible and transparent antenna was explored in [4], where a water-based unidirectional antenna was developed. In this design, an off-centre-fed dipole was used as the main radiator and a water reflector was used underneath it to transform the bidirectional radiation pattern to a unidirectional radiation pattern mimicking a microstrip patch antenna. Pure water is almost 100% transparent [53] and has high permittivity. The high permittivity of water was utilized in this design to realize a transparent flexible reflector plane. The water layer was held inside a hollow rectangular PDMS bolus. The main dipole radiator was made from transparent conductive mesh, VeilShield, which had an area of 43 mm × 2 mm. The area taken up by VeilShield is approximately 2.5% of the total antenna area, this section has nearly 70% transparency [48] and the remaining 97.5% of the area was taken up by water and PDMS, whose transparency was nearly the same as that of PDMS (94%). Thus this water-based unidirectional antenna has very high transparency and excellent flexibility.

Figure 19.14 presents a schematic diagram of the fabrication process of the antenna. The proposed fabrication method of this water antenna [4] is cost-effective, simple, and environmentally friendly. The antenna was manufactured by using three 3D printed moulds whose geometries were shown in figure 19.14. The moulds were fabricated from photopolymer resin in a stereolithography (SLA) based 3D printer

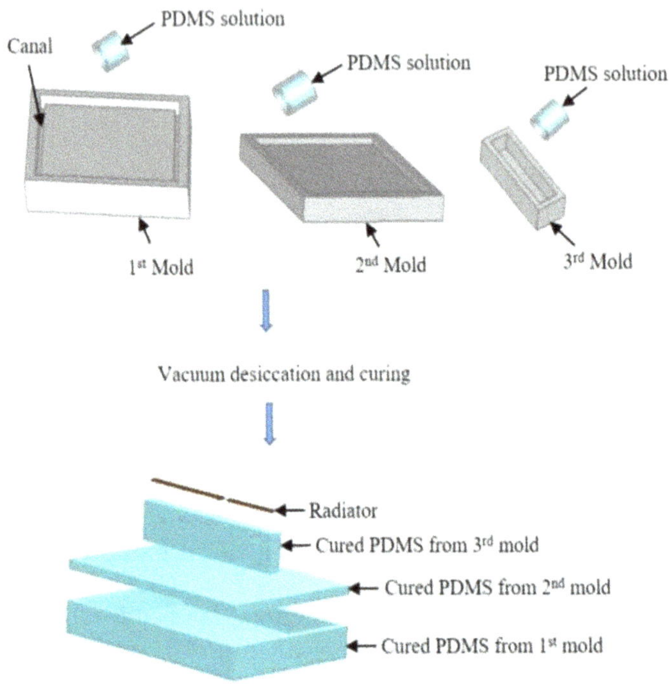

Figure 19.14. Multi-stage fabrication process of the transparent and flexible antenna, using water as the reflector. (Reproduced with permission from [4]. Copyright 2020 IEEE.)

(Form 2 from Formlabs). After mould fabrication, the PDMS solution was prepared using a Dow Corning Sylgard 184 Silicone Elastomer Kit with a 10:1 base to curing agent mixing ratio. Then PDMS solution was poured in the moulds, degassed through a vacuum desiccator, and cured in an oven (UF55, Memmert) at 65 °C for two hours. After curing the PDMS, they were carefully taken out from the moulds and PDMS layers from moulds 1 and 2 were attached to form a hollow box and then PDMS from mould 3 was attached on top of the hollow box to contain the radiator. The dipole radiator was made by cutting VeilShield and then attached on the PDMS holder using a small amount of uncured PDMS and cured in the oven at 85 °C for 30 min. In the next stage, an SMA connector was connected to the radiator using silver conductive epoxy which was cured in the oven (UF55, Memmert) at 65 °C for two hours. In the last stage, pure water was injected inside the hollow PDMS box using a syringe. Figure 19.15 shows a photograph of the VeilShield and fabricated prototype. Figure 19.15(b) reveals the high transparency of the prototype and figure 19.15(c) reveals its flexibility through a bent view with a very small radius.

Figure 19.16 shows the predicted and measured $|S_{11}|$ of the antenna. The predicted result is the result computed numerically using CST Microwave Studio 2017. The measurements were conducted in Agilent N5242A PNA-X. The antenna was tested in free space and on flat and human-hand-shaped phantoms (the same phantoms as shown in figure 19.12) to examine its suitability for near body area networks and on curved surfaces. From figure 19.16 it is seen that the antenna

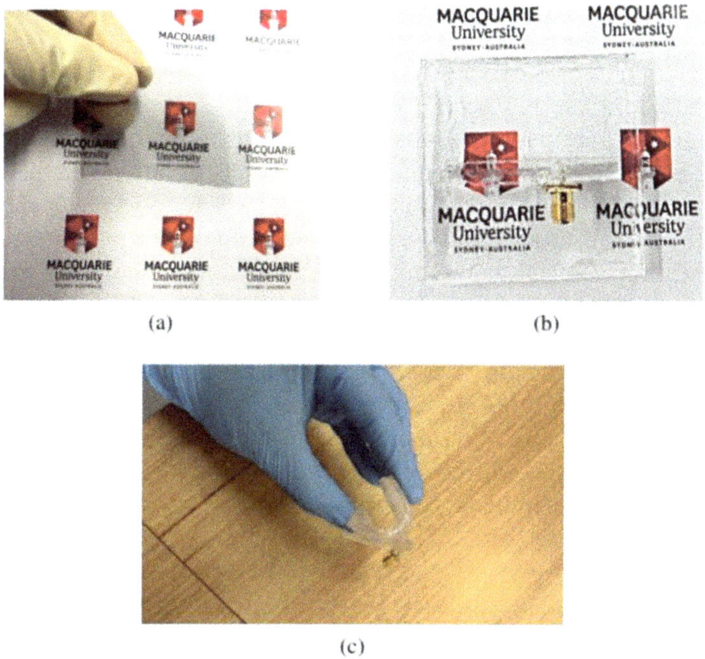

Figure 19.15. (a) A VeilShield sample demonstrating its transparency. (b) The water antenna prototype. (c) Water antenna bent to a very small radius. (Reproduced with permission from [4]. Copyright 2020 IEEE.)

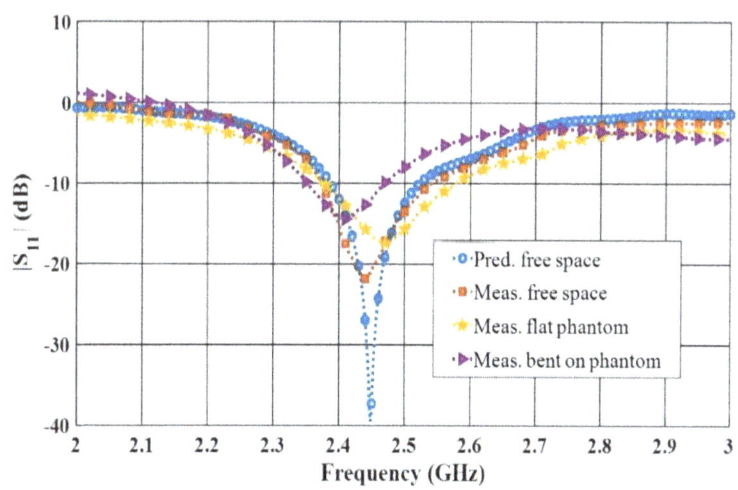

Figure 19.16. Predicted and measured $|S_{11}|$ of the water antenna in free space, on a flat phantom, and under bending. (Reproduced with permission from [4]. Copyright 2020 IEEE.)

maintains a 170 MHz 10 dB return-loss bandwidth in free space, which is affected slightly when it operates on the phantom. However, the effect of the phantom is not very significant, which demonstrates the effectiveness of the shielding provided by the water layer.

19.4 Future wearable electronics: higher frequencies and flexible antennas

In the last decade, there has been enormous progress in wireless communications. Modern communication technology is moving towards the millimetre wave (mmW) spectrum. It is predicted that 5G will connect billions of devices in the near future. To meet this demand, research has been devoted to effectively designing 5G networks. There is no doubt that wearable technology will play a significant role in next-generation wireless communications. Wearable antennas operating in the mmW band can be attractive candidates for short-to-medium range high-speed applications. However, the design and characterization of wearable antennas in the mmW spectrum are associated with some inevitable challenges. For example, the sizes of human body features are electrically much larger than the small and fine features of antennas at mmW frequencies and, thus, traditional full-wave computations become inefficient. Moreover, realization of mmW wearable antennas is a challenging task. Another major concern is the biological safety of the electromagnetic exposure in this band.

In the literature, some research endeavours have been devoted to the development of wearable antennas in the mmW frequency spectrum. In [53–57] the propagation of a wave at 60 GHz was numerically and experimentally investigated on a skin-equivalent phantom. A 2×2 microstrip patch antenna array operating at 60 GHz and realized with 0.07 mm thick copper foils and cotton-woven fabric substrate was explored in [57]. An inkjet-printed Yagi–Uda antenna operating at 24.5 GHz was reported in [58]. The printing was conducted in the Dimatix DMP-2831 inkjet printer with Cabot CCI-300 silver nanoparticle ink and flexible liquid-crystalline polymer (LCP) laminate. In [59] a printed Yagi-Uda antenna at 60 GHz was demonstrated. The antenna was fabricated using copper foil and fabric substrate ($\varepsilon_r = 2.5$, $\tan\delta = 0.016$) and achieved a gain of 9.2 dBi and an efficiency of 48%. A polymer-based 4×2 microstrip patch antenna array operating in the V-band (50–60 GHz) was reported in [60]. In this band the PDMS was characterized using a 50 Ω microstrip line and a relative permittivity of 2.68 and loss tangent of 0.04 were obtained. The demonstrated antenna achieved a measured gain of 12.3 dBi and a radiation efficiency of 28%.

19.5 Conclusion

This chapter has described the characterization, realization, and functionalization of flexible and conformal antennas for wireless communication and sensing applications. Starting with an overview of the strong demand for flexible antennas and the challenges associated with their development, a non-exhaustive summary of the design, fabrication, and experimental testing of wearable antennas has been presented, including some research approaches towards the development of flexible and conformal antennas in ultrawideband, optically transparent, and higher-frequency applications.

References

[1] Mohamadzade B, Hashmi R M, Simorangkir R B V B, Gharaei R, Rehman S and Abbasi Q H 2019 Recent advances in fabrication methods for flexible antennas in wearable devices: state of the art *Sensors* **19**

[2] Abbas S M, Desai S C, Esselle K P, Volakis J L and Hashmi R M 2018 Design and characterization of a flexible wideband antenna using polydimethylsiloxane composite substrate *Int. J. Antennas Propag.*

[3] Simorangkir R B V B, Yang Y, Hashmi R M, Björninen T, Esselle K P and Ukkonen L 2018 Polydimethylsiloxane-embedded conductive fabric: characterization and application for realization of robust passive and active flexible wearable antennas *IEEE Access* **6** 48102–12

[4] Sayem A S M, Simorangkir R B V B, Esselle K P, Hashmi R M and Liu H 2020 A method to develop flexible robust optically transparent unidirectional antennas utilizing pure water, PDMS and transparent conductive mesh *IEEE Trans. Antennas Propag.*

[5] Zhou Y, Bayram Y, Du F, Dai L and Volakis J L 2010 Polymer–carbon nanotube sheets for conformal load bearing antennas *IEEE Trans. Antennas Propag.* **58** 2169–75

[6] Song L *et al* 2014 Stretchable and reversibly deformable radio frequency antennas based on silver nanowires *ACS Appl. Mater. Interfaces* **6** 4248–53

[7] Kiourti A and Volakis J L 2014 Stretchable and flexible E-fiber wire antennas embedded in polymer *IEEE Antennas Wirel. Propag. Lett.* **13** 1381–84

[8] Alqadami A S M *et al* 2016 Assessment of PDMS technology in a MIMO antenna array *IEEE Antennas Wirel. Propag. Lett.* **15** 1939–42

[9] Simorangkir R B V B, Yang Y, Matekovits L and Esselle K P 2017 Dual-band dual-mode textile antenna on PDMS substrate for body-centric communications *IEEE Antennas Wirel. Propag. Lett.* **16** 677–80

[10] Jang T, Zhang C, Youn H, Zhou J and Guo L J 2017 Semitransparent and flexible mechanically reconfigurable electrically small antennas based on tortuous metallic micromesh *IEEE Trans. Antennas Propag.* **65** 150–8

[11] Zhang L, Wang Z and Volakis J L 2012 Textile antennas and sensors for body-worn applications *IEEE Antennas Wirel. Propag. Lett.* **11** 1690–3

[12] Ivsic B, Bonefacic D and Bartolic J 2013 Considerations on embroidered textile antennas for wearable applications *IEEE Antennas Wirel. Propag. Lett* **12** 1708–11

[13] Wise P L, Reaney I M, Lee W E, Price T J, Iddles D M and Cannell D S 2001 Structure–microwave property relations of Ca and Sr titanates *J. Eur. Ceram. Soc.* **21** 2629–32

[14] Rajesh S, Murali K P, Rajani K V and Ratheesh R 2009 $SrTiO_3$-filled PTFE composite laminates for microwave substrate applications *Int. J. Appl. Ceram. Tech.* **6** 553–61

[15] Johnston I D, McCluskey D K, Tan C K L and Tracey M C 2014 Mechanical characterization of bulk Sylgard 184 for microfluidics and microengineering *J. Micromech. Microeng.* **24** 035017-1–7

[16] Tiercelin N, Coquet P, Sauleau R, Senez V and Fujita H 2006 Polydimethylsiloxane membranes for millimeter-wave planar ultra flexible antennas *J. Micromech. Microeng.* **16** 2389–95

[17] Trajkovikj J, Zürcher J-F and Skrivervik A K 2013 PDMS a robust casing for flexible W-BAN antennas *IEEE Antennas Propag. Mag.* **55** 287–97

[18] Colas A 1990 *Silicones: Preparation Properties and Performances* (Midland, MI: Dow Corning)

[19] Sebastian M T and Namitha L K 2017 Rubber–ceramic composites *Microwave Materials and Applications* (New York: Wiley) pp 537–74

[20] Bodö P and Sundgren J-E 1986 Titanium deposition onto ion-bombarded and plasma-treated polydimethylsiloxane: surface modification interface and adhesion *Thin Solid Films* **136** 147–59

[21] Bayram Y *et al* 2010 E-textile conductors and polymer composites for conformal lightweight antennas *IEEE Trans. Antennas Propag.* **58** 2732–6

[22] Le Duc C and Zhadobov M 2017 Impact of antenna topology and feeding technique on coupling with human body: application to 60-GHz antenna arrays *IEEE Trans. Antennas Propag.* **65** 6779–87

[23] Takimoto T, Onishi T, Saito K, Takahashi M, Uebayashi S and Ito K 2007 Characteristics of biological tissue equivalent phantoms applied to UWB communications *Electron. Commun. Part I* **90** 48–55

[24] Kobayashi T, Nojima T, Yamada K and Uebayashi S 1993 Dry phantom composed of ceramics and its application to SAR estimation *IEEE Trans. Microw. Theory Tech* **41** 136–40

[25] Ito K, Furuya K, Okano Y and Hamada L 2001 Development and characteristics of a biological tissue-equivalent phantom for microwaves *Electron. Commun. Part I* **84** 67–77

[26] Mobashsher A T and Abbosh A M 2015 Artificial human phantoms: human proxy in testing microwave apparatuses that have electromagnetic interaction with the human body *IEEE Microw. Mag.* **16** 42–62

[27] Hartsgrove G, Kraszewski A and Surowiec A 1987 Simulated biological materials for electromagnetic radiation absorption studies *Bioelectromagnetics* **8** 29–36

[28] Mohamadzade B, Simorangkir R B V B, Maric S, Lalbakhsh A, Esselle K P and Hashmi R M 2020 Recent developments and state of the art in flexible and conformal reconfigurable antennas *Electronics* **9** 1375

[29] Salleh S M, Jusoh M, Ismail A H, Kamarudin M R, Nobles P, Rahim M K A, Sabapathy T, Osman M N, Jais M I and Soh P J 2018 Textile antenna with simultaneous frequency and polarization reconfiguration for WBAN *IEEE Access* **6** 7350–8

[30] Kumar J, Basu B, Talukdar F A and Nandi A 2017 Graphene-based multimode inspired frequency reconfigurable user terminal antenna for satellite communication *IET Commun.* **12** 67–74

[31] Tahir F A and Javed A 2015 A compact dual-band frequency reconfigurable textile antenna for wearable applications *Microw. Opt. Technol. Lett.* **57** 2251–7

[32] Kasirga T S, Ertas Y N and Bayindir M 2009 Microfluidics for reconfigurable electromagnetic metamaterials *Appl. Phys. Lett.* **95** 214102

[33] King A J, Patrick J F, Sottos N R, White S R, Huff G H and Bernhard J T 2013 Microfluidically switched frequency reconfigurable slot antennas *IEEE Antennas Wirel. Propag. Lett.* **12** 828–31

[34] Simorangkir R B, Yang Y, Esselle K P and Zeb B A 2017 A method to realize robust flexible electronically tunable antennas using polymer-embedded conductive fabric *IEEE Trans. Antennas Propag.* **66** 50–8

[35] So J-H, Thelen J, Qusba A, Hayes G J, Lazzi G and Dickey M D 2009 Reversibly deformable and mechanically tunable fluidic antennas *Adv. Funct. Mater.* **19** 3632–7

[36] Mazlouman S J, Jiang X J, Mahanfar A, Menon C and Vaughan R G 2011 A reconfigurable patch antenna using liquid metal embedded in a silicone substrate *IEEE Trans. Antennas Propag.* **59** 4406–12

[37] Moghadas H, Zandvakili M, Sameoto D and Mousavi P 2016 Beam reconfigurable aperture antenna by stretching or reshaping of a flexible surface *IEEE Antennas Wirel. Propag. Lett.* **16** 1337–40

[38] Alharbi S, Chaudhari S, Inshaar A, Shah H, Zou C, Harne R L and Kiourti A 2018 E-textile origami dipole antennas with graded embroidery for adaptive RF performance *IEEE Antennas Wirel. Propag. Lett.* **17** 2218–22

[39] Shah S I H and Lim S 2017 A dual band frequency reconfigurable origami magic cube antenna for wireless sensor network applications *Sensors* **17** 2675

[40] Liu C-X, Zhang Y-J and Li Y 2019 Precisely manipulated origami for consecutive frequency reconfigurable antennas *Smart Mater. Struct.* **28** 095020

[41] Shah S I H and Lim S 2019 Thermally beam-direction-and beamwidth switchable monopole antenna using origami reflectors with smart shape memory polymer hinges *IEEE Antennas Wirel. Propag. Lett.* **18** 1696–700

[42] Foerster J *et al* 2001 Ultra-wideband technology for short-or medium-range wireless communications *Int. Technol. J.* **2** 1–11

[43] Fort A, Desset C, De Doncker P, Wambacq P and Van Biesen L 2006 An ultra-wideband body area propagation channel model—from statistics to implementation *IEEE Trans. Microw. Theory Techn* **54** 1820–6

[44] Koohestani M, Zurcher J-F, Moreira A A and Skrivervik A K 2014 A novel low-profile vertically-polarized UWB antenna for WBAN *IEEE Trans. Antennas Propag.* **62** 1888–94

[45] Mohamadzade B, Simorangkir R B V B, Hashmi R M and Lalbakhsh A 2020 A conformal ultrawideband antenna with monopole-like radiation patterns *IEEE Trans. Antennas Propag.* **68** 6383–8

[46] Simorangkir R B V B, Kiourti A and Esselle K P 2018 UWB wearable antenna with a full ground plane based on PDMS-embedded conductive fabric *IEEE Antennas Wireless Propag. Lett.* **17** 493–6

[47] Mohamadzade B, Simorangkir R B V B, Hashmi R M, Chao-Oger Y, Zhadobov M and Sauleau R 2020 A conformal band-notched ultrawideband antenna with monopole-like radiation characteristics *IEEE Antennas Wirel. Propag. Lett.* **19** 203–7

[48] Sayem A S M, Simorangkir R B V B, Esselle K P and Hashmi R M 2019 Development of robust transparent conformal antennas based on conductive mesh–polymer composite for unobtrusive wearable applications *IEEE Trans. Antennas Propag.* **67** 7216–24

[49] Sayem A S M, Esselle K P, Hashmi R M and Liu H 2020 Experimental studies of the robustness of the conductive-mesh–polymer composite towards the development of conformal and transparent antennas *Smart Mater. Struct.* **29** 085015

[50] Sayem A S M, Esselle K P and Hashmi R M 2020 Increasing the transparency of compact flexible antennas using defected ground structure for unobtrusive wearable technologies *IET Microw. Antennas Propag.* **14** 1869–77

[51] Sayem A S M, Esselle K P and Hashmi R M 2020 Robustness analysis of the polymer conductive-mesh composite for the realization of transparent and flexible wearable antennas *14th European Conf. on Antennas and Propagation (Copenhagen, Denmark, 15–20 March 2020)*

[52] Sayem A S M, Le D, Simorangkir R B V B, Bjórninen T, Esselle K P, Hashmi R M and Zhadobov M 2020 Optically transparent flexible robust circularly polarized antenna for UHF RFID tags *IEEE Antennas Wirel. Propag. Lett.* **19** 2334

[53] Saidur R, Meng T, Said Z, Hasanuzzaman M and Kamyar A 2012 Evaluation of the effect of nanofluid-based absorbers on direct solar collector *Int. J. Heat Mass Transfer* **55** 5899–907

[54] Chahat N, Valerio G, Zhadobov M and Sauleau R 2013 On-body propagation at 60 GHz *IEEE Trans. Antennas Propag.* **61** 1876–88

[55] Guraliuc A R, Zhadobov M, Valerio G, Chahat N and Sauleau R 2014 Effect of textile on the propagation along the body at 60 GHZ *IEEE Trans. Antennas Propag.* **62** 1489–94

[56] Chahat N, Zhadobov M, Le Coq L, Alekseev S I and Sauleau R 2012 Characterization of the interactions between a 60-GHz antenna and the human body in an off-body scenario *IEEE Trans. Antennas Propag.* **60** 5958–65

[57] Chahat N, Zhadobov M, Muhammad S A, Le Coq L and Sauleau R 2013 60-GHz textile antenna array for body-centric communications *IEEE Trans. Antennas Propag.* **61** 1816–24

[58] Tehrani B K, Cook B S and Tentzeris M M 2016 Inkjet printing of multilayer millimeter-wave Yagi–Uda antennas on flexible substrates *IEEE Antennas Wirel. Propag. Lett.* **15** 1

[59] Sabri S F, Sam S M, Kamardin K, Daud S M and Salleh N 2016 Review of the current design on wearable antenna in medical field and its challenges *J. Teknol.* **78** 111–17

[60] Hage-Ali S, Tiercelin N, Coquet P, Sauleau R, Fujita H, Preobrazhensky V and Pernod P 2009 A millimeter-wave microstrip antenna array on ultra-flexible micromachined poly-dimethylsiloxane (PDMS) polymer *IEEE Antennas Wirel. Propag. Lett.* **8** 1306–9

www.ingramcontent.com/pod-product-compliance
Ingram Content Group UK Ltd.
Pitfield, Milton Keynes, MK11 3LW, UK
UKHW051908100926
8556IPUK00005B/812

9 780750 334372